*Physiological Ecology of
North American
Plant Communities*

W. DWIGHT BILLINGS, Wyoming, 1978. Photograph by Kim Peterson

Physiological Ecology of North American Plant Communities

EDITED BY

BRIAN F. CHABOT

Section of Ecology and Systematics
Cornell University

AND

HAROLD A. MOONEY

Department of Biological Sciences
Stanford University

NEW YORK   LONDON

CHAPMAN AND HALL

First published 1985 by
Chapman and Hall
733 Third Avenue, New York NY 10017
Published in Great Britain by
Chapman and Hall Ltd
11 New Fetter Lane, London EC4P 4EE
© *1985 Chapman and Hall*

Printed in Great Britain at the
University Press, Cambridge

ISBN   0 412 23240 5

Library of Congress Cataloging in Publication data

Main entry under title:

Physiological ecology of North American plant communities.

   Includes bibliographies and index.
   1. Botany – North America – Ecology. 2. Plant
communities North America. 3. Plant physiology.
1. Chabot, Brian F. II. Mooney, Harold A.
QK110.P49 1985   581.5′097   84-9586
ISBN 0-412-23240-5

British Library Cataloguing in Publication Data

Physiological ecology of North American
   Plant communities.
   1. Botany—North America—Ecology
   I. Chabot, Brian F.   II. Mooney, H.A.
   581.5′097   QK110

   ISBN   0-412-23240-5

Contents

Contents

Preface

Although, as W.D. Billings notes in his chapter in this book, the development of physiological ecology can be traced back to the very beginnings of the study of ecology it is clear that the modern development of this field in North America is due in the large part to the efforts of Billings alone. The foundation that Billings laid in the late 1950s came from his own studies on deserts and subsequently arctic and alpine plants, and also from his enormous success in instilling enthusiasm for the field in the numerous students attracted to the plant ecology program at Duke University.

Billings' own studies provided the model for subsequent work in this field. Physiological techniques, normally confined to the laboratory, were brought into the field to examine processes under natural environmental conditions. These field studies were accompanied by experiments under controlled conditions where the relative impact of various factors could be assessed and further where genetic as opposed to environmental influences could be separated. This blending of field and laboratory approaches promoted the design of experiments which were of direct relevance to understanding the distribution and abundance of plants in nature. Physiological mechanisms were studied and assessed in the context of the functioning of plants under natural conditions rather than as an end in itself.

This approach was utilized by a large number of Billings' students as they moved out across North America, through his support and encouragement, to engage in research. They did not confine their activities to a single vegetation type, but rather as a group they covered most of the plant communities on the continent. It is because of this research breadth that the information was available to assemble this book. Not all of the authors of this book were students of Billings; however, they all felt his influence in one way or another.

It is inevitable that certain plant communities received more attention than others as is evident from the book structure and contents. In particular, deserts and the arctic and alpine, have received a disproportionate share of concentrated studies. Important vegetation types, such as tropical forests, are just beginning to attract researchers to their varied problems.

As noted earlier, knowledge from physiological ecology is necessary for understanding the patterning of plants on the landscape. Of equal importance, this knowledge is essential for a rational management of ecosystems. Studies from physiological ecology quantify the resource base of individual species, elucidate the specific mechanisms that have evolved to acquire and allocate these resources, and provide insights into the nature of species interactions. This knowledge can be utilized to specify the impact of various perturbations and stresses which are currently being imposed on ecosystems. Such information is currently being used to predict the effects of SO_2, acid rain, CO_2 enhancement, etc. We know that species differ in their sensitivity to these factors. Knowledge of the precise differences can help us predict how species composition in a community might shift as the environment is modified or at what seasons or times species will be least susceptible to the impact of pollutants.

In its modern configuration plant physiological ecology is an exciting developing field. The pace of its development has accelerated in recent years as sophisticated instrumentation has become available to quantify precisely the microenvironment of plants and their physiological responses in nature. The enormous diversity of the plant types which occur in North America provides the material for the development of a comprehensive comparative picture of the mechanisms which plants have evolved to cope with some of the hottest, coldest, wettest, and most arid habitats on earth. This book is an attempt to bring together some of the initial efforts in this field and it is our tribute to the leadership and inspiration of W.D. Billings.

Contributors

M.G. Barbour   Department of Botany, University of California, Davis, Davis
CA 95616

W.D. Billings   Department of Botany, Duke University, Durham NC 27706

L.C. Bliss   Department of Botany, University of Washington, Seattle
WA 98195

M.M. Caldwell   Department of Range Science, Utah State University, Logan
UT 84322

B.F. Chabot   Section of Ecology and Systematics, Cornell University, Ithaca
NY 14853

F.S. Chapin III   Institute of Arctic Biology, University of Alaska, Fairbanks
AL 99701

T.M. De Jong   Department of Pomology, University of California, Davis, Davis
CA 95616

E.L. Dunn   School of Biology, Georgia Institute of Technology, Atlanta
GA 30602

J.R. Ehleringer   Department of Biology, University of Utah, Salt Lake City
UT 84112

C.C. Grier   College of Forest Resources, University of Washington, Seattle
WA 98195

B.L. Haines   Department of Botany, University of Georgia, Athens
GA 30602

D.J. Hicks   Section of Ecology and Systematics, Cornell University, Ithaca
NY 14853

T.M. Hinckley College of Forest Resources, University of Washington, Seattle WA 98195

J.P. Lassoie Department of Natural Resources, Cornell University, Ithaca NY 14853

W.T. Lawrence Centre for Energy and Environment Research, GPO Box 3682, San Juan, Puerto Rico 00936

P.C. Miller Department of Biology, San Diego State University, San Diego CA 92182

H.A. Mooney Department of Biological Sciences, Stanford University, Stanford CA 94305

P.S. Nobel Department of Biology, University of California Los Angeles, Los Angeles CA 90024

W.C. Oechel Systems Ecology Research Group, San Diego State University, San Diego CA 92182

B.M. Pavlik Department of Botany, Iowa State University, Ames IA 50011

R.W. Pearcy Department of Botany, University of California, Davis, Davis CA 95616

P.G. Risser Illinois Natural History Survey, Natural Resources Laboratory, 607 East Peabody Drive, Champaign IL 61820

R.H. Robichaux Department of Botany, University of California, Berkeley, Berkeley CA 94720

G.R. Shaver The Ecosystems Center, Marine Biological Center, Woods Hole MA 02543

W.K. Smith Department of Botany, University of Wyoming, Laramie WY 82071

1

The historical development of physiological plant ecology

W.D. BILLINGS

'Those who cannot remember the past are condemned to repeat it' – Santayana (*The Life of Reason*, 1905–1906, Vol. 1. Reason in Common Sense)

During the last 25 years there has been a renaissance of an aspect of ecology designated either as 'physiological plant ecology' or 'plant ecophysiology'. These areas of research are concerned with the broad continuum where ecology and physiology overlap and join forces to solve problems.

Certain problems define the roles of physiological ecology. Among these are:

1. attempts to explain the reasons for the present geographical distributions of populations of wild plant taxa and crop plant cultivars by measuring their physiological, morphological, and reproductive adaptations of environmental factors acting in concert;

2. the definition of the potential environmental tolerance ranges of plant taxa (native, adventive, and agricultural) and the evolution of the responsible character traits;

3. the determination of how ecosystems operate in a changing biosphere; in this regard, physiological ecology works with its companion sciences: community ecology, systems ecology, and also the physical sciences.

Physiological ecology is not a science in isolation but an integral part of the whole of ecology. If I did not realize it earlier, it was brought home to me by F.H. Bormann in 1964 during a midnight walk around the countryside at Brookhaven while we were both spending some time in George Woodwell's laboratory. His arguments were logical and persuasive; one should realize that physiological ecology is just one of the bonds that holds ecology together in its attempt to understand the biosphere.

1.1 The beginnings

As befits an interdisciplinary science, one can trace the roots of physiological plant ecology to multiple sources. There are three main ones: plant geography, plant physiology, and ecology. Beyond these, geology, climatology, soil science, and agronomy all had influences on the origin and development of physiological ecology.

In 1935, B.E. Livingston reminded us that the field of physiological ecology existed among crop scientists long before there was 'ecology' itself. Actually, as one might suppose, it dates back over 2000 years to the Greeks and Romans. Evenari (1980/81) calls Theophrastus 'the first seed physiologist and seed ecologist'. Theophrastus knew, for example, that during seed maturation, plants deposit food reserves in seeds; these reserves are needed in seedling growth. Theophrastus also noted that, in several kinds of plants, seeds of wild plant populations did not all germinate at once in the Mediterranean climate of Greece but over a period of several years.

Farmers and naturalists have been making useful observations on plant growth and distribution for all the centuries between the time of Theophrastus and now. However, I shall be concerned only with certain observations and discoveries of the last 200 years.

In the latter half of the 18th century, both in Europe and North America, there began a blossoming of interest in the relationships between plants, climates, atmosphere, and soils. From the standpoint of physiological ecology, the most important discovery during this time was that of photosynthesis through the observations of Priestley, Ingenhousz, Senebier, Lavoisier, de Saussure, and others. According to Egerton (1976), Benjamin Franklin learned in 1774 from Priestley himself of the purification, by plants, of air 'spoiled' by animals. Franklin was pleased since this fitted in nicely with his ideas on natural cycles and balances in nature. The ecological importance of Priestley's discovery was not lost on the Marquis Francois Jean de Chastellux (1787) who spent 1780–1782 in America. Chastellux was one of the 40 members of the French Academy, a Major-General in the American Revolution, and friend of Washington. He was concerned about the effects on the atmosphere of the increasingly rapid destruction of the eastern American deciduous forest for wood and cropland. Today, this has a familiar and timely ring in regard to the increase in biospheric CO_2. Yet, Chastellux's comments were made not only before photosynthesis was known but even before O_2 and CO_2 were identified as the gases concerned.

While some European scientists of the 18th century were concerned with the relationships of plants and the atmosphere, eastern North America was still forested and almost unknown as compared to Europe; it was a gold mine of new species for the plant explorers. Foremost among these naturalists were Pehr Kalm, from Sweden (see Benson, 1937) and William Bartram of Philadelphia (see Harper 1958). Kalm, in particular, in his travels in North America between 1748 and 1751, observed and described the relationships between native plants and soils. He knew that certain species could be used as indicators of the suitability of the various soil types for agriculture. This technique was used far later, with some precision, in the arid mountain valleys of western North America in the first decades of the twentieth century by H.L. Shantz and his colleagues (Shantz and Piemeisel, 1924).

Samuel Williams (1794) was also considerably ahead of his time. Among his interests were the water and temperature relations of maple forests in Vermont. By measuring water loss from a maple twig and its leaves, cutting down the trees, counting its leaves, and determining the number of trees per acre, he calculated the transpiration per acre of forest per day.

Among those 18th century Americans scientifically inclined toward plants, one must mention the writer of the Declaration of Independence whose talents and accomplishments were many: Thomas Jefferson. Long before the term 'phenology' appeared in Europe in the mid-19th century, Jefferson was recording the growth, blooming dates, fruiting, etc. of his garden and farm plants in the *Garden Book* starting in 1766 at his birthplace of Shadwell and resumed in 1771 at his new home. Monticello, both near Charlottesville, Virginia. With a few gaps, this continued until 1824, two years before his death. When he was away, on government service in New York and Philadelphia, he even attempted with the help of his daughter and son-in-law at Charlottesville to make comparative phenological observations on the same kinds of garden plants on a latitudinal gradient, and with some success. Even though busy with the Revolutionary War, he continued phenological and weather records; on the day of independence itself, 4 July 1776, in Philadelphia, he recorded the temperature at 6 a.m., 9 a.m., 1 p.m., and 9 p.m. with a newly purchased thermometer. In 1782, Jefferson drew a calendar of the bloom of flowers at Monticello 'planted this spring and the season was very backward' (Betts, 1944).

From 1784 until 1789, Jefferson was Minister Plenipotentiary to France. He had to abandon temporarily his Monticello Garden Book but he was busy with plants and travels in Europe. In a letter to William Drayton dated 30 July 1787, he noted that crossing three high mountains in the Alps: '... enabled me to form a scale of the tenderer plants, and arrange them according to their different powers of resisting cold...'. 'In ascending, we lose these plants, one after another, as we rise, and find them again in the contrary order as we descend on the other side; and this is repeated three times.' The fact that these were cultivated plants does not detract from his observations on differential cold tolerance nor on the individualistic distribution of species. Jefferson's observations were solid and reflect the mental breadth and scientific influence, on men such as Humboldt, of this widely travelled scientist and intellectual.

1.2 The 19th century

It is difficult to discuss simultaneously the roles of the three main sources of plant physiological ecology: plant geography, plant physiology, and ecology. So, I shall follow each line of development until their interactions near the end of the 19th century became so numerous and powerful that it is nearly impossible to separate their effects on the development of physiological plant ecology.

There have been times in the history of civilization that are remarkable for the simultaneous appearance of groups of highly intelligent leaders with such broad interests from politics to philosophy and science that their impact on the future outweighed their numbers. The latter decades of the 18th century and first decades of the 19th was one of these times. In North America, there were Franklin, Jefferson, John Adams, James Madison, and many others. They were the leaders in the American Revolution. All had scientific abilities, as well as political. Franklin, for example, founded the American Philosophical Society in Philadelphia in 1743. Jefferson later served several years as President of the Society. John Adams, even while the war was still going on around him, similarly founded the American Academy of Arts and Sciences.

The same thing was happening in England and Europe within the Royal Society and in the various Academies. They, too, had their leaders: Banks, Priestley, Lavoisier, de Saussure, Gay-Lussac, Goethe, Willdenow, and many others. It was an age of world exploration for plants and animals, a continuation of the work started more than half a century before by Linnaeus and his students (Kalm, Osbeck, Solander, *et al.*). It was Sir Joseph Banks who sent Captain James Cook and his naturalists, including himself, on the great voyages of exploration of the Pacific and the Polar regions. At the same time, experimenters were discovering elements, the gas laws, the composition of the atmosphere, and gas exchanges of plants and animals. The stage was set near the end of the century for the entrance of the incomparable Alexander von Humboldt, the first great plant geographer.

Before Humboldt, Willdenow had travelled far, even to North America. His observation that similar climates in different parts of the earth resulted in vegetations with similar physiognomies relatively independent of their floras was a step toward the basics of life forms, comparative community ecology – and eventually to the modern comparative physiological ecology of widely separated regions of similar vegetations. One immediately thinks of the five widely separated broad-sclerophyll regions occurring in mediterranean-type climates, and the studies in physiological ecology which have done so much to explain this phenomenon of convergent evolution.

Willdenow introduced Humboldt to botany (Botting, 1973). But, after Humboldt's move to the University at Göttingen, he met George Forster, who had been the naturalist on Cook's second voyage. Humboldt was determined to become an explorer for plants. Before getting a chance at this, however, he moved on to Freiberg and studied mining; in 1792, he was appointed Assistant Inspector in the Department of Mines. Even while inspecting mines, he wrote a flora of Freiberg and carried out advanced experiments in physiology. Meanwhile, he met Goethe who had a great influence on his scientific thinking. In 1797, he resigned from the Department of Mines and with his ample inheritance was ready to see the world and its plants.

But first, there were scientific observations to be made in the European mountains and a stay in France, where he met the famous old explorer, Bougainville, who gave him excellent additional training. Finally, with his friend Aimée Bonpland, he convinced the King of Spain, Carlos IV, to grant them passports for scientific exploration in Spanish America, most of which was unknown scientifically. In June 1799, well-equipped with instruments, they set out from Spain on the great adventure.

Eventually, the ship reached the Venezuelan coast where Humboldt and Bonpland debarked at Cumaná and were visually and mentally overwhelmed by the richness and diversity of tropical plants. They collected in and described the coastal tropical rainforest. From Caracas, the wide llanos of the interior were crossed to the headwaters of the Orinoco. They followed the Orinoco down to the coast. Humboldt and Bonpland had made the first scientific expedition through this almost unknown region and collected 12 000 plant specimens. They had learned more about the natural history of the tropical rainforest, its environment, and people than had been known in South America up to that time.

By way of Cuba, Humboldt reached northern Colombia and started southward in the Andes to Bogotá and Quito. It was from Quito that he studied relationships between plants and high Andean environments, first on Pichincha which rises above Quito to an elevation of 15 672 ft (4777 m). Later he measured the upper limits of plants on the higher Ecuadorian volcanoes: Cotopaxi, Antisana, and eventually on the highest of all, Chimborazo, which rises to 20 702 ft (6310 m). Here, Humboldt, Bonpland, Carlos Montúfar and an Indian reached, without any real climbing equipment, an elevation of 19 286 ft (5878 m) but were stopped by a deep ravine too wide to cross in the soft snow. This climb in June 1802, set, by far, a new high elevation record for mountain-climbing which was not broken for 35 years. Humboldt thought Chimborazo to be the highest mountain on earth and was greatly disappointed to hear that higher peaks were found later in the Himalayas and in the Andes themselves. Botanically, the higher slopes of Chimborazo and Cotopaxi are quite barren even below snow-line. The highest plants that they found on Chimborazo were some crustose lichens above snow-line at an elevation of 5157 m; the highest moss seen was at about 4400 m. Actually, even

vascular plants occur higher up on both Chimborazo and Cotopaxi (Diels, 1937; Mulroy, 1979).

A pair of remarkable diagrams of the distribution of plants with elevation on tropical mountains resulted from Humboldt's many months of travel in the Andes. These appeared in Humboldt and Bonpland (1805) *Essai sur la Geographie des Plantes*. The first is a diagrammatic cross-section of Ecuador from east to west showing vegetational zones and the presence of genera and families of plants at different elevations. With data on barometric pressure, temperatures, and gravitational attraction (!) alongside elevation one could make a comparative assessment of the tolerance ranges of the different plant taxa of the Andean region. This diagram was, and is, a useful tool in the comparative ecology of tropical plants. The other diagram shows the lowering of elevational zones along a latitudinal gradient from the tropics to Lapland. Humboldt was far ahead of his time.

In 1804, before returning to Europe, Humboldt came to the United States to visit Jefferson, by then the President of the United States. After sometime in Philadelphia, in company with Bonpland, Montúfar, Fothergill, and Charles Willson Peale, Humboldt travelled to Washington by stagecoach. So, two great men met, Humboldt invited by Jefferson to be his house-guest. While Jefferson had already sent Lewis and Clark on their expedition to explore the newly purchased Louisiana Territory and on to the Pacific, he was quick to probe Humboldt's mind and records not only for science but for other information as well.

In a few days, Humboldt and Bonpland were off to Europe and home after an absence of several years during which more was learned of the ecological relationships in tropical America than had ever been known before. His influence on the scientific exploration of the American West is difficult to judge. In a way, it was unfortunate that he did not arrive before the departure of Lewis and Clark – he certainly would have gone along. One can imagine what could have been learned of the environmental relationships of the biota in the North American prairies and mountains. As he said in *Cosmos*: 'In the great chains of causes and effects, no thing and no activity should be regarded in isolation'. It seems particularly appropriate in regard to North American ecosystems that this statement which is the essence of the holocoenotic approach to ecology appears in more direct words in John Muir's diary of 27 July 1869, written in the High Sierra (Muir, 1911).

The great biological explorers of the rest of the nineteenth century: Schouw, Darwin, Bates, Wallace, Spruce, Gray, Grisebach, DeCandolle were influenced by Humboldt's work and ideas. It was particularly the British naturalists who early in life had, or took, the opportunity to travel around the earth in the middle decades of the 19th century studying plants and animals. The biological wealth of the tropics, in particular, opened the eyes of these 'cold-winter' Europeans, as it had to Humboldt, to new and diverse kinds of organisms – and similarly opened their minds to questions, and to new and unsuspected answers. Again, as with Humboldt, they returned home after their travels, often never to travel again. The years were spent going over their collections, thinking about what they had seen, writing their books, and proposing their theories. Where would we be scientifically if neither Darwin nor Wallace had seen the world in this way? Certainly, others eventually would have come up with evolutionary ideas and the data upon which they must be based. But the work of these naturalists had to be done then, in the days of sailing ships and plenty of time. It had to be done before modern laboratories, fast transportation by airplane, rapid communication, and the 'need-to-produce' could influence young ecologists into today's 'get there, get the data, and get out' existence. While this latter approach has been productive of many new ideas and much information, and may be necessary in answering some of today's biospheric problems, there is also need for the long-term approach of the 19th century in modern form.

The impact of evolutionary theory, which emerged in large part from 19th century biogeography, on ecological thought and progress cannot be overestimated. If Humboldt advanced biogeography, both Wallace and Darwin have also influenced ecology considerably in the mid-20th century (Harper, 1967).

Simultaneously with the work of Darwin and Wallace, other naturalists continued their work. The phytogeographers Alphonse DeCandolle, Kerner, Asa Gray, Joseph Hooker, Grisebach, Drude, Warming, and A.F.W. Schimper tried to see and explain the earth as a phytogeographical whole by using the collected results of the great explorations including those of their own. It is not possible here to include all of their contributions to the development of ecology, and in particular to physiological ecology. Modern ecology was bound to emerge from their thinking. By the time of Kerner's publication of *Die Pflanzenleben der Donauländer* in 1863, ecology had arrived, even though still without a name. Kerner's later experimental plant ecology also anticipated physiological ecology.

The change from phytogeography to holistic ecology came gradually with much overlap. Kerner was essentially an ecologist while Grisebach (1877), with his relationships between world vegetation and climate, was a phytogeographer, and yet Grisebach's emphasis on life-forms indicated an ecological approach to causes. Drude (1890), again, was primarily a phytogeographer, both floristic and vegetational, but ecology shows up even more in his writing. His proposal of the phenomenon of 'acclimatization' drew phytogeography and plant physiology even closer together into an ecology of plants. But it was Eugen Warming in Denmark in his *Plantesamfund* (1895) and Andreas Franz Wilhelm Schimper's great *Pflanzengeographie auf Physiologische Grundlage* (1898) who really began to bring things together. These two men had by far the greatest influence not only on the science of ecology itself but on looking at the earth from an ecological point of view. Here, especially in Schimper's work, is a valid beginning of modern physiological ecology. To quote, in translation, from Schimper (1898) 'The ecology of plant-distribution will succeed in opening out new paths on condition only that it leans closely on experimental physiology, for it presupposes an accurate knowledge of the conditions of the life of plants which experiment alone can bestow'.

Neither Warming nor Schimper could have reached this point without their earlier associations with the botanists and plant physiologists of the latter third of the 19th century. Most of them were German: Sachs, Klebs, F. Haberlandt, Pringsheim,

Strasburger, and others. Justus von Liebig must be included here because of his pioneer work in agricultural chemistry, the nutritional ecology of crop plants, and the concept of limiting factors. These physiologists had an ecological approach and breadth of knowledge; it is no wonder that ecology was emerging as a holistic science. One wonders if ecologists are becoming too specialized today.

Returning briefly to Europe at the turn of the century, plant ecology and physiological ecology, in addition to the work of Warming and Schimper, were emerging in many places. Kerner was continuing his observations of the growth of plants in the Alps and Schröter was beginning his *Pflanzenleben der Alpen* which appeared in first edition in 1908. Schröter's classic was the start of an understanding of relationships between the growth of high mountain plants and their patterned and severe environments. V.H. Blackman (1919), in England, suggested and developed the technique of mathematical growth analysis which has proved so useful in recent years in phytotron studies (Patterson *et al.*, 1978). Raunkiaer (1934), in Denmark, as early as 1903 was attempting to characterize plant geographical regions by a floristic analysis of the life-forms present. Great importance was placed on leaf size and the positions of the perennating bud; very good ideas, indeed. And, in far away New Zealand, the self-trained Leonard Cockayne (1897) made the first experiments on the effects of freezing on survival and growth of alpine plants using a mutton refrigeration plant in Christchurch!

Forest ecologists, particularly in Germany, Russia, and Scandinavia, were studying growth, physiology, and site requirements of forest trees and also of the understory plants. Gutorovich as early as 1897 found that the use of ground cover within the taiga was an excellent way to differentiate stand types in this kind of forest. Morosov (1904) brought such information together in a classic paper which set up a classification of northern forests based on the use of ground cover plants and reproductive capacity of the trees as important indicators of soil conditions and, thus, of forest site quality. Much work continued along these lines in Russia up to World War I.

Meanwhile, in Finland, A.K. Cajander arrived at a similar plant indicator system for forest types based upon ground vegetation. His hypothesis, as published in *Ueber Waldtypen* (1909) was that these forest floor indicator plants characterized biologically equivalent sites. He found that growth of trees of a given species was quite different depending upon forest type, as defined by ground cover, but varied relatively little *within* a single forest type, as so defined. This is an early and practical union of phytosociology and physiological ecology. Cajander was later to become Prime Minister of Finland.

Studies on forest tree roots, their growth and distribution were also well-started by the turn of the century. One needs only to read the early works of Büsgen (1897), Engler (1903), and Wieler (1894) to know that the physiological ecology of roots was already an active subject of research in central Europe.

1.3 The spread of physiological ecology to North America

With all the activity in plant geography and ecology in Europe in that latter half of the 19th century, it could not be long before its influence was felt in North America. Asa Gray at Harvard had been close to Darwin's work in phytogeography since, mid-century. The real surprise was the sudden jump to the Midwest, particularly to the prairies, only recently settled. This was primarily due to the work of one man, Charles E. Bessey (Tobey, 1981). He was graduated from the Michigan Agricultural College in 1869, taught at Ames, Iowa, and eventually came under the influence of Asa Gray. He spent some time working with Gray and Farlow at Harvard. He moved to the University of Nebraska at Lincoln in 1884. His success there with students was phenomenal. Although he was a mycologist interested in plant diseases, he was the first real teacher of plant ecology in the United States. Among his ecology students were Roscoe Pound (later the famous dean of the Harvard Law School who remained, also, a life-long botanist) and Frederic E. Clements.

How did Bessey establish such an outstanding group of young ecologists in the 1880s and 1890s on the American frontier so isolated from Europe? It was quite an accomplishment; I believe that it was due to Bessey's wide knowledge, and to his kind, enthusiastic personality and ability as a great teacher. Who else could have inspired two young scientists, under 25 years of age, to tackle the research to do a *Phytogeography of Nebraska* only 16 years after the Battle of the Little Bighorn? In the book, published in 1897, the two authors (Pound and Clements) acknowledge not only Bessey but the inspiration provided by Drude's (1896) *Deutschlands Pflanzengeographie* and his earlier (1890) *Handbuch der Pflanzengeographie*. So, the European ecological influence extended to the still wild prairies and plains of North America.

Another, and independent, beginning of plant ecology in America was essentially contemporary with that at Nebraska. It, too, was in the Midwest but in the more urban surroundings of the University of Chicago. John M. Coulter, who had botanized in the Rocky Mountain West with the Hayden Expedition to the Yellowstone Plateau in 1872 and later in Colorado in the mid-1870s, had a young student, Henry Chandler Cowles, interested in the relationships between geological substrate and vegetational change. Cowles had been influenced by still another European source: Warming's (1895) *Plantesamfund*. Out of this enthusiasm emerged Cowles's classic series of papers in 1899 on succession on the sand dunes at the southern end of Lake Michigan. In 1901, these were followed by a series on physiographic ecology of the Chicago region.

Neither Clements nor Cowles appeared at first to be headed in the direction of physiological ecology. However, they both were, in their own research and writings, and later by the directions taken by a number of their students. For example, Clements in his 1907 book *Plant Physiology and Ecology* says 'The proper task of physiology is the study of the external factors of the environment in which the plant lives, and of the activities and structure which these factors call forth'. And, again, 'physiology was originally understood to be an inquiry into the nature of plants. This is the view that pervades the following pages, *and in accordance with this the subject matter of ecology is merged with that of physiology*' (italics mine). In contrast to the general view of

many ecologists, Clements was not all succession and climax. In fact, he maintained his interest and work in physiological ecology for the rest of his life, as in his controlled work on transpiration rates in coastal dune plants (Martin and Clements, 1939). His transplant gardens on Pike's Peak, and with H.M. Hall in California, led to the classic studies by Clausen, Keck, and Hiesey.

Clements had few doctoral students, but at least two were outstanding: John E. Weaver and Homer Leroy Shantz. Both maintained a life-long interest in the prairies, plains, and, in the case of Shantz, in the deserts to the west. Both men understood physiological ecology in its broadest sense. Weaver and his students studied grassland root systems in relation to soils and laid the foundations for our understanding of prairie ecosystems. He and Clements collaborated in using the plant itself as an instrument: the phytometer method (Clements and Weaver, 1918; Clements and Goldsmith, 1924). Shantz became an authority also on the relationships between wild plants and soils in arid climates. He perfected the use of plants as indicators of soil chemical and physical conditions (Shantz and Piemeisel, 1924). It is apparent, at once, that this group had their focus on the use of plants directly in ecological studies rather than using an indirect approach through the environment. In this way, they were the forerunners (with some other people) of modern physiological ecology. Also, since they worked on plants of grassland and desert ranges, they were among the pioneers in the science of range management.

One must return again to the late 19th century to find another and important step toward physiological ecology. In 1889, C. Hart Merriam, influenced by Humboldt and Bonpland's (1805) *Essai sur la Geographie des Plantes*, studied the elevational distribution of vegetation on the unexplored San Francisco Peaks which rise above the deserts of northern Arizona. The results of this expedition are in Merriam (1890). Merriam and his colleague, Vernon Bailey, discerned seven life zones extending from the desert, through forests, to cold alpine screes. Being much influenced by Humboldt, Merriam correlated these elevational zones with temperature, especially summer temperature, relating these to latitudinal equivalents. This rather tenuous relationship between elevation and latitude can be carried too far. Merriam also realized that desert plants in full sun and forest herbs in shade grew in microclimates quite different from the general zonal climates. He suggested that long-term climatic studies be set up in each of his zones, a suggestion that has never been followed up to this day on San Francisco Peaks, that ideal mountain for such measurements. Sixty years later environmental research of this kind was started, and still continues, on other mountains in the West particularly in the Front Range of Colorado (Marr, 1961).

The American Southwest, with its floristic diversity and environmental stress was a region ready for ecological studies; physiological ecology was to have its share (Billings, 1980). In addition to Merriam, other pioneers were Andrew Ellicott Douglass, an astronomer, who became interested in the relationships between climate, tree rings, and tree growth. He started his tree ring work on the Kaibab and Coconino Plateaus within sight of San Francisco Peaks during 1901 (Douglass, 1909) and continued it until well after mid-century. He founded the Tree-Ring Laboratory at the University of Arizona which, today, is the main center for the study of tree growth and physiology as influenced by weather and climate.

A little earlier, Frederick V. Coville began describing the vegetation and flora of the western American deserts and their mountains. Coville had received an A.B. degree from Cornell in 1887, taught there briefly, and went to work in 1888 as a botanist for the US Department of Agriculture, the same organization which put Merriam on San Francisco Peaks. Coville was sent on an expedition to explore the plant life of the Death Valley region. The results he reported in great detail (Coville, 1892).

The Death Valley work had a great influence on Coville. He could see the possibilities of desert research, and though his systematic and physiological work as Curator of the National Herbarium kept him pretty closely tied to Washington, he continued his interest in the desert. Not one to let an opportunity slip by, Coville suggested to the young Carnegie Institution of Washington that it set up a desert botanical laboratory in the Southwest to 'study the relation of plants to an arid climate, and to

substrata of unusual composition' (Coville and MacDougal, 1903). The Carnegie Institution immediately appointed Coville and Daniel T. MacDougal as a committee to look into this idea and to suggest a suitable place for the laboratory. After traveling through the Southwest during January and February 1903, the two men chose the city of Tucson which reciprocated by donating land on Tumamoc Hill west of town for a laboratory site and natural preserve in the Sonoran Desert. The Carnegie Institution soon built the laboratory and appointed W.A. Cannon as the first resident investigator.

It is difficult to overestimate the effects of this laboratory on desert ecological research and on the development of plant physiological ecology (Billings, 1980; McGinnies, 1981). It attracted many great botanists. With Cannon and MacDougal working on roots and water relations of desert plants, Forrest Shreve on the distributions of species and vegetations in relation to temperature and moisture, and Clements on indicator plants and the general plant ecology and paleoecology of the southwestern deserts, the Desert Laboratory was a most active center of physiological ecology in the first decades of the 20th century. It was here that Shreve recognized that each plant species had a different way of coping with desert environments thus anticipating Gleason's 'individualistic concept' and also Good's 'Theory of Tolerance' by many years (Shreve, 1915, and subsequent papers). Among the young physiological ecologists attracted to the Laboratory was Heinrich Walter in 1925; Walter was never to lose his interests in the water and temperature relations of desert plants. Although the old Desert Laboratory is no longer supported by the Carnegie, it is still being utilized by such ecologists as Paul Martin of the University of Arizona, Ray Turner of the US Geological Survey, and others. Also, it seems appropriate that the people of the Carnegie Institution Plant Biology Laboratory at Stanford and the Stanford University ecology group now carry on modern desert physiological ecology in Coville's old haunts in Death Valley.

By no means all plant physiological ecology in the United States during the first two decades of the 20th century was confined to the prairies, deserts, and western mountains. Burton E. Livingston at Johns Hopkins studying evapotranspiration, Edgar N. Transeau measuring the productivity of corn crops and the effects of drought stress on leaf structure, William S. Cooper, the ultimate geobotanist, and Robert F. Griggs working on timberlines were all active. They did not confine their work to physiological ecology alone. In fact, much of their research concerned vegetation, environments, and change. There was a strong stamp of independence on these men.

1.4 The middle decades

Until World War I, physiological ecology was primarily a field science with not many people involved: they were, however, very good ones. The equipment was often makeshift, logistics in the field were difficult and slow, but the observations were excellent.

The War took a tremendous toll of young scientists, particularly in Europe. But there were other reasons, too. The pioneers, especially in America, grew old, retired to their laboratories, became administrators, died. The cost of field work, with little financial support, became prohibitive. This lack of money also applied to instrumentation which, in general, because of the questions being asked, needed to be portable, reliable, yet reasonably priced. The 1920s were expensive times. The great teachers and theorists were mostly turning to the problems of vegetation and succession which were not only important, and centers of controversy, but also such research was relatively cheap to do. The few young people going into ecology were, with their teachers, doing exciting work on the structure and dynamics of vegetation.

The Depression of the 1930s meant even less money for research in ecology. With the great Midwestern drought in the middle of the Depression, the thoughts and work of plant ecologists were aimed toward the application of ecological principles to the solution of practical problems in grassland and forest management, and toward conservation. This trend, ably led by people such as Paul Sears, Aldo Leopold, Stanley Cain, H.J. Oosting, and Murray Buell set the stage for the solution of many of

the large ecological problems which face us today. Unfortunately, World War II devastated much of Europe and kept most young American ecologists busy at other things for three or four years. The combination of these events combined to create a noticeable gap in the development of physiological ecology.

The years from the 1920s until the late 1940s, however, were not completely lost in regard to physiological ecology. The young teachers and researchers of the 1920s and 1930s kept ecology alive during such discouraging times. Also, 'whole plant' physiologists and agronomists were continuing to direct their attention to plant–environment relationships. These included E.N. Transeau, Paul Kramer, Frits Went, Walter Loomis, William Hiesey and James Bonner in the United States, and V.H. Blackman, G.E. Blackman, H. Lundegårdh, Otto Stocker, N.A. Maximov, B. Huber, A. Pisek, Heinrich Walter, and others in Europe and Britain. All of these people had, in addition to their physiology, strong ecological interests. In fact, almost all of them published ecological papers. Went, for example, not only discovered auxin but also published papers on the sociology of epiphytes in tropical rainforests and on the ecology of desert annuals in the Death Valley region. In sum, they had the breadth of training and knowledge to bring physiology and ecology together. They and a few ecologists, notably Thorvald Sørensen and Tyge Böcher in Denmark and Charles Olmsted in the United States were the immediate ancestors of physiological ecology in its modern renaissance after World War II.

During the 1920s and 1930s, there was another great series of researches, first in Scandinavia and later in the United States, that combined the study of evolution with ecology, taxonomy, physiology, and genetics into the new and exciting field of genecology. It was Göte Turesson (1922) in Sweden who first recognized the role of 'ecotypes' and 'ecological races' as the genetically based populational units resulting from selection along environmental gradients or discontinuities. Such ecotypes within a species allow fitness to the different environments within the geographical range of the species. If the environments are not sharply defined but occur gradually along a gradient, the result of environmental selection is a genetic gradient or 'ecocline' (Gregor and Watson, 1954; Langlet, 1959).

Even before Turesson's classic and definitive research, F.E. Clements, taking his cue from Bonnier (1895) in the Alps, had begun transplant studies along an elevational gradient from 1830 m to 4260 m on Pike's Peak in Colorado (Clements and Hall, 1921). Hall, in 1922, began a similar series of transplant gardens along a transect from San Fransisco Bay across the Sierra Nevada. This, too, was supported by the Carnegie Institution. After Hall's death in 1932, the work was taken over by William M. Hiesey, Jens Clausen, and David D. Keck. This team, members of the Division of Plant Biology, Carnegie Institution, Stanford, with their colleagues in more recent years, Malcolm Nobs and Olle Björkman, worked together on the 'nature of species' along this transect. Their combination of physiology, cytogenetics, taxonomy, and ecology using 'ecological races' (ecotypes and ecolines) of a number of species has given physiological ecology tremendous impetus. All of this can be traced back to the early work of Clements, Hall, and particularly to Turesson, working by himself, in Sweden during the early 1920s.

One cannot overstate the impact of the work of Turesson and of Clausen, Keck, and Hiesey on the renewal of physiological ecology. Turesson, himself, realized that many of the answers to his ecotypes lay in physiology, but he was not equipped to do this kind of work. Olmsted (1944), at the University of Chicago, using growth chambers, showed that the grass *Bouteloua curtipendula* had photoperiodic ecotypes in relation to flowering along a latitudinal gradient from Arizona to Canada. In the same year, Hiesey began physiological experiments in the Earhart phytotron at Cal Tech with *Achillea* populations from the Carnegie transect across the Sierra Nevada. Both Olmsted and Hiesey were pioneers and mainly measured growth and flowering in the controlled environments; good instruments to measure metabolism and water potential were not available then.

1.5 The post-war revival of physiological ecology

About 1950, there was a renewal of interest in plant physiological ecology. This has continued at an ever-accelerating pace through the 1960s and 1970s until the present time all over the world. Why? The reasons are so many and so intertwined that it is not easy to answer this question. As one who was on the scene, I can only give my opinions.

First of all, much of this growth was due to the increasing numbers of young people coming out of the armed forces, interested in ecology, who were better trained in the physical sciences, mathematics, and physiology than their predecessors of the 1930s.

Secondly, the ecology teaching centers and leaders of the 1930s were ready to receive these students. Perhaps some of these older ecologists were not well-trained in physiological ecology, but they were receptive to new approaches. Also, an attempt was made to review the possibilities for modern research in the region of overlap between ecology and plant physiology (Billings, 1957). During the 1950s, the principal centres for this kind of training in the United States were Duke, Cal Tech, Emory, Wisconsin, Rutgers, and very few other places.

Thirdly, field logistics were greatly improved. With automobiles, good roads, airplanes that could land in out-of-the-way places, it was possible to get instrument systems into the field in places that were inaccessible before.

The rapid development of very good physiological and meteorological instrument systems also aided this research tremendously. Among these systems were infrared gas analysers which could be used to measure net photosynthesis and dark respiration continuously. Some physiologists were using these, particularly the British models, to great advantage especially in Kramer's laboratory at Duke. Some of us who were Oosting's students (Bormann, Bourdeau, Woodwell, and I) prevailed on Kramer's students to show us how they worked. But still, they were mostly laboratory instruments. In 1958, Mooney, Clebsch, and I took a battery-powered one to 3400 m in the alpine region of the Medicine Bow Mountains of Wyoming. We knew very little of cuvette cooling (we used snow) and many other aspects of IRGA operation but surprisingly got good photosynthetic data on alpine plants *in situ* (Billings *et al.*, 1966). Certainly, the German and Austrian physiological ecologists used such systems in their mobile laboratories about the same time (as did Went in his mobile laboratory), but not at such high elevations – and ours was definitely *not* a mobile laboratory, the pieces being assembled on the open alpine tundra with its wind, low temperatures, and lightning. Since then, Walter Moser has backpacked his IRGA to an even more inaccessible peak in the Tyrolean Alps. Improved instrument systems followed improved instrument systems. A good mobile laboratory now amazes anyone who measured gas exchange and water relations under the relatively primitive conditions of 25 years ago.

The development of small computers whose ancestors in the 1950s filled whole rooms, has aided data acquisition and reduction even in the field. One can carry in a pocket a mini-computer more powerful than the giants of 25 years ago.

Even with improvements in field instrumentation it would be impossible to do controlled experiments in physiological ecology without phytotrons and controlled environment growth chambers. It was in the 1940s that Frits Went built the first phytotron, the Earhart Laboratory at Cal Tech. Since most of us could not afford to build a phytotron, and even if we could, we would not have known how to operate it, we tried building small growth chambers in the 1950s, and succeeded fairly well. Among our early attempts were three which were built from supermarket display freezers. These were used for alpine and arctic plants and environments; they still are being used. Soon, refrigeration companies saw a ready and expanding market. Now, many laboratories have commercially available growth chambers. Many phytotrons, such as the one at Duke, use modular commercial growth chambers which can be modified to fit particular problems. The phytotron, of which there are now quite a number in different parts of the world, is a powerful tool in controlled physiological ecology studies.

None of this improvement in instrumentation and field logistics, nor the research, could have been done, then or now, without money. Such financial aid became available about 1950 from the National Science Foundation which had been founded in the late 1940s by an Act of Congress. Without the Foundation, the field of physiological ecology would

have developed much more slowly in the last 30 years. There was some help also from the US Atomic Energy Commission, Department of Defense, and the National Institutes of Health, and, more recently, the Department of Energy, and the Environmental Protection Agency. There was continued within-house support of some physiological ecology at the Department of Agriculture including the US Forest Service. Fortunately, this continues but not at the pace required if important biospheric problems are to be solved. The Carnegie Institution also finances some research in physiological ecology.

The impacts of funding and hardware on the rise of physiological ecology in the last 30 years have been mentioned. These were necessary, but without advances in the companion sciences of genecology, ecosystem ecology, and plant physiology, physiological ecology would have made little progress.

The long latitudinal gradient from the tropics to the Arctic in North America drew several of us into a combination of genecology and physiological ecology. We did indeed find ecotypes that differed in metabolism and other physiological characteristics within every widespread species no matter how long or short the environmental gradient or whether it went from warm to cold, wet to dry, or short day to long day.

Physiological ecology was aided also by the fact that, even though much of plant physiology was becoming oriented toward the molecular level, there was renewed interest by many plant physiologists in ecology and the effects of the environment. This group even included a number of molecular physiologists. The contributions of these environmental physiologists led to the building of the phytotrons. Ecology and physiology were approaching each other, one from the field and one from the laboratory. This natural synthesis of physiology, ecology, taxonomy, and genetics has been demonstrated nicely by Osmond *et al.* (1980) using the genus *Atriplex*, a taxon studied intensively by Hall and Clements (1923) in their pioneering work in biosystematics.

The last 25 years have also seen advances in plant physiology which have important ecological implications. These include the working out of the C_3 or Calvin cycle photosynthetic pathway and photorespiration, the C_4 or Hatch and Slack pathway, the complexities of the Crassulacean acid metabolism (CAM) pathway, and the discrimination against certain stable and unstable carbon isotopes in carbon fixation. All of these have been useful in helping to explain the relationships of certain plant taxa to summer temperature, to drought, and to ecosystem productivity. Gates, Knoerr, and others have worked out leaf energy budgets in different climates. The clarification of plant and cell water relationships by Kramer, Slatyer, and others together with the development by Scholander and others of techniques for measuring leaf water potentials, leaf diffusion resistance and conductance have been of considerable help to ecologists. Progress has been made in understanding mineral nutrition in wild plants. Chapin, Shaver and other ecologists have been quick to apply this knowledge to working out nutrient cycling patterns in natural ecosystems.

The rapid development of physiological ecology in the last 30 years has also gone hand in hand with the greatly increased understanding of ecosystem operation. This was actually begun by Raymond Lindemann in the early 1940s but was not really followed up in a powerful way until the early 1950s when the Odums and others became involved. Since then, physiological ecology has been an integral part of ecosystem and biosphere research. The International Biological Program of the late 1960s and 1970s occupied many of us in the physiological ecology of photosynthesis, plant growth, reproduction, and primary productivity. Without the knowledge and techniques of physiological ecology, the program in the various biomes would have been far less successful. Excellent examples of this kind of contribution to the understanding of ecosystem operation may be seen in Bliss (1977) and in Tieszen (1978).

1.6 Some thoughts about the future of physiological ecology

A history is no more than a story if it cannot be of help in guiding the future. Where should *we* take physiological ecology? The following are some opinions and suggestions.

We should continue the use of physiological ecology to help us toward a better understanding of the evolution of taxa at any level. We need more comparative studies of related species within a genus or within subgenera. These studies should not only be focused on maintenance (C_3 or C_4 photosynthesis, for example) of individuals and populations but also on the physiological ecology of reproduction, be it sexual or vegetative.

This logically leads to another need which is that of explaining species distributions: 'Why does that plant grow where it does?' What one should be aiming at here is the determination of tolerance ranges which make predictions of future geographical ranges possible.

Physiological ecology has an important role to play also in deciphering the structure and operation of ecosystems; this includes crop ecosystems. In addition to the usual techniques, there is a new one which is proving to be very useful. This concerns the use of stable carbon isotopes. Plants with C_3 photosynthesis show a greater discrimination against the heavier carbon isotope ^{13}C. They show a $\delta^{13}C$ value of about -26.5 to -28 while C_4 plants have a $\delta^{13}C$ value of approximately -12.5 (Tieszen *et al.*, 1979). This provides a label for the kinds of organic material derived from each of the photosynthetic types and can be followed into the herbivores of an ecosystem. By determining the $\delta^{13}C$ value for the carbon in such an animal, one can tell the nature of the grasses or browse upon which it has fed, as Tieszen and his colleagues have done. Recently, van der Merwe (1982) has used $\delta^{13}C$ values of prehistoric human skeletons to discover whether the diets of these people consisted of C_3, C_4 plants, or a mixture. He and Vogel (1978) have utilized the $\delta^{13}C$ values of such skeletal material to trace the northeastward movement of maize (C_4) from the prairies to the forests as a diet staple among prehistoric American Indians.

Boutton *et al.* (1980), and Harrison (personal communication, 22 April 1981) have shown that C_4 plant biomass decreases with elevation in eastern Wyoming while C_3 plant biomass increases. This is largely a function of decreasing summer temperatures and shortening of the growing season with increasing elevation. Harrison (personal communication, 22 April 1981) has carried this one step further. In a single excavation pit in Wyoming, the $\delta^{13}C$ values of bison bones nearer the surface of the present soil reflect a bison diet with a substantial proportion of C_4 grasses. With increasing depth in the pit, the older bison bones show $\delta^{13}C$ values that indicate a bison diet composed almost entirely of C_3 grasses about 8000 years ago when the climate was cooler and before many C_4 grasses had migrated into the ecosystem from warmer climate refugia.

Finally, the earth's biosphere is also an ecosystem. Owing to both natural and man-caused events, the biosphere is changing rapidly. Examples are the known and continuing increase in atmospheric carbon dioxide, the problem of acid precipitation, and the postulated decrease in the stratospheric ozone layer due to chlorofluoromethane release from industry and oxides of nitrogen from stratospheric flight. If the ozone depletion becomes serious, there is the danger of increased ultraviolet-B irradiation at the surface of the earth with possible damage to plants, people, and other living organisms. The possible effects of such increased UV-B irradiation have been reviewed by Caldwell (1979). It is apparent that physiological ecology has an important role to play here. Caldwell *et al.* (1982) have demonstrated, for example, that arctic taxa and ecotypes are more sensitive to photosynthetic inhibition by UV-B radiation than their alpine counterparts in the middle latitudes or in tropical mountains.

The carbon dioxide question is much more complex, more people are involved, and the effects likely to be more far-reaching. The great amount of literature on this problem is somewhat misleading since except for the knowledge that the carbon dioxide content of the atmosphere is increasing and that this is largely the result of the combustion of fossil fuels, most of the effects are postulated and not proven. There is considerable disagreement among scientists on what is going to happen to the biosphere as a result of the CO_2 increase. Here, physiological ecology can be of great assistance in actual field and phytotron experiments on the effects of increased CO_2 concentrations on photosynthesis, plant growth, primary production, and reproduction. At long last, such work is getting started using individual plants and also microcosm samples of real ecosystems (Billings *et al.*, 1982). It is becoming more obvious that physiological ecology has a job to do in regard to the carbon dioxide problem as well as on other problems posed by biospheric changes.

Physiological ecologists have an important and unique role to play in the understanding of ecosystemic and biospheric problems. The reason is relatively simple and straightforward: such ecologists deal with the mechanisms by which environment and organisms interact. If we are to understand how communities, ecosystems, and the biosphere function, with or without stress, these interactions must be known. Normal interactions are basic to the stability of the system. Because of their sensitivity, they are the first processes to show rate changes. These changes *can* be the mark of fitness for the individual or the local population. Conversely, if carried to extremes, they *may* be the first signals of instability in the ecosystem.

References

Benson, A.B. (ed.) (1937) *Peter Kalm's Travels in North America*, Revised from original Swedish, Wilson-Erickson, Inc., New York. 2 volumes.

Betts, E.M. (ed.) (1944) *Thomas Jefferson's Garden Book 1766–1824*, American Philosophical Society, Philadelphia.

Billings, W.D. (1957) Physiological ecology. *Annual Review of Plant Physiology*, **8**, 375–92.

Billings, W.D. (1980) American deserts and their mountains: An ecological frontier. *Bulletin of the Ecological Society of America*, **61**, 203–9.

Billings, W.D., Clebsch, E.E.C. and Mooney, H.A. (1966) Photosynthesis and respiration rates of Rocky Mountain alpine plants under field conditions. *American Midland Naturalist*, **75**, 34–44.

Billings, W.D., Luken, J.O., Mortensen, D.A. and Peterson, K.M. (1982) Arctic tundra: A source or sink for atmospheric carbon dioxide in a changing environment? *Oecologia*, **53**, 7–11.

Blackman, V.H. (1919) The compound interest law and plant growth. *Annals of Botany*, **33**, 353–60.

Bliss, L.C. (ed.) (1977) *Truelove Lowland, Devon Island,*

Canada: A High Arctic Ecosystem, University of Alberta Press, Edmonton.

Bonnier, G. (1895) Recherches experimentales sur l'adaptation des plantes au climat alpin. *Annales des sciences naturelles, Botanique, 7th series*, **20**, 217–358.

Botting, D. (1973) *Humboldt and the Cosmos*, Harper and Row, Publishers, New York.

Boutton, T.W., Harrison, A.T. and Smith, B.N. (1980) Distribution of biomass of species differing in photosynthetic pathway along an altitudinal transect in southeastern Wyoming grassland. *Oecologia*, **45**, 287–98.

Büsgen, M. (1897) *Bau und Leben unserer Waldbäume*, Gustav Fischer, Jena.

Cajander, A.K. (1909) Ueber Waldtypen. *Acta Forestalia Fennica*, **1**.

Caldwell, M.M. (1979) Plant life and ultraviolet radiation: Some perspective in the history of the earth's UV climate. *BioScience*, **29**, 520–5.

Caldwell, M.M., Robberecht, R., Nowak, R.S. and Billings, W.D. (1982) Differential photosynthetic inhibition by ultraviolet radiation in species from the arctic–alpine life zone. *Arctic and Alpine Research*, **14**, 195–202.

Chastellux, Marquis Francois Jean de (1787) *Travels in North America in the Years 1780, 1781, and 1782*, translated from French by 'An English Gentleman' in 1827, White, Callaher, and White, New York.

Clements, F.E. (1907) *Plant Physiology and Ecology*, Henry Holt and Co, New York.

Clements, F.E. and Goldsmith, G.W. (1924) *The Phytometer Method in Ecology: The Plant and Community as Instruments*, Carnegie Institution, Washington.

Clements, F.E. and Hall, H.M. (1921) Experimental taxonomy. *Carnegie Institution Washington Year Book*, **20**, 395–6.

Clements, F.E. and Weaver, J.E. (1918) The phytometer method. *Carnegie Institution Washington Year Book*, **17**, 288.

Cockayne, L. (1897) On the freezing of New Zealand alpine plants. Notes of an experiment conducted in the freezing-chamber, Lyttleton. *Transactions and Proceedings of the New Zealand Institute*, **30**, 435–42.

Coville, F.V. (1892) Botany of the Death Valley Expedition. *Contributions from the United States National Herbarium*, **4**, 1–363, plus 21 plates and map.

Coville, F.V. and MacDougal, D.T. (1903) Desert Botanical Laboratory of the Carneigie Institution. *Carnegie Institution of Washington Publication* No. 6, 1–58, plus plates.

Cowles, H.C. (1899) The ecological relations of the vegetation on the sand dunes of Lake Michigan. *Botanical Gazette*, **27**, 95–117, 167–202, 281–308, 361–91.

Cowles, H.C. (1901) The physiographic ecology of Chicago and vicinity. *Botanical Gazette*, **31**, 73–108, 145–81.

Diels, L. (1937) Beiträge zur Kenntnis der Vegetation und Flora von Ecuador. *Bibliotheca Botanica*, **29**(116), 1–190.

Douglass, A.E. (1909) Weather cycles in the growth of big trees. *Monthly Weather Review*, **37**, 225–37.

Drude, O. (1890) *Handbuch der Pflanzengeographie*, J. Engelhorn, Stuttgart.

Drude, O. (1896) *Deutschlands Pflanzengeographie*, J. Engelhorn, Stuttgart.

Egerton, F.N. (1976) Ecological studies and observations before 1900, in: *Issues and Ideas in America*. (eds. B.J. Taylor and T.J. White), University of Oklahoma Press, Norman, pp. 311–51.

Engler, A. (1903) Untersuchungen über das Wurzelwachstum der Holzarten. *Mitteilungen der schweizer Zentralanstalt für das forstliche Versuchswesen*, **7**, 247–317.

Evenari, M. (1980/81) The history of germination research and the lesson it contains for today. *Israel Journal of Botany*, **29**, 4–21.

Gregor, J.W. and Watson, P.J. (1954) Some observations and reflexions concerning the patterns of intraspecific differentiation. *New Phytologist*, **53**, 291–300.

Grisebach, A. (1877) *La Vegetation du Globe*, two volumes, Librarie J.-B. Bailliere et Fils, Paris.

Gutorovich, I.I. (1897) Notes of a northern forester. (in Russian). *Lesnoi Zhurnal*, **27**.

Hall, H.M. and Clements, F.E. (1923) The phylogenetic method in taxonomy. *Carneigie Institution Washington*, **326**, 1–355 plus 58 plates.

Harper, F. (ed.) (1958) *Bartram, William. Travels*, edited with commentary and an annotated index by Francis Harper, Naturalist's edition, Yale University Press, New Haven.

Harper, J.L. (1967) A Darwinian approach to plant ecology. *Journal of Ecology*, **55**, 247–70.

Humboldt, A. and Bonpland, A. (1805) *Essai sur la Geographie des Plantes*, Chez Levrault, Schoell et Compagne, Paris.

Kerner, A. (1863) *Die Pflanzenleben der Donauländer*, Vienna, Translated by H.S. Conard (1951) as *The Background of Plant Ecology*, Iowa State College Press, Ames.

Langlet, O. (1959) A cline or not a cline – a question of Scots pine. *Silvae Genetica*, **8**, 1, 13–36.

Livingston, B.E. (1935) Atmometers of porous porcelain and paper, their use in physiological ecology. *Ecology*, **16**, 438–72.

Marr, J.W. (1961) Ecosystems of the east slope of the Front Range in Colorado. *University of Colorado Studies, Series in Biology*, **8**, 1–134.

Martin, E.V. and Clements, F.E. (1939) *Adaptation and Origin in the Plant World. I. Factors and Functions in Coastal Dunes*, Carnegie Institution, Washington.

McGinnies, W.G. (1981) *Discovering the Desert: Legacy of the Carnegie Desert Botanical Laboratory*, University of Arizona Press, Tucson.

Merriam, C.H. (1890) Results of a biological survey of the San Francisco Mountain region and desert of the Little Colorado, Arizona, *North American Fauna*, No. 3, 1–136 plus plates and maps.

Morosov, G.F. (1904) On stand types and their importance in forestry (in Russian). *Lesnoi Zhurnal*, **34**, 6–25.

Muir, J. (1911) *My First Summer in the Sierra*, Houghton Mifflin Company, Boston.

Mulroy, J.C. (1979) *Contributions to the Ecology and Biogeography of the Saxifraga cespitosa L. Complex in the Americas*, PhD dissertation, Duke University, Durham, North Carolina.

Olmsted, C.E. (1944) Growth and development in range grasses. IV. Photo-periodic responses in twelve geographic strains of side-oats grama. *Botanical Gazette*, **106**, 46–74.

Osmond, C.B., Björkman, O. and Anderson, D.J. (1980) *Physiological Processes in Plant Ecology: Toward a Synthesis with Atriplex*, Ecological Studies, 36, Springer-Verlag, New York.

Patterson, D.T., Meyer, C.R. and Quimby, P.C. Jr (1978) Effects of irradiance on relative growth rates, net assimilation rates, and leaf area partitioning in cotton and three associated weeds. *Plant Physiology*, **62**, 14–17.

Pound, R. and Clements, F.E. (1897) *The Phytogeography of Nebraska*, Botanical Seminar, Lincoln.

Raunkiaer, C. (1934) *The life forms of plants and statistical plant geography*, Clarendon Press, Oxford.

Schimper, A.F.W. (1898) *Pflanzengeographie auf Physiologische Grundlage*, Verlag von Gustav Fischer, Jena.

Schröter, C. (1908) *Pflanzenleben der Alpen*, Verlag von Albert Raustein, Zürich.

Shantz, H.L. and Piemeisel, R.L. (1924) Indicator significance of the natural vegetation of the south-western desert region. *Journal of Agricultural Research*, **28**, 721–802. 14 pl.

Shreve, F. (1915) *The Vegetation of a Desert Mountain Range as Conditioned by Climatic Factors*, Publication 217, Carnegie Institution of Washington.

Tieszen, L.L. (ed.) (1978) *Vegetation and Production Ecology of an Alaskan Arctic Tundra*, Ecological Studies, 29, Springer-Verlag, New York.

Tieszen, L.L., Hein, D., Qvortrup, S.A., Troughton, J.H. and Imbamba, S.K. (1979) Use of $\delta^{13}C$ values to determine vegetation selectivity in East African herbivores. *Oecologia*, **37**, 351–9.

Tobey, R.C. (1981) *Saving the Prairies: The Life Cycle of the Founding School of American Plant Ecology, 1895–1955*, University of California Press, Berkeley.

Turesson, G. (1922) The genotypic response of the plant species to the habitat. *Hereditas*, **3**, 211–350.

van der Merwe, N.J. (1982) Carbon isotopes, photosynthesis, and archeology. *American Scientist*, **70**, 596–606.

van der Merwe, N.J. and Vogel, J.C. (1978) ^{13}C content of human collagen as a measure of diet in Woodland North America. *Nature*, **276**, 815–16.

Warming, E. (1895) *Plantesamfund*, Copenhagen.

Wieler, A. (1894) Ueber die Periodicität in der Wurzelbildung der Pflanzen. *Forstwissenschaftliche Zentralblatt*, **16**, Heft 7.

Williams, S. (1794) *The Natural and Civil History of Vermont*, Walpole, New Hampshire.

2

Arctic

F. STUART CHAPIN III AND GAIUS R. SHAVER

2.1 Introduction

The Arctic has received considerable attention from plant physiological ecologists, because the extreme nature of the physical environment suggests dramatic physiological adjustment by organisms that live there. In North America, Dwight Billings and his associates have played a major role in describing the nature of adaptations of arctic organisms to their environment (e.g. Bliss, 1956, 1962a, 1971; Billings and Mooney, 1968; Mooney and Billings, 1961). These and other studies (e.g. Lewis and Callaghan, 1976; Savile, 1972; Warren Wilson, 1966) suggest that low temperature is a major factor limiting growth and activity of arctic plants but that arctic plants exhibit many adaptations allowing effective metabolism at low temperature.

It is common for ecotypes of a single species to be adapted to extemely different thermal regimes, whereas one must go to the generic or family level to find plants from radically different light or moisture regimes (Mooney, 1975). This suggests that adaptation to low temperature occurs readily in the course of evolution. The temperature response of growth changed after only 20 years of selection under a new thermal regime in one Georgia annual (Christy and Sharitz, 1980). If plants adapt so readily to low temperature, how important is low temperature in influencing the growth of arctic plants? Recent work suggests that other environmental factors also strongly influence the physi-

ology and growth of arctic plants. In this chapter we explore the direct and indirect consequences of low temperature and other aspects of the arctic environment for the growth of arctic plants. We attempt to pinpoint (a) attributes that distinguish arctic from temperate plants, and (b) environmental factors responsible for variation in growth within the arctic. Many of our conclusions derive from studies conducted under the International Biological Program. Syntheses of this work should be consulted for additional information (Bliss, 1977a; Bliss *et al.*, 1973, 1981; Brown *et al.*, 1980; Heal and Perkins, 1978; Sonesson, 1980; Wielgolaski, 1975a). Excellent detailed reviews of earlier work in arctic physiological ecology are also available (Billings and Mooney, 1968; Bliss, 1962a, 1971; Savile, 1972).

2.2 Environment

The Arctic is the treeless zone with a southern boundary approximately coincident with the mean summer position of the arctic front (Fig 2.1, Barry *et al.*, 1981; Bliss *et al.*, 1973; Hare, 1968). The polar air mass responsible for the arctic climate is cold, due to low annual radiation input. This results in negative (°C) mean annual air temperature and therefore presence of permafrost (permanently frozen ground), a short snow-free season for plant growth, low atmospheric moisture content, and low precipitation.

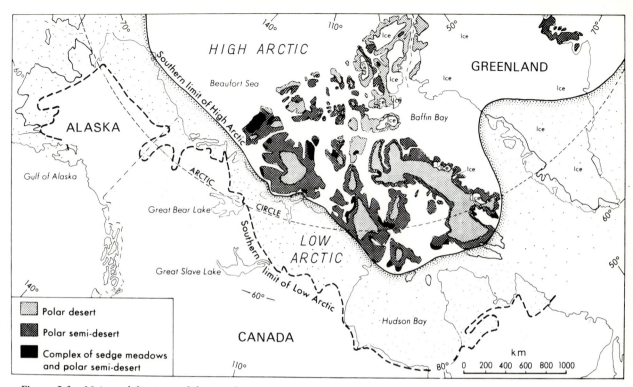

Figure 2.1 Major subdivisions of the North American and Greenland arctic (from Bliss, 1981).

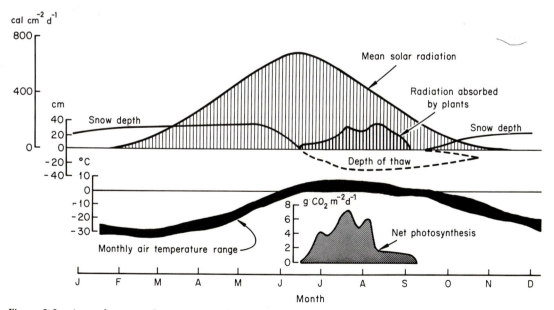

Figure 2.2 Annual pattern of environmental variables and net photosynthesis at Barrow, Alaska (after an article in *Mosaic*, 1980).

The 24-h photoperiod of the arctic summer (from snow-melt until late July or mid-August) compensates for low intensity of radiation, so that the daily total radiation input in July is similar in arctic and temperate regions (Barry *et al.*, 1981; LeDrew and Weller, 1978; Fig. 2.2, Table 2.1). Radiation is maximal at snow-melt and decreases through the growing season. Coastal areas and islands, which comprise a large proportion of the vegetated arctic, experience frequent summer cloudiness, reducing incoming solar radiation by 50% or more (Weller and Holmgren, 1974).

Air temperatures during the growing season are low (Table 2.1), and frosts or temporary snow cover can occur at any time. The end of the growing season is particularly variable from year to year (Myers and Pitelka, 1978). The growing season includes all the time that plants are snow-free and thawed, approximately 10–14 weeks in the low arctic and 6–10 weeks at higher latitudes. Annual precipitation is generally 50–250 mm, similar to rainfall in temperate deserts, but less than half of this falls during the growing season (Fig. 2.3, Dingman *et al.*, 1980; Ryden 1977, 1981). Water from winter snow-melt runs off in 3–10 days over the soil surface and through the upper few centimeters of thawed soil (Outcalt *et al.*, 1975; Courtin and Labine, 1977; Rydèn, 1981). Despite low precipitation, soils in the rooting zone are generally wet (> 40% volumetric water content), because permafrost prevents drainage, except in polar deserts and certain well-drained microsites (e.g. rock fellfield, sand dunes, and dry ridges) where thaw depth is greater (Dingman *et al.*, 1980; Rydèn, 1977). Polar deserts are wet only during the period immediately following snow-melt, because there is low precipitation input and deep thaw due to lack of surface soil organic mat which would otherwise insulate the soil (Bliss, 1981).

In wet tundra about 70% of summer net radiation is dissipated in evaporating water, leaving little energy to melt and warm the soil (2–9%) or warm the air (18–32%) (Weller and Holmgren, 1974; Barry *et al.*, 1981). In polar semidesert there is less evaporation from the soil surface, but little vegetation to prevent large convective heat losses, so there is only slightly greater heat input to soil (4–16% of total) (Addison and Bliss, 1980; Courtin and Labine, 1977). In shrub tundra about 20% of net radiation is dissipated in heating soil (Haag and Bliss, 1974). Thus, there is a relatively small soil heat flux in arctic tundra, and soils are colder than

Table 2.1 Environmental characteristics of polar semidesert, arctic lowland, temperate alpine, and temperate grassland sites

	Polar semidesert Devon Island[*]	Arctic lowland Barrow[†]	Alpine Niwot Ridge[‡]	Temperate grassland Pawnee[§]
Radiation				
July mean global solar radiation (MJ m^{-2}d^{-1})[¶]	23.9	18.0	20.1	24.1
July mean daytime radiation (MJ m^{-2}h^{-1})[¶]	0.87	0.85	1.34	1.61
Maximum photoperiod (h)	24	24	15	15
Temperature (°C)				
Air (July mean)	3.0	3.7	12.5	22.0
(January mean)	−34.4	−25.9	−10.1	−3.8
Soil (July mean at 10 cm)	8.8	2.1	14.3	—
Maximum thaw depth (cm)	60	32	∞	∞
Length of growing season (d)	53	70	80	193
Precipitation (mm)				
Annual	185	170	1020	311
Summer	42	63	207	217
Soil moisture (% by volume)	—	80	—	—
Wind (July mean) (m s^{-1})	3.1	4.8	5.8	—

[*]Beach Ridge; Courtin and Labine, 1977; Bliss, 1977b; Labine, personal communication; Rydèn, 1977.
[†]Billings and Mooney, 1968; Chapin, 1981; Dingman *et al.*, 1980.
[‡]Billings and Mooney, 1968; Barry, 1973; Chapin, 1981.
[§]Lauenroth personal communication; Sims *et al.*, 1978.
[¶]1 W m^{-2} = 0.086 MJ m^{-2}d^{-1} = 2.06 cal cm^{-2}d^{-1} = 2.06 ly d^{-1} = 6.34 μE m^{-2}s^{-1} at 760 nm = 58 footcandles at 555 nm = 650 lux.

Figure 2.3 Arctic plants are adapted to withstand snow and low temperatures at any time of year as with this July snowstorm in Alaskan tussock tundra. Tundra is characterized by absence of trees and may be dominated by plants from any of several growth forms, in this case by a mixture of evergreen and deciduous shrubs, single-shooted and tussock-forming graminoids, mosses, and lichens.

summer air temperature and radiation regimes might suggest.

Local topographic gradients in plant production are closely tied to gradients in soil conditions (Bell and Bliss, 1978; Billings and Mooney, 1968; Shaver *et al.*, 1979) and are almost as pronounced as gradients spanning the latitudinal range of arctic tundra. This suggests that the soil environment is more important than aerial environment in determining the range of productivity in tundra (Bell and Bliss, 1978; Shaver *et al.*, 1979), although microhabitat differences in wind speed and therefore in leaf temperature may also be important (Bliss, 1956; Warren Wilson, 1959).

The nature of the soil environment is largely conditioned by the presence of permafrost, a layer of frozen soil and bedrock up to 400 m thick (Bliss *et al.*, 1973). Growing-season soil temperatures are low and decrease exponentially from the soil surface to permafrost (Table 2.1). During winter the soil is frozen completely. Soil thaw proceeds quickly immediately following snowmelt, but does not reach maximum depth (20–80 cm) until late August (Addison and Bliss, 1980; Everett *et al.*, 1981; Fig. 2.2). In autumn, soils remain partially thawed, and biological activity continues even after snow and freezing air temperatures halt aboveground plant growth. Soil freezes from the top and bottom toward the center of the thawed zone (Fig. 2.2; Outcalt *et al.*, 1975), perhaps concentrating soluble nutrients at the freezing front in soils with uniformly high soil water content. The freeze-thaw process *per se* increases concentrations of soluble phosphate and ammonium in organic soils (Saebø, 1969; Summerfield and Rieley, 1973).

Nutrient availability is generally low in arctic soils because (a) low precipitation limits atmospheric input, (b) low temperature virtually eliminates chemical weathering of parent material, (c) low temperature minimizes nitrogen fixation rates, and (d) low temperature and frequently poor aeration (low soil oxygen content) reduce rates of nutrient release from soil organic matter (Babb and Whitfield, 1977; Chapin *et al.*, 1978, 1980b; Dowding *et al.*, 1981; Ellis, 1980). Nitrogen is available almost entirely as ammonium rather than nitrate (Everett *et al.*, 1981; Flint and Gersper, 1974). Although nutrient availability is generally low, it is highly heterogeneous, with concentrations varying by an order of magnitude between adjacent microhabitats (Barèl and Barsdate, 1978). Concentrations of soil oxygen and available nutrients and rates of nutrient input and turnover decrease substantially with depth, and 90% of the root and microbial biomass and biological activity occur within the top 5 cm of soil (Barèl and Barsdate, 1978; Chapin *et al.*, 1978, 1979; Flint and Gersper, 1974). Thus, most biotic activity in tundra occurs in a narrow band from 5 cm below to 10 cm above the soil surface. As in many ecosystems, there is a flush of nutrient availability in spring (perhaps 40% of annual turnover) followed by large fluctuations throughout the rest of the summer (Barèl and Barsdate, 1978; Chapin *et al.*, 1978; Everett *et al.*, 1981).

Frost features such as ice-wedge polygons, frost scars, and stone stripes play a key role in shaping the mosaic of communities that make up arctic tundra and in determining patterns of seedling establishment and succession (Billings and Mooney, 1968; Bliss, 1971; Britton, 1966; Churchill and Hanson, 1958; Hopkins and Sigafoos, 1951; Shaver *et al.*, 1979). Frost features create micro-environments with radical differences in soil moisture, soil nutrients, and soil stability (op cit.). The development of these frost features is a consequence of low autumn and winter temperatures acting upon soil that frequently has a high water content (e.g. Lachen-

bruch, 1962). Other sources of disturbance that differ in importance among arctic areas include fire (Wein and Bliss, 1973; Racine, 1981) and animal activity (McKendrick *et al.*, 1980).

2.3 Vegetation and growth forms

Arctic vegetation is highly variable. In contrast to most other major vegetation types, which are defined by the dominant plant growth form (e.g. 'deciduous forest' or 'grassland'), arctic vegetation is defined by latitude, climate, and the absence of trees (Fig. 2.3). The dominant growth form may be a vascular plant, moss, or lichen; woody or herbaceous; erect, prostrate or caespitose; deciduous, wintergreen, or evergreen (Bell and Bliss, 1977; Bliss, 1981; Wielgolaski, 1975b). Growth form composition and vegetation vary dramatically along short mesotopographic gradients as well as regionally and latitudinally (Billings and Mooney, 1968; Bliss, 1956, 1981; Miller, 1982; Webber, 1978; Wielgolaski, 1975b).

Bliss (1981) has divided North American arctic vegetation into two major types: (a) tundra with 100% cover and wet-to-moist soil and (b) polar desert or semidesert with frequently less than 5% cover and dry soil (Fig. 2.1). With the decrease in precipitation to the north, there is a general increase in proportion of semidesert and desert vegetation and a progressive restriction of the wetter tundras to lowlands. The mosaic distribution of arctic vegetation types is thus superimposed upon a latitudinal gradient of frequency of each type (Table 2.2).

The lack of trees and the low stature of arctic vegetation have been ascribed both to winter desiccation and ice abrasion and to a low annual carbon gain that is inadequate to build and maintain large woody support structures (Billings and Mooney, 1968; Savile, 1972; Tranquillini, 1979). Morphological adaptations to wind include the prostrate or caespitose growth habit, branch layering (rooting) in shrubs, and protection of the perennating bud by resinous scales or dead leaves. Shrubs greater than 1 m tall usually grow only where snow cover protects them from wind. However, the lack of krummholz at the arctic treeline and the occurrence of deciduous trees near year-round warm (0–10 °C)

Table 2.2 Major North American arctic vegetation types, after Bliss (1981). Biomass, production and root: shoot values are estimates compiled from Bliss (1977), Chapin and Shaver (unpublished), Chapin *et al.*, (1979), Miller (1982), Miller *et al.*, (1982), Rosswall and Heal (1975), Shaver and Cutler (1979), Tieszen (1978a), Wein and Bliss (1974).

	Deserts and semideserts		Tundras				
	Polar desert	Polar semidesert	Wet sedge-moss	Tussock	Low shrub	Tall shrub	Heath
Dominant growth form	Cushion plants, rosettes	Caespitose graminoids, cushion plants, rosettes, lichens	Rhizomatous graminoids, mosses	Tussock graminoids, low deciduous and evergreen shrubs, mosses	Low deciduous and evergreen shrubs	Tall deciduous shrubs	Low evergreen shrub, lichen
North Amer. distribution	Mid- to high-Arctic uplands	Discontinuous in south, continuous in mid- to high-Arctic	Continuous in south, patchy in north	N. slope Alaska and mainland Canada	Mainly N. slope Alaska and mainland Canada	River bottoms in Alaska and mainland Canada	Mountains and dry ridges
Topography/ soils	Slopes and well-drained plains	Slopes and well-drained plains	Flat, poorly drained lowlands, fine soils	Lowlands and gentle slopes, fine soils	Gentle slopes, coarser soils	Coarse, well-drained soils	Coarse, well-drained soils
Soil moisture	Moist (spring), then very dry	Moist–dry	Very wet; standing water common	Wet–moist	Moist	Wet–moist	Dry–moist
Vegetation height	0–5 cm	0–10 cm	10–20 cm	10–40 cm	20–60 cm	up to 2 m	0–10 cm
Vascular cover (%)	< 1	10–20	80–100	80–100	80–100	50–100	10–70
Aboveground primary production ($g\,m^{-2}\,yr^{-1}$)	< 2	2–20	30–120	40–100	40–80	n.d.	10–50
Aboveground vascular biomass ($g\,m^{-2}$)	5–20	20–100	50–250	200–400	150–500	n.d.	50–200
Vascular root:shoot ratio	< 0.6:1	0.5–2:1	8–12:1	1–5:1	2–5:1	n.d.	n.d.
Aerial extent ($10^6\,km^2$)	0.80	1.50	1.00	0.90	1.28	0.23	n.d.

springs in Alaskan North Slope tundra (Murray, 1980) suggest that soil temperature and related factors may be equally as important as wind in explaining the general absence of trees in tundra. Warmer soil temperatures may allow a longer growing season, greater nutrient absorption, and therefore improved carbon and nutrient balance. Alternatively, the presence or absence of trees may be determined largely by controls over present and past reproduction and seedling establishment rather than by current controls over photosynthesis and growth (Black and Bliss, 1980). For plants with perennating buds or major storage organs beneath the soil surface, soil temperature and rate of thaw in spring may strongly limit both growth rate and the effective length of the growing season (Bell and Bliss, 1978).

The arctic flora has very few annual plants, presumably due to the difficulty of completing the entire life cycle in a single cold growing season of 6–12 weeks (Billings and Mooney, 1968; Bliss, 1971; Savile, 1972). Desert annuals, on the other hand, reproduce in a similar length of time but at much warmer air and soil temperatures.

Mosses are major components of arctic vegetation. Moss biomass may be several times the vascular plant biomass, particularly in wet to mesic areas (Bliss, 1977b; Britton, 1966; Clarke *et al.*,

1971; Rastorfer, 1978; Vitt and Pakarinen, 1977; Webber, 1978). However, moss productivity rarely exceeds vascular plant productivity except in local microhabitats, and the widespread moss mats characteristic of antarctic sites are not found in the Arctic (Bliss, 1979). Lichens are relatively abundant on dry ridges with low snow cover (Vitt and Pakarinen, 1977), and in tussock tundra but contribute little to annual productivity (Wielgolaski *et al.*, 1981).

In addition to the diversity of plant growth forms above ground there is considerable diversity in rooting patterns, rooting depths, and root:shoot ratios both among sites (e.g. Bliss, 1977b; Webber 1977, 1978; Wein and Bliss, 1974; Wielgolaski, 1975b), and among species within a site (Bell and Bliss, 1978; Shaver and Billings, 1975; Shaver and Cutler, 1979). Root:shoot ratio and rooting depth are not well-correlated with above-ground growth form, although there is a trend toward decreased root:shoot ratios in drier sites (Table 2.2; Webber, 1977). Despite the shallow depth of thawed soil (usually 20–50 mm), the dominant species partition the soil environment through differing root distributions and seasonality of root growth (Bell and Bliss, 1978; Shaver and Billings, 1975; Shaver and Cutler, 1979).

2.4 Phenology and growth

Patterns of growth in arctic plants are similar to those of temperate species, but growth occurs at lower temperatures, responds to different environmental cues, and is condensed into a shorter time period. Leaf and flower buds of arctic plants overwinter in an advanced state of development, and spring growth begins as soon as bud temperatures increase to 0 °C for even a few hours each day (Billings and Mooney, 1968; Bliss, 1962a, 1971; Sørensen, 1941). As in most ecosystems, above-ground spring growth of arctic deciduous forbs and shrubs is rapid due to rapid and nearly synchronous leaf expansion, but graminoids initiate leaves throughout the growing season (Bliss, 1971; Johnson and Tieszen, 1976; Tieszen, 1978a). Graminoid leaves produced late in the season continue elongation the following year. Leaves of deciduous species gain weight until early August, then lose weight as breakdown products are trans-

located to stems or rhizomes for winter storage (Bliss, 1956; Johnson and Tieszen, 1976; Muc, 1977; Tieszen, 1972). In polar semidesert, leaves of many 'deciduous' species are wintergreen, continue expanding and functioning in a second growing season (Bell and Bliss, 1977), and then are shed. In contrast, leaves of evergreen shrubs are initiated 2–4 weeks following snowmelt, gain weight throughout the first growing season, and then continue to photosynthesize for two or more years, (Bliss, 1956; Johnson and Tieszen, 1976; Svoboda, 1977). Similar seasonal patterns of growth over longer seasons have been documented in trees, shrubs, forbs, and graminoids of the temperate zone (Grigal *et al.*, 1976; Kozlowski, 1971; Sims and Singh, 1978a). Growth of mosses and lichens has no distinct seasonal pattern but occurs whenever conditions are suitable (Callaghan *et al.*, 1978; Oechel and Sveinbjörnsson, 1978).

In vascular deciduous species above-ground growth occurs most rapidly just following snowmelt, when solar radiation and air temperature are most favorable (Fig. 2.1). In contrast, root growth continues at substantial rates until late July as the soil continues to warm; there is an August decrease in root elongation that is photoperiodically controlled (Kummerow and Russell, 1980; Shaver and Billings, 1977). Growth of primary roots begins as soon as the soil thaws, but fine lateral roots generally grow most actively during the latter part of the growing season (Bell and Bliss, 1978; Shaver and Billings, 1975). The concentration of above-ground growth in early spring and of below-ground growth in mid- and late summer enables growth of each plant part to occur when it experiences most favorable conditions.

Not surprisingly, tundra plants are well adapted to grow at low temperatures. Arctic and alpine plants have a lower optimum temperature (15–20 °C) for shoot growth than temperate species (25–30 °C) (Kummerow *et al.*, 1980; Scott, 1970; Tieszen, 1978b). Similarly, root growth by arctic plants is less temperature-sensitive and in some studies has a lower optimum temperature than that of temperate species (Bell and Bliss, 1978; Billings *et al.*, 1978; Chapin, 1974; Ellis and Kummerow, 1982; Shaver and Billings, 1975). In *Eriophorum angustifolium* at Barrow, the most rapid root elongation rates occur deep in the thawed layer, at the lowest soil temperatures (1–4 °C), suggesting that root growth in the field is controlled more by allocation patterns than by temperature (Shaver and Billings, 1977).

To a limited extent arctic plants ameliorate their thermal environment. The small stature and compact nature of arctic plants, particularly tussock and cushion growth forms, maximize the thickness of the layer of still air (boundary layer) between plant and bulk air, so that leaf temperatures can be 3–8 °C (occasionally as much as 20 °C) warmer than bulk air (Bliss, 1962a; Miller *et al.*, 1976; Mølgaard, 1982; Oechel and Sveinbjörnsson, 1978; Warren Wilson, 1957a). Willow stems propped above the ground surface to minimize boundary layer effects produced 30–40% less biomass than controls (Warren Wilson, 1959). Similarly, the tussock

growth form common in arctic and alpine environments raises soil temperature, enhancing root growth and nutrient release within the tussock (Chapin *et al.*, 1979). The parabolic shape and heliotropic nature of many arctic flowers raise ovary temperature 2–10 °C, thus speeding development and increasing the temperature and activity of pollinators (Hocking and Sharpin, 1965; Kjellberg *et al.*, 1982; Mølgaard, 1982).

Despite these physiological and morphological adaptations common to arctic and alpine plants, shoot temperature is generally less than that required for maximum growth. Growth and phenological development are more rapid in warm microsites (Bliss, 1956; Sørensen, 1941; Warren Wilson, 1959) and when plants are transplanted to a warmer climate (Chapin and Chapin, 1981). The maximum and the average relative growth rate $(gg^{-1}d^{-1})$ of tundra graminoids growing *in situ* is at

Table 2.3 Maximum relative growth rate (RGR) and annual production of shoots of tundra and temperate graminoids. Rates were calculated from seasonal changes in above-ground biomass. Rates chosen were maximum rates, before water limitation or retranslocation reduced growth rate. The seasonal average relative growth rate showed similar differences between sites. From Chapin (1983).

Species	Site	Latitude	RGR $(mg\,g^{-1}d^{-1})$	Annual production $(g\,m^{-2})$	Reference
Arctic					
Carex aquatilis	Barrow, Alaska	71°18'	67	—	Tieszen (1972)
Duponita fischeri			92	—	
Eriophorum angustifolium			128	—	
Wet meadow tundra			93	101	
Eriophorum vaginatum	Atkasook, Alaska	70°27'	78	—	Chapin *et al.* (1980a)
Carex meadow	Devon Is, Canada	75°33'	37	38	Muc (1977)
Alpine					
Sedge Meadow	Mt. Washington, N.H.	44°16'	64	187	Bliss (1966)
Heath-rush meadow			110	117	
Sedge-rush meadow			57	104	
Temperate Grassland					
Andropogon grassland	Illinois	40°20'	44	240	Baier *et al.* (1972)
Mountain grassland	Bridger, MT	45°47'	60	249	Sims and Singh (1978a, b)
Mountain grassland	Bison, MT	47°19'	54	272	
Northwest bunchgrass	ALE, WA	46°24'	16	98	
Mixed grass prairie	Dickenson, ND	46°54'	47	351	
Mixed grass prairie	Cottonwood, SD	43°57'	47	249	
Mixed grass prairie	Hays, KA	38°52'	50	363	
Shortgrass prairie	Pawnee, CO	40°49'	42	172	
Shortgrass prairie	Pantex, TX	35°18'	45	257	
Tallgrass prairie	Osage, OK	36°57'	62	346	
Desert grassland	Jornada, NM	32°36'	32	148	

least as high as that of temperate graminoids despite a 15–20 °C difference in average air temperature during the growing season (Table 2.3, Bliss, 1962b; Chapin, 1983; Chapin *et al.*, 1980c). The two-fold greater production of temperate compared to arctic graminoid-dominated vegetation thus must be due to their longer growing season and greater quantity of overwintering green material rather than to higher *in situ* growth rate. We suggest that tundra plants have their observed growth rates, not because low temperature prevents evolution of a more rapid growth rate, but because the present *in situ* growth rate is close to the maximum that can be supported by the available environmental resources (e.g. light and nutrients).

There are striking differences in growth rate among growth forms of arctic plants, with forbs and deciduous shrubs growing most rapidly and evergreen shrubs growing most slowly (Bliss, 1956; Johnson and Tieszen, 1976). Communities dominated by the most rapidly growing species (e.g. tall shrub tundra) typically have high productivity and occur on nutrient-rich sites, whereas communities dominated by slowly growing evergreens have lower productivity and occur on less fertile sites (Miller, 1982). However, most tundra communities contain a mixture of growth forms, and an increase in nutrient availability associated with fire or disturbance causes an increase in dominance of rapidly growing species (Chapin and Shaver, 1981; Wein and Bliss, 1973). The growth of rapidly growing species is supported by high capacity for photosynthesis (Johnson and Tieszen, 1976; Limbach *et al.*, 1982) and nutrient absorption (Chapin and Tryon, 1982).

Differences in growth rate and/or production among communities within the arctic correlate more closely with variation in soil moisture, aeration, and nutrient availability than with air temperature (Miller, 1982; Shaver *et al.*, 1979; Warren Wilson, 1957b; Webber, 1978). This implies that arctic plants have been so successful in adapting to low temperature that other ecological factors are more important in explaining patterns of community structure and production within the arctic. These other factors such as low nutrient availability, poor soil aeration, and short growing season are indirect

consequences of low temperature. Fertilization studies have consistently shown production in the arctic to be limited by nitrogen and/or phosphorus availability (Babb and Whitfield, 1977; Haag, 1974; McKendrick *et al.*, 1978; Shaver and Chapin, 1980; Warren Wilson, 1966). Within a given community, growth of most species may be limited by the same nutrients (Shaver and Chapin, 1980).

In polar semideserts water availability directly limits plant growth during parts of the year (Teeri, 1973), but in other arctic communities where soil water content is high the strong positive correlation between soil moisture and plant growth (Shaver *et al.*, 1979; Webber, 1978) is not well understood. Water may be more important in moving nutrients to the root surface or in changing redox conditions and nutrient availability than in its direct effect upon plant growth (Chapin and Shaver, 1981; Shaver *et al.*, 1979).

2.5 Photosynthesis and respiration

Maximum rates of photosynthesis by arctic plants measured in the field are similar to those of temperate-zone species, but are reached at lower temperature (Lechowicz, 1982; Tieszen, 1978a; Tieszen and Wieland, 1975; Tieszen *et al.*, 1980; Wager, 1941). For vascular plants the range is from 20–250 nmol $CO_2 g^{-1} s^{-1}$, with the sequence: forb > grass > deciduous shrub > evergreen shrub (Johnson and Tieszen, 1976; Limbach *et al.*, 1982; Shvetsova and Voznesenskii, 1970; Tieszen *et al.*, 1980). Maximum moss photosynthetic rates are usually in the range 3–30 nmol $g^{-1} s^{-1}$, and lichens from 2–20 nmol $g^{-1} s^{-1}$. Photosynthetic rate is loosely correlated with maximum growth rate across species and growth forms, but does not correlate with growth rate of an individual species across a range of habitats (Kallio and Heinonen, 1975; Oechel, 1976; Oechel and Sveinbjörnsson, 1978; Tieszen, 1978a).

Photosynthetic rate of arctic plants is generally greatest at 10–15 °C and is relatively insensitive to changes in temperature down to about -4 °C (Billings *et al.*, 1971; Johnson and Tieszen, 1976; Limbach *et al.*, 1982; Mayo *et al.*, 1977; Oechel and Sveinbjörnsson, 1978; Skre, 1975; Tieszen, 1973,

1975, 1978a; Tieszen *et al.*, 1980). At 0 °C moss and lichen photosynthesis can be over half of maximum (Kallio and Heinonen, 1971; 1975; Oechel and Collins, 1976). Berry and Björkman (1980) suggest that the low temperature optimum of photosynthesis in arctic plants results from high RuBP carboxylase concentrations, allowing a substantial photosynthetic rate at low temperature. However, RuBP carboxylase also reacts with oxygen, so that high levels of this enzyme would result in a high photorespiration rate, reducing net photosynthesis at higher temperatures, leading to a low temperature optimum. Thermolability and temperature-dependent kinetic characteristics of RuBP carboxylase are less important than a large enzyme quantity in explaining the low temperature optimum of arctic plants (Chabot, 1979; Chabot *et al.*, 1972; Tieszen and Sigurdson, 1973).

Although temperature optima of arctic plants are 10–30 °C lower than those of temperate plants, they are still higher than the average summer leaf temperature (6–10 °C). Simulations suggest that further lowering the temperature optimum of photosynthesis to the average leaf temperature would increase annual carbon gain by only 5% (Miller *et al.*, 1976; Tieszen, 1978b). This lower temperature optimum could presumably be achieved through a higher RuBP carboxylase concentration (Berry and Björkman, 1980) at the cost of increased root biomass to acquire the additional nitrogen. Thus, the observed 10–15 °C temperature optimum of photosynthesis in tundra plants apparently results from a balance between selection for effective function at low temperature and selection for moderate protein requirement.

Because of low light intensity, photosynthetic rate in arctic vascular plants is rarely light-saturated and more closely parallels daily and seasonal patterns of light intensity than temperature (Mayo *et al.*, 1977; Miller *et al.*, 1976; Shvetsova and Voznesenskii, 1970; Tieszen, 1975, 1978a). Both the light compensation point (5.6–7.0 W m^{-2}) and the light intensity at which photosynthesis becomes light-saturated (280 W m^{-2}) are lower than comparable values for most temperate vascular plants (Mayo *et al.*, 1977; Tieszen, 1978a). Consequently, positive photosynthetic rates may be maintained 24 h a day

during much of the growing season, if the night is cloudless (Mayo *et al.*, 1977; Tieszen, 1978a). Mosses have a similar light compensation point but saturate at a lower light intensity (100 W m^{-2}) than vascular plants, so that moss photosynthesis is closely tied to irradiance only in the shade of a well developed vascular canopy (Miller *et al.*, 1978; Oechel, 1976; Oechel and Sveinbjörnsson 1978). High photosynthetic rates at low light levels are due in part to high chlorophyll concentrations and low chlorophyll a to b ratio (Mooney and Billings, 1961; Tieszen, 1970, 1972, 1978a), just as in temperate plants from low light (shade) environments (Boardman, 1977). Like shade plants, arctic plants also have low Hill reaction rates (electron transport in light reaction) (Billings *et al.*, 1971; Tieszen and Helgager, 1968). Thus in arctic plants the light reaction of photosynthesis has many characteristics in common with shade plants, whereas the dark reaction is characterized by a high RuBP carboxylase concentration, as in sun plants. There are no C_4 arctic plants.

In the field, photosynthetic potential per unit leaf area is strongly correlated with carboxylase activity and other factors associated with 'mesophyll' resistance (Tieszen and Wieland, 1975; Tieszen *et al.*, 1980). However, on a community basis the major limiting factors to photosynthesis are the available leaf area and its state of development. These are limited by the length of the growing season and never reach levels that would allow maximum carbon gain (Miller *et al.*, 1976; Tieszen, 1975). Peak leaf area index in wet tundra is about 0.5–0.6 (Caldwell *et al.*, 1978), and permits only 30% of the carbon gain that would be possible with 'optimal' development of leaf area (Miller *et al.*, 1976). Community production is thus strongly dependent on the date of snowmelt, when leaves become photosynthetically competent and begin expansion (Tieszen, 1974, 1978b). Evergreens have an advantage in that they start the season with a full complement of leaves, but their maximum photosynthetic rates are lower than for graminoids, forbs, or deciduous shrubs (Johnson and Tieszen, 1976; Limbach *et al.*, 1982).

Despite large differences in production between sites of differing nutrient availability, photosynthetic rate of vascular plants in the field is not correlated with concentration of any mineral nutrient in leaves or soil (Tieszen, 1978a). In other cases there is a strong correlation between leaf nitrogen and maximum photosynthetic capacity (Limbach *et al.*, 1982). Fertilization either does not affect or reduces photosynthesis, while stimulating growth (Bigger and Oechel, 1982; Oechel and Sveinbjörnsson, 1978; Tieszen, 1978a; Warren Wilson, 1960). This suggests that nutrients stimulate growth processes rather than directly affecting photosynthesis.

Arctic plants have high mitochondrial oxidative rates and therefore high respiration rates for both growth and maintenance processes (Lechowicz *et al.*, 1980) at low temperatures (Billings *et al.*, 1971; Mayo *et al.*, 1977; Tieszen, 1973; Tieszen and Wieland, 1975; Wager, 1941). This high respiratory capacity is probably necessary to support observed rapid growth rates and can be attained through high enzyme levels, again suggesting a high nutrient cost for effective growth at low temperature. Arctic plant enzymes may have a low energy of activation which would maximize the effectiveness of each unit of enzyme under prevailing low temperature (Simon, 1979). The high respiratory capacity of arctic plants leads to exhaustion of carbohydrate reserves and increased mortality when arctic plants are grown in a warm environment (Chapin and Chapin, 1981).

Arctic vascular plants acclimate photosynthetically to changes in temperature, but not as rapidly or as completely as do alpine species (Billings *et al.*, 1971; Chabot, 1979; Tieszen and Helgager, 1968). In *Oxyria digyna*, acclimation to lower growth temperature involves higher respiration rates, higher mitochondrial oxidative capacity, increased Hill reaction activity, and increased carboxylation activity, but lower chlorophyll efficiency (Billings *et al.*, 1971; Chabot *et al.*, 1972). In many cases (but not in *Oxyria*) the respiration response dominates, shifting photosynthetic maximum to a lower temperature and lower rate. Maximum photosynthetic rate in lichens may not decrease with cold acclimation (Larson, 1979; but see Kallio and Heinonen, 1971).

The importance of photosynthetic temperature acclimation to vascular plants in the field has not

been assessed. In some arctic mosses there is little seasonal temperature acclimation, but in polytrichaceous mosses seasonal temperature acclimation may play a major role in maintaining high photosynthetic rates over normal seasonal changes in temperature (Hicklenton and Oechel, 1976; Oechel, 1976 ; Oechel and Sveinbjörnsson, 1978; Stoner *et al.*, 1978).

In sum, arctic plants have adapted to the arctic environment by producing photosynthetically efficient tissues with high levels of metabolic activity. These tissues are efficient principally because they have high enzyme levels to fix and metabolize carbon. These adaptations suggest a high nitrogen requirement per unit of leaf and in a nitrogen-limited environment must limit the potential productivity of tundra plants.

2.6 Water relations

In wet sedge and tussock tundra, vascular plant water stress is rare and transient. Water potentials (xylem tension) less than -1.5 to -2.0 MPa are infrequently recorded, with minor effect on photosynthesis due to stomatal closure (Miller *et al.*, 1980; Oberbauer and Miller, 1979; Stoner and Miller, 1975; Stoner *et al.*, 1978). In wet sedge tundra, leaf water potential recovers to -0.1 to -0.3 MPa most nights, with normal midday minima of -0.5 to -1.2 MPa. A similar pattern prevails in tussock tundra, with less recovery (to -0.4 to -0.6 MPa) at night and midday minima of -1.0 to -1.5 MPa. Photosynthesis occurs even at 'night', so that stomates are not fully closed then, and even greater nighttime recovery might be possible in full dark.

The major reason for the lack of water stress in vascular plants of tussock and wet sedge tundras is the high soil moisture (Oberbauer and Miller, 1979; Shaver *et al.*, 1979). Moreover, wet sedge tundra plants also have low root resistances even at the low ambient soil temperatures, and root resistance is unrelated to soil temperature (Stoner and Miller, 1975). In contrast, most temperate plants fail to absorb water and become drought-stressed at low soil temperatures (Kramer, 1942). Vapor-pressure deficits are also low in wet sedge and tussock tundras due to low air temperatures and frequently foggy or humid weather.

In polar semidesert, vascular plant water stress may reduce photosynthesis due to stomatal closure. Water potentials of -3.5 to -4.4 MPa have been recorded for *Dryas* spp. and *Saxifraga oppositifolia*, with stomatal closure in the saxifrage at -2.1 to -2.9 MPa (Courtin and Mayo, 1975; Teeri, 1973). In *Dryas integrifolia* turgor pressure is maintained over a wide range of total water potentials, suggesting that this polar semidesert species is capable of osmotic adjustment to soil moisture deficit (Hartgerink and Mayo, 1976).

Water stress clearly affects plant distribution and productivity along local mesotopographic gradients. In a transect running from a wet sedge-moss community to a *Dryas*-dominated fellfield, average maximum water potential ranged from about -0.5 MPa to less than -3.0 MPa, with the lowest water potentials recorded in the site with the driest soils and highest air temperatures (Oberbauer and Miller, 1979). Effects of water stress on productivity may be minimized, however, by the fact that those species with lower threshold water potentials for stomatal closure grow in sites where lowest water potentials occur (Johnson and Caldwell, 1975, 1976; Oberbauer and Miller, 1979; Stoner and Miller, 1975; Teeri, 1973).

Many of the effects of water stress on plant distribution and production may be exerted through winter desiccation (Savile, 1972), although there have been no studies of the winter water relations of arctic plants. The osmotic component of total water potential, which would reflect freezing-point depression and thus freezing resistance, increases from -6.05 MPa in winter to -1.7 MPa during active growth in *Dryas integrifolia* (Hartgerink, cited in Courtin and Mayo, 1975).

Mosses and lichens are particularly affected by water availability and show little control over tissue water uptake and loss (Lechowicz, 1981; Oechel and Sveinbjörnsson, 1978). Moss and lichen photosynthesis thus is dependent upon availability of water at or near the soil surface, for capillary uptake or by direct absorption of rain or mist (Miller *et al.*, 1980). Moss resistance to water loss is a function of shoot morphology and shoot density; as the moss dries, resistance increases due to infolding and rolling of the leaves (Oechel and Sveinbjörnsson, 1978). In general, mosses from wetter habitats have higher water contents and less resistance to water loss at low water content (Oechel and Collins, 1976; Oechel and Sveinbjörnsson, 1978). Maximum photosynthetic rates also occur at higher water content in wet-site moss species. Water stress does limit moss photosynthesis frequently in wet sedge tundra, yet positive net photosynthesis is possible at most times in the growing season due to recurrent fog, dew, and rainfall. In lichens, wetting after a dry period is followed by a period of high respiration and negative carbon balance, as the lichen redevelops its photosynthetic capacity (Lechowicz, 1981). If drying occurs before the lichen has had time for sufficient photosynthesis, its carbon balance for the entire wetting–drying cycle is negative. Thus lichen photosynthesis is closely linked to daily and seasonal cycles of moisture availability (Moser and Nash, 1978).

2.7 Mobile carbon pools

Total nonstructural carbohydrate (TNC) pools in arctic plants are large, as high as 40–60% dry weight in storage tissues or rapidly growing parts (Archer and Tieszen, 1980; Mooney and Billings, 1961; Russell, 1940; Shaver and Billings, 1976; Skre, 1975) although values vary considerably with method of analysis. In the spring, concentrations typically are in the 15–40% range (Muc, 1977; Shaver *et al.*, 1979; Svoboda, 1977). These high TNC concentrations are determined both genetically and environmentally (Eagles, 1967; Mooney and Billings, 1961; Stewart and Bannister, 1973; Warren Wilson, 1954) but the reasons for such high concentrations are unclear. Possible explanations include: (1) a high storage requirement to support rapid spring growth, (2) an imbalance between controls on carbon acquisition versus carbon utilization in response to temperature or nutrients ('sink limitation'; Evans, 1975), (3) reserves for regrowth after grazing, or (4) osmotic protection against frosts and freezing.

Rapid spring growth in arctic plants is associated with a decline in TNC (McCown, 1978; Muc, 1977; Russell, 1940; Shaver and Billings, 1976; Svoboda, 1977) suggesting that energy storage to support spring growth is one of the roles played by the large TNC reserves of arctic plants. Evidence discussed below suggests that these reserves support below-ground growth and maintenance to a greater extent than shoot growth. TNC reserves are gradually replenished during the growing season, reaching a peak in late July and August. TNC often declines slightly prior to dormancy (Skre *et al.,* 1975; Svoboda, 1977), perhaps supporting late season root growth and activity at a time when photosynthetic capacity is low. Seasonal and daily TNC fluctuations are less pronounced, and minimal TNC levels are higher in arctic than in alpine (Fonda and Bliss, 1966; Mooney and Billings, 1960, 1961) or temperate (White, 1973) plants, suggesting that in the Arctic the availability of carbohydrates may not directly limit plant growth (Mooney and Billings, 1961; Russell, 1940; Shaver and Billings, 1976; Warren Wilson, 1960, 1966).

Fertilization of tundra results in a decrease in TNC in most arctic species, as storage carbohydrates are utilized in growth (McKendrick *et al.,* 1978; Shaver and Chapin, 1980; Warren Wilson, 1966). This TNC reduction with fertilization indicates that nutrient limitation is among the factors responsible for the large TNC reserves of tundra plants. Growth chamber studies indicate that the TNC accumulation due to low nutrient availability becomes more pronounced at low soil temperatures (McCown, 1978). Temperate plants from low-nutrient habitats also accumulate TNC, but not to the levels observed in tundra plants (White, 1973), suggesting that nutrient limitation cannot entirely explain the high TNC levels of tundra plants.

TNC concentration is positively correlated with latitude and inversely correlated with temperature (Eagles, 1967; McCown, 1978; Mooney and Billings, 1961; Stewart and Bannister, 1973) suggesting that low temperature may be responsible for carbohydrate accumulation either by inhibiting translocation or by reducing carbon utilization relative to photosynthesis (Warren Wilson, 1966). Arctic and alpine plants translocate ^{14}C-labelled carbohydrates to roots at 2 °C or less (Allessio and Tieszen, 1975; Dadykin, 1954; Wallace and Harrison, 1978), suggesting that temperature inhibition of translocation is not a serious limitation in

tundra. There have been no critical tests of Warren Wilson's (1966) hypothesis that carbon utilization is more restricted by low temperature than is photosynthesis. Tundra plants do have a high capacity to respire and grow at low temperature, as discussed above, but there are suggestions that nutrients are either unavailable or cannot be converted to organic form at low temperature (Haag, 1974). The stimulatory effect of nutrients upon growth in absence of a stimulatory effect upon photosynthesis is consistent with Warren Wilson's hypothesis.

Intensive grazing could result in occasional heavy demands upon carbohydrate reserves. However, repeated severe clipping reduces TNC to only about 50–70% of control concentration, or 4–20% dry weight (Chapin and Slack, 1979; Mattheis *et al.,* 1976). These reductions in TNC would not be considered severe in temperate species, and are of similar magnitude to normal annual fluctuations.

Nonstructural carbohydrates could serve as 'antifreeze' (McCown, 1978). However, observed sugar concentrations of arctic plants would reduce the freezing point of cytoplasm only 1–2 °C and cannot by themselves explain why the flowers and leaves of arctic plants survive summer frosts and continue development without apparent damage (Billings and Mooney, 1968; Bliss, 1962a; Sørensen, 1941).

Lipid concentrations are higher in arctic than in temperate plants (Bliss, 1962c; Hadley and Bliss, 1964; Svoboda, 1977), but again the causes are unclear. Because of their high energy content and compact storage, fats may be important energy sources for rapid growth (Bliss, 1962c; Hadley and Bliss, 1964; Svoboda, 1977). However, lipids require three times more energy to produce than do polysaccharides (Chapin *et al.,* 1980c; Penning de Vries, 1975) and may be disadvantageous if carbon and energy are at all limiting to growth. Alternatively, the high lipid content of tundra plants may be associated with membranes needed to maintain their high metabolic capacity. A larger proportion of the total lipid of plants was membrane-associated in a permafrost-dominated marsh than in an adjacent hot spring (Kedrowski and Chapin, 1978). A third possibility is that phenolic resins account for much of the high total 'lipid' measured in most methods, and that those resins are important as anti-herbivore defenses (Bryant and Kuropat, 1980; Jung *et al.*, 1979). The role of 'lipids' in arctic plants will remain uncertain until the 'lipid' fraction is separated into storage, membrane, and defensive components.

2.8 Nutrient absorption, storage, and loss

Absorption of both ammonium and phosphate, the two most limiting nutrients in most arctic communities, is much less temperature-sensitive in tundra than in temperate plants (Chapin, 1981; Chapin and Bloom, 1976; McCown, 1978). For example, the major graminoids at Barrow maintain 20–60% of their 20 °C phosphate absorption rates at 1 °C (Chapin and Bloom, 1976), whereas phosphate absorption by temperate graminoids is strongly inhibited below 10 °C (Carey and Berry, 1978; Sutton, 1969). There is no difference in temperature optimum of phosphate absorption between tundra and temperate plants examined to date (Chapin and Bloom, 1976). Effective nutrient absorption by tundra plants at low temperature is in part a consequence of membranes which maintain their fluidity and function at low temperature (McCown, 1978), due to the many double bonds present in fatty acids of membrane phospholipids (Kedrowski and Chapin, 1978).

Because nutrient absorption is temperature dependent even in tundra, one would expect low *in situ* rates of nutrient absorption. However, when grown and measured under similar conditions, tundra species and ecotypes have a higher phosphate absorption capacity than their temperate counterparts (Chapin, 1974, 1981; McNaughton *et al.*, 1974). Thus, the phosphate absorption process of tundra graminoids has adapted to low temperature by a decrease in temperature sensitivity below optimum temperature, by increased affinity of roots for phosphate at low temperature, and by an increase in uptake rate at all temperatures, but not by any change in temperature optimum.

Growth of plants at high root temperature results in a compensatory decrease in capacity of roots to absorb phosphate, when measured under standard conditions (Chapin, 1974). This may partially explain the seasonal decline in absorption capacity of roots collected in the field (Chapin and Bloom, 1976). The extent of temperature acclimation of phosphate absorption is less developed in arctic plants than in ecotypes and species from more thermally fluctuating environments (Chapin, 1974).

Although arctic plants have effectively compensated for low temperature both evolutionarily and physiologically, nutrient absorption rates are still low due to very low concentrations of nutrients in the soil solution. In wet tundra at Barrow, the calculated seasonal pattern of phosphate absorption is much more closely correlated to phosphate availability ($r^2 = 0.92$) than to soil temperature ($r^2 = 0.004$) (Chapin *et al.*, 1978). Thus, the seasonal pattern of nutrient release by decomposition, particularly during the spring flush of nutrients, may be much more important than seasonal changes in thaw depth or soil temperature in determining patterns of nutrient absorption. Absorption continues actively as long as the soil is thawed – as much as a month after leaves have senesced (Chapin and Bloom, 1976). Soil aeration may also indirectly limit nutrient absorption by reducing abundance of mycorrhizal fungi (Antibus and Linkins, 1978).

Arctic plants, like other plants from ammonium-dominated soils, grow more effectively with ammonium as a nitrogen source than do temperate plants from nitrate-dominated soils, particularly at low soil temperatures (McCown, 1978). The importance of minerals other than nitrogen and phosphorus has received little attention in the Arctic, although plant surveys at Barrow indicate that all other elements are present in plants at concentrations normally considered adequate for plant growth. Certain elements such as manganese reach levels that prove toxic in crops (Ulrich and Gersper, 1978), but there have been no critical studies of potential toxicity of iron, aluminum, manganese, etc., in the Arctic.

There is considerable variation in nutrient absorption capacity among arctic plants. Deciduous shrubs with their large annual nutrient requirement have the highest root absorption capacity and evergreen shrubs the lowest; graminoids are intermediate (Chapin and Tryon, 1982). Among populations of *Carex aquatilis* at Barrow, plants with the most rapid growth have the highest capacity to absorb nutrients (Shaver *et al.*, 1979).

In autumn, abscised leaves of arctic deciduous plants contain 40–60% of maximum pre-senescent nitrogen, phosphorus and potassium content (Chapin *et al.*, 1975, 1980a; Muc, 1977; Wielgolaski *et al.*, 1975) a proportion similar to that observed in other ecosystems (Chapin, 1980). Most of the remaining nitrogen and phosphorus is apparently retranslocated to storage organs, whereas potassium may be largely leached (Chapin *et al.*, 1980a; Morton, 1977).

2.9 Allocation

The seasonal patterns of growth described earlier are a consequence of predictable patterns of allocation of newly acquired and stored resources. The pronounced drop in carbohydrate concentrations in stems and rhizomes of arctic deciduous shrubs, forbs, and graminoids coincides with and has been assumed to support directly early spring leaf production (McCown, 1978; Mooney and Billings, 1960; Muc, 1977; Skre, 1975). However, Barrow graminoid shoots become photosynthetically self-sufficient almost immediately following leaf initiation (Tieszen, 1975, 1978a) and do not translocate labelled carbohydrate from rhizomes into shoots (Allessio and Tieszen, 1975). Computer simulations suggest that the spring decline in stem or rhizome reserve carbohydrate concentrations serves primarily to support growth and maintenance of below-ground tissue at a time when most photosynthetic carbon is retained by rapidly growing leaves (Stoner *et al.*, 1978). However, ^{14}C studies indicate substantial translocation of carbohydrates from stems to growing shoots in some evergreen and deciduous shrubs (Campbell, 1983).

In contrast to carbohydrates, stored nutrients play a major role in directly supporting spring leaf growth (Chapin *et al.*, 1980a). Nutrients absorbed from soil also move directly into new shoots (Chapin and Bloom, 1976) but serve primarily to replenish nutrient stores in stems and rhizomes in late summer and autumn (Chapin *et al.*, 1980a). Thus fertilization affects tundra plant growth more in subsequent years than in the year of application.

Because of the lack of light saturation of photosynthesis, arctic plants absorb a high proportion ($> 1\%$) of incoming solar radiation. Also, long daylight hours result in low net respiratory carbon loss by leaves, so that about 80% of net photosynthesis is converted to above-ground plant biomass or translocated below-ground, a higher net production efficiency than in most temperate communities. Because of the large below-ground biomass in arctic plants, half of the carbon translocated below ground is lost in respiration, whereas in temperate grasslands 15% of carbon translocated below ground is respired and the remainder converted to new biomass (Coleman *et al.*, 1976). Thus, due to long daylight hours and high below-ground biomass, arctic plants produce above-ground tissue efficiently but expend much of acquired carbon in respiration below ground (Babb and Whitfield, 1977, Chapin *et al.*, 1980b; Peterson and Billings, 1975).

Environmental factors exert a strong effect upon patterns of allocation in arctic plants. For example, both low root temperature and low nutrient availability generally result in high root:shoot ratios (e.g. Brouwer, 1966) and could explain the generally high root:shoot ratios of arctic plants (Chapin and Chapin, 1981). Even in tundra, root:shoot ratio declines in response to increased nutrient availability (Dennis *et al.*, 1978). However, the longevity of tundra roots may be equally important in explaining high root:shoot ratios (Bell and Bliss, 1978; Shaver and Billings, 1975). Polar desert plants have lower root:shoot ratios than other arctic plants (Bell and Bliss, 1978).

Both genetic and environmental factors determine rates of root and leaf turnover and therefore allocation pattern (Chapin and Chapin, 1981). For example, *Eriophorum* species have an annual or nearly annual pattern of root turnover, whereas *Carex* roots live 8–10 years or more (Shaver and Billings, 1975). Leaf longevity may vary from less than one year in some grasses (Tieszen, 1978a) to 6–8 years or more in some evergreens (Shaver and Chapin, unpublished). Nutrient availability in particular exerts a strong control over turnover rates, with higher rates of leaf (Shaver, 1981) or tiller (Fetcher and Shaver, 1982) turnover at high nutrient levels. In general leaf turnover rates are lowest in the most exposed, infertile, or unproductive habitats, as Ewers and Schmid (1981) found for subalpine conifers.

2.10 Reproduction

Arctic plant populations are maintained and dispersed both by seed and by vegetative means. Most arctic species are long-lived perennials, and many depend heavily on clonal growth by tillering, branch layering, etc., in undisturbed vegetation (Billings and Mooney, 1968; Bliss, 1971; Savile, 1972; Sørenson, 1941). In *Oxyria digyna*, arctic populations show greater rhizome development than alpine populations, suggesting that clonal growth may be selected for as an adaptation to the arctic (Mooney and Billings, 1961). On the other hand, both natural and anthropogenic disturbances are quickly colonized by seedlings of native species, at least if the soil is stable and wet-to-mesic (Bliss and Wein, 1972; Chapin and Chapin, 1980; Gartner *et al.*, 1983; Hernandez, 1973; Wein and Bliss, 1973), and much of the natural patterning in the vegetation is due to recurrence of frost-related disturbance (Britton, 1966; Hopkins and Sigafoos, 1951).

Arctic soils have a substantial resident population of buried seeds (Leck, 1980; McGraw, 1980; Gartner, *et al.*, 1983), similar in number to those of temperate grasslands (Lippet and Hopkins, 1950; Miles, 1973). Buried seed can account for most of the seedlings present in organic disturbances (Chester and Shaver, 1982a; Gartner *et al.*, 1983). In temperate zones the seed pool is poorly represented by the late successional species (Miles, 1979). However, in tundra many dominant climax species are also important colonizers and have abundant buried seed (McGraw,. 1980; Gartner *et al.*, 1983). The large buried seed pool in tundra is due in part to low soil temperatures that prolong seed viability (Billings and Mooney, 1968).

Flowering in arctic vegetation varies greatly from year to year. Even in a good flowering year reproductive allocation is lower than in comparable temperate-zone species (Chester and Shaver, 1982b). However, the proportion of reproductive 'effort' accounted for by viable seeds was somewhat higher than in temperate species, suggesting that arctic plants may produce more seed per unit of reproductive biomass (Chester and Shaver, 1982b). Flowering is initiated by long photoperiod and prevented by short photoperiod, as in temperate long-day plants (Teeri and Tonsor, 1981).

Viablility and/or germinability of seed of arctic plants is fairly high, 50–100% in most species (Bliss, 1958; Mooney and Billings, 1961; Wein and MacLean, 1973). Intrinsic or constitutive dormancy mechanisms are generally absent or poorly developed except in a few genera such as *Carex* (Billings and Mooney, 1968; Bliss, 1971). In some cases light may be necessary or may stimulate germination above rates in the dark (Bliss, 1958; Mooney and Billings, 1961; Wein and MacLean 1973). Optimum temperature for germination may be as high as 25–30 °C (Wein and MacLean, 1973), with arctic ecotypes germinating better at high temperatures than alpine ecotypes. These light and temperature requirements would be met in disturbances or favorable microsites where seedlings are more likely to establish. In the absence of such favorable conditions, seeds may enter the buried seed pool.

Seedling growth rate is often low under natural conditions owing to inability to compete effectively for resources. However, in disturbances where there is little competition such as stabilized frost boil margins (Hopkins and Sigafoos, 1951), recently drained thaw lakes (Dennis, 1968), burned areas (Racine, 1981; Wein and Bliss, 1973) and areas of human impact (Chapin and Chapin, 1980; Chester and Shaver, 1982a; Gartner *et al.*, 1983), seedlings readily establish and grow rapidly. Slow (100–10 000 year) cycles of disturbance and seedling establishment may be extremely important in the maintenance of tundra vegetation. Seedling establishment of certain species also occurs in established vegetation (McGraw and Shaver, 1982; Wager, 1938). In wet coastal tundra the carbon cost of producing a new tiller by sexual reproduction is about 10 000 times greater than by vegetative reproduction (Chapin *et al.*, 1980c). Thus, the substantial allocation to sexual reproduction that occurs in the Arctic indicates the importance of a long-term evolutionary framework within which to view the physiological ecology of arctic plants.

2.11 Summary

Recent studies have added considerable detail and perspective to our understanding of the physiological ecology of arctic plants, as outlined by Billings and his associates. Arctic plants are so well adapted to low temperature that they photosynthesize, respire, absorb nutrients, and grow at rates similar to those of temperate plants, but at much lower ambient temperatures. Thus, adaptation to low temperature is a major factor distinguishing arctic from temperate plants. However, owing to effective temperature adaptation by all arctic organisms, it is primarily factors other than temperature that directly determine the variation in growth rate and distribution of plants within the arctic. The low temperature of the Arctic indirectly results in (a) a short growing season, (b) low nutrient availability, and (c) permafrost with restricted drainage and frequently poor aeration. It is these factors that directly limit growth most strongly within the arctic and determine the mosaic of arctic vegetation.

Tundra plants grow as rapidly as their temperate counterparts but have lower annual production due to the shorter growing season and small quantity of overwintering green material. The rapid growth of shoots in spring is largely supported by concurrent photosynthesis, and the decline in carbohydrates that coincides with spring growth may be largely due to below-ground growth and respiration. Successful reproduction by seed is infrequent in the arctic. However, tundra plants allocate a substantial proportion of reserves to sexual reproduction, and seedlings generally can be found. Within the time frame of arctic succession (hundreds or thousands of years) seedling establishment is probably at least as important in the Arctic as in other ecosystems in the establishment and maintenance of vegetation.

Acknowledgements

We thank Dwight Billings and the many other arctic workers with whom we have discussed the quirks of arctic plants. J. Fox and M. Chapin critically reviewed the manuscript. Work leading to these generalizations was funded by the National Science Foundation (GV-29343, DPP-76-80642, and DEB-79-05842), the Department of Energy, and the Army Research Office.

References

Addison, P.A. and Bliss, L.C. (1980) Summer climate, microclimate, and energy budget of a polar semidesert on King Christian Island, Northwest Territory, Canada. *Arctic and Alpine Research*, **12**, 161–70.

Allessio, M.L. and Tieszen, L.L. (1975) Patterns of carbon allocation in an arctic tundra grass, *Dupontia fischeri* (Gramineae), at Barrow, Alaska. *American Journal of Botany*, **62**, 797–807.

Antibus, R.K. and Linkins, A.E. (1978) Ectomycorrhizal fungi of *Salix rotundifolia* Trautv. I. Impact of surface applied Prudhoe Bay crude oil on mycorrhizal structure and composition. *Arctic*, **31**, 366–80.

Archer, S. and Tieszen, L.L. (1980) Growth and physiological responses of tundra plants to defoliation. *Arctic and Alpine Research*, **12**, 531–52.

Babb, T.A. and Whitfield, D.W.A. (1977) Mineral nutrient cycling and limitation of plant growth in the Truelove Lowland ecosystem, in *Truelove Lowland, Devon Island, Canada: A High Arctic Ecosystem* (ed. L.C. Bliss), University Alberta Press, Edmonton, Canada, pp. 589–606.

Baier, J.D., Bazzaz, F.A., Bliss, L.C. and Boggess, W.R. (1972) Primary production and soil relations in an Illinois sand prairie. *American Midland Naturalist*, **88**, 200–8.

Barèl, D. and Barsdate, R.J. (1978) Phosphorus dynamics of wet coastal tundra soils near Barrow, Alaska, in *Environmental Chemistry and Cycling Processes* (eds D.C. Adriano and I.L. Brisbin), DOE Symposium Series, Conf-760429. Washington, D.C., pp. 516–37.

Barry, R.G. (1973) A climatological transect along the east slope of the Front Range, Colorado. *Arctic and Alpine Research*, **5**, 89–110.

Barry, R.G., Courtin, G.M. and Labine, C. (1981) Tundra climates, in *Tundra Ecosystems: A Comparative Analysis* (eds L.C. Bliss, O.W. Heal and J.J. Moore), Cambridge University Press, Cambridge, pp. 81–114.

Bell, K.L. and Bliss, L.C. (1977) Overwinter phenology of plants in a polar semidesert. *Arctic*, **30**, 118–21.

Bell, K.L. and Bliss, L.C. (1978) Root growth in a polar semidesert environment. *Canadian Journal of Botany*, **56**, 2470–90.

Berry, J. and Björkman, O. (1980) Photosynthetic response and adaptation to temperature in higher plants. *Annual Review of Plant Physiology*, **31**, 491–543.

Bigger, C.M and Oechel, W.C. (1982) Nutrient effect on maximum photosynthesis in arctic plants. *Holarctic Ecology*, **5**, 158–63.

Billings, W.D. and Mooney, H.A. (1968) The ecology of arctic and alpine plants. *Biological Reviews*, **43**, 481–529.

Billings, W.D., Godfrey, P.J., Chabot, B.F. and Bourque, D.P. (1971) Metabolic acclimation to temperature in arctic and alpine ecotypes of *Oxyria digyna*. Arctic and Alpine Research, **3**, 277–89.

Billings, W.D., Peterson, K.M. and Shaver, G.R. (1978) Growth, turnover and respiration rates of roots and tillers in tundra graminoids, in *Vegetation and Production Ecology of an Alaskan Arctic Tundra* (ed. L.L. Tieszen), Springer-Verlag, New York, pp. 415–34.

Black, R.A. and Bliss, L.C. (1980) Reproductive ecology of *Picea mariana* (Mill.) BSP., at treeline near Inuvik, Northwest Territories, Canada. *Ecological Monographs*, **50**, 331–54.

Bliss, L.C. (1956) A comparison of plant development in microenvironments of arctic and alpine tundras. *Ecological Monographs*, **26**, 303–37.

Bliss, L.C. (1958) Seed germination in arctic and alpine species. *Arctic*, **11**, 180–8.

Bliss, L.C. (1962a) Adaptations of arctic and alpine plants to environmental conditions. *Arctic*, **15**, 117–44.

Bliss, L.C. (1962b) Net primary production of tundra ecosystems, in *Die Stoffproduktion der Pflanzendecke* (ed. H. Lieth), Gustav Fischer Verlag, Berlin, pp. 35–46.

Bliss, L.C. (1962c) Caloric and lipid content in alpine tundra plants. *Ecology*, **43**, 753–7.

Bliss, L.C. (1966) Plant productivity in alpine microenvironments on Mt. Washington, New Hampshire. *Ecological Monographs*, **36**, 125–55.

Bliss, L.C. (1971) Arctic and alpine plant life cycles. *Annual Review of Ecology and Systematics*, **2**, 405–38.

Bliss, L.C. (Ed.) (1977a) *Truelove Lowland, Devon Island, Canada: A High Arctic Ecosystem*, University Alberta Press, Edmonton, 714 pp.

Bliss, L.C. (1977b) General summary Truelove Lowland ecosystem, in *Truelove Lowland, Devon Island, Canada: A High Arctic Ecosystem* (ed. L.C. Bliss), University Alberta Press, Edmonton, pp. 657–75.

Bliss, L.C. (1979) Vascular plant vegetation of the Southern Circumpolar Region in relation to antarctic, alpine, and arctic vegetation, *Canadian Journal of Botany*, **57**, 2167–78.

Bliss, L.C. (1981) North American and Scandinavian tundras and polar deserts, in *Tundra Ecosystems: A Comparative Analysis* (eds L.C. Bliss, O.W. Heal and J.J. Moore), Cambridge University Press, Cambridge, pp. 8–24.

Bliss, L.C. and Wein, R.W. (1972) Plant community responses to disturbances in the western Canadian Arctic. *Canadian Journal of Botany*, **50**, 1097–109.

Bliss, L.C., Courtin, G.M., Pattie, D.L., Riewe, R.R., Whitfield, D.W.A. and Widden, P. (1973) Arctic tundra ecosystems. *Annual Review of Ecology and Systematics*, **4**, 359–99.

Bliss, L.C., Heal, O.W. and Moore, J.J. (eds) (1981) *Tundra Ecosystems. A Comparative Analysis*, Cambridge University Press, Cambridge, 813 pp.

Boardman, N.K. (1977) Comparative photosynthesis of sun and shade plants. *Annual Review of Plant Physiology*, **28**, 355–77.

Britton, M.E. (1966) Vegetation of the arctic tundra, in *Arctic Biology* (ed. H.P. Hanson), Oregon State University Press, Corvallis, pp. 67–130.

Brouwer, R. (1966) Root growth of grasses and cereals, in *The Growth of Cereals and Grasses* (eds F.L. Milthorpe and J.D. Ivins), Butterworths, London, pp. 153–66.

Brown, J., Miller, P.C., Tieszen, L.L and Bunnell, F.L. (eds) (1980) *An Arctic Ecosystem. The Coastal Tundra at Barrow, Alaska*, Dowden, Hutchinson and Ross, Stroudsburg, Pennsylvania, 571 pp.

Bryant, J.P. and Kuropat, P.J. (1980) Selection of winter forage by subarctic browsing vertebrates: The role of plant chemistry. *Annual Review of Ecology and Systematics*, **11**, 261–85.

Caldwell, M.M., Johnson, D.A. and Fareed, M. (1978) Constraints on tundra productivity: Photosynthetic capacity in relation to solar radiation utilization and water stress in arctic and alpine tundras, in *Vegetation and Production Ecology of an Alaskan Arctic Tundra* (ed by L.L. Tieszen), Springer-Verlag, New York, pp. 323–42.

Callaghan, T.V., Collins, N.J. and Callaghan, C.H. (1978) Photosynthesis, growth, and reproduction of *Hylocomium splendens* and *Polytrichum commune* in Swedish Lapland. *Oikos*, **31**, 73–88.

Campbell, J.S. (1983) Growth, biomass partitioning and carbon allocation in three dwarf shrubs in the Canadian Low Arctic. PhD Thesis, University of Alberta, Edmonton.

Carey, R.W. and Berry, J.A. (1978) Effects of low temperature on respiration and uptake of rubidium ions by excised barley and corn roots. *Plant Physiology*, **61**, 858–60.

Chabot, B.F. (1979) Metabolic and enzymatic adaptations to low temperature in *Comparative Mechanisms of Cold Adaptation* (eds L.S. Underwood, L.L. Tieszen, A.B. Callahan and G.E. Folk), Academic Press, New York, pp. 283–301.

Chabot, B.F., Chabot, J.F. and Billings, W.D. (1972) Ribulose-1, 5-diphosphate carboxylase activity in arctic and alpine populations of *Oxyria digyna*. Photosynthetica, **6**, 364–9.

Chapin, F.S. III (1974) Morphological and physiological mechanisms of temperature compensation in phosphate absorption along a latitudinal gradient. *Ecology*, **55**, 1180–98.

Chapin, F.S. III (1980) The mineral nutrition of wild plants. *Annual Review of Ecology and Systematics*, **11**, 233–60.

Chapin, F.S. III (1981) Field measurements of growth and phosphate uptake in *Carex aquatilis* along a latitudinal gradient. *Arctic and Alpine Research*, **13**, 83–94.

Chapin, F.S. III (1983) Direct and indirect effects of temperature on arctic plants. *Polar Biology*, **2**, 47–52.

Chapin, F.S. III and Bloom, A.J. (1976) Phosphate absorption: Adaptation of tundra graminoids to a low temperature, low phosphorus environment. *Oikos*, **26**, 111–21.

Chapin, F.S. III and Chapin, M.C. (1980) Revegetation of an arctic disturbed site by native tundra species. *Journal of Applied Ecology*, **17**, 449–56.

Chapin, F.S. III and Chapin, M.C. (1981) Ecotypic differentiation of growth processes in *Carex aquatilis* along latitudinal and local gradients. *Ecology*, **62**, 1000–9.

Chapin, F.S. III and Shaver, G.R. (1981) Changes in soil properties and vegetation following disturbance of Alaskan arctic tundra. *Journal of Applied Ecology*, **18**, 605–17.

Chapin, F.S. III and Slack, M. (1979) Effect of defoliation upon root growth, phosphate absorption, and respiration in nutrient-limited tundra graminoids. *Oecologia (Berlin)*, **42**, 67–79.

Chapin, F.S. III and Tryon, P.R. (1982) Phosphate absorption and root respiration of different plant growth forms from northern Alaska. *Holarctic Ecology*, **5**, 164–71.

Chapin, F.S. III, Barsdate, R.J. and Barèl, D. (1978) Phosphorus cycling in Alaskan coastal tundra: A hypothesis for the regulation of nutrient cycling. *Oikos*, **31**, 189–99.

Chapin, F.S. III, Johnson, D.A. and McKendrick, J.D. (1980a) Seasonal movement of nutrients in plants of differing growth form in an Alaskan tundra ecosystem: implications for herbivory. *Journal of Ecology*, **68**, 189–209.

Chapin, F.S. III, Miller, P.C., Billings, W.D. and Coyne, P.I. (1980b) Carbon and nutrient budgets and their control in coastal tundra, in *An Arctic Ecosystem: The Coastal Tundra at Barrow, Alaska* (eds J. Brown, P.C. Miller, L.L. Tieszen and F.L. Bunnell), Dowden, Hutchinson and Ross, Stroudsburg, Pennsylvania, pp. 458–82.

Chapin, F.S. III, Tieszen, L.L., Lewis, M.C., Miller, P.C. and McCown, B.H. (1980c) Control of tundra plant allocation patterns and growth, in *An Arctic Ecosystem: The Coastal Tundra of Northern Alaska* (eds J. Brown, P.C. Miller, L.L. Tieszen, and F.L. Bunnell), Dowden, Hutchinson and Ross, Stroudsburg, Pennsylvania, pp. 140–85.

Chapin, F.S. III, Van Cleve, K. and Chapin, M.C. (1979) Soil temperature and nutrient cycling in the tussock growth form of *Eriophorum vaginatum*. *Journal of Ecology*, **67**, 169–89.

Chapin, F.S. III, Van Cleve, K. and Tieszen, L.L. (1975) Seasonal nutrient dynamics of tundra vegetation at Barrow, Alaska. *Arctic and Alpine Research*, **7**, 209–26.

Chester, A.L. and Shaver, G.R. (1982a) Seedling dynamics of some cotton grass tussock tundra species during the natural revegetation of small disturbed areas. *Holarctic Ecology*, **5**, 207–11.

Chester, A.L. and Shaver, G.R. (1982b) Reproductive effort in cotton grass tussock tundra. *Holarctic Ecology*, **5**, 200–6.

Christy, E.J. and Sharitz, R.R. (1980) Characteristics of three populations of a swamp annual under different temperature regimes. *Ecology*, **61**, 454–60.

Churchill, E.D. and Hanson, H.C. (1958) The concept of climax in arctic and alpine vegetation. *Botanical Review*, **24**, 127–91.

Clarke, G.C.S., Greene, S.W. and Greene, D.M. (1971) Productivity of bryophytes in polar regions. *Annals of Botany*, **35**, 99–108.

Coleman, D.C., Andrews, R., Ellis, J.E. and Singh, J.S. (1976) Energy flow and partitioning in selected man-managed and natural ecosystems. *Agro-Ecosystems*, **3**, 45–54.

Courtin, G.M. and Labine, C.L. (1977) Microclimatology studies on Truelove Lowland, in: *Truelove Lowland, Devon Island, Canada: A High Arctic Ecosystem* (ed. L.C. Bliss), University Alberta Press, Edmonton, pp. 73–106.

Courtin, G.M and Mayo, J.M. (1975) Arctic and alpine plant water relations, in *Physiological Adaptations to the Environment* (ed. F.J. Vernberg), Intext Educational Publishers, New York, pp. 201–24.

Dadykin, V.P. (1954) Peculiarities of plant behavior in

cold soils. *Voprosy Botanicheski*, **2**, 455–72 (in Russian); 473–89 (in French).

Dennis, J.G. (1968) Growth of tundra vegetation in relation to arctic microenvironments at Barrow, Alaska, PhD Dissertation, Duke University, Durham, North Carolina, 289 pp.

Dennis, J.G., Tieszen, L.L. and Vetter, M.A. (1978) Seasonal dynamics of above and belowground production of vascular plants at Barrow, Alaska, in *Vegetation and Production Ecology of an Alaskan Arctic Tundra* (ed. L.L. Tieszen), Springer-Verlag, New York, pp. 113–40.

Dingman, S.L., Barry, R.G., Weller, G., Benson, C., LeDrew, E.F. and Goodwin, C.W. (1980) Climate, snow cover, microclimate, and hydrology, in *An Arctic Ecosystem. The Coastal Tundra at Barrow, Alaska* (eds J. Brown, P.C. Miller, L.L. Tieszen and F.L. Bunnell), Dowden, Hutchinson and Ross, Stroudsburg, Pennsylvania, pp. 30–65.

Dowding, P., Chapin, F.S. III, Wielgolaski, F.E. and Kilfeather, P. (1981) Nutrients in tundra ecosystems, in: *Tundra Ecosystems: A Comparative Analysis* (eds L.C. Bliss, O.W. Heal and J.J. Moore), Cambridge University Press, Cambridge, pp. 647–83.

Eagles, C.F. (1967) Variation in the soluble carbohydrate content of climatic races of *Dactylis glomerata* (Cocksfoot) at different temperatures. *Annals of Botany*, **31**, 645–51.

Ellis, B.A. and Kummerow, J. (1982) Temperature effect on growth rates of *Eriophorum vaginatum* roots. *Oecologia (Berlin)*, **54**, 136–7.

Ellis, S. (1980) An investigation of weathering in some arctic–alpine soils on the northeast flank of Oksskolten, North Norway. *Journal of Soil Science*, **31**, 371–85.

Evans, L.T. (1975) The physiological basis of crop yield, in: *Crop Physiology* (ed. L.T. Evans), Cambridge University Press, Cambridge, pp. 327–55.

Everett, K.B., Vassiljevskaya, V.D., Brown, J. and Walker, B.D. (1981) Tundra and analogous soils, in: *Tundra Ecosystems: A Comparative Analysis* (ed. L.C. Bliss, O.W. Heal and J.J. Moore), Cambridge University Press, Cambridge, pp. 139–79.

Ewers, F.W. and Schmid, R. (1981) Longevity of needle fascicles of *Pinus longaeva* (Bristlecone pine) and other North American pines. *Oecologia (Berlin)*, **51**, 107–15.

Fetcher, N. and Shaver, G.R. (1982) Growth and tillering patterns within tussocks of *Eriophorum vaginatum*. *Holarctic Ecology*, **5**, 180–6.

Flint, P.S. and Gersper, P.L. (1974) Nitrogen nutrient levels in arctic tundra soils, in *Soil Organisms and Decomposition in Tundra* (eds A.J. Holding, O.W. Heal, S.F. MacLean Jr and P.W. Flanagan), Tundra Biome Steering Committee, Stockholm, pp. 375–87.

Fonda, R.W. and Bliss, L.C. (1966) Annual carbohydrate cycle of alpine plants on Mt. Washington, New Hampshire. *Bulletin of the Torrey Botanical Club*, **93**, 268–77.

Gartner, B.L., Chapin, F.S. III and Shaver, G.R. (1983) Demographic patterns of seedling establishment and growth of native graminoids in an Alaskan tundra disturbance. *Journal of Applied Ecology*, **20**, 965–80.

Grigal, D.F., Ohmann, L.F. and Brander, R.B. (1976) Seasonal dynamics of tall shrubs in northeastern Minnesota: Biomass and nutrient element changes. *Forest Science*, **22**, 195–208.

Haag, R.W. (1974) Nutrient limitations to plant production in two tundra communities. *Canadian Journal of Botany*, **52**, 103–16.

Haag, R.W. and Bliss, L.C. (1974) Energy budget changes following surface disturbance to upland tundra. *Journal of Applied Ecology*, **11**, 355–74.

Hadley, E.B. and Bliss, L.C. (1964) Energy relationships of alpine plants on Mt. Washington, New Hampshire. *Ecological Monographs*, **34**, 331–57.

Hare, F.K. (1968) The arctic. *Quarterly Journal of the Royal Meteorological Society*, **58**, 439–59.

Hartgerink, A.P. and Mayo, J.M. (1976) Controlled-environment studies on net assimilation and water relations of *Dryas integrifolia*. *Canadian Journal of Botany*. **54**, 1884–95.

Heal, O.W. and Perkins, D.F. (eds) (1978) *Production Ecology of some British Moors and Montane Grasslands*, Springer-Verlag, New York, 426 pp.

Hernandez, H. (1973) Natural plant recolonization of surficial disturbances, Tuktoyaktuk Peninsula Region, Northwest Territories. *Canadian Journal of Botany*, **51**, 2177–96.

Hicklenton, R.R. and Oechel, W.C. (1976) Physiological aspects of the ecology of *Dicranum fuscescens* in the subarctic. I. Acclimation and acclimation potential of CO_2 exchange in relation to habitat, light, and temperature. *Canadian Journal of Botany*, **10**, 1104–19.

Hocking, B. and Sharpin, C.D. (1965) Flower basking by arctic insects. *Nature*, **206**, 215.

Hopkins, D.M. and Sigafoos, R.S. (1951) Frost action and vegetation patterns on Seward Peninsula, Alaska. *U.S. Geological Survey Bulletin*, **974-C**, 51–100.

Johnson, D.A. and Caldwell, M.M. (1975) Gas exchange of four arctic and alpine tundra plant species in relation to atmospheric and soil moisture stress. *Oecologia (Berlin)*, **21**, 93–108.

Johnson, D.A. and Caldwell, M.M. (1976) Water potential components, stomatal function, and liquid phase water transport resistances of four arctic and alpine species in relation to moisture stress. *Physiologia Plantarum*, **36**, 271–8.

Johnson, D.A. and Tieszen, L.L. (1976) Aboveground biomass allocation, leaf growth, and photosynthesis patterns in tundra plant forms in arctic Alaska. *Oecologia (Berlin)*, **24**, 159–73.

Jung, H.G., Batzli, G.O. and Seigler, D.S. (1979) Patterns in the phytochemistry of arctic plants. *Biochemical Systematics and Ecology*, **7**, 203–9.

Kallio, P. and Heinonen, S. (1971) Influence of short-term low temperature on net photosynthesis in some subarctic lichens. *Reports of the Kevo Subarctic Research Station*, **8**, 63–72.

Kallio, P. and Heinonen, S. (1975) CO_2 exchange and growth of *Rhacomitrium lanuginosum* and *Dicranum elongatum*, in *Fennoscandian Tundra Ecosystems, Part 1: Plants and Microorganisms* (ed. F.E. Wielgolaski), Springer-Verlag, Berlin, pp. 138–48.

Kedrowski, R.A. and Chapin, F.S. III (1978) Lipid properties of *Carex aquatilis* from hot spring and permafrost-dominated sites in Alaska: Implications for nutrient requirements. *Physiologia Plantarum*, **44**, 231–7.

Kjellberg, B., Karlsson, S. and Kerstensson, I. (1982) Effects of heliotropic movements of flowers of *Dryas octopetala* L. on gynoecium temperature and seed development. *Oecologia (Berlin)*, **54**, 10–13.

Kozlowski, T.T. (1971) *Growth and Development of Trees. Vol. 1, Seed Germination, Ontogeny and Shoot Growth*, Academic Press, New York, 443 pp.

Kramer, P.J. (1942) Species differences with respect to water absorption at low soil temperatures. *American Journal of Botany*, **29**, 828–32.

Kummerow, J., McMaster, G.S. and Krause, D.A. (1980) Temperature effects on growth and nutrient contents in *Eriophorum vaginatum* under controlled environmental conditions. *Arctic and Alpine Research*, **12**, 335–41.

Kummerow, J. and Russell, M. (1980) Seasonal root growth in the arctic tussock tundra. *Oecologia (Berlin)*, **47**, 196–9.

Lachenbruch, A.H. (1962) Mechanics of thermal contraction cracks and ice-wedge polygons in permafrost. *Geological Society of America Special Paper*, **70**, 1–69.

Larson, D.W. (1979) Whole plant gas exchange and acclimation in lichens, in *Comparative Mechanisms of Cold Adaptation* (eds L.S. Underwood, L.L. Tieszen, A.B. Callahan and G.E. Folk), Academic Press, New York, pp. 303–10.

Lechowicz, M.J. (1981) The effects of climatic pattern on lichen productivity: *Cetraria cucullata* (Bell.) Ach. in the arctic tundra in northern Alaska. *Oecologia (Berlin)*, **50**, 210–16.

Lechowicz, M.J. (1982) Ecological trends in lichen photosynthesis. *Oecologia (Berlin)*, **53**, 330–6.

Lechowicz, M.J., Hellens, L.E. and Simon, J.-P. (1980) Latitudinal trends in the responses of growth respiration and maintenance respiration to temperature in the beach pea, *Lathyrus japonicus. Canadian Journal of Botany*, **58**, 1521–4.

Leck, M.A. (1980) Germination in Barrow, Alaska, tundra soil cores. *Arctic and Alpine Research*, **12**, 343–9.

LeDrew, E.F. and Weller, G. (1978) A comparison of the radiation and energy balance during the growing season for an arctic and alpine tundra. *Arctic and Alpine Research*, **10**, 665–78.

Lewis, M.C. and Callaghan, T.V. (1976) Tundra, in *Vegetation and the Atmosphere*, Vol. 2 (ed. J.L. Monteith), Academic Press, London, pp. 399–433.

Limbach, W.E., Oechel, W.C. and Lowell, W. (1982) Photosynthetic and respiratory responses to temperature and light of three Alaskan tundra growth forms. *Holarctic Ecology*, **5**, 150–7.

Lippet, R.D. and Hopkins, H.H. (1950) Study of viable seeds in various habitats in mixed prairies. *Transactions of the Kansas Academy of Science*, **53**, 355–64.

Mattheis, P.J., Tieszen, L.L. and Lewis, M.C. (1976) Responses of *Dupontia fischeri* to simulated lemming grazing in an Alaskan arctic tundra. *Annals of Botany*, **40**, 179–97.

Mayo, J.M., Hartgerink, A.P., Despain, D.G., Thompson, R.G., van Zinderin Bakker, E.M. and Nelson, S.D. (1977) Gas exchange studies of *Carex* and *Dryas*, Truelove Lowland, in *Truelove Lowland, Devon Island, Canada: A High Arctic Ecosystem* (ed. L.C. Bliss), University Alberta Press, Edmonton, pp. 265–80.

McCown, B.H. (1978) The interactions of organic nutrients, soil nitrogen, and soil temperature and plant growth and survival in the arctic environment, in *Vegetation and Production Ecology of an Alaskan Arctic Tundra* (ed. L.L. Tieszen), Springer-Verlag, New York, pp. 435–56.

McGraw, J.B. (1980) Seed bank size and distribution of seeds in cottongrass tussock tundra, Eagle Creek, Alaska. *Canadian Journal of Botany*, **58**, 1607–11.

McGraw, J.B. and Shaver, G.R. (1982) Seedling density and seedling survival in Alaskan cotton grass tussock tundra. *Holarctic Ecology*, **5**, 212–17.

McKendric, J.D., Batzli, G.O., Everett, K.R. and Swanson. J.C. (1980) Some effects of mammalian herbivores and fertilization on tundra soils and vegetation. *Arctic and Alpine Research*, **12**, 565–78.

McKendrick, J.D., Ott, V.J. and Mitchell, G.A. (1978) Effects of nitrogen and phosphorus fertilization on carbohydrate and nutrient levels in *Dupontia fisheri* and *Arctagrostis latifolia*, in *Vegetation and Production Ecology of an Alaskan Arctic Tundra* (ed. L.L. Tieszen), Springer-Verlag, New York, pp. 509–37.

McNaughton, S.J., Campbell, R.S., Freyer, R.A., Mylroie, J.E. and Rodland, K.D. (1974) Photosynthetic properties and root chilling responses of altitudinal ecotypes of *Typha latifolia* L. *Ecology*, **55**, 168–72.

Miles, J. (1973) Natural recolonization of experimentally bared soil in *Callunetum* in north-east Scotland. *Journal of Ecology*, **61**, 399–412.

Miles, J. (1979) *Vegetation Dynamics*, Chapman and Hall, London, 80 pp.

Miller, P.C. (1982) Environmental and vegetational variation across a snow accumulation area in montane tundra in central Alaska. *Holarctic Ecology*, **5**, 85–98.

Miller, P.C., Mangan, R. and Kummerow, J. (1982) Vertical distribution of organic matter in eight vegetation types near Eagle Summit, Alaska. *Holarctic Ecology*, **5**, 117–24.

Miller, P.C., Stoner, W.A. and Tieszen, L.L. (1976) A model of stand photosynthesis for the wet meadow tundra at Barrow, Alaska. *Ecology*, **57**, 411–30.

Miller, P.C., Stoner, W.A., Tieszen, L.L., Allessio, M.L., McCown, B.H., Chapin, F.S. III and Shaver, G. (1978) A model of carbohydrate, nitrogen, phosphorus allocation and growth in tundra production, in *Vegetation and Production Ecology of an Alaskan Arctic Tundra* (ed. L.L. Tieszen), Springer-Verlag, New York, pp. 577–98.

Miller, P.C., Webber, P.J., Oechel, W.C. and Tieszen, L.L. (1980) Biophysical processes and primary production, in *An Arctic Ecosystem: The Coastal Tundra at Barrow, Alaska* (eds J. Brown, P.C. Miller, L.L. Tieszen and F.L. Bunnell), Dowden, Hutchinson and Ross, Stroudsburg, Pennsylvania, pp. 66–101.

Mølgaard, P. (1982) Temperature observations in high arctic plants in relation to microclimate in the vegetation of Peary Land, North Greenland. *Arctic and Alpine Research*, **14**, 105–15.

Mooney, H.A. (1975) Plant physiological ecology – a synthetic view, in *Physiological Adaptation to the Environment* (ed. F.J. Vernberg), Intext Educational Publishers, New York, pp. 19–36.

Mooney, H.A. and Billings, W.D. (1960) The annual carbohydrate cycle of alpine plants as related to growth. *American Journal of Botany*, **47**, 594–8.

Mooney, H.A. and Billings, W.D. (1961) Comparative physiological ecology óf arctic and alpine populations of *Oxyria digyna*. *Ecological Monographs*, **31**, 1–29.

Morton, A.J. (1977) Mineral nutrient pathways in a *Molinietum* in autumn and winter. *Journal of Ecology*, **65**, 993–9.

Moser, T.J. and Nash, T.H., III (1978) Photosynthetic

patterns of *Cetraria cucullata* (Bell.) Ach. at Anaktuvuk Pass, Alaska. *Oecologia (Berlin)*, **34**, 37–43.

Muc, M. (1977) Ecology and primary production of Sedge-moss Meadow communities, Truelove Lowland, in *Truelove Lowland, Devon Island, Canada: A High Arctic Ecosystem* (ed. L.C. Bliss), University Alberta Press, Edmonton, Canada, pp. 157–80.

Murray, D.F. (1980) Balsam poplar in arctic Alaska. *Canadian Journal of Anthropology*, **1**, 29–32.

Myers, J.P. and Pitelka, F.A. (1978) Variations in summer temperature patterns near Barrow, Alaska: Analysis and ecological interpretation. *Arctic and Alpine Research*, **11**, 131–44.

Oberbauer, S. and Miller, P.C. (1979) Plant water relations in montane and tussock tundra vegetation types in Alaska. *Arctic and Alpine Research*, **11**, 69–81.

Oechel, W.C. (1976) Seasonal patterns of temperature response of CO_2 flux and acclimation in arctic mosses growing *in situ*. *Photosynthetica*, **10**, 447–56.

Oechel, W.C. and Collins, N.J. (1976) Comparative CO_2 exchange patterns in mosses from two tundra habitats at Barrow, Alaska. *Canadian Journal of Botany*, **54**, 1355–69.

Oechel, W.C. and Sveinbjörnsson, B. (1978) Primary production processes in arctic bryophytes at Barrow, Alaska, in *Vegetation and Production Ecology of an Alaskan Arctic Tundra* (ed. L.L. Tieszen), Springer-Verlag, New York, pp. 269–98.

Outcalt, S.I., Goodwin, C., Weller, G. and Brown, J. (1975) Computer simulation of the snowmelt and soil thermal regime at Barrow, Alaska. *Water Resources Research*, **11**, 709–15.

Penning de Vries, F.W.T. (1975) The cost of maintenance processes in plant cells. *Annals of Botany*, **39**, 77–92.

Peterson, K.M. and Billings, W.D. (1975) Carbon dioxide flux from tundra soils and vegetation as related to temperature at Barrow, Alaska. *American Midland Naturalist*, **94**, 88–98.

Racine, C.H. (1981) Tundra fire effects on soils and three plant communities along a hill-slope gradient in the Seward Peninsula, Alaska. *Arctic*, **34**, 71–84.

Rastorfer, J.R. (1978) Composition and bryomass of the moss layers of two wet-tundra-meadow communities near Barrow, Alaska, in *Vegetation and Production Ecology of an Alaskan Arctic Tundra* (ed. L.L. Tieszen), Springer-Verlag, New York, pp. 169–83.

Rosswall, T. and Heal, O.W. (eds) (1975) *Structure and function of tundra ecosystems*, Ecological Bulletin no. 20, Swedish Natural Science Research Council, Stockholm.

Russell, R.S. (1940) Physiological and ecological studies on an arctic vegetation. III. Observations on carbon assimilation, carbohydrate storage and stomatal movement in relation to the growth of plants on Jan Mayen Island. *Journal of Ecology*, **28**, 289–309.

Rydèn, B.E. (1977) Hydrology of Truelove Lowland, in *Truelove Lowland, Devon Island, Canada: A High Arctic Ecosystem* (ed. L.C. Bliss), University Alberta Press, Edmonton, pp. 107–36.

Rydèn, B.E. (1981) Hydrology of northern tundra, in *Tundra Ecosystems: A Comparative Analysis* (eds L.C. Bliss, O.W. Heal, and J.J. Moore), Cambridge University Press, Cambridge, pp. 115–38.

Saebø, S. (1969) On the mechanism behind the effect of freezing and thawing on dissolved phosphorus in *Sphagnum fuscum* peat. *Scientific Reports of the Agricultural College of Norway*, **48(14)**, 1–10.

Savile, D.B.O. (1972) *Arctic adaptations in plants*, Canada Department of Agriculture Research Branch. Monograph No. 6, 81 pp.

Scott, D. (1970) Relative growth rates under controlled temperatures of some New Zealand indigenous and introduced grasses. *New Zealand Journal of Botany*, **8**, 76–81.

Shaver, G.R. (1981) Mineral nutrition and leaf longevity in an evergreen shrub, *Ledum palustre* ssp. *decumbens*. *Oecologia (Berlin)*, **49**, 362–5.

Shaver, G.R. and Billings, W.D. (1975) Root production and root turnover in a wet tundra ecosystem, Barrow, Alaska. *Ecology*, **56**, 401–9.

Shaver, G.R. and Billings, W.D. (1976) Carbohydrate accumulation in tundra graminoid plants as a function of season and tissue age. *Flora*, **165**, 247–67.

Shaver, G.R. and Billings, W.D. (1977) Effects of day length and temperature on root elongation in tundra graminoids. *Oecologia (Berlin)*, **28**, 57–65.

Shaver, G.R. and Chapin, F.S. III (1980) Response to fertilization by various plant growth forms in an Alaskan tundra: Nutrient accumulation and growth. *Ecology*, **61**, 662–75.

Shaver, G.R. and Cutler, J.C. (1979) The vertical distribution of live vascular phytomass in cottongrass tussock tundra. *Arctic and Alpine Research*, **11**, 335–42.

Shaver, G.R., Chapin, F.S. III and Billings, W.D. (1979) Ecotypic differentiation of *Carex aquatilis* on ice-wedge polygons in the Alaskan coastal tundra. *Journal of Ecology*, **67**, 1025–46.

Shvetsova, V.M. and Voznesenskii, V.L. (1970) Diurnal and seasonal variations in rate of photosynthesis in some plants of Western Taimyr. *Botanicheskii Zhurnal*, **55**, 66–76.

Simon, J.-P (1979) Adaptation and acclimation of higher plants at the enzyme level: latitudinal variations of thermal properties of NAD malate dehydrogenase in *Lathyrus japonicus* Willd. (Leguminosae). *Oecologia (Berlin)*, **39**, 273–87.

Sims, P.L. and Singh, J.S. (1978a) The structure and function of ten western North American grasslands. II. Intra-seasonal dynamics in primary producer compartments. *Journal of Ecology*, **66**, 547–72.

Sims, P.L. and Singh, J.S. (1978b) The structure and function of ten western North American grasslands. III. Net primary production, turnover and efficiencies of energy capture and water use. *Journal of Ecology*, **66**, 573–97.

Sims, P.L., Singh, J.S. and Lauenroth, W.K. (1978) The structure and function of ten western North American grasslands. I. Abiotic and vegetational characteristics. *Journal of Ecology*, **66**, 251–85.

Skre, O. (1975) CO_2 exchange in Norwegian tundra plants studied by infrared gas analyser technique, in *Fennoscandian Tundra Ecosystems. Part I: Plants and Microorganisms* (ed. F.E. Wielgolaski), Springer-Verlag, Berlin, pp. 168–83.

Skre, O., Berg, A. and Wielgolaski, F.E. (1975) Organic compounds in alpine plants, in *Fennoscandian Tundra Ecosystems, Part I. Plants and Microorganisms* (ed. F.E. Wielgolaski), Springer-Verlag, Berlin, pp. 339–50.

Sonesson, M. (ed.) (1980) *Ecology of a Subarctic Mire*, Ecological Bulletins 30, Swedish Natural Sciences Research Council, Stockholm, 313 pp.

Sørensen, T. (1941) Temperature relations and phenology of the northeast Greenland flowering plants. *Meddeleser om Grønland*, **125**, 1–305.

Stewart, W.S. and Bannister, P. (1973) Seasonal changes in carbohydrate content of three *Vaccinium* spp. with particuliar reference to *V. uliginosum* L. and its distribution in the British Isles. *Flora*, **162**, 134–55.

Stoner, W.A and Miller, P.C. (1975) Water relations of plant species in the wet coastal tundra at Barrow, Alaska. *Arctic and Alpine Research*, **7**, 109–24.

Stoner, W.A., Miller, P.C. and Oechel, W.C. (1978) Simulation of the effect of the tundra vascular plant canopy on the productivity of four moss species, in *Vegetation and Production Ecology of an Alaskan Arctic Tundra* (ed. L.L. Tieszen), Springer-Verlag, New York, pp. 371–87.

Summerfield, R.J. and Rieley, J.O. (1973) Substrate freezing and thawing as a factor in the mineral nutrient status of mire ecosystems. *Plant and Soil*, **38**, 557–66.

Sutton, C.D. (1969) Effect of low soil temperature on phosphate nutrition of plants – a review. *Journal of Science of Food and Agriculture*, **20**, 1–3.

Svoboda, J. (1977) Ecology and primary production of Raised Beach communities, Truelove Lowland, in

Truelove Lowland, Devon Island, Canada: A High Arctic Ecosystem (ed. L.C. Bliss), University Alberta Press, Edmonton, pp.185–216.

Teeri, J. (1973) Polar desert adaptations of a high arctic plant species. *Science*, **179**, 496–7.

Teeri, J.A. and Tonsor, S.J. (1981) Variability in photoperiod and the inhibition of flowering in a high latitude population of *Saxifraga rivularis*. *Canadian Journal of Botany*, **59**, 388–91.

Tieszen, L.L. (1970) Comparisons of chlorophyll content and leaf structure in arctic and alpine grasses. *American Midland Naturalist*, **83**, 238–53.

Tieszen, L.L. (1972) The seasonal course of aboveground production and chlorophyll distribution in a wet arctic tundra at Barrow, Alaska. *Arctic and Alpine Research*, **4**, 307–24.

Tieszen, L.L. (1973) Photosynthesis and respiration in arctic tundra grasses, field light intensity and temperature responses. *Arctic and Alpine Research*, **5**, 239–51.

Tieszen, L.L. (1974) Photosynthetic competence of the subnivean vegetation of an arctic tundra. *Arctic and Alpine Research*, **6**, 253–6.

Tieszen, L.L. (1975) CO_2 exchange in the Alaskan arctic tundra: seasonal changes in the rate of photosynthesis of four species. *Photosynthetica*, **9**, 376–90.

Tieszen, L.L. (1978a) Photosynthesis in the principal Barrow, Alaska, species: A summary of field and laboratory responses, in *Vegetation and Production Ecology of an Alaskan Arctic Tundra* (ed. L.L. Tieszen), Springer-Verlag, New York, pp. 241–68.

Tieszen, L.L. (1978b) Summary, in *Vegetation and Production Ecology of an Alaskan Arctic Tundra* (ed. L.L. Tieszen), Springer-Verlag, New York, pp. 621–45.

Tieszen, L.L. and Helgager, J.A. (1968) Genetic and physiological adaptation in the Hill reaction of *Deschampsia caespitosa*. *Nature*, **219**, 1066–7.

Tieszen, L.L. and Sigurdson, D.C. (1973) Effect of temperature on carboxylase activity and stability in some Calvin cycle grasses from the arctic. *Arctic and Alpine Research*, **5**, 59–66.

Tieszen, L.L. and Wieland, N.K. (1975) Physiological ecology of arctic and alpine photosynthesis and respiration, in *Physiological Adaptation to the Environment* (ed. F.J. Vernberg), Intext Educational Publishers, New York, pp. 157–200.

Tieszen, L.L., Miller, P.C. and Oechel, W.C. (1980) Photosynthesis, in *An Arctic Ecosystem: The Coastal Tundra at Barrow, Alaska* (eds J. Brown, P.C. Miller, L.L. Tieszen and F.L. Bunnell), Dowden, Hutchinson and Ross, Stroudsburg, Pennsylvania, pp. 102–39.

Tranquillini, W. (1979) *Physiological Ecology of the Alpine Timberline*, Springer-Verlag, New York.

Ulrich, A. and Gersper, P.L. (1978) Plant nutrient limitations of tundra plant growth, in *Vegetation and Production Ecology of an Alaskan Arctic Tundra* (ed. L.L. Tieszen), Springer-Verlag, New York, pp. 457–81.

Vitt, D.H. and Pakarinen, P. (1977) The bryophyte vegetation, production, and organic compounds of Truelove Lowland, in *Truelove Lowland, Devon Island, Canada: A High Arctic Ecosystem* (ed. L.C. Bliss), University Alberta Press, Edmonton, pp. 225–44.

Wager, H.G. (1938) Growth and survival of plants in the arctic. *Journal of Ecology*, **26**, 390–410.

Wager, H.G. (1941) On the respiration and carbon assimilation rates of some arctic plants as related to temperature. *New Phytologist*, **40**, 1–19.

Wallace, L.L. and Harrison, A.T. (1978) Carbohydrate mobilization and movement in alpine plants. *American Journal of Botany*, **65**, 1035–40.

Warren Wilson, J. (1954) The influence of 'midnight sun' conditions on certain diurnal rhythms in *Oxyria digyna*. *Journal of Ecology*, **42**, 81–94.

Warren Wilson, J. (1957a) Observations on the temperatures of arctic plants and their environment. *Journal of Ecology*, **45**, 499–531,

Warren Wilson, J. (1957b) Arctic plant growth. *Advancement of Science*, **13**, 383–8.

Warren Wilson, J. (1959) Notes on wind and its effects in arctic–alpine vegetation. *Journal of Ecology*, **47**, 415–27.

Warren Wilson, J. (1960) Observations on net assimilation rates in arctic environments. *Annals of Botany*, **24**, 372–81.

Warren Wilson, J. (1966) An analysis of plant growth and its control in arctic environments. *Annals of Botany*, **30**, 383–402.

Webber, P.J. (1977) Belowground tundra research: A commentary. *Arctic and Alpine Research*, **9**, 105–11.

Webber, P.J. (1978) Spatial and temporal variation of the vegetation and its productivity, Barrow, Alaska, in *Vegetation and Production Ecology of an Alaskan Arctic Tundra* (ed. L.L. Tieszen), Springer-Verlag, New York, pp. 37–112.

Wein, R.W. and Bliss, L.C. (1973) Changes in arctic *Eriophorum* tussock communities following fire. *Ecology*, **54**, 845–52.

Wein, R.W. and Bliss, L.C. (1974) Primary production in arctic cottongrass tussock tundra communities. *Arctic and Alpine Research*, **6**, 261–74.

Wein, R.W. and MacLean, D.A. (1973) Cottongrass (*Eriophorum vaginatum*) germination requirements and colonizing potential in the Arctic. *Canadian Journal of Botany*, **51**, 2509–13.

Weller, G. and Holmgren, B. (1974) The microclimates of the arctic tundra. *Journal of Applied Meteorology*, **13**, 854–62.

White, L.M. (1973) Carbohydrate reserves of grasses: a review. *Journal of Range Management*, **26**, 13–18.

Wielgolaski, F.E. (ed.) (1975a) *Fennoscandian Tundra Ecosystems, Part I: Plants and Microorganisms*, Springer-Verlag, Berlin.

Wielgolaski, F.E. (1975b) Vegetation types and plant biomass in tundra. *Arctic and Alpine Research*, **4**, 291–305.

Wielgolaski, F.E., Bliss, L.C., Svoboda, J. and Doyle, G. (1981) Primary production of tundra, in *Tundra Ecosystems: A Comparative Analysis* (eds L.C. Bliss, O.W. Heal and J.J. Moore), Cambridge University Press, Cambridge, pp. 187–225.

Wielgolaski, F.E., Kjelvik, S. and Kallio, P. (1975) Mineral content of tundra and forest tundra plants in Fennoscandia, in *Fennoscandian Tundra Ecosystems, Part I: Plants and Microorganisms* (ed. F.E. Wielgolaski), Springer-Verlag, Berlin, pp. 316–32.

3

Alpine

L.C. BLISS

3.1 Introduction

Alpine regions hold a special fascination for biologists because of their stressful environments, their similarity with the Arctic, their aesthetic setting, and their island-like isolation. Because alpine habitats extend over a great range of latitude, from the highlands and volcanoes of Central America and Mexico to the northern mountains of the Yukon Territory and Alaska, their environments are very diverse. Consequently, there is a greater range in photoperiod, length of the growing season, total and net radiation, diurnal temperature, and precipitation than in the Arctic. The diversity of geologic substrates, presence of permafrost in only the northern mountains, soil drainage patterns, and size of these lands above the climatic limit of tree growth often add to the diversity in patterning of physiologically adapted ecotypes and the general patterning of species and plant communities.

As used here, alpine refers to those lands above the climatic limit of upright trees, although it may include islands of krummholz and the often associated local patches of subalpine herbs and shrubs. Zonation within the subalpine and alpine has been variously defined (Bliss, 1969, 1981; Clausen, 1965; Douglas and Bliss, 1977; Löve, 1970).

While early alpine studies centered upon descriptions of plant communities and floristics (e.g. Cooper, 1908; Cox, 1933; Holm, 1927; Rydberg, 1914), these workers and others often commented on environmental influences. More recent studies have integrated meso- and microenvironmental data in helping to explain community pattern (Bamberg and Major, 1968; Bliss, 1963, 1969; Douglas and Bliss, 1977; Hrapko and La Roi, 1978; Johnson and Billings, 1962) and plant production (Bliss, 1956, 1966; Scott and Billings, 1964; Webber and May, 1977).

While some conclusions presented here result from studies undertaken in the International Biological Programme, most North American emphasis on arctic–alpine systems was conducted in the Arctic (Barrow, Alaska and Truelove Lowland, Devon island, N.W.T.). Syntheses of these studies should be consulted for further detail (Bliss, 1977a; Brown *et al.*, 1980; Hobbie, 1980; Tieszen, 1978a; Tieszen *et al.*, 1981). Earlier reviews which summarize information on the physiological ecology of alpine plants include Billings (1974a), Billings and Mooney (1968), Bliss (1962a, 1971) and for timberline studies Tranquillini (1978).

The major objectives of this chapter are: (1) to summarize environmental similarities and dissimilarities within the alpine in relation to arctic and temperate environments, and (2) to discuss the physiological responses of plants adapted to these environments.

3.2 Environment

For many mountain ranges in North America, photoperiod during much of the growing season averages 15–18 h, though the range is from 13 to

Table 3.1 Partitioning of net radiation at alpine tundra sites in mid-summer

Location	Date	sample days no.	R_n (MJ m^{-2} d^{-1})	Energy flux (%)			Bowen ratio (B)	Reference
				LE	H	G		
Mount Washington, NH								
(44 °N) 1660 m*	July 1965	1	11.30	31	50	19	1.61	Courtin (1968)
1840 m†	July 1965	1	9.29	48	46	06	0.96	Courtin (1968)
Niwot Ridge, CO								
(40 °N, 3500 m)	July 1973	19	12.43	38	50	12	1.31	Le Drew and Weller (1978)
Piute Pass, CA								
(37 °N, 3540 m)	July 1968	1	20.35	5	88	7	17.6	Chabot and Billings (1972)

*Diapensia cushion.
†Sedge-moss.

14 h in Central American alpine to 24 h in northern Alaskan and Yukon mountains. While many alpine environments are characterized by summer sunshine, the maritime climate within the mountains of New England and the Gaspé Peninsula has much summer cloudiness. The result is a considerable range in the amount of total incoming radiation and how it is dissipated. Data in Table 3.1 are for sunny days in these three contrasting environments. They show that two alpine sites in the generally wet and humid Mt Washington, NH environment dissipate net radiation (R_n) mainly as sensible heat flux (H) in the *Diapensia lapponica* cushion plant community or as equal amounts of sensible heat and latent heat of evapotranspiration (LE) in the sedge-moss site. In the latter system, soil heat flux (G) is quite minor. It is surprising that in the Sierran alpine, where soil temperatures are high, G is such a small percentage of the total. The very high levels of net radiation are dissipated almost solely as sensible heat (note the high Bowen ratio). There is need for more studies on radiation balance within alpine environments.

Short-wave radiation is quite variable in different mountain massifs. Maximum daily values reach 33.5–35.6 MJ m^{-2} d^{-1} on clear days in the normally cloudy Presidential Range, NH as well as in the Rocky Mountains, the Sierra Nevada, CA and the Olympia Mountains, WA. Daily means, expressed on a monthly basis, range from 12 to 17 on Mount Washington, and 21 to 25 in the Rocky Mountains and Pacific Northwest Mountains, to 25 to

28 MJ m^{-2} d^{-1} in the clear, high elevation alpine of the Sierra Nevada (Bell and Bliss, 1979; Bliss, 1966, 1969; Chabot and Billings, 1972; Douglas and Bliss, 1977; Hadley and Bliss, 1964; Harter, 1981; Klikoff, 1965a; Table 3.2). A generalized pattern of radiation, temperature, and snow cover is presented in Fig. 3.1 for Niwot Ridge, a representative alpine site in the central Rocky Mountains.

Summer air and soil temperatures within the growing environment (-20 to $+20$ cm) of most alpine plants are considerably higher than the general temperature data, recorded at standard Weather Bureau height (1.2 m), would indicate. Soil temperatures within the rooting zone (-5 to -20 cm) are generally 8–13 °C, but are much higher in the Sierra Nevada with higher solar radiation (Table 3.2). Mean daily air temperatures generally range from 9 to 12 °C at 10–15 cm. Mean daily maxima at 10 cm are often 15–20 °C and daily minima 4–8 °C in the Presidential Range (Bliss, 1966). In comparable sites at 3450 m in the alpine of the Sierra Nevada, daily maxima average 14–18 °C and minima 5–8 °C (Chabot and Billings, 1972). Diurnal temperature oscillations average 10–12 °C in many alpine areas, a much greater range than in the constant light but low sun angle environment of the Arctic. In most alpine sites, outside of the tropics, nightly freezing is uncommon in spite of strong reradiation. Temperatures of 0 °C or lower at 10 cm were recorded on only 4% of the 72 nights in the Presidential Range, 27% of the 62

Table 3.2 Mean daily environment data for select alpine sites in North America

Location	Solar radiation $(MJ\,m^{-2}\,d^{-1})$ July	August	Temperature (C°) −10 to −15 cm July	August	+10 to +15 cm July	August	Wind (40–60 cm) $(m\,s^{-1})$ July–August	Precipitation (cm) July–August	VPD (kPa) July–August
Mt Washington, NH[a]									
1664 m (heath rush)	15.4	17.1	9.0	11.6	11.3	11.4	2.9	14.8	0.11
1840 m (sedge moss)	12.1	16.5	8.8	8.7	9.3	9.3	3.7	15.6	0.09
Trail Ridge Road, CO[b]									
3500 m (Kobresia meadow)	24.9	21.2	—	—	7.0[h]	10.0[h]	2.0	10.0	0.39
Medicine Bow Mts, WY[c]									
3320 m (sedge meadow)	—	—	8.1	9.4	10.8	8.8	1.8	10.6	—
3350 m (cushion plants)	—	—	12.6	10.7	12.5	10.3	4.4	10.3	—
Signal Mt Jasper Nat. Park[d]									
2193 m (heath shrub)	19.6	16.0	5.8	4.1	9.5	7.2	3.1	9.7	0.41
Piute Pass, Sierra Nevada, CA[e]									
3540 m (herb fellfield)	28.3	25.4	22.0[i]	18.0[i]	10.5[i]	5.0[i]	2.3	6.7	0.61
Elk Mt Olympic Mts, WA[f]									
1960 m (herb meadow)	21.2	17.9	9.0	10.0	9.5	9.2	3.5	5.8	0.33
1972 m (cushion plants)	—	—	—	12.7	—	9.8	3.5	5.5	—
Grouse Ridge, Mt Baker, WA[g]									
1785 m (herb fellfield)	22.0	20.9	—	—	10.7	11.2	1.6	7.7	0.15

[a]Bliss, 1969 (3 yr). [b] Bell and Bliss, 1979 (1 yr). [c]Bliss, 1956 (2 yr). [d]Harter, 1981 (3 yr). [e]Chabot and Billings, 1972 (1 yr). [f]Bliss, 1969 (2 yr). [g]Douglas and Bliss, 1977 (2 yr).
[h]Extrapolated data at 120 cm.
[i]Extrapolated data.

nights in the Olympic Mountains (Bliss, 1969), and 5% of the 65–75 nights on Mt Baker, Northern Cascades (Douglas and Bliss, 1977).

For years, high ultraviolet radiation was assumed to be important in the alpine environment. Caldwell (1968) showed that UV-B (2800–3150 nm) was 26% greater at 4350 m than at 1670 m on a clear day. Under natural conditions, the epidermal pigments (flavonoids) filter UV-B irradiance and photoreactivation further reduces the potential damaging effects. Red leaves also reduce UV transmission, thus the presence of high leaf anthocyanin content acts as a UV filter (Caldwell, 1968).

In a study of UV-B irradiance from equatorial alpine to high latitude arctic environments, Caldwell et al. (1980) report that maximum daily total shortwave radiation varied by only 60% and total daily shortwave radiation by only 15%, yet UV-B

radiation varied seven-fold, all on clear days. This seven-fold difference on a DNA effective basis, however, is not reflected in significant differences in UV-B transmittance to leaf mesophyll (Robberecht et al., 1980). Tropical plants exhibit a much greater epidermal attenuation of UV-B, the result of flavonoid and alkaloid compounds, than their arctic counterparts. In general UV-B transmittance to the mesophyll tissue is > 2% in the species along this arctic–alpine gradient. Although UV-B irradiance is probably not a major factor today in reducing alpine plant production, the consistently greater photosynthetic damage between ecotypes of *Oxyria digyna* and species pairs to *Taraxacum* and *Lupinus* in arctic vs. alpine populations, the more rapid epidermal leaf damage of some arctic populations (Caldwell et al., 1982) and the consistently higher UV epidermal transmittance at high latitudes (Robberecht et al.,

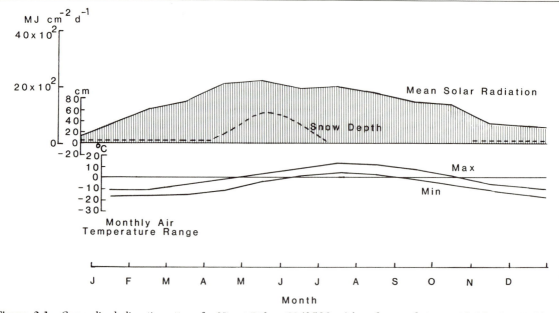

Figure 3.1   Generalized climatic pattern for Niwot Ridge, CO (3500 m) based upon data provided by P.J. Webber.

1980) all suggest an evolutionary selective force is operating in this arctic-alpine gradient.

Plant temperatures of broad-leaved herbs are generally elevated above ambient by 5–15 °C in the sunny Rocky Mountain alpine (Salisbury and Spomer, 1964). Narrow-leaved species, such as *Kobresia bellardii*, have leaf temperatures generally within 1 °C of ambient and never more than 2.8 °C higher (Bell and Bliss, 1979). In winter, leaf temperatures of *Kobresia* were the same as ambient. In the Presidential Range, the cushion plant *Diapensia lapponica*, growing in well-drained soils, had leaf temperatures often 3–5 °C above ambient on cloudy days and 12–15 °C higher on sunny days. In contrast the sedge, *Carex bigelowii*, growing in cold, wet soils with near constant wind had leaf temperatures only 0.5–1.5 °C above ambient (Courtin, 1968). Clearly, the cushion plant growth form is an adaptation for maintaining a thicker plant boundary layer. The total impact of microclimate on plant form and physiology will be discussed later.

Wind is an omnipresent component of alpine environments. Wind speeds in summer at 40 to 60 cm are generally low (1–2 m s⁻¹) in meadows and considerably higher on ridges and cols (3–5 m s⁻¹) (Bell and Bliss, 1979; Bliss, 1956, 1966; Chabot and

Billings, 1972; Douglas and Bliss, 1977; Klikoff, 1965a). The highest winds on a monthly basis have been reported in the New Zealand alpine where values of 5–8 m s⁻¹ are common (Mark, 1962; Bliss and Mark, 1974). In the New Zealand mountains cushion plants, representing diverse plant families, predominate in these habitats.

Wind profiles measured within a windy fellfield and a moderately protected *Kobresia* meadow in the Rocky Mountain alpine show that winter winds can be 2–5 times higher than in summer. In winter, gusts in the fellfield exceeded 7 m s⁻¹ at 2 cm and 35 m s⁻¹ at 200 cm. In summer, measurements at 2 cm within the plant canopy averaged 0.08 m s⁻¹ in a snowbed protected site and 0.5 m s⁻¹ in the fellfield. Wind speed near the top of the plant canopy averaged 1–1.5 m s⁻¹ at 10 cm in the *Kobresia* meadow and 2–2.3 m s⁻¹ at the same height above the cushion plants in the fellfield. Summer winds were strong enough to induce a 3–6 °C cooling at the plant surface of the fellfield compared with protected snowbed sites (Bell and Bliss, 1979). High winds (6.8–10.8 m s⁻¹) at 60 cm have also been effective in reducing net assimilation of four alpine species 1 or 2 d following storms on Mt Washington (Hadley and Bliss, 1964). From this discussion it is evident that wind velocity is greatly influenced by micro and mesotopography, and is greatly reduced at the plant canopy, but with important differences between growth forms.

Summer precipitation is a highly variable factor, and increases from western to eastern mountain ranges. Atmospheric vapor-pressure deficit (VPD) decreases along the same west to east transect (Table 3.2). This is strongly reflected in the plant communities; *Carex bigelowii, Juncus trifidus*, and heath species predominate in the New England mountains, herbs in the Rocky Mountains, and scattered dwarf herbs and cushion plants in the dry summer climate of the Olympic Mountains and Sierra Nevada.

Soil moisture stress (< -1.5 MPa) occurs in many alpine plant communities in August and September. Klikoff (1965a) reported that soil water potential (Ψ_s) at -10 cm dropped below -1.5 MPa by midsummer in upland dry meadows, while still above -1.5 MPa in moist meadows (*Saxifraga–Antennaria*) and above -0.03 MPa in wet meadows (*Calamagrostis*). Data from Trail Ridge Road, Rocky Mountain National Park, CO indicate that Ψ_s of fellfield soils at -10 cm was generally < -1.5 MPa, and that at -2 cm was always < 4.0 MPa except after a rain (Bell and Bliss, 1979). In *Kobresia* meadows, Ψ_s was seldom below -2.0 MPa and usually > -1.0 MPa. Although problems occur with winter measurements of Ψ_s using porous cup thermocouple psychrometers, soil water potential was usually < -4.0 MPa, though values of -0.5 to -2.5 MPa were obtained for short periods of warmer winter weather.

Oberbauer and Billings (1981) found that leaf water potentials (Ψ_l) were lowest in plants on a ridgetop and highest in those in a wet meadow in the Medicine Bow Mountains, WY. In general, deep rooted species (*Trifolium parryi*) showed maximum leaf conductance at plant water potentials as low as -1.7 MPa. Minimum Ψ_l was -2.18 MPa on a windward site. In contrast the shallow-rooted species (*Paronychia pulvinata, Artemisia scopulorum, Hymenoxys grandiflora*) survived Ψ_l of -3.5 to -4.0 MPa. Leaves of *Polygonum bistortoides* showed high leaf conductances near -1.0 MPa, had a sharp decline in conductance < 1.5 MPa as they wilted, and leaves died at potentials below -2.3 MPa.

During a drought period in the North Cascades, WA and BC, Ψ_s reached -1.4 to -2.7 MPa with Ψ_l generally -0.5 to -1.0 MPa lower. *Lupinus latifolius* wilted under these soil conditions (-2.7 MPa) and the plants died back within a few days (Douglas and Bliss, 1977). Comparably low Ψ_l and Ψ_s have been reported in high subalpine herbs and soils in the Olympic Mountains (Kuramoto and Bliss, 1970) and on Niwot Ridge, CO (Ehleringer and Miller, 1975).

3.3 Floristics

Alpine floras are highly variable in species richness. This is the product of areal extent of each mountain range, its proximity to other high mountains, past floristic migration routes during glacial and interglacial periods, and the relative environmental severity of each alpine area, in relation to adjacent lowland environments.

Alpine floras probably first evolved in the central Asian Highlands and the Cordillera during the period of mountain building in late Miocene and early Pliocene (Billings, 1974b; Dorf, 1960; Wolfe and Hopkins, 1967; Yurtsev 1972). From these centers the floras spread in the Late Tertiary and Pleistocene. They also were enriched by upward migration and evolution from species adapted to lower mountain slopes (Billings, 1978; Chabot and Billings, 1972).

All of the floras have a significant arctic component, though this is greatest in the Rocky Mountain ranges and the New England mountains (Table 3.3). Alpine floras of the Intermountain Region and the Sierra Nevada have fewer species in common with the northern mountains and the Arctic (Billings, 1978). Many mountain massifs have a vascular plant flora of 150 to 200 + species.

3.4 Plant communities and environmental gradients

Because alpine areas are 'islands' within scattered larger mountain massifs, and because the floras and plant community patterns are so varied throughout

Table 3.3 Vascular plant floristics of North American alpine regions

Location	Latitude N°	No. species	Floristic affinities (%)					
			Regional or local endemics	Arctic–alpine	Boreal	Rocky Mountains	Cascade Mountains	Sierra Nevada
Signal Mt, Jasper N.P.[a]	52	151	7	49	14	30	—	—
Lewis Range, MT[b]	50	185	6	55		37	2	—
Flint Creek, Range, MT[b]	46	165	7	45	—	46	2	—
Beartooth Mts, WY[c]	45	194	2	42	2	51	2	2
Trail Ridge Road, CO[d]	40	157	14	56	30	—	—	—
North Cascades, WA[e]	48	82	9	20	26	45	—	—
Wassuk Range, NV[f]	40	70	1	17	—	66[j]	—	16
Mt Humphreys, CA[g]	37	72	17	14	3	30	21	15
San Francisco Peak, AZ[h]	35	49	10	41	—	49	—	—
Mt Washington NH[i]	40	110	4	64	32	—	—	—

[a]Hrapko and La Roi (1978). [b]Bamberg and Major (1968). [c]Johnson and Billings (1962). [d]Willard (1979). [e]Douglas and Bliss (1977). [f]Bell and Johnson (1980). [g]Chabot and Billings (1972). [h]Little (1941). [i]Bliss (1962). [j]Includes Great Basin ranges.

the continent, the vegetation has only been classified into very generalized units (Bliss, 1979, 1981).

Some authors have related plant community patterns to environmental gradients. Johnson and Billings (1962) classified seven plant community types along a three-dimensional gradient of wind exposure, topographic location, and depth of snow cover on the Beartooth Plateau, MO. They further described, in detail, the importance of cryopedogenic patterns and processes relative to soil and rock movement. Active mass wasting processes occur in many alpine areas today (Bell and Bliss, 1973; Billings and Mooney, 1959; Fahey, 1973; Thompson, 1961), albeit at a reduced rate relative to early

Holocene time. The environmental gradients best correlated with the patterning of nine plant communities in the Presidential Range were wind, snow depth, and atmospheric moisture (Bliss, 1963). Broad gradients of plant communities and atmospheric moisture result from the forced topographic rise of warm air masses in summer, while narrow gradients of wind and depth of snow cover often result from local patterned ground (solifluction terrace) features. The importance of meso- and microtopography and meso- and microclimatology in relation to plant community patterns has been described in the Rocky Mountains for vascular plant communities (Billings and Bliss, 1959; Bliss, 1956;

Figure 3.2a *Abies lasiocarpa* (foreground) and *Salix brachycarpa* (centre) established in the lee of rocks. Alpine meadow herbs include *Geum rossii, Polygonum bistortoides, Carex drummondiana* and *Hymenoxys grandiflora*, Snowy Range, Medicine Bow Mountains, 3350 m.

Hrapko and La Roi, 1978; Johnson and Billings, 1962; Marr, 1961) and for cryptogam communities (Flock, 1978; See and Bliss, 1980); in the Cascades and Olympic Mountains (Bliss, 1969; Douglas and Bliss, 1977); and in the Sierra Nevada (Klikoff, 1965a). From these studies, it is evident that soil pH, soil water potential, snow cover and time of melt, and wind exposure, as influenced by topography, are controlling factors in the meso- and micropatterning of plant species in diverse alpine environments (Fig. 3.2a,b,c).

3.5 Seed germination and seedling establishment

Based upon studies of species from diverse alpine environments (Amen, 1965, 1968; Amen and Bonde, 1964; Bliss, 1958; Bonde, 1965; Chabot and Billings, 1972; Marchand and Roach, 1980; Sayers and Ward, 1966), it is evident that many species produce viable seed. Seed dormancy is characteristic of only 20–40% of alpine and arctic species, mostly seed coat inhibition (Amen and Bonde, 1964; Bell and Amen, 1970; Rocow, 1970), chilling, after-ripening, and light requirements (Amen, 1966; Bliss, 1958). Most species require high temperatures (20–30 °C) with only a small percentage of seeds germinating at < 10 °C; germination is often stimulated by oscillating temperatures (Mooney and Billings, 1961; Sayers and Ward, 1966). Having optimum germination at relatively high temperatures is probably an important selective mechanism to deter fall and spring germination when high frequency of frosts would select against seedling establishment.

3.2b   Patterning of alpine herbs and dwarf shrubs in a snowbed site (*Antennaria lanata, Carex nigricans, Erigeron peregrinus*) with dark colored patches of *Phyllodocae glanduliflora* on slope. Soil boils with *Dryas octopetala* cushion plants on wind-exposed bench beyond; Sunshine Basin, Banff National Park, Alberta, 2425 m.

Seedling establishment, while not common in many alpine environments, does occur under favorable conditions of bare and moist soil. Marchand and Roach (1980) reported that three species common to abandoned trails were also species producing large quantities of seed that germinated readily. *Juncus trifidus* germinated best at a 12–15 °C oscillating night–day temperature regime and *Potentilla tridentata* at 21–26 °C and 23–28 °C regimes. *Arenaria groenlandica* germination was above 60% at the three highest temperature regimes used. As expected, field germination percentages were lower than under laboratory conditions but did occur in all three species.

Seedling mortality is often the result of soil drought and needle ice formation (Bonde, 1968; Osburn, 1961), yet in the Sierra Nevada annuals

predominate in these sites (Jackson and Bliss, 1981). The presence of other species within clumps of cushion plants led Griggs (1956) to believe that the clumps act as 'nurse plants'. While this may occur, cushion plants also grow around previously existing plants. Bonde (1968) attributed mortality of seedlings within cushion plants to competition for moisture, nutrients, and light, with soil moisture stress most important in bare soil sites.

3.6 Growth forms and plant growth

3.6.1 ANNUALS

While annuals are a minor floristic component of all alpine and high subalpine environments, their greatest species diversity is in the Sierra Nevada

3.2c Alpine herb meadow at weather station dominated by *Carex phaeocephala, Phlox diffusa, Potentilla diversifolia,* and *Draba lonchocarpa.* On the exposed slope beyond *Lupinus lepidus* and *Phlox diffusa* dominate with *Carex nigricans* and *Antennaria lanata* occurring at the snowbed site; Elk Mountain, Olympic National Park, WA 1960 m.

(Went, 1953). The total number of annual species reaching the alpine is probably only 10 (Major and Taylor, 1977) and in most alpine studies only 2–5 species have been reported for specific areas (Chabot and Billings, 1972; Jackson and Bliss, 1981; Klikoff, 1965a). Went (1953) attributed the greater number of annuals to a physiological and ecological segregation of species adapted to the hot, dry environments of the valleys and lower slopes. Chabot and Billings (1972) added to this in pointing out that alpine annuals are in an energy-rich and droughty environment as are their desert counterparts.

Jackson and Bliss (1982) found that along a transect from subalpine open meadows to the alpine in the central Sierra Nevada, annuals decreased from 15 to 4 species. The requirement for higher

temperatures and longer water stress-free periods are believed to greatly limit the extension of annuals and ephemerals (perennial species that function for only part of the growing season) into the alpine, even though their short life span of 2–3 months might be hypothesized to be an adaptation of these species to a short alpine growing season. Annuals are found only in open sites of loose gravel or soil, nearly devoid of other species. The same is true for *Koenigia islandica* in the Rocky Mountains (Dahl, 1963).

3.6.2 CUSHION AND ROSETTE FORMS

Cushion and rosette-forming species are selected for exposed ridges and slopes where there is little winter snow (5–10 cm maximum), well-drained soils, and

Figure 3.3 Cushion plants in a wind-exposed site, Trail Ridge Road, Rocky Mountain National Park, CO 3730 m. The species include *Trifolium nanum* (above coin), *Minuartia obtusiloba* (right of coin), *Silene acaulis* (dark patch below coin), and *Paronychia pulvinata* (below and left of coin).

strong winds with reduced boundary layer conditions. Species from diverse plant families have evolved this growth from: *Diapensia lapponica –* Diappensiaceae and *Loiseleuria procumbens–* Ericaceae in the New England mountains; *Silene acaulis, Arenaria obtusiloba –* Carophyllaceae, *Trifolium nanum, T. dasyphyllum –* Leguminoseae, *Paronychia pulvinata –* Boraginaceae, *Phlox caespitosa –* Polemoniaceae, in the central Rocky Mountains (Fig. 3.3); and *Lupinus brewerii –* Leguminoseae, *Potentilla brewerii –* Rosaceae, *Eriogonum marifolium, E. ochrocephalum –* Polygonaceae, and *Arenaria nuttallii –* Carophyllaceae in the Sierra Nevada.

Although these growth forms are common in alpine and arctic areas, few species have been studied in detail. *Diapensia lapponica* (Courtin, 1968; Hadley and Bliss, 1964) from the Presidential Range and *Dryas integrifolia* (Bliss, 1977b) from the High Arctic show the same physiological response pattern. The species have winter green or evergreen leaves, low rates of shoot and root growth, great longevity, low rates of net assimilation, low carbohydrate reserves, low conductance of water vapor, and leaf temperatures 15–20 °C above ambient on sunny days and even values of ΔT of 5–8 °C on cloudy days (Table 3.4). Relatively large amounts of seed are produced per plant, the seeds are generally viable, but few seedlings are observed. This is the ultimate K-selected group of alpine species, although most other growth forms also fit this category. This growth form is also common on warmer and drier soils, regardless of soil

Table 3.4 Plant growth and physiological responses of different growth forms within alpine environments

Plant growth and Physiological response	Plant growth forms				Shrubs	
	Annuals	Cushion and rosette	Forbs	Graminoids	Deciduous	Evergreen
Plant age (yr)	1	20–50+	10–50	25–250	20–50+	20–50+
Leaves/shoot/year	2–10	3–10	3–20	2–10	5–15	5–15
Net assimilation (μmol CO_2 g^{-1} s^{-1})	32–95	13–32	13–63	13–158	13–93	6–32
Carbohydrate reserves	nil	intermediate	high	high	intermediate	low
Aboveground production	low	low	high	high	intermediate	low
Leaf resistance	low	high	intermediate	low	low	high
Leaf water potential	high	low	high	intermediate	intermediate	low

References: Bell and Bliss (1979), Courtin (1968), Ehleringer and Miller (1975), Hadley and Bliss (1964), Jackson and Bliss (1984), Johnson and Caldwell (1974, 1975).

texture and parent material, in much of the polar semi-deserts of the High Arctic (Bliss, 1979, 1981).

3.6.3 GRAMINOIDS

Grasses, sedges, and rushes are present in all alpine environments, though one or more groups may predominate in specific habitats. The species used to illustrate growth and development patterns are *Carex bigelowii* for a moist to wet site (Bliss, 1966; Courtin, 1968; Fonda and Bliss, 1966; Hadley and Bliss, 1964) and *Kobresia bellardii* for a dry site (Caldwell *et al.*, 1978; Bell and Bliss, 1979).

Growth is rapid as soon as snow melts and the soils begin to warm. This results from large carbohydrate reserves that are mobilized from the rhizomes in *Carex*, and from shoot bases in *Kobresia*. These species also carry over winter, 5–10% biomass of green tissue that probably contains high concentrations of nitrogen and phosphorus as do arctic graminoids (Chapin, 1978). Shoots live 3–5 yr but a rhizome system (*Carex*) or clump (*Kobresia*) may live 50–100 yr. Plant production is considerable for the graminoids, but rates of net assimilation may be low in a cloudy, wet coastal maritime or alpine environment (19–25 μmol CO_2 g^{-1} s^{-1}) in contrast with *Kobresia* from a sunny, relatively dry continental environment (10–13 μmol CO_2 m^{-2} s^{-1} (Table 3.4). Leaf water potentials can be quite low in *Kobresia* (summer and winter) and leaf temperatures of both *Kobresia* and

Carex are little elevated above ambient temperatures in their respective 'dry' and 'wet' alpine environments.

3.6.4 DECIDUOUS AND EVERGREEN SHRUBS

While low *Salix* shrubs (30–50 cm) are fairly common along streams, in cirques, and just above timberline in the Rocky Mountains and eastern alpine areas, *Salix* and *Betula* are absent or very minor in the Sierra Nevada, Cascade Ranges, and Olympic Mountains. Dwarf evergreen shrubs (10–20 cm) occur in nearly all mountains but dwarf deciduous shrubs are also minor in the western mountains. Low and dwarf shrubs occur in microhabitats where the shoots are covered with snow most of the winter. The evergreen heath species, exemplified by *Cassiope tetragona*, always occur in snowbed sites or at least where snow does not melt until early to mid-June.

Data from *Vaccinium uliginosum* (Bliss, 1966; Hadley and Bliss, 1964) and *Cassiope tetragona* (Harter, 1981) indicate these plants are rather long lived ($\simeq$ 20 to 50 yr), produce 8–10 leaf pairs per year, grow slowly, and have rather low net assimilation rates (Table 3.4). It is assumed that carbohydrate storage is limited in this growth form, but data are not available. Belowground biomass and production are generally less in shrubs than aboveground biomass and net production in contrast with graminoids and forbs where the root and rhizome component is large. Rates of net assimilation are low relative to forbs and graminoids and leaf conductances are low, especially in evergreen leaved species.

3.7 Plant phenology and growth

3.7.1 ABOVEGROUND SYSTEMS

As with other stressful environments, alpine plants are adapted to initiate growth as soon as environmental conditions become favorable. Leaf and flower buds overwinter in an advanced stage of development and begin expanding as soon as bud temperatures reach 0 °C or above for a few hours per day (Bliss, 1962a, 1971; Mark, 1970; Spomer, 1964). Numerous species overwinter with wintergreen or evergreen leaves (Bell, 1974) and are able to initiate photosynthesis immediately following snowmelt. For most forbs and deciduous shrubs, leafing is synchronous, but graminoids initiate leaves throughout the growing season (Bliss, 1956, 1966; Fareed and Caldwell, 1975; Johnson and Caldwell, 1974). Flowering is generally synchronous, reaching a peak 2–3 wk after snowmelt. Some cushion plants such as *Diapensia lapponica* and *Saxifraga oppositifolia* flower within 1–3 d of snowmelt. Plant phenology is strongly influenced by variation in microenvironments relative to microtopography (Bliss, 1956, 1966; Fareed and Caldwell, 1975).

Most shrub and forb species complete 70–80% of their shoot elongation in 20–30 d, but graminoids continue leaf production and inflorescence elongation until late in the season (Bliss, 1966). Three years of data from Mt Washington showed that 47–63% of shoot growth occurred within 34 d of the start of the growing season in *Carex bigelowii*. The fourth year had much higher air and soil temperatures and solar radiation in June; 90% of total shoot growth occurred in 34 d that year. Air temperature and air plus soil temperature were significantly correlated with both shoot growth and shoot production in *Carex bigelowii* and *Juncus trifidus* (Bliss, 1966).

In *Carex exserta* meadow communities near treeline in the summer-dry Sierra Nevada, soil moisture is high after snowmelt but decreases in mid-season. Plants of three non-graminoid life forms differ in the timing of leaf growth, phenology and rates of vegetative and reproductive growth. Flowering begins in the annual *Polygonum minimum*, after a month of rapid leaf growth, and flower buds continue to emerge until the plants die. Death occurs when Ψ_s drops below about $-3.0\,\text{MPa}$. *Saxifraga aprica*, an ephemeral perennial forb, emerges under snow and flowers 2 wk later for only 1 wk. Seasonal leaf growth has ceased by this time. Dormancy occurs 7 wk or less after snowmelt, depending on Ψ_s (-1.0 to $-1.5\,\text{MPa}$). A wintergreen perennial, *Penstemon heterodoxus*, flowers after a month of rapid leaf growth for a period of 2 wk. During late summer in dry years ($< 3.0\,\text{MPa}$), all the leaves can die, despite their overwintering capability. Leaf growth in these species is essentially over by the onset of mid-season soil moisture stress, but flowering and fruiting strategies of each life form differ in relation to their ability to cope with soil water stress (Jackson and Bliss, 1984).

Working in the Medicine Bow Mountains, WY, Scott and Billings (1964) correlated 55 environmental variables with the standing crop (aboveground) of species at 50 sites. They found frequent correlations ($P = < 0.1$) with altitude, winter snow cover, soil moisture regime, soil movement, per cent clay, potassium, and available water in the subsoil. Shoot and root net production at a mesic site were significantly correlated with solar radiation, mean daily wind run, soil temperature, and soil moisture. At a xeric site, only solar radiation accounted for any significant variation in growth. These results show that no single environmental factor predominates in determining plant growth.

Data from several alpine studies indicate that aboveground net annual production averages $25–75\,\text{g m}^{-2}$ in snowbed, cushion plant, and dry herb fellfield communities (Billings and Bliss, 1959; Bliss, 1956, 1966), and $100–300\,\text{g m}^{-2}$ in a broad range of forb-rich and graminoid meadows, and dwarf shrub (health) communities (Bliss, 1956, 1966; Klikoff, 1965a).

Alpine plants tend to have lower chlorophyll contents and higher chlorophyll *a* to *b* ratios than their arctic counterparts (Mooney and Billings, 1961; Tieszen, 1970). The peaking of chlorophyll content before net plant production (Muc, 1977; Tieszen, 1972) indicates that photosynthetic efficiency reaches a peak before plant biomass, but that maximum community uptake of CO_2 occurs later in the season with maximum leaf area (Tieszen and Wieland, 1975). Leaf area indices of alpine communities on Mt Washington ranged from 0.94 in heath-rush and 1.15 in cusion plant communities to 1.95 in dwarf heath shrub and 3.30 in an herb-rich snowbed community (Bliss, 1970).

3.7.2 BELOWGROUND SYSTEMS

Tundra plants concentrate their aboveground and belowground biomass near the soil surface in the most favorable microclimate (Bliss, 1962a, 1971; Billings, 1974a). In a detailed study of alpine plant root systems, Daubenmire (1941) pointed out the important functions of absorption, storage of reserves, anchorage, and vegetative reproduction by these belowground organs, but added that these organs do not have a uniform morphology in an alpine environment. Studies have shown that variation in belowground biomass occurs between plant communities and along environmental gradients (Bliss, 1966; Holch et al., 1941; Scott and Billings, 1964; Webber and May, 1977).

While the increase in belowground biomass has often been interpreted as a response to a more severe environment (Bliss, 1970; Scott and Billings, 1964), Webber and May (1977) found that the shifts in aboveground:belowground ratio appear to be controlled more by plant growth form, rather than a product of an environmental gradient. On Niwot Ridge, ratios of alive aboveground:belowground biomass were 1:14 for a *Trifolium dasyphyllum – Silene acaulis* (tap roots) fellfield, 1:17 for a *Kobresia bellardii* (fibrous roots) dry meadow, 1:20 for a *Caltha leptosepala* (rhizomes) wet meadow, and 1:25 for a *Deschampsia caespitosa* (fibrous roots) moist meadow (Webber and May, 1977). They assumed that belowground net annual production equaled that aboveground. This assumption needs verification.

In most alpine sites, the majority of roots and rhizomes are in the upper 30 cm of soil (Bliss, 1956, 1966; Daubenmire, 1941; Holch et al., 1941; Retzer, 1956; Scott and Billings, 1964; Webber and May, 1977). In the Presidential Range, many herbaceous species and the dwarf heath species have root systems confined to the upper 10–15 cm (Bliss, 1966), though *Carex bigelowii* and *Juncus trifidus* have abundant roots to 30–40 cm and frequent roots to 50–80 cm. In the intensive study of root systems on Niwot Ridge, Webber and May (1977) reported 50–80% of the belowground biomass occurs in the upper 10 cm although both fellfield and shrub tundra plants have considerable biomass at depths of 50–80 cm.

There are a few estimates of living belowground biomass for different alpine environments. Values range from $\simeq 2500$ to $4500 \, g \, m^{-2}$ in six alpine communities on Niwot Ridge (Webber and May, 1977); $2270 \, g \, m^{-2}$ in a fellfield to $4485 \, g \, m^{-2}$ in a *Caltha* dominated wet meadow. Data from the Medicine Bow Mountains provide estimates of $1060 \, g \, m^{-2}$ in a *Geum rossii* dominated turf (Mooney and Billings, 1960) and $1100 \, g \, m^{-2}$ in a mesic meadow dominated by *Geum rossii*, *Trifolium parryi*, *Artemisia scopulorum* and *Polygonum bistortoides*. Bliss (1966) reported $540 \, g \, m^{-2}$ in a rush-heath meadow and $3630 \, g \, m^{-2}$ in a sedge-moss meadow in the Presidential Range.

Rates of turnover for belowground biomass are limited in the literature. Webber and May (1977) provide estimates of 16 yr for shrub tundra, 19 yr for dry meadow and 22–26 yr for fellfield, wet meadow, moist meadow and snowbed communities. Turnover rates were lower in two cushion plant communities (3.7, 4.7 yr) compared with two herbfield communities (4.6, 10.8 yr) in the New Zealand alpine; the dominant species had fibrous roots (Bliss and Mark, 1974).

In contrast with data available for the Arctic (Chapin and Slack, 1979; Shaver and Billings, 1975, 1977), there are no detailed studies of root growth and root respiration in alpine environments. It may well be that maximum root growth occurs in mid to late season with warmer soils, as in the Arctic, but data are lacking.

3.8 Physiological responses

3.8.1 PHOTOSYNTHESIS AND RESPIRATION

Alpine plants generally photosynthesize by the C_3 pathway though a very few C_4 system species reach the alpine in tropical regions (Tieszen *et al.*, 1979). Utilizing the C_3 system, we would expect alpine plants to:

1. possess low temperature requirements for growth and photosynthesis;
2. have light compensation and saturation points at high intensities;
3. have lower rates of photosynthesis, partly due to photorespiration;
4. have reduced r_m (mesophyll resistance) to compensate for higher r_s (stomate resistance); and
5. have relatively high chlorophyll *a* to *b* ratios (Tieszen and Wieland, 1975).

The general pattern of net assimilation in alpine species is quite similar to that reported for plants of other environments. Rates of net assimilation are lower in evergreen leaved shrubs than in herbs, and species adapted to wet sites reduce their rates of net assimilation more rapidly with the onset of soil water stress than do species from dry sites. There are, however, numerous ways in which alpine plants differ in net assimilation and respiration compared with low elevation species. Growth of alpine and arctic plants is restricted to a 2–4 mon snow-free period when they can capture, store, and utilize energy within these low temperature environments. This has resulted in certain adaptations and constraints.

Alpine plants begin growth rapidly following snowmelt. Depending upon the species, flowering may precede leaf and shoot elongation (some cushion plant species) or may follow shoot elongation (most graminoids and forbs). Early in the season, rates of net assimilation may be quite low or even negative due to high rates of respiration associated with rapid plant growth (Hadley and Bliss, 1964). With reduced shoot growth and completion of flowering, respiration rates drop and therefore net assimilation increases. This pattern was found in graminoids, forbs, evergreen cushion plants, and

dwarf heath shrub species (Hadley and Bliss, 1964). The recycling of respiratory CO_2 released from roots and stems through hollow stems of a rapidly growing subalpine species (*Mertensia ciliata*) may makeup for part of its early season deficit (Billings and Godfrey, 1967). Species released from snowbeds after mid-July take much less time to reach maturity, are smaller in size (Billings and Bliss, 1959; Rocow, 1969), and have a shorter period of time for photosynthesis. Whether this results in species with higher rates of net assimilation occupying these snowbeds is not known. Eventually if snow melts late enough on the average (late July–early August), vascular plants are eliminated or greatly reduced in numbers, leaving only cryptogams.

Combining field and laboratory studies we have a reasonable understanding of the photosynthetic response of a diverse group of alpine and arctic species to temperature, light, and water potential. The response measured is influenced by the regime in which species have been grown (acclimation) (Billings *et al.*, 1971; Mooney and West, 1964; Pearcy, 1969) and the ecotype being tested (Mooney and Billings, 1961). Alpine populations of *Oxyria digyna* (Mooney and Billings, 1961), *Thalictrum alpinum* (Mooney and Johnson, 1965), and *Trisetum spicatum* (Clebsch, 1960) have photosynthetic optima at higher temperatures than do arctic ecotypes. Working with seventeen populations of *Oxyria digyna* from widely distributed arctic and alpine environments, Billings *et al.* (1971) found that alpine populations showed a high degree of homeostasis in net assimilation compared with arctic ecotypes. In only a few alpine populations was the temperature at which maximum net photosynthesis occurred increased by warm acclimation, and maximum net assimilation rates were not lowered by warm acclimation as occurred in all arctic populations. Alpine plants appear to be preadapted to the alternating diurnal temperature regime in contrast to arctic plants which grow in a more temperature stable environment. Maximum rates of net assimilation were reached at lower temperature in alpine and subalpine species compared with low elevation species from the White Mountains, CA when grown under uniform conditions. The authors, however, caution that much of the plant response is due to environmentally conditioned plasticity rather than to a genetically fixed response (Mooney *et al.*, 1964).

Hill reaction data on isolated chloroplasts of *Deschampsia caespitosa* showed that an alpine population had a higher temperature optimum and greater phenotypic flexibility than did an arctic population (Tieszen and Helgager, 1968). Hill reaction activity increases when species are grown at low temperatures and activity is higher in ecotypes from cold environments. The temperature optima will shift when individuals are grown at different temperatures (Pearcy, 1969).

In contrast with ecotypic responses of *Oxyria digyna* (Billings *et al.*, 1961) in photosynthetic electron transport, a diverse group of species from the White Mountains, CA, showed no relationship between photosynthetic efficiency at reduced CO_2 levels and elevation origin of a plant (Mooney, *et al.*, 1964).

Alpine plants grow in high light intensity environments and, as a consequence, these species have high light compensation and saturation points. Mooney and Billings (1961) reported that alpine populations of *Oxyria digyna* reached light saturation at higher intensities than did arctic populations. Similar findings were reported for populations of *Trisetum spicatum* (Clebsch, 1960). Hadley and Bliss (1964) reported that high light saturation and increased photosynthetic efficiency at higher temperatures (20–25 °C) for *Geum peckii* were in contrast with other herbaceous alpine species (*Carex bigelowii, Juncus trifidus, Potentilla tridentata*). *Geum* is restricted to lower, sunny alpine slopes in the Presidential Range while the other herbs extend upslope into the cooler and cloudy upper alpine areas.

Scott *et al.* (1970) found that light intensity was highly correlated with the net assimilation rates in five subalpine species in the Medicine Bow Mountains. Scott and Billings (1964) showed that for nine species light compensation point increased with temperature rise and that maximum net assimilation rates were generally 38–50 μmol CO_2 g^{-1} s^{-1}.

As with photosynthesis, respiration rates tend to be higher at various temperatures in alpine than in arctic populations (Clebsch, 1960; Mooney and Billings, 1961) and lower elevation populations (Mooney, 1963; Mooney *et al.*, 1964). The studies by Klikoff (1966, 1968) on altitudinal populations of *Sitanion hystrix* from the Sierra Nevada and by Billings *et al.* (1971) for latitudinal populations of *Oxyria digyna* show that rates of mitochondrial activity are genetically based. Rates of activity were higher in alpine than in subalpine and montane populations of *Sitanion* and higher in arctic than in alpine populations of *Oxyria*. This helps to explain the relatively high rates of respiration of tundra plants, especially at low temperatures. These data also suggest that lower elevation species could not maintain sufficient mitochondrial oxidative rates to survive the lower temperatures of alpine sites (Klikoff, 1968).

3.8.2 WATER RELATIONS

It is well established that both arctic and alpine species are subjected to water stress, especially those plants of wind exposed, well-drained soils (Bell and Bliss, 1979; Bliss, 1956; Courtin, 1968; Ehleringer and Miller, 1975; Mooney *et al.*, 1965). Species from contrasting alpine habitats frequently show different responses to water stress. For example, species from wet habitats show a sharp decrease in water conductance as leaf water content decreases, while species from drier habitats have a slower stomatal response (Miller *et al.*, 1978). Peterson and Billings (1982) found that 7 of 11 alpine species grown under high moisture conditions (daily watering) produced more dry weight than when grown under a low moisture regime (weekly watering). In general species normally growing in habitats with daily meltwater had a greater reduction in growth than plants from dry habitats. The effect was most pronounced in *Deschampsia caespitosa*.

Geum rossii and *Kobresia bellardii* from dry sites, *Deschampsia caespitosa* and *Bistorta bistortoides* from moist sites, and *Caltha leptosepala* from wet sites on Niwot Ridge (3500 m) respond quite differently to soil moisture stress (Ehleringer and Miller, 1975; Johnson and Caldwell, 1974, 1975). As soil water potential (Ψ_s) drops there is a more rapid decline in turgor potential (Ψ_p) in *Deschampsia* than in *Geum*. Further, *Geum* has more elastic cell walls than *Deschampsia* and leaf expansion shows less reduction in *Geum* as Ψ_s drops. Under conditions of high Ψ_s, *Deschampsia* maintains higher Ψ_p and greater leaf expansion. Under conditions of low soil moisture stress, *Deschampsia* maintains higher rates of net assimilation (11 as opposed to $7.6\,\mu mol$ $CO_2\ m^{-2}\ s^{-1}$), and higher water use efficiency (photosynthesis/transpiration, P_sT_r) than *Geum*. However, *Geum* has higher P_s/T_r than *Deschampsia* as water stress increases (Johnson and Caldwell, 1975). As Ψ_1 decreased from -0.8 to -2.1 MPa, net assimilation dropped 40% in *Geum*, but only 25% in *Deschampsia*. Minimum leaf water potentials of -2.3 MPa in *Geum rossii* and -3.1 MPa in *Deschampsia caespitosa* were measured in 1972 (Ehleringer and Miller, 1975), a year when precipitation was within 10% of the 9 yr average.

Alpine and subalpine plants from moist and dry meadows show a very different photosynthetic response to the onset of drought. Klikoff (1965b) reported that the photosynthetic rate of *Calamagrostis breweri* (wet meadow species) decreased to near zero at $\Psi_s < -1.0$ MPa, but *Carex exserta* (dry meadow species) operated near maximum at the same soil water potential and at 25% of maximum at -2.0 MPa. Similar patterns of photosynthetic response were reported for *Caltha leptosepala* and *Saussaurea americana* in wet sites and *Festuca idahoensis*, *Lupinus latifolius*, and *Eriophyllum lanatum* from mesic grass and dry grass–forb subalpine communities. Maximum rates of net assimilation were 63–70 $\mu mol\ CO_2\ g^{-1}\ s^{-1}$ at light saturation in the latter three species but only 32 $\mu mol\ CO_2\ g^{-1}\ s^{-1}$ in the wet site species (Kuramoto and Bliss, 1970).

Billings and Bliss (1959) hypothesized that soil drought speeded fall dormancy. Holway and Ward (1965) found this to be true in *Geum rossii* and Spomer and Salisbury (1968) reported that decreasing soil temperatures also induced dormancy in this species.

Based upon research from the Austrian Alps, there is an increase in diffusivity of gases from leaves with increased elevation. With increasing elevation maximum leaf conductance increased as did the daily sum of leaf conductance. These results help explain the relatively high rates of evapotranspiration measured in alpine plants of mesic sites (Körner and Mayr, 1980).

In a summer and winter study of *Kobresia bellardii* on Trail Ridge Road, CO (3600 m), Ψ_1 was near 0 MPa during snowmelt, but generally ranged from -0.8 to -1.5 MPa (midday measurements) in summer. Both summer and winter minima reached -2.2 MPa; winter values steadily increased from the February low to -0.5 MPa in April. Winter and summer diurnal changes in Ψ_1 averaged 0.2–0.4 MPa (Bell and Bliss, 1979). This species maintains a high concentration of oligosaccharides in winter shoots (frost hardiness?) and is able to elongate its leaves in winter (February–April) during periods when surface temperatures were 0° to -3 °C. When Ψ_s dropped below -2.0 MPa for 7 d or more, leaves died back or did not change length (Bell and Bliss, 1979). Rates of net assimilation are relatively high (5.7–12.6 $\mu mol\ CO_2\ m^{-2}\ s^{-1}$) for individual leaves in midseason (Johnson and Caldwell, 1974). Leaf water potentials of -2.4 MPa appeared to have little effect in reducing net assimilation (Johnson *et al.*, 1974). This amazing species is clearly very well adapted to survive winter conditions with moderate snow cover, produce short bursts of leaf growth in winter, initiate quite rapid leaf elongation in April and May when other plants are dormant, and maintain net assimilation under quite severe soil drought conditions in summer.

In a recent study of *Cassiope mertensiana* at treeline and *C. tetragona* in the adjacent alpine, Harter (1981) found that both species maintain low Ψ_1 and $\Psi_{\pi+\gamma}$ (osmotic plus matric potential) at snow release and early in the growing season (-2.0 to 5.0 MPa). Maximum Ψ_1 values (-0.5 to -1.5 MPa) were reached in mid-summer. Osmotic plus matric potentials declined to -2.5 MPa in *C. mertensiana* and to -5.0 MPa in *C. tetragona* by November. *Cassiope tetragona* (new leaves) showed greater cell wall elasticity than in *C. mertensiana* and thus a more constant Ψ_p.

3.8.3 CARBON ALLOCATION AND PLANT GROWTH

Alpine species are adapted to a rapid burst of growth following snowmelt, generally at the expense of belowground root and rhizome carbohydrate reserves (Fonda and Bliss, 1966; Mooney and Billings,

1960). The general pattern of the non-structural carbohydrate (TNC) cycle appears similar for plants of warm, sunny alpine environments in the Sierra Nevada (Mooney and Billings, 1965) and the central Rocky Mountains (Mooney and Billings, 1960). Carbohydrate reserves are generally a higher percentage of dry weight in species with tap roots and rhizomes than in species with fibrous roots, dicotyledonous than in monocotyledonous species, and forbs than in woody species. Lipid and protein contents are generally higher in evergreen or wintergreen-leaved alpine plants (Bliss, 1962b; Bliss and Mark, 1974; Hadley and Bliss, 1964). The wet site plant, *Caltha leptosepala*, had higher sugar levels in roots and shoots of plants released from snowbed sites in late June than in plants released in mid-July (Rocow, 1969). The latter plants were not able to increase their sugar levels significantly owing to the shortened growing season. There also appears to be a reduction in carbohydrate content in some species with increased elevation (Mooney and Billings, 1965).

Sugar content and cold resistance are linked in alpine species as they are in plants from other cold-dominated climates. The fall increase in TNC appears universal in alpine plants. The relatively high concentrations of oligosaccharides (low molecular mass carbohydrates) in alpine (Bell and Bliss, 1979) and arctic plants (Muc, 1977; Svoboda, 1977) are believed to be associated with freezing tolerance and the winter growth of *Kobresia bellardii* noted previously.

Several studies of *Carex bigelowii*, from the Presidential Range, showed that this species is quite well 'programmed' by current summer climatic conditions. Rhizome reserves were mobilized in June to support a burst of shoot growth (Fonda and Bliss, 1966). At this time, respiration exceeded photosynthesis (Hadley and Bliss, 1964). The negative carbon balance was maintained until 75–90% of shoot growth was completed. Respiration then dropped while net assimilation rates were maintained and carbohydrate reserves were then replaced. Dry matter production of shoots may be increased 30–40% during warm, sunny summers, but only a 5–10% increase occurred in cool, cloudy summers (Bliss, 1966). During a cool, cloudy period in August, shoot carbohydrates dropped from 22 to 13% and rhizomes from 22 to 18%, followed by a rise to former levels with the return of sunny days (Fonda and Bliss, 1966). Weather conditions in August can be rather critical in terms of the levels of the TNC that are stored for the next growing season.

3.8.4 ENZYMATIC AND NUTRIENT ADAPTATIONS

While a good deal is known about general patterns of net assimilation, respiration, and water balance in alpine species, the underlying processes that determine plant adaptations to low temperatures are still poorly known. Improved metabolic activity at low temperatures may be primarily a function of resistance of chloroplast and mitochondrial membranes to low temperature phase changes and to quantitative increase in enzyme levels (Chabot, 1979). In alpine and arctic populations of *Oxyria digyna*, rates of net assimilation and the activity of RuBP carboxylase were highest in low-temperature grown plants (Chabot *et al.*, 1972). Increased enzyme activity appeared due to changes in the amount of RuBP carboxylase enzyme present. There does not appear to be a thermal optima for carboxylase activity below 35–40°C in several arctic species (Chabot *et al.*, 1972; Tieszen and Sigurdson, 1973).

In the Arctic, Tieszen (1978b) has shown that the general pattern for leaf nitrogen content and RuBp carboxylase activity are closely parallel. However, when carboxylase activity is expressed on a unit nitrogen basis, much of the early and late season nitrogen is not directly associated with net assimilation (Chapin *et al.*, 1980). The pattern of lipid changes with high spring values observed by McCown (1978) are interpreted by Chapin (1979) as possibly reflecting biosynthesis and rapid growth rather than as a pool of storage lipid (Bliss, 1962b).

3.9 Timberline

The physiological ecology of timberline has been most intensively studied in Europe (Tranquillini, 1978), although in recent years several North American studies have added significantly to our understanding of the climatic limits of tree growth.

Timberline as used here refers to the broad ecotone between forests and alpine tundra and includes tree islands and krummholz (Tranquillini, 1978; Wardle, 1974).

Seven hypotheses outlined by Daubenmire (1954) to explain the position and occurrence of timberline, have been reduced to two major ones (Tranquillini, 1978; Wardle, 1974):

1. winter and spring desiccation of leaves and buds determines the limit to tree growth; and

2. summer environmental conditions as they influence carbon fixation, shoot growth, and shoot maturation prevent tree growth, beyond approximately the position of the 10 °C isotherm for the mean temperature of the warmest month. It is believed that these hypotheses explain the limit for alpine and northern timberlines formed by a variety of species.

Comparative studies on the carbon balance of *Larix decidua* and *Pinus cembra* at timberline near Obergurgl, Austria showed that while net assimilation rates are nearly double in *Larix*, the shorter growing season (59%) (Tranquillini, 1962, 1964) and smaller needle mass of *Larix* result in *Pinus cembra* being able to fix 50% more carbon on a tree to tree basis (Tranquillini and Schütz, 1970).

Richards (1980) summarized the literature on the physiological response of evergreen and deciduous timberline species. He concluded that summer environmental conditions which reduce net assimilation rates at timberline compared with lower elevation populations do not differentially affect deciduous and evergreen species. Answers must therefore come from comparative studies of water stress during periods of winter and spring desiccation.

Richards (1980) compared summer and winter physiological responses of deciduous *Larix lyallii* with the sympatric evergreens *Abies lasiocarpa* and *Picea engelmannii*. He concluded that the deciduous needle habit permits *Larix* to maintain great tolerance of and resistance to winter desiccation. The overwintering buds were tolerant of midwinter water potentials of -4.0 to -5.0 MPa and water contents of 100–150%. A large resistance to water flux between buds and xylem developed in October and maintained the buds as relatively isolated units until February. The sharp decreases in xylem pressure potentials that developed in late winter and spring were not transmitted to the buds. As a result there was 30–64% less winter desiccation damage to *Larix* than to adjacent *Abies lasiocarpa*. The winter advantages are balanced by summer disadvantages; *Larix* requires more light for net assimilation and there are large reductions in assimilation in response to relatively low atmospheric VPD demands. In order to maintain a positive annual carbon gain, rates of net assimilation are high, there is a high leaf area to leaf weight ratio which permits a high carbon gain from a low carbon leaf investment. The trees grow in open stands for low Ψ_s also reduces net assimilation, which influences tree density. Richards (1980) stated that both a moist summer (reduced atmospheric and soil-water deficits) and desiccating winter conditions favor deciduous trees at timberline. The frequently observed upright tree form in *Larix*, compared with the krummholz form of *Picea* and *Abies* at timberline, clearly reflects differential survival of buds.

Winter studies of leaf water content in conifers have given varying results across the continent. Lindsay (1971) studied *Abies lasiocarpa* and *Picea engelmannii* in the Medicine Bow Mountains and found that Ψ_1 varied from -1.5 MPa in summer to -3.0 to -3.4 MPa in winter in both species. At timberline (3300 m), during a late summer drought, Ψ_1 dropped to equal winter values. In the closed forest (3050 m), Ψ_1 seldom dropped below -2.0 MPa in winter, the result of deep snow cover, probably more available soil water, and reduced transpiration stress at lower wind speeds. Hansen and Klikoff (1972) reported Ψ_1 as low as -8.0 to -9.0 MPa in timberline *Picea engelmannii*. This was attributed to low soil water potentials in the Wasatch Mountains, UT. On Mt Washington, Marchand and Chabot (1978) reported minimum Ψ_1 of -2.8 MPa for *Picea mariana* and -1.75 MPa for *Abies balsamea* over a two-winter period. They concluded that timberline position in these eastern mountains might best be explained by reproductive biology and demography rather than by classical desiccation theory. The Presidential Range has high levels of summer and winter precipitation and therefore winter desiccation is less likely than in the western North American mountains. Similar findings have been reported for *Picea rubens* on Mount Monadnock, NH where average water potential in winter was − 1.1 MPa. More variation was found within a single tree than between juvenile and adult trees, and among trees on an altitudinal gradient or comparing seasons (Kincaid and Lyons, 1981).

At timberline in the Arctic, *Picea mariana* trees and saplings showed little water stress in summer and winter; osmoregulation maintained a constant turgor, permitting needle expansion and growth in summer. Combined $\Psi_{\pi+\gamma}$ decreased with the onset of winter while Ψ_l remained stable, reflecting changes in water relations associated with winter hardiness. Seedlings, however, were severely affected by summer drought. Temperatures $< 15\,°C$, the lower cardinal limit for seed germination, were probably the final control of this species at its northern limit and of stand reproduction south of forest limit (Black and Bliss, 1980).

3.10 The role of snow

As discussed earlier, snow depth in winter and its time of melt in spring and summer are important in the habitat selection of species and plant communities in many alpine areas. Snow cover reduces near soil surface fluctuations of temperature, enables some species to maintain wintergreen leaves, and may permit some photosynthesis to occur in springtime (Richardson and Salisbury, 1977; Salisbury, 1976). The absence of red or far-red light under deep snow (2 m) and its reappearance under shallow snow (40 to 50 cm) may provide a signal for regulating growth responses in both higher and lower plants (Richardson and Salisbury, 1977). They also reported seeds of nine species of 23 tested germinated under deep snow.

Two studies that detail the physiological response of diverse species to winter snow have recently been completed. The absence of *Kobresia bellardii* in fellfields and deep snow sites ($> 140\,$cm) was explained on the basis of fall and winter rather than summer conditions. *Kobresia* did not survive the fellfield environment with little or no snow cover because of mechanical damage by windblown snow and sand. Low Ψ_s led to leaf wilting in both summer and winter and resulted in early leaf dieback.

Moderate ($\simeq 75\,$cm) and deep snowbed sites ($\simeq 120\,$cm) permitted early snow accumulation (September-October), preventing completion of dieback and frost hardening. Low winter temperatures resulted in death of unhardened leaves. Delayed spring growth shortened the photosynthetic season. Reduced carbohydrate reserves, associated with early summer leaf growth, led to wilting which further reduced net assimilation. Shallow snow accumulation sites ($\simeq 15\,$cm) are clearly the preferred habitat of this alpine sedge, based upon both the experimental transplants and the functioning of natural populations (Bell and Bliss, 1979).

Cassiope mertensiana and *C. tetragona* are sympatric species in the Alberta Rocky Mountains where they occur in deep ($> 1\,$m) and shallow ($< 50\,$cm) snow accumulation sites, respectively. Snow cover $> 15\,$cm was present in the *C. mertensiana* sites by late October and plants seldom experienced winter temperature $< − 15$ to $− 20\,°C$. *Cassiope tetragona* was snow covered 3–6 wk later, was often exposed in winter and the plants experienced temperatures $< − 25$ to $− 30\,°C$. Beneath deep snow, minimum temperatures were $− 2$ to $− 4\,°C$ and $− 6$ to $− 8\,°C$ under shallow snow (*C. tetragona* sites). Summer flowering, plant growth, microclimatic conditions, and summer trends in Ψ_l and $\Psi_{\pi+\gamma}$ were similar. Leaf water potentials and osmotic plus matric potentials were $− 2.0$ to $− 3.5\,$MPa in *C. mertensiana* and $− 2.5$ to $− 5.0\,$MPa in *C. tetragona* at snow release. Seasonal maxima occur in mid-summer ($\Psi_l \simeq − 0.5$ to $− 1.0\,$MPa and $\Psi_{\pi+\gamma}$ $− 1.0$ to $− 1.5\,$MPa) and the potentials drop rapidly in autumn. Rapidly expanding new leaves of *C. tetragona* have elastic cell walls and are hydrolabile while *C. mertensiana* new leaves are hydrostable with rigid cell walls. Growth and photosynthesis are favored on mesic sites in the snowbed species while *C. tetragona* is tolerant of higher VPD and soil moisture stress. In winter *C. mertensiana* was cold hardy to $− 26\,°C$ while *C. tetragona* showed little injury to $− 36\,°C$ (Harter, 1981).

Cassiope mertensiana is restricted to snow protected sites where early and deep snow reduces winter desiccation and low-temperature injury. *Cassiope tetragona* occupies more exposed sites because of greater tolerance to summer atmospheric and soil moisture stress, and to winter desiccation and low temperatures. Dwarf evergreen shrub species occur in the warmer microenvironments and some can range above treeline. The evergreen habit provides adaptive advantages related to carbon allocation and low nutrient demand, but requires at least limited snow protection with the advantages this brings.

3.11 Summary

Alpine plants are adapted to a low-temperature environment through a variety of mechanisms. This is exemplified by rates of photosynthesis, respiration, and growth that equal those of temperate plants operating under less severe conditions. The alpine environment is also characterized by high irradiance and often soils with low water potentials. Soil nutrients, especially nitrogen are generally low, but the dynamics of nutrient absorption and distribution within the plant have been little investigated.

Most plants produce abundant seed and seed germination occurs in many species, yet seedlings are uncommon. The growth forms of cushion plants, rosettes, graminoids, and dwarf shrubs have adaptive significance in different microtopographic sites. Rapid growth following snowmelt results from utilization of carbohydrate reserves and current photosynthates, especially in forbs and graminoids with wintergreen leaves.

Snow depth and time of snowmelt appear more critical in the distribution of alpine species and the resultant plant communities and the physiological response of species than in the Arctic.

Acknowledgements

I thank Dwight Billings who guided my start into cold stressed environments and to my many graduate students who have taught me more than they often realized. The review comments of Louise Jackson and John Harter have been most helpful.

Research on which many of these studies were based has been funded by the National Science Foundation, the National Research Council of Canada and the Canadian petroleum industry.

References

Amen, R.D. (1965) Seed dormancy in the alpine rush, *Luzula spicata* L. *Ecology*, **45**, 361–4.

Amen, R.D. (1966) The extent and role of seed dormancy in alpine plants. *Quarterly Review Biology*, **41**, 271–81.

Amen, R.D. (1968) A model of seed dormancy. *Botanical Review*, **34**, 1–31.

Amen, R.D. and Bonde, E.K. (1964) Dormancy and germination in alpine *Carex* from the Colorado Front Range. *Ecology*, **45**, 881–4.

Bamberg, S.A. and Major, J. (1968) Ecology of the vegetation and soils associated with calcareous parent materials in three alpine regions of Montana. *Ecological Monographs*, **38**, 127–67.

Bell, K.L. (1974) Autumn, winter, and spring phenology of some Colorado Plants. *American Midland Naturalist*, **91**, 460–4.

Bell, K.L. and Amen, R.D. (1970) Seed dormancy in *Luzula spicata* and *L. parviflora*. *Ecology*, **51**, 492–6.

Bell, K.L. and Bliss, L.C. (1973) Alpine disturbance studies: Olympic National Park, U.S.A. *Biological Conservation*, **5**, 28–35.

Bell, K.L. and Bliss, L.C. (1979) Autecology of *Kobresia bellardii*: Why winter snow accumulation limits local distribution. *Ecological Monographs*, **49**, 377–402.

Bell, K.L. and Johnson, R.E. (1980) Alpine flora of the Wassuk Range, Mineral County, Nevada. *Madroño*, **27**, 25–35.

Billings, W.D. (1974a) Arctic and alpine vegetation: plant adaptations to cold summer climates, in *Arctic And Alpine Environments* (eds J.D. Ives and R.G. Barry) Methuen, London, pp. 403–43.

Billings, W.D. (1974b) Adaptations and origins of alpine plants. *Arctic Alpine Research*, **6**, 129–42.

Billings, W.D. (1978) Alpine phytogeography across the Great Basin, in *Intermountain Biogeography: A Symposium. Great Basin Naturalist Memoirs* No. 2. Brigham Young University, Provo, Utah, pp. 105–17.

Billings, W.D. and Bliss, L.C. (1959) An alpine snowbank environment and its effects on vegetation, plant development and productivity. *Ecology*, **40**, 389–97.

Billings, W.D., and Godfrey, P.J. (1967) Photosynthetic utilization of internal carbon dioxide by hollow-stemmed plants. *Science*, **158**, 121–3.

Billings, W.D. and Mooney, H.A. (1959) An apparent frost

hummock–sorted polygon cycle in the alpine tundra of Wyoming. *Ecology*, **40**, 16–20.

Billings, W.D. and Mooney, H.A. (1968) The ecology of arctic and alpine plants. *Biological Review*, **43**, 481–529.

Billings, W.D., Clebsch, E.E.C. and Mooney, H.A. (1961) Effect of low concentrations of carbon dioxide on photosynthesis rates of two races of *Oxyria*. *Science*, **133**, 1834.

Billings, W.D., Godfrey, P.J., Chabot, B.F. and Bourque, D.P. (1971) Metabolic acclimation to temperature in arctic and alpine ecotypes of *Oxyria digyna*. *Arctic Alpine Research*, **3**, 277–89.

Black, R.A. and Bliss, L.C. (1980) Reproductive ecology of *Picea mariana* (Mill.) BSP., at tree line near Inuvik, Northwest Territories, Canada. *Ecological Monographs*, **50**, 331–54.

Bliss, L.C. (1956) A comparison of plant development in microenvironments of arctic and alpine tundras. *Ecological Monographs*, **26**, 303–37.

Bliss, L.C. (1958) Seed germination in arctic and alpine species. *Arctic*, **11**, 180–8.

Bliss, L.C. (1962a) Adaptations of arctic and alpine plants to environmental conditions. *Arctic*, **15**, 117–44.

Bliss, L.C. (1962b) Caloric and lipid content of alpine tundra plants. *Ecology*, **43**, 753–7.

Bliss, L.C. (1963) Alpine plant communities of the Presidential Range, New Hampshire. *Ecology*, **44**, 678–97.

Bliss, L.C. (1966) Plant productivity in alpine microenvironments on Mount Washington, New Hampshire. *Ecological Monographs*, **36**, 125–55.

Bliss, L.C. (1969) Alpine plant community patterns in relation to environmental parameters, in *Essays In Plant Geography And Ecology* (ed. K.N.H. Greenridge), Nova Scotia Museum, Halifax, N.S. pp. 167–84.

Bliss, L.C. (1970) Primary production within arctic tundra ecosystems, in *Productivity And Conservation In Northern Circumpolar Lands* (eds W.A. Fuller and P.G. Kevan), IUCN New Ser. No. 16. Morges, Switzerland, pp. 77–85.

Bliss, L.C. (1971) Arctic and alpine plant life cycles. *Annual Review Ecology and Systematics*, **2**, 405–38.

Bliss, L.C. (ed.) (1977a) *Truelove Lowland, Devon Island, Canada: A High Arctic Ecosystem*, University of Alberta Press, Edmonton.

Bliss, L.C. (1977b) General summary Truelove Lowland ecosystem, in *Truelove Lowland, Devon Island, Canada: A High Arctic Ecosystem*, University of Alberta Press, Edmonton, pp. 657–75.

Bliss, L.C. (1979) Vascular plant vegetation of the Southern Circumpolar Region in relation to Antarctic, alpine, and arctic vegetation. *Canadian Journal of Botany*, **57**, 2167–78.

Bliss, L.C. (1981) North American and Scandinavian tundras and polar deserts, in *Tundra Ecosystems: A Comparative Analysis* (eds L.C. Bliss, D.W. Heal, and J.J. Moore), Cambridge University Press, London, pp. 8–24.

Bliss, L.C. and Mark, A.F. (1974) High alpine environments and primary production on the Rock and Pillar Range, Central Otago, New Zealand. *New Zealand Journal Botany*, **13**, 445–83.

Bonde, E.K. (1965) Further studies on the germination of seeds of Colorado alpine plants. *University of Colorado Studies, Series In Biology*, No. 18, 1–30.

Bonde, E.K. (1968). Survival of seedlings of an alpine clover (*Trifolium nanum* Tou.). *Ecology*, **49**, 1193–5.

Brown, J., Miller, P.C., Tieszen, L.L., Bunnell, F.L. and Maclean, S.F.Jr (eds) (1980) *An Arctic Ecosystem: The Coastal Tundra at Barrow, Alaska*, Dowden, Hutchinson and Ross. Stroudsburg, PA.

Caldwell, M.M. (1968) Solar ultraviolet radiation as an ecological factor for alpine plants. *Ecological Monographs*, **38**, 243–68.

Caldwell, M.M., Johnson, D.A. and Fareed, M. (1978) Constraints on tundra productivity: Photosynthetic capacity in relation to solar radiation, utilization and water stress in arctic and alpine tundras, in *Vegetation and Production Ecology of An Alaskan Arctic Tundra*, (ed. L.L. Tieszen), Springer, Verlag, N.Y. pp. 323–42.

Caldwell, M.M., Robberecht, R. and Billings, W.D. (1980) A steep latitudinal gradient of solar ultraviolet-B radiation in the arctic-alpine zone. *Ecology*, **61**, 600–11.

Caldwell, M.M. Robberecht, R., Nowak, R.S. and Billings, W.D. (1982) Differential photosynthetic inhibition by ultraviolet radiation in species from the arctic-alpine life zone. *Arctic Alpine Research*, **14**, 195–202.

Chabot, B.F. (1979) Metabolic and enzymatic adaptations to low temperature, in *Comparative Mechanisms of Cold Adaptation* (eds L.S. Underwood, L.L. Tieszen, A.B. Callahan, G.E. Folk), Academic Press, N.Y., pp. 283–301.

Chabot, B.F. and Billings, W.D. (1972) Origins and ecology of the Sierran alpine flora and vegetation. *Ecological Monographs*, **42**, 163–99.

Chabot, B.F., Chabot, J.F. and Billings, W.D. (1972) Ribulose-1, 5-diphosphate carboxylase activity in arctic and alpine populations of *Oxyria digyna*. *Photosynthetica*, **6**, 364–9.

Chapin, F.S. III (1978) Phosphate uptake and nutrient utilization by Barrow tundra vegetation, in *Vegetation And Production Ecology of An Alaskan Arctic Tundra* (ed. L.L. Tieszen), Springer-Verlag, N.Y., pp. 483–507.

Chapin, F.S. III (1979) Nutrient uptake and utilization by tundra plants, in *Comparative Mechanisms of Cold Adaptation* (eds L.S. Underwood, L.L. Tieszen, A.B. Callahan, G.E. Folk), Academic Press, N.Y., pp. 215–34.

Chapin, F.S. III and Slack, M. (1979) Effect of defoliation upon root growth, phosphate absorption, and respiration in nutrient-limited tundra graminoids. *Oecologia*, **42**, 67–79.

Chapin, F.S. III, Tieszen, L.L., Lewis, M.C., Miller, P.C. and McCown, E.H. (1980) Control of tundra plant allocation patterns and growth, in *An Arctic Ecosystem: The Coastal Tundra of Northern Alaska* (eds J. Brown, P.C. Miller, L.L. Tieszen, F.L. Bunnell and S.F. MacLean). Dowden, Hutchinson and Ross, Stroudsberg, PA, pp. 140–85.

Clausen, J. (1965) Population studies of alpine and subalpine races of conifers and willows in California high Sierra Nevada. *Evolution*, **19**, 56–68.

Clebsch, E.E.C. (1960) Comparative morphology and physiological variation in arctic and alpine populations of *Trisetum spicatum*, PhD thesis, Duke University, Durham, NC

Cooper, W.S. (1908) Alpine vegetation in the vicinity of Long's Peak, Colorado. *Botanical Gazette*, **45**, 319–37.

Courtin, G.M. (1968) Evapotranspiration and energy budgets of two alpine microenvironments, Mt Washington, New Hampshire, PhD thesis, University Illinois, Urbana.

Cox, D.F. (1933) Alpine plant succession on James Peak, Colorado. *Ecological Monographs*, **3**, 300–72.

Dahl, E. (1963) On the heat exchange of a wet vegetation surface and the ecology of *Koenegia islandica*. *Oikos*, **14**, 190–211.

Daubenmire, R.F. (1941) Some ecological features of the subterranean organs of alpine plants. *Ecology*, **22**, 370–8.

Daubenmire, R.L. (1954) Alpine timberlines in the Americas and their interpretation. *Butler University Botanical Studies*, **11**, 119–36.

Douglas, G.W. and Bliss, L.C. (1977) Alpine and high subalpine plant communities of the North Cascades Range, Washington and British Columbia. *Ecological Monographs*, **47**, 113–50.

Dorf, E. (1960) Climatic changes of the past and present. *American Scientist*, **48**, 341–54.

Ehleringer, J. and Miller, P.C. (1975) Water relations of selected plant species in the alpine tundra, Colorado. *Ecology*, **56**, 370–80.

Fahey, D.B. (1973) An analysis of diurnal freeze-thaw and frost heave cycles in the Indian Peaks Region of the Colorado Front Range. *Arctic Alpine Research*, **5**, 269–81.

Fareed, M. and Caldwell, M.M. (1975) Phenological patterns of two alpine tundra plant populations on Niwot Ridge, Colorado. *Northwest Science*, **49**, 17–23.

Flock, J.W. (1978) Lichen-bryophyte distribution along a snow-cover-soil moisture gradient, Niwot Ridge, Colorado. *Arctic Alpine Research*, **10**, 31–47.

Fonda, R.W. and Bliss, L.C. (1966) Annual carbohydrate cycle of alpine plants on Mt Washington, New Hampshire. *Bulletin Torry Botanical Club*, **93**, 268–77.

Griggs, R.F. (1956) Competition and succession in a Rocky Mountain fellfield. *Ecology*, **37**, 8–20.

Hadley, E.B. and Bliss, L.C. (1964) Energy relationships of the alpine plants on Mt Washington, New Hampshire. *Ecological Monographs*, **34**, 339–57.

Hansen, D.H. and Klikoff, L.G. (1972) Water stress in krummholz, Wasatch Mountains, Utah. *Botanical Gazette*, **133**, 392–4.

Harter, J.E. (1981) Comparative Autecology of Cassiope Species at Treeline in Jasper National Park, Alberta, PhD Thesis, University of Alberta, Edmonton.

Hobbie, J.E. (ed.) (1980) *Limnology of Tundra Ponds, Barrow, Alaska.* Dowden, Hutchinson and Ross, Stroudsburg, PA.

Holch, H.E., Hertel, E.W., Oakes, W.C. and Whitewell, H.H. (1941) Root habits of certain plants of the foothills and alpine belts of Rocky Mountain National Park. *Ecological Monographs*, **3**, 327–45.

Holm, H.T. (1927) Vegetation of the alpine region of the Rocky Mountains of Colorado. *National Academy Science Memoirs*, **3**, 1–45.

Holway, J.G. and Ward, R.T. (1965) Phenology of alpine plants in northern Colorado. *Ecology*, **46**, 73–83.

Hrapko, J.O. and La Roi, G.H. (1978) The alpine tundra vegetation of Signal Mountain, Jasper National Park. *Canadian Journal of Botany*, **56**, 309–32.

Jackson, L.E. and Bliss, L.C. (1982) Distribution of ephemeral herbaceous plants near treeline in the Sierra Nevada, California, U.S.A. *Arctic Alpine Research*, **14**, 33–42.

Jackson, L.E. and Bliss, L.C. (1984) Phenology and water relations of three plant life-forms in a dry treeline meadow. *Ecology*, **65**, 1302–14.

Johnson, D.A. and Caldwell, M.M. (1974) Field measurements of photosynthesis and leaf growth rates of three alpine plant species. *Arctic Alpine Research*, **6**, 245–51.

Johnson, D.A. and Caldwell, M.M. (1975) Gas exchange of four arctic and alpine tundra plant species in relation to atmospheric and soil moisture stress. *Oecologia*, **21**, 93–108.

Johnson, D.A., Caldwell, M.M. and Tieszen, L.L. (1974) Photosynthesis in relation to leaf water potential in three alpine plant species, in *Primary Production and*

Production Processes, Tundra Biome (eds L.C. Bliss and F.E. Wielgolaski), Tundra Biome Steering Committee, Edmonton, pp. 205–10.

Johnson, P.L. and Billings, W.D. (1962) The alpine vegetation of the Beartooth Plateau in relation to cryopedogenic processes and patterns. *Ecological Monographs*, **32**, 105–35.

Kincaid, D.T. and Lyons, E.E. (1981) Winter water relations of red spruce on Mount Monadrock, New Hampshire. *Ecology*, **621**, 1155–69.

Klikoff, L.G. (1965a) Microenvironmental influence on vegetational pattern near timberline in the Central Nevada. *Ecological Monographs*, **35**, 187–211.

Klikoff, L.G. (1965b) Photosynthetic response to temperature and moisture stress of three timberline meadow species. *Ecology*, **46**, 516–17.

Klikoff, L.G. (1966) Temperature dependence of the oxidative rates of mitochondria in *Danthonia intermedia*, *Penstemon davidsonii* and *Sitanion hystrix*. *Nature*, **212**, 529–30.

Klikoff, L.C. (1968) Temperature dependence of mitochondrial oxidative rates of several plant species of the Sierra Nevada. *Botanical Gazette*, **129**, 227–30.

Körner, C. and Mayr, R. (1980) Stomatal behavior in alpine plant communities between 600 and 2600 metres above sea level, in *Plants and their Atmospheric Environment* (eds J. Grace, E.D. Ford and P.G. Jarvis), Blackwell, London, pp. 205–18.

Kuramoto, R.T. and Bliss, L.C. (1970) Ecology of subalpine meadows in the Olympic Mountains, Washington. *Ecological Monographs*, **40**, 317–47.

Le Drew, E.F. and Weller, G. (1978) A comparison of the radiation and energy balance during the growing season for an arctic and alpine tundra. *Arctic Alpine Research*, **10**, 655–78.

Lindsay, J.H. (1971) Annual cycle of leaf water potential in *Picea engelmannii* and *Abies lasiocarpa* at timberline in Wyoming. *Arctic Alpine Research*, **3**, 131–8.

Little, E.L. (1941) Alpine flora of San Francisco Mountain, Arizona. *Madroño*, **6**, 65–81.

Löve, D. (1970) Subarctic and subalpine: where and what? *Arctic Alpine Research*, **2**, 63–73.

McCown, B.H. (1978) The interactions of organic nutrients, soil nitrogen, and soil temperature and plant growth and survival in the arctic environment, in *Vegetation and Production Ecology of an Alaskan Arctic Tundra* (ed. L.L. Tieszen), Springer-Verlag, N.Y., pp. 435–56.

Major, J. and Taylor, D.W. (1977) Alpine, in *Terrestrial Vegetation of California* (eds M.G. Barbour and J. Major), John Wiley and Sons, N.Y. pp. 601–75.

Marchand, P.J. and Chabot, B.L. (1978) Winter water relations of timberline plant species on Mt. Washington, New Hampshire. *Arctic, Alpine Research*, **10**, 105–16.

Marchand, P.J. and Roach, D.A. (1980) Reproductive strategies of pioneering alpine species: seed production, dispersal and germination. *Arctic Alpine Research*, **12**, 137–46.

Mark, A.F. (1962) Zonation of vegetation and climate on the Old Man Range, Central Otago. *Otago University Science Record*, **12**, 5–8.

Mark, A.F. (1970) Floral initiation and development in New Zealand alpine plants. *New Zealand Journal Botany*, **81**, 67–75.

Marr, J.W. (1961) Ecosystems of the Front Range of Colorado, *University Colorado Studies, Series in Biology*, No. 8, University Colorado Press, Boulder.

Miller, P.C., Stoner, W.A. and Ehleringer, J.R. (1978) Some aspects of water relations of arctic and alpine regions, in *Vegetation and Production Ecology of an Alaskan Arctic Tundra* (ed. L.L. Tieszen), Springer-Verlag, N.Y., pp. 343–57.

Mooney, H.A. (1963) Physiological ecology of coastal, subalpine, and alpine populations of *Polygonum bistortoides*. *Ecology*, **44**, 812–16.

Mooney, H.A. and Billings, W.D. (1960) The annual carbohydrate cycle of alpine plants as related to growth. *American Journal Botany*, **47**, 594–8.

Mooney, H.A. and Billings, W.D. (1961) Comparative physiological ecology of arctic and alpine populations of *Oxyria digyna*. *Ecological Monographs*, **31**, 1–29.

Mooney, H.A. and Billings, W.D. (1965) Effects of altitude on carbohydrate content of mountain plants. *Ecology*, **46**, 750–1.

Mooney, H.A. and Johnson. A.W. (1965) Comparative physiological ecology of an arctic and an alpine population of *Thalictrum alpinum* L. *Ecology*, **46**, 721–7.

Mooney, H.A. and West, M. (1964) Photosynthetic acclimation of plants of diverse origin. *American Journal Botany*, **51**, 825–7.

Mooney, H.A., Hillier, R.D. and Billings, W.D. (1965) Transpiration rates of alpine plants in the Sierra Nevada of California. *American Midland Naturalist*, **74**, 374–86.

Mooney, H.A., Wright, R.D. and Strain, B.R. (1964) The gas exchange capacity of plants in relation to vegetation zonation in the White Mountains of California. *American Midland Naturalist*, **72**, 281–97.

Muc, M. (1977) Ecology and primary production of the Truelove Lowland sedge-moss meadow communities, in: *Truelove Lowland, Devon Island Canada: A High Arctic Ecosystem* (ed. L.C. Bliss) University of Alberta Press, Edmonton, pp. 157–84.

Oberbauer, S. and Billings, W.D. (1981) Drought tolerance and water use by plants along an alpine topographic gradient. *Oecologia*, **50**, 325–31.

Osburn, W.S. Jr (1961) Successional potential resulting from differential seedling establishment in alpine tundra stands. *Bulletin Ecological Society America*, **42**, 146–7.

Pearcy, R.W. (1969) Physiological and varied environmental studies of ecotypes of *Deschampsia caespitosa* (L.) Beauv, PhD thesis, Colorado State University, Fort Collins.

Peterson, K.M. and Billings, W.D. (1982) Growth of alpine plants under controlled drought. *Arctic Alpine Research*, **14**, 189–194.

Retzer, J.L. (1956) Alpine soils of the Rocky Mountains. *Journal Soil Science*, **7**, 22–32.

Richards, J.H. (1980) *Ecophysiology of a Deciduous Timberline Tree, Larix lyallii Parl*, PhD thesis, University of Alberta, Edmonton.

Richardson, S.G. and Salisbury, F.B. (1977) Plant responses to the light penetrating snow. *Ecology*, **58**, 1152–8.

Robberecht, R., Caldwell, M.M. and Billings, W.D. (1980) Leaf ultraviolet optical properties along a latitudinal gradient in the arctic-alpine zone. *Ecology*, **61**, 611–19.

Rocow, T.F. (1969) Growth, caloric content and sugars in *Caltha leptosepala* in relation to alpine snowmelt. *Bulletin Torrey Botanical Club*, **96**, 689–98.

Rocow, T.F. (1970) Ecological investigations of *Thalspi alpestre* L. along an elevational gradient in the central Rocky Mountains. *Ecology*, **51**, 649–56.

Rydberg, P.A. (1914) Formation of the alpine zone. *Bulletin Torrey Botanical Club*, **41**, 457–74.

Salisbury, F.B. (1976) Snow flowers. *Utah Science*, **37**, 35–41.

Salisbury, F.B. and Spomer, G.G. (1964) Leaf temperatures of alpine plants in the field. *Planta*, **60**, 497–505.

Sayers, R.L. and Ward, R.T. (1966) Germination responses in alpine species. *Botanical Gazette*, **127**, 11–16.

Scott, D. and Billings, W.D. (1964) Standing crop and productivity of an alpine tundra. *Ecological Monographs*, **34**, 243–70.

Scott, D., Hiller, R.D. and Billings, W.D. (1970) Correlation of CO_2 exchange with moisture regime and light in some Wyoming subalpine meadow species. *Ecology*, **51**, 701–2.

See, M.G. and Bliss, L.C. (1980) Alpine lichen-dominated communities in Alberta and Yukon. *Canadian Journal Botany*, **58**, 2148–70.

Shaver, G.R. and Billings, W.D. (1975) Root production

and root turnover in a wet tundra ecosystem, Barrow, Alaska. *Ecology*, **56**, 401–9.

Shaver, G.R. and Billings, W.D. (1977) Effect of daylength and temperature on root elongation in tundra graminoids. *Oecologia*, **28**, 57–65.

Spomer, G.G. (1964) Physiological ecology studies of alpine cushion plants. *Physiological Plantarum*, **17**, 717–24.

Spomer, G.G. and Salisbury, F.B. (1968) Eco-physiology of *Geum turbinatum* and implications concerning alpine environments. *Botanical Gazette*, **129**, 33–49.

Svoboda, J. (1977) Ecology and primary production of raised beach communities Truelove Lowland. *Truelove Lowland, Devon Island, Canada: A High Arctic Ecosystem* (ed. L.C. Bliss), University of Alberta Press, Edmonton, pp. 185–216.

Thompson, W.F. (1961) The shape of New England mountains. Part II. *Appalachia*, **44**, 316–35.

Tieszen, L.L (1970) Comparisons of chlorophyll content and leaf structure in arctic and alpine grasses. *American Midland Naturalist*, **83**, 238–53.

Tieszen, L.L. (1972) The seasonal course of aboveground production and chlorophyll distribution in a wet arctic tundra at Barrow, Alaska. *Arctic Alpine Research*, **4**, 307–24.

Tieszen, L.L. (ed.) (1978a) *Vegetation and Production Ecology of an Alaskan Arctic Tundra*, Springer-Verlag, N.Y.

Tieszen, L.L. (1978b) Photosynthesis in the principal Barrow, Alaska species: a summary of field and laboratory responses, in *Vegetation and Production of an Alaskan Arctic Tundra* (ed. L.L. Tieszen), Springer-Verlag, N.Y., pp. 241–68.

Tieszen, L.L. and Helgager, J.A. (1968) Genetic and physiological adaptation in the Hill reaction of *Deschampsia caespitosa*. *Nature*, **219**, 1066–7.

Tieszen, L.L. and Sigurdson, D.C. (1973) Effect of temperature on carboxylase activity and stability in some Calvin cycle grasses from the arctic. *Arctic Alpine Research*, **5**, 59–66.

Tieszen, L.L. and Wieland, N.K. (1975) Physiological ecology of arctic and alpine photosynthesis and respiration, in *Physiological Adaptation to the Environment* (ed. F.J. Vernberg), Intex Educational Publishers, N.Y., pp. 157–200.

Tieszen, L.L., Lewis, M.C., Miller, P.C., Mayo, J., Chapin, F.S. III and Oechel, W. (1981) An analysis of processes of primary production in tundra growth forms, in *Tundra, Ecosystems: A Comparative Analysis* (eds L.C. Bliss, D.W. Heal, J.J. Moore), Cambridge University Press, London, pp. 285–355

Tieszen, L.L., Senyimba, M.M., Imbamba, S.K. and Troughton, J.H. (1979) The distribution of C_3 and C_4

grasses and carbon isotope discrimination along an altitudinal and moisture gradient in Kenya. *Oecologia*, **37**, 337–40.

Tranquillini, W. (1962) Beitrag zur Kausalanalyse des Wettbewerbs ökologisch verchiedener Holzarten. *Berichte der Deutschen Botanishen Gesellschaft*, **75**, 353–64.

Tranquillini, W. (1964) Photosynthesis and dry matter production of trees at high altitudes, in *The Formation of Wood in Forest Trees* (ed. M.H. Zimmerman), Academic Press, N.Y., pp. 505–18.

Tranquillini, W. (1978) *Physiological Ecology of the Alpine Timberline*, Springer-Verlag, N.Y.

Tranquillini, W. and Schütz, W. (1970) Über der Rindenatmung einiger Bäume an der Waldgrenze. *Centralblatt fur das Gasamte Forstwessen*, **87**, 42–60.

Wardle, P. (1974) Alpine timberlines, in *Arctic and Alpine Environments* (eds J.D. Ives and R.G. Barry), Methuen, London, pp. 371–402.

Webber, P.J. and May, D.E. (1977) The magnitude and distribution of belowground plant structures in the alpine tundra of Niwot Ridge, Colorado. *Arctic Alpine Research*, **9**, 157–74.

Went, F.W. (1953) Annual plants at high altitudes in the Sierra Nevada, California. *Madroño*, **12**, 109–14.

Willard, B.E. (1979) Plant sociology of alpine tundra, Trail Ridge, Rocky Mountain in National Park, Colorado. *Colorado School Mines Quarterly*, **74**(4), 1–119.

Wolfe, J.A. and Hopkins, D.M. (1967) Climatic changes recorded by Tertiary land floras in northwestern North America, in *Tertiary Correlations and Climatic Changes in the Pacific* (ed. K. Hatai), Saski, Sendi, pp. 67–76.

Yurtsev, B.A. (1972) Phytogeography of northeastern Asia and the problems of Transberingian floristic interrelations, in *Floristics and Paleofloristics of Asia and Eastern North America* (ed. A. Graham) Elsevier, Amsterdam, pp. 19–54.

4

Taiga

WALTER C. OECHEL AND
WILLIAM T. LAWRENCE

4.1 Introduction

The taiga is predominantly a forested ecosystem which covers a vast expanse of North America roughly north of 50 °N latitude (Fig. 4.1). It occurs over a wide geographical area and with diverse climatological conditions (Larsen, 1980). Its northern limit is tree line in the Arctic, which generally occurs along the July 13 °C isotherm which is also roughly the southern extent of the Arctic front

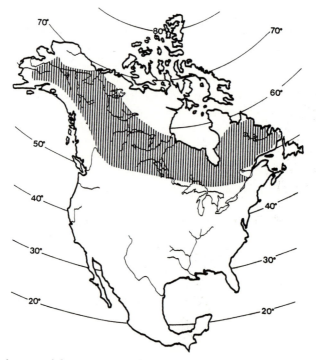

Figure 4.1   General range of the taiga in North America as abstracted from several sources (see text).

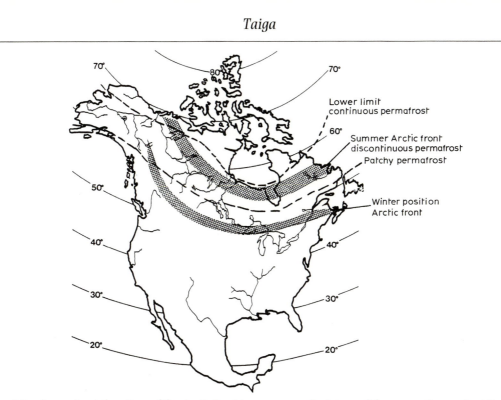

Figure 4.2 Approximate locations of the Arctic front in summer and winter and the range of permafrost throughout the taiga (Bryson and Hare, 1974; Hare and Hay, 1974).

during the summer season (Fig. 4.2). The southern limit of the taiga vegetation is less abrupt and usually occurs on the July 18 °C isotherm which is the average location of the arctic front during the winter (Bryson and Hare, 1974; Hare and Hay, 1974). Only recently has the taiga been so extensive. The area presently included in the taiga was largely glaciated. The ice retreated about 12 500 years before present in the northwest Canadian Arctic but remained even longer further east (Andrews, 1975). Plant populations survived the glaciation in refugia, which included areas of central Alaska, or by retreating in advance of the ice (Hultén, 1937). Species recolonized areas as the glaciers retreated and were limited only by their physiological tolerances and reproductive capabilities.

4.2 Vegetation zones of the taiga

Detailed classification schemes for the taiga vegetation were proposed by Hanson (1953) who dif-ferentiates between six major and twenty-two minor vegetation zones for northwest Alaska and by Viereck and Dyrness (1980) who produced a hierarchical classification for all of Alaska, which included one aquatic and four terrestrial vegetation formations. Viereck and Dyrness (1980) also recognize over 400 communities based on the dominant plant species.

In this chapter we define four major vegetation zones:

1. the forest–tundra ecotone,
2. the open boreal woodland,
3. the main boreal forest, and
4. the boreal–mixed forest ecotone.

All four vegetation zones are species impoverished. Taiga forests are dominated by four genera of conifers. *Picea*, *Pinus*, *Abies*, and *Larix*, and by two hardwood genera, *Populus* and *Betula* (Hare, 1950; Hare and Ritchie, 1972; La Roi, 1967). The dominant forest types are muskeg, closed canopy forest, and lichen or feather moss woodlands (Hare, 1950; Hare and Ritchie, 1972; Hustich, 1949).

The forest–tundra ecotone is the most northerly region of the taiga. It is composed of a lichen- or moss-dominated woodland with an open tree canopy of white spruce, *Picea glauca*, and black spruce, *Picea mariana* (see Hultén, 1968, for full nomenclature). The spruce are often small and slow growing and have low reproductive potentials (Elliot, 1979). Larger trees occupy more topographically protected sites; stunted or prostrate individuals occur in exposed areas. However, in some instances white spruce may be tree-like and black spruce prostrate at the same site (Hustich, 1953). Hardwoods persist in the far north but are mostly shrubby and are confined to margins of water courses where there is fresh alluvium (Bliss and Cantlon, 1957). The largest hardwoods are in sites free of permafrost (Lawrence, 1983). The forest–tundra ecotone integrates with arctic tundra throughout its range; some spruce stands impinge on coastal tundra sites near Hudson Bay and the Beaufort Sea.

To the south of the forest–tundra ecotone, the height and density of the spruce increases, and the area is classified as an open boreal woodland (Fig. 4.3). The canopy is open and in most cases dominated by black spruce, with larch (*Larix laricina*) restricted to wetter sites. Shrub cover is discontinuous. In the western part of Canada the forest floor is covered with feather mosses. Lichens (*Cladonia* spp.) predominate on the forest floor in the mid and eastern longitudes across Canada (Fig. 4.4). By far the most common forest type in the open boreal woodland is lichen woodland.

The main boreal forest vegetation zone is composed of continuous stands of coniferous trees with a closed canopy and a moss and sub-shrub understory. Three hardwood species, aspen (*Populus tremuloides*), balsam poplar (*Populus balsamifera*), and

Figure 4.3 View of a lichen–spruce woodland near Schefferville, Quebec. Black spruce dominate the tree canopy and lichens cover the forest floor.

Figure 4.4 Close view of the forest floor within a lichen woodland at Schefferville, Quebec. Lichen is *Cladonia alpestris*, the shrub *Ledum groelandicum*.

paper birch (*Betula papyrifera*), may become important after disturbance. The main boreal forest forms a truly transcontinental crescent from Alaska to Newfoundland, with *P. glauca,* and *L. laricina* occurring throughout. Balsam fir (*Abies balsamea*) and jack pine (*Pinus banksiana*) occur in the eastern and southern parts of this vegetation zone (Hare, 1950; Hustich, 1953; La Roi, 1967). The main boreal forest occurs on sites that are free of water stress but are not boggy.

The boreal–mixed forest ecotone occurs along the southern border of the taiga. It is a mixture of the dominant taiga forest species and southern non-taiga species. In the eastern part of Canada, the southern limit of the taiga intergrades with red and white pines (*P. resinosa* and *P. strobus*), sugar maple (*Acer saccharum*), arbor-vitae (*Thuja occidentalis*), black ash (*Fraxinus nigra*), and bigtooth aspen (*Populus grandidentata*) (Hare, 1950). The mid-

continent limit is a transition to aspen parklands and grassland; in the west the limit is a transition to subalpine and coastal conifer forests.

The forest tundra ecotone and open boreal woodland vegetation zones are often referred to as the subarctic. When this is done, the term boreal forest refers only to the main boreal forest and boreal–mixed forest ecotone. The entire taiga vegetation complex spans a latitudinal range exceeding 20°, a north/south distance of more than 2200 km. Taiga stands occur from north of 69 °N and near the Beaufort Sea to the northern shore of Lake Superior at 49 °N (Hosie, 1969; Porsild, 1957; Viereck and Little, 1972; Walter, 1979).

4.3 Environment

The climate in the taiga is characterized by a strong seasonal variation in solar radiation. Summers are cool and winters are extemely cold. Precipitation occurs more frequently in the summer when precipitation exceeds evaporation. The snow cover lasts more than half the year. Discontinuous permafrost occurs throughout most of the taiga's range (Brown, 1970).

Subarctic and boreal environments are climatologically quite different from temperate subalpine environments. In subarctic areas, solar radiation is usually significantly lower, day length longer, diurnal temperature fluctuations reduced, and summer temperatures colder than in subalpine areas. In subalpine areas water stress may affect photosynthesis. Water stress is not usually a limiting factor in the taiga except on thin canopied southern exposures.

Total radiation received and the amount absorbed and lost are limiting factors which affect production in the taiga. Maximum day length increases from 16 h at the southern edge of the taiga to 24 h in the north, but the total energy received decreases to the north. Mean annual solar radiation ranges between about 120 kilolangleys (kly) at the boreal mixed forest ecotone to about 80 kly at the arctic treeline (Hare and Hay 1974; Hare and Ritchie, 1972; Landsberg, 1961). Of this total incident radiation, 62–75% is absorbed. Yearly net radiation is over 40 kly yr^{-1} at the southern ecotone but decreases to about 15 kly yr^{-1} at the arctic treeline. The major factor accounting for the low net radiation in the extreme north is the strongly negative energy balance for several months in the winter when little or no solar radiation is received (Hare and Hay, 1974; Hare and Ritchie, 1972). Winter energy losses are accentuated in the northern taiga by snow cover and lack of a well-developed canopy, both of which increase albedo and decrease interception and capture of solar radiation (Hay, 1969; Pruit, 1978).

Throughout the taiga seasonal temperature fluctuations are great; monthly mean temperature at some sites can vary seasonally almost 90 °C, (Fig. 4.5). Diurnal fluctuations are between 6 and 14 °C, with the least daily temperature excursion occurring during winter and the greatest during summer. Mean yearly temperatures increase toward the south, with local anomalies, from a

minimum of −6.2 °C at Norman Wells near the forest-tundra ecotone, to a high of 2.5 °C at Winnipeg which is on the southern edge of the boreal forest. Interestingly, Winnipeg has the highest annual mean temperature but also has the greatest recorded extremes in temperature.

Key temperature parameters have been identified as being closely related to vegetation in the taiga. Of special importance is the number of days with a mean temperature above 10 °C (Fig. 4.5). Hopkins (1959) used this index to separate tundra, boreal, and coastal forests in Alaska. Walter (1979) and Walter *et al.* (1975) found taiga to be bounded by $120 \geq 10$ °C days on the south and $30 \geq 10$ °C days on the north. The lichen woodland at Schefferville has the fewest days with a mean temperature above 10 °C. Sites in the main boreal forest zone have more days above 10 °C. The number of days above 10 °C increases to the south to a maximum of 138 at Winnipeg.

One of the most important of ecological factors in the taiga is the presence or absence of permafrost. Changes in the energy budget of a site may change it from a permafrost-free to permafrost-dominated area or vice versa, depending on whether energy gain is reduced or increased by changes in canopy cover due to disturbance or succession. If the depth of thaw decreases, the volume of soil available for nutrient and water exploitation by the plant roots is reduced. Nutrient absorption by roots is also reduced due to cold soil temperatures. Plants may be limited to sites with soil depths compatible with their rooting patterns. A deeply tap-rooted species with few lateral roots may not survive in soils with a high permafrost level. Pulling (1918) observed that black spruce and larch which have shallow, widespread roots in permafrost-dominated sites had exactly the same rooting pattern in deep soils. Tap-rooted species, such as jack and white pine, had few side branches regardless of soil depth. White spruce and balsam poplar had flexible rooting patterns; they had tap roots in deep soils and more mat-like roots in shallow soils. Freeze/thaw cycles in permafrost zones can fracture roots due to ice formation. In most soils where permafrost is present, roots are restricted to the drier, ice-free organic and humus layers.

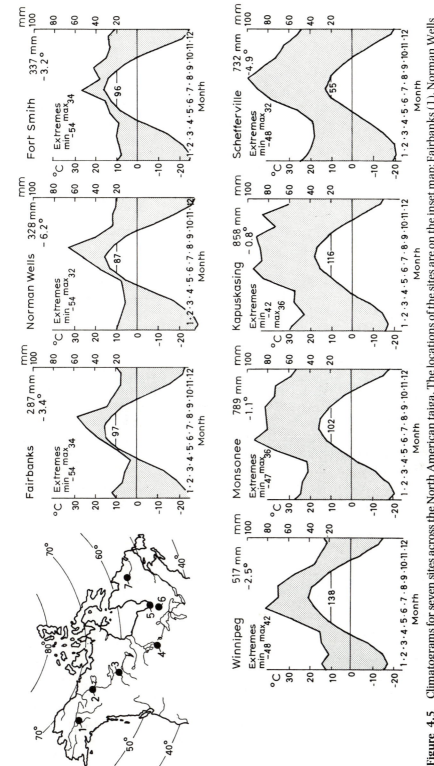

Figure 4.5 Climatograms for seven sites across the North American taiga. The locations of the sites are on the inset map: Fairbanks (1), Norman Wells (2), Fort Smith (3), Winnipeg (4), Monsonee (5), Kapuskasing (6), and Schefferville (7). Data from Hare and Hay (1974) and after Walter (1973). Lower line is monthly mean temperature (scale on left) and upper line is monthly total precipitation (mm scale on right). The number under the temperature curve is total days with mean temperature over 10 °C. Total yearly precipitation is in the upper right corner, with yearly mean temperature immediately below. The extreme minimum and maximum temperatures are noted below the site name in the upper left corner. Months are indicated along the lower margin.

4.4 Growth forms and phenology

Although the taiga is depauperate in species, it is rich in growth forms. In mature communities the tree overstory is usually composed of an evergreen conifer, but deciduous conifers or hardwoods also occur in the overstory. The understory of evergreen or deciduous shrubs usually includes low bush cranberry (*Vaccinium vitis-idaea*), blueberry (*Vaccinium uliginosum*), *Ledum groenlandicum* or *Ledum palustre*, and *Betula nana* or *Betula gladulosa* (Fig. 4.6). More closed forests tend to have a thick, continuous cover of mosses. In better drained areas of central Alaska feather mosses, *Pleurozium schreberi* and *Hylocomium splendens*, predominate; *Polytrichum commune* and *Ptillium crista-castensia* are also often present. Wetter or more poorly drained areas are usually dominated by *Sphagnum* species. In open forest areas, such as the woodlands

near Schefferville, Quebec, similar growth forms are present except that the open canopy permits development of dense lichen mats, especially of *Cladonia alpestris (Cladina stellaris)*.

Phenological development is generally rapid and growth is compressed into the few summer months of the short growing season. At Schefferville leaf bud break in *Betula glandulosa* occurs in late June, senescence in early August, and leaf abscission in mid-September. There is only a 5–6 week period between bud break and the beginning of leaf senescence.

The photosynthetic period is a month longer for evergreen species such as *Ledum groenlandicum* compared to the period available for photosynthesis in deciduous species. Leaf duration is also longer in evergreen shrubs. Leaves of *Ledum groenlandicum* may persist for $1\frac{1}{2} - 2\frac{1}{2}$ yr, which increases the photosynthetic return of the leaf (Prudhomme,

Figure 4.6 The forest floor in a 120 year old black spruce forest. Feather moss cover is 100%, with scattered low shrubs. The organic layer can exceed 30 cm in depth, and the soil thaws less than 85 cm in this type of stand.

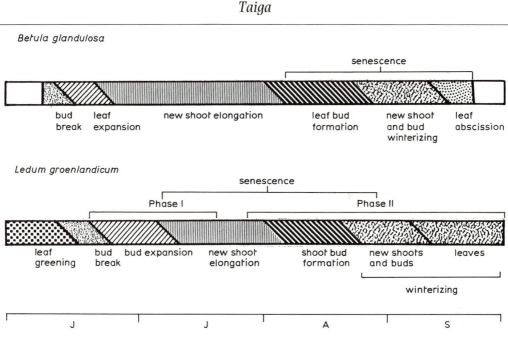

Figure 4.7 Seasonal phenology of two dominant shrub layer components of the lichen woodland near Schefferville, Quebec. *Betula* is a deciduous shrub, and *Ledum* evergreen. For both species the months of June through September comprise the entire growing season (from Prudhomme, 1983).

1983; Fig. 4.7). In central Alaska, *Picea mariana* may maintain leaves in excess of 15 yr (Hom and Oechel, 1983), thereby maximizing the carbon return per unit of carbon and nutrient invested in leaves.

4.5 Controls on carbon balance

Physiological data are available for relatively few taiga species. Among the dominant tree species, maximum rates of photosynthesis (P_{max}) are highest in deciduous hardwood species, which occur in the post-disturbance successional stages of moist sites. Photosynthetic rates in conifers are lower than those of hardwoods. Of the conifers, white spruce have higher P_{max} values than do black spruce. The lower photosynthetic rates of conifers may reflect the general pattern of nutrient impoverishment and cold rooting zone temperatures in later successional stages, which are the result of the permafrost mediated reduction in the depth of the active layer and rate reduction in cold-sensitive nutrient uptake processes (Lawrence and Oechel, in preparation).

Black spruce, *Picea mariana*, which is the dominant tree in extensive areas of the taiga, has

maximum photosynthetic rates measured in the field of about $0.02–0.14\,\mu mol\,CO_2\,g^{-1}$ dry wt s^{-1} (Table 4.1) (Vowinckel *et al.*, 1975; Lawrence and Oechel, in prep.). Rates of $0.03\,\mu mol\,g^{-1}\,s^{-1}$ were measured for greenhouse grown seedlings (Black and Bliss, 1980). Early in the season, black spruce had lower photosynthetic rates at treeline sites than in the boreal forest. This may reflect the effect of the harsh winter conditions, wind damage, and desiccation which are prevalent at treeline and may also reflect the need for a recovery period following dormancy before full photosynthetic capacity is reached. Maximum summer rates at treeline, even in krummholz forms, were similar to those measured in the forest (Vowinckel *et al.*, 1975). Other than an early season depression in photosynthesis, there is little seasonal change in maximum photosynthesis rates between June and August in *P. mariana*. In response to nutrient limitation, black spruce needles may be maintained for 13 years. At this time, photosynthetic activity can still be 40% of the maximum rate (Hom and Oechel, 1983).

The deciduous conifer, *Larix laricina*, has a maximum photosynthetic rate of $0.14\,\mu mol\,CO_2\,g^{-1}\,s^{-1}$ (Auger, 1974), but as leaves senesce photosynthetic

Table 4.1 Maximum photosynthetic rates (P_{max}, μmol CO_2 g^{-1} dry wt s^{-1}) of various growth forms in boreal regions

Growth form	Species	P	Location	Reference	Notes
Trees					
Evergreen conifer	Picea mariana	0.02	Schefferville, Quebec	Vowinckel et al. (1975)	Field
	Picea mariana	0.03	McKenzie Delta	Black and Bliss (1980)	Greenhouse grown
Deciduous conifer	Larix laricina	0.13–0.14	Schefferville, Quebec	Auger (1974)	Field
Deciduous hardwood	Populus tremuloides	0.07	Fairbanks, Alaska	Oechel et al., unpub.	Field
	Populus tremuloides	0.17	Fairbanks, Alaska	Lawrence and Oechel (1983b)	Growth chamber grown
	Populus balsamifera	0.09	Fairbanks, Alaska	Lawrence and Oechel (1983b)	Growth chamber grown
	Betula papyrifera	0.03	Fairbanks, Alaska	Lawrence and Oechel (1983b)	Growth chamber grown
	Alnus	0.11	Fairbanks, Alaska	Lawrence and Oechel (1983b)	Growth chamber grown
$\bar{x}$		0.08			
Deciduous shrubs	Betula nana	0.18	Eagle Creek, Alaska	Bigger and Oechel (1982)	Tundra above spruce
	Rubus chamaemorus	0.10	Eagle Creek, Alaska	Bigger and Oechel (1982)	Tundra above spruce
	Vaccinium uliginosum	0.09	Eagle Creek, Alaska	Bigger and Oechel (1982)	Tundra above spruce
$\bar{x}$		0.12			
Evergreen shrubs	Ledum palustre	0.06	Eagle Creek, Alaska	Bigger and Oechel (1982)	Tundra above spruce
	Ledum groenlandicum	0.03	Eagle Creek, Alaska	Smith and Hadley (1974)	Growth chamber grown
	Vaccinium vitis-idaea	0.03	Eagle Creek, Alaska	Bigger and Oechel (1982)	Tundra above spruce
	Empetrum nigrum	0.01	Eagle Creek, Alaska	Bigger and Oechel (1982)	Tundra above spruce
$\bar{x}$		0.03			
Sedges	Carex bigelowii	0.09	Eagle Creek, Alaska	Bigger and Oechel (1982)	Tundra above spruce
	Eriophorum vaginatum	0.03	Eagle Creek, Alaska	Bigger and Oechel (1982)	Tundra above spruce
$\bar{x}$		0.06			
Mosses	Dicranum fuscescens	0.005	Schefferville, Quebec	Hicklenton and Oechel (1976)	Field grown
	Polytrichum commune	0.02	Washington Creek, AK	Skre and Oechel (1981)	Field grown
	Pleurozium schreberi	0.008	Washington Creek, AK	Skre and Oechel (1981)	Field grown
	Hylocomium splendens	0.008	Washington Creek, AK	Skre and Oechel (1981)	
	Sphagnum subsecundum	0.004	Washington Creek, AK	Skre and Oechel (1981)	Field grown
$\bar{x}$		0.009			
Lichens	Cladina stellaris	0.01	Schefferville, Quebec	Carstairs and Oechel (1978)	Field grown

rates fall by about 30% from August to September. Leaf respiration rates at 14 °C in midseason are about 25% of maximum photosynthetic rates.

In deciduous hardwood species, photosynthesis rates from $0.07\,\mu mol\,CO_2\,g^{-1}s^{-1}$ in trembling aspen (*Populus tremuloides*) to $0.03\,\mu mol\,CO_2\,g^{-1}s^{-1}$ in paper birch (*Betula papyrifera*) (Table 4.1). Maximum photosynthetic rates measured in growth chamber and field studies can be quite different, as is shown for *Populus tremuloides* (Table 4.1).

Deciduous shrubs which are characteristic of the taiga but which were measured at a site above timberline at Eagle Creek, Alaska, had a mean P_{max} of $0.12\,\mu mol\,CO_2\,g^{-1}s^{-1}$. This rate was higher than the mean rate measured for deciduous trees (Table 4.1). Average photosynthetic rate in the evergreen shrub species was about 25% of the mean deciduous shrub rate. Sedges fall between the evergreen and deciduous shrubs in photosynthetic rate and averaged about $0.06\,\mu mol\,CO_2\,g^{-1}s^{-1}$.

Measured moss photosynthesis values average $0.009\,\mu mol\,CO_2\,g^{-1}s^{-1}$, ranging from 0.004 for *Sphagnum subsecundum* to $0.02\,\mu mol\,CO_2\,g^{-1}s^{-1}$ for *Polytrichum commune* (Table 4.1). Maximum photosynthetic rates vary in different vascular plant communities. Maximum photosynthetic rates in *Pleurozium schreberi* are $0.004\,\mu mol\,CO_2\,g^{-1}$ dry wt s^{-1} in a birch stand and $0.008\,\mu mol\,CO_2\,g^{-1}$ dry wt s^{-1} in a white spruce stand. Respiration rates measured in the top portion of the thallus are about three-fold those of the middle portion.

4.6 Temperature response

The temperatue optimum for photosynthesis in black spruce growing under natural conditions in the field at Schefferville, Quebec, is around 15 °C. 90% of the maximum photosynthetic rate is maintained from 9 to 23 °C (Vowinckel *et al.*, 1975). Black spruce seedlings grown in a greenhouse for 12 months had a broad photosynthetic temperature response which retained 31% of P_{max} at 0 °C and 41% of P_{max} at 30 °C. Photosynthesis is positive from less than 0 °C to greater than 40 °C. This suggests that photosynthesis is not likely to be strongly limited by temperature under ordinary climatic conditions

during the summer period from June to August (Black and Bliss, 1980). However, photosynthesis may be limited by low temperature at other times of the year (Vowinckel *et al.*, 1975).

Photosynthetic acclimation was well developed in *Ledum groenlandicum* (Smith and Hadley, 1974) especially in populations found in the northern boreal forest of central Alaska and in alpine populations. Plants of a central Alaska population showed a change in photosynthetic rate of only about 7% when grown and measured at 15 °C with 10 °C night temperatures compared with plants grown and measured at 30 °C with 25 °C night temperatures. Homeostasis of respiration was also highest in the central Alaska population; respiration rates shifting only 20% between cool and warm grown plants when measured at the growth temperature. Acclimation of respiration was complete within two weeks, whereas acclimation of photosynthesis was not complete within the two week period.

Acclimation shifts the optimum temperature for photosynthesis from 0–10 °C early in June to 5–25 °C in August. Temperature compensation points for photosynthesis also shift with acclimation. The upper compensation point of photosynthesis is at 15–20 °C early in the season and increases to > 30 °C in late August. Photosynthesis at 0 °C is 62% of maximum in early June and 11% of maximum in October. Obviously, any discussion of the temperature response of photosynthesis is complicated by the ability of photosynthesis to acclimate during the growing season in response to changing environmental conditions.

In the lichen *Cladina stellaris* (Opiz) Brodo, which was formerly called *Cladonia alpestris* (L.) Rabh, the

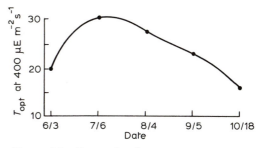

Figure 4.8 Seasonal changes in termerature optima for photosynthesis (T_{opt}) in *Cladina stellaris* at $400\,\mu E\,m^{-2}s^{-1}$.

Table 4.2 Seasonal characteristic of the temperature response surface of *Cladina stellaris* at $800\,\mu E\,m^{-2}\,s^{-1}$ and optimal water contents. Characteristics include the maximum photosynthetic rate observed (P_{max}) (mg $CO_2\,g^{-1}$ h^{-1}), the temperature optimum (T_{opt}) (°C), and the percentage of P_{max} at 5 and 30 °C. These values were generated from second power polynomial regression. The coefficient of determination is R^2.

	P_{max}	T_{opt}	% P_{max} @ 5°C	% P_{max} @ 30°C	Coeff. of Deter. (R^2)
June	0.86	28	51	100	0.98
July	2.15	25	28	95	0.82
August	2.30	21	22	79	0.80
September	2.04	20	49	74	0.83
October	1.48		28	41	0.95

temperature range for photosynthesis is broad and high rates of photosynthesis occur from 5 to 30 °C. In June, for example, 51% of the maximum photosynthetic rate can occur at 5 °C, but by August only 22% of the maximum rate can occur at 5 °C (Table 4.2). However, the temperature optima for photosynthesis are consistently greater than the mean temperature for the period of measurement (Fig. 4.8). Temperature optima may exceed mean temperatures by as much as 27 °C, indicating incomplete acclimation (Carstairs and Oechel, 1978; Oechel, 1976; Oechel *et al.*, 1975). These temperature optima for photosynthesis are generally higher than those reported for arctic (Larsen and Kershaw, 1975), antarctic (Lange and Kappen, 1972), and some subarctic lichen populations (Kershaw, 1977) but are similar to those reported for *Cladonia* species from a northern temperate area (Lechowicz and Adams, 1974).

The dark respiration rates of *C. stellaris* change less during the growing season than did the photosynthetic rates which suggests that the shift in the temperature optima of photosynthesis in this lichen is a consequence of increasing P_{max} only (Berry and Bjorkman, 1980).

4.7 Light response

Net daily carbon gain by the foliage of black spruce is between 15 and 30 mg $CO_2\,g^{-1}$ dry wt d^{-1}. During the summer months the environmental factor with the greatest potential to limit photosynthesis is light intensity. Cloud cover is frequent and extensive. However, overcast skies increase diffuse radiation which results in a higher percentage of incident photosynthetically active radiation (PAR) at the moss surface than on cloudless days (Skre *et al.*, 1983). The low light compensation and light saturation points of many taiga species allow daily net carbon gains. Photosynthesis is only minimally limited by PAR during the summer (Vowinckel *et al.*, 1975).

Light saturation in *Picea mariana* under field conditions occurs about $800\,\mu E\,m^{-2}\,s^{-1}$ at 13–20 °C, when photosynthesis is 90% of the maximum rate (Fig. 4.9). Saturation was even lower, $450–500\,\mu E\,m^{-2}\,s^{-1}$, in laboratory-grown seedlings (Black and Bliss, 1980; van Zindern Bakker, 1974). Light compensation occurs between 12 and $35\,\mu E\,m^{-2}\,s^{-1}$ for different growth conditions (Black and Bliss, 1980; van Zindern Bakker, 1974; Vowinckel *et al.*, 1975). Light saturation in larch also occurs at about $800\,\mu E\,m^{-2}\,s^{-1}$, which is almost 36% of full sunlight. However, the light compensation point in larch decreases through the season from about $80\,\mu E\,m^{-2}\,s^{-1}$ in June to about $50\,\mu E\,m^{-2}\,s^{-1}$ in August.

In mosses light compensation and saturation points for photosynthesis also vary seasonally. At one location in Schefferville, Quebec, compensation in *Dicranum fuscescens* drops from

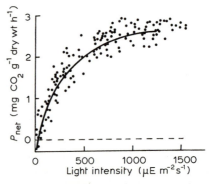

Figure 4.9 Relationship between net photosynthesis and light intensity in black spruce at Schefferville, Quebec, at air temperatures of 13–20 °C. (From Vowinckel *et al.*, 1975.)

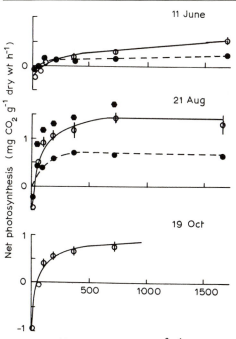

Figure 4.10 Net photosynthesis at 5 (●) and 20 °C (○) in the moss *Pleurozium schreberi* as a function of light intensity in excised 2 year old shoots from the intensive black spruce site in central Alaska (Washington Creek). Also shown are moss shoots from a white spruce site (●). Confidence limits are ± 2 s.e. and are plotted at each light step. (From Skre and Oechel, 1981).

$30–60 \, \mu E \, m^{-2} s^{-1}$ in early June to $15–30 \, \mu E \, m^{-2} s^{-1}$ in late August. Light saturation increases from $60–125 \, \mu E \, m^{-2} s^{-1}$ to $125–250 \, \mu E \, m^{-2} s^{-1}$ over the same period. Similar decreases in light compensation over the growing season were seen for four moss species in central Alaska, although light saturation points were less variable over the course of the season (Fig. 4.10) (Skre and Oechel, 1981). At light intensities above $1000 \, \mu E \, m^{-2} s^{-1}$ photo inhibition of photosynthesis has been reported in one moss species (Oechel and Collins, 1976).

Light compensation points and light saturation points in lichens are temperature sensitive. At 22 °C photosynthesis in *C. stellaris* light saturates at $500 \, \mu E \, m^{-2} s^{-1}$. Higher and lower temperatures reduced light saturation values. At 5 °C light saturation points were about $200 \, \mu E \, m^{-2} s^{-1}$. Carstairs and Oechel (1978) found light saturation levels at most temperatures to be greater than $400 \, \mu E \, m^{-2} s^{-1}$ for all months of the year except June. For most months, light compensation points were above $40 \, \mu E \, m^{-2} s^{-1}$ at temperatures of 15 °C and higher. Below 15 °C, compensation points were less than $40 \, \mu E \, m^{-2} s^{-1}$.

Kimball *et al.* (1973) found that 10% of the incident radiation penetrated a snow cover of approximately 3.5 cm. The reported light compensation points for photosynthesis in lichens indicate that there may be periods when positive photosynthesis occurs under the snow such as has been reported for *C. stellaris* (Barashkova, 1971). Lichens are commonly snow covered until the middle of June which is the month with highest light intensities (Tout, 1963, 1964); they could have positive net assimilation under snow approximately 5 cm deep. Because of low solar angles and light intensity, lichen photosynthesis under snow in October is much less likely.

High light intensities may interact with water stress to limit photosynthesis in lichens. High light intensities are generally correlated with high evaporation potentials, especially at high temperatures which are common in lichen mats (Carstairs and Oechel, unpublished; Kershaw and Field, 1975). Since lichens dry out very rapidly in warm. dry air and are unable to photosynthesize at very low water contents, it is unlikely that extended periods of positive net assimilation are possible under conditions of high light and high temperature. However, under lower light intensities, the lichen will generally desiccate more slowly and may be able to carry on photosynthesis for a longer period of time.

There appears to be a lag in the response of lichen photosynthesis to new environmental conditions. Lichens may adapt primarily to long-term changes rather than to day-to-day variations in climatic conditions. Long-term adaptations are important in a plant which can photosynthesize only for limited periods when sufficient moisture is present in the thallus. Analysis using multiple regressions indicated that 97% of the seasonal variation in the maximum photosynthetic rate (P_{max}) was explained in terms of temperature and radiation conditions during the month prior to the field sampling period (Carstairs and Oechel, 1978).

4.8 Water stress

Despite generally shallow root systems, black spruce tend to show little water stress, which may be related to wet soils and low atmospheric vapor-pressure deficits associated with areas where black spruce occur. In Schefferville, dawn water potentials range from about -0.2 to -0.35 MPa and minimum water potentials measured were -1.5 MPa. Shaded portions of the trees had water potentials as much as 0.3 MPa higher than sunlight portions. Minimum and maximum water potentials for black spruce in the Mackenzie Delta area varied from -2.1 MPa to -0.6 MPa. Generally, water stress was lower in the larger trees studied (Black and Bliss, 1980).

However, conifers are sensitive to drought and low water potentials. Water stress of -1.5 MPa results in a 50% reduction in net photosynthesis in greenhouse-grown seedlings of black spruce and a decrease in leaf water potential to -2.5 MPa further reduces net photosynthesis to the compensation point (Black and Bliss, 1980). Even though low water potentials can severely curtail photosynthesis, the prevailing environmental situation in the field seldom induces water stress in black spruce. Osmoregulation by the trees during water content changes maintained a constant turgor of about 1 MPa which allows continued growth and cell expansion (Black and Bliss, 1980).

In contrast to the vascular plants, normal moisture conditions in the field may impose a major limitation on photosynthesis in mosses and lichens (Table 4.3). Since they have no vascular tissue or roots to tap soil moisture, many taiga moss species depend on precipitation for moisture input. With only a few days of dry, sunny conditions, mosses can become desiccated and brittle despite a saturated soil. Feather mosses are especially susceptible to desiccation, but desiccation also occurs in other species such as *Dicranum fuscescens*. *Sphagnum* species have high water retention capabilities; they have the capacity to wick water up on the stem surface from belowground sources and have a compact surface from which water is lost very slowly. *Polytrichum* species transport water internally through the stem structure (Lawrence and Oechel, in prep.; Skre *et al.*, 1983) and apparently regulate water loss by altering leaf orientation. Polytrichaceous and sphagneous species, therefore, are better buffered against drought than are many other species and are capable of sustaining photosynthesis during what would be severe drought conditions for other moss species.

Water contents for maximum photosynthesis in mosses show a remarkable range, from a low of 100% for *Polytrichum* to a 725% minimum for *Sphagnum subsecundum* (Table 4.3). Compensation points for net photosynthetic carbon gain range from 20–30% water content for *P. commune, P. schreberi*, and *H. splendens*, 62% for *S subsecundum*, to 75% in *Polytrichum alpinum* (Table 4.3). In most moss species there is a gradual decrease in both net

Table 4.3 Per cent water content yielding optimal net photosynthesis, maximal respiration and photosynthetic compensation. Data from drying experiments conducted in July 1976 at 15 °C and at 720 μE m^{-2} s^{-1} or in darkness in plants from Washington Creek, Alaska. Confidence limits are ± 2 s.e. $N = 6$. (Skre and Oechel, 1981.)

	Species			
	Polytrichum commune	*Pleurozium schreberi*	*Hylocomium splendens*	*Sphagnum subsecundum*
Date Sampled	16 July	12 July	18 July	3 August
Optimum for photosynthesis	100 ± 20*	678 ± 180	425 ± 100	725 ± 140
Compensation for photosynthesis	25 ± 8	32 ± 8	20 ± 5	62 ± 10
Optimum for respiration	—	764 ± 250	1120 ± 300	925 ± 270

*Estimated from field measurements on intact plants (August 1975).

Table 4.4 Time for establishment of stable photosynthesis as well as maximum photosynthesis rates and dark respiration at $15°C$ (mg CO_2 produced $g^{-1} h^{-1}$) at stable state after cutting and desiccation, compared with excised moist samples. Data for Washington Creek, central Alaska. Confidence limits ± 2 S.E. indicated $N = 6$. (Skre and Oechel, 1981.)

Species	Desiccation time (d)	Stabilization time (h)	Maximum photosynthesis	Dark respiration at $15°C$
Polytrichum	0	2	2.40 ± 0.12	0.60 ± 0.12
commune	8	2	0.60 ± 0.14	0.44 ± 0.10
	16	3	0.30 ± 0.12	0.44 ± 0.10
Sphagnum	0	3	0.76 ± 0.10	0.50 ± 0.08
subsecundum	8	8	0.32 ± 0.10	0.52 ± 0.10
	16	10	-0.08 ± 0.08	0.62 ± 0.10
Hylocomium	0	3	1.36 ± 0.16	0.52 ± 0.12
splendens	4	5	0.50 ± 0.18	0.65 ± 0.16
	8	6	0.40 ± 0.20	0.45 ± 0.18
Pleurozium	0	2	0.72 ± 0.15	0.56 ± 0.10
schreberi	4	3	0.56 ± 0.18	0.64 ± 0.12
	8	6	0.20 ± 0.20	0.62 ± 0.10

photosynthesis and dark respiration as leaf water content drops. Above the moisture content for maximum photosynthesis, there is a plateau or, in some cases, a slight decrease in photosynthesis. Stålfelt (1937) explained this behavior as increased diffusion resistance in the external water film surrounding the moss. A similar reaction is also attributed to dark respiration as a function of leaf moisture content in *Hypnum* sp. (Fraymouth, 1928) where oxygen limitation was noted at high moisture levels.

When mosses are rehydrated following desiccation there may be a rapid increase of respiration and a depression of normal photosynthetic rates (Peterson and Mayo, 1975). Within 12 h of rehydration, respiration rates decrease to their original level while photosynthesis stabilizes at a level considerably lower than that of hydrated mosses (Table 4.4). the greatest effect of desiccation was found in *Polytrichum*, where photosynthesis decreased from 2.4 mg $CO_2 g^{-1} h^{-1}$ before desiccation to 0.3 mg $CO_2 g^{-1} h^{-1}$ upon rewetting after 16d desiccation. The depression in photosynthetic rates is also dependent on the length of the desiccation. In *Pleurozium* the photosynthetic rates upon rehydration decreased from 0.72 to 0.56 mg $CO_2 g^{-1} h^{-1}$

after 4 d desiccation and to 0.20 mg $CO_2 g^{-1} h^{-1}$ after 8 d desiccation. The stablization time varied from 2–3 h in *Polytrichum* to 8–10 h in *Sphagnum* depending on desiccation time. Peterson and Mayo (1975) reported that 8 h were required for restoration of photosynthesis when rewetting *Dicranum* after 8 days of drying. Stålfelt (1937) reported recovery times in *Hylocomium* varying from 6 h after 6 d desiccation to 11 h after a 17 d period. Thus, the period of hydration necessary to re-establish maximal photosynthesis rates after a rain varies with the length of the preceding dry period.

Cladina stellaris at Schefferville shows no seasonal photosynthetic response to environmental water conditions. The optimal tissue moisture content was found to fall at a point between 35 and 55% of saturation under various light and temperature conditions at each sampling period. This is much lower than the optimal tissue moisture content of 75% reported by Lechowicz (1978) for *Cladina stellaris* and *C. evansii*. However, because of the relative insensitivity of photosynthesis to water content, these differences in optimal water content may have less impact on carbon uptake than would at first be assumed. Under a favorable light and temperature environment, the lichen maintained positive net

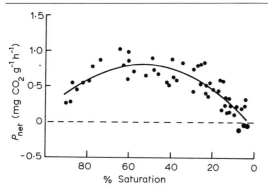

Figure 4.11 Net carbon dioxide assimilation pattern in the photosynthetically active section of *Cladina stellaris* as it dries from saturation at 5 °C and 400 $\mu E\,m^{-2}\,s^{-1}$ in August. The line is a second degree polynomial regression fit to the data (Carstairs and Oechel, 1978).

carbon dioxide assimilation down to moisture contents of 5–10% of saturation (Fig. 4.11). Thus a positive carbon balance is possible until the lichen is almost air dry which occurs at 3–4% saturation in the field (Carstairs and Oechel, 1978 and unpublished). Although no change in the photosynthetic response to moisture conditions was detected, a trend in the relationship between photosynthesis and the water holding capacity and drying rate of the lichen was measured over a period of approximately 1 h at the beginning of all assimilation experiments. At the beginning and end of the growing season, when moisture is more available, the lichen appears to have a greater water holding capacity, but it also dries out more rapidly in dry air. During the mid-summer period the lichen became noticeably tougher and less pliable even when moistened. Water conductance decreases during the period when maximum photosynthesis is greatest. This implies that either the intercellular space diffusion resistance and the liquid phase transfer resistance (Sestak *et al.*, 1971) for carbon dioxide and water vapor vary differentially through the season or that intercellular metabolic processes (e.g. enzymatic activity) are responsible for changes in the maximum photosynthetic rate.

4.9 Nutrient relations

The taiga is a nutrient-poor region with heavily podzolized soils. The podzols occur where precipi-

tation exceeds evaporation, which is common throughout the taiga (Dyrness and Grigal, 1979; Larsen, 1980; Pruit, 1978). Owing to cold soil temperatures, decomposition is slow, and nutrients are tied up as undecomposed organic matter. As communities develop after disturbance, the soil is more and more shaded and insulated; and organic matter is sequestered in frozen soils or is immobilized in peat.

Stands which have a moss understory are under a unique type of nutrient stress. Not only is decomposition slowed by permafrost and chronically cold soils, but living feather moss acts as a remarkable nutrient 'sponge'. Laboratory and field work has shown that practically any nitrogenous or ionic nutrient that falls on the moss surface is absorbed by the moss and immobilized (Oechel *et al.*, in prep.). Even litter is trapped in the moss layer, which can be over 30 cm deep. *Polytrichum* can mobilize water and nutrients from the mineral layer, so it need not rely solely on input from above the moss surface. In many taiga ecosystems the forest floor contains the largest pool of nitrogen above the mineral soil (Van Cleve *et al.*, 1983).

On permafrost-dominated sites the roots of trees and shrubs are found throughout the unfrozen organic layers of the soil, but few roots penetrate into the nutrient-poor mineral soil. Practically all plants have mycorrhizal associations. A survey of 29 vascular plant species showed that 4 were ectomycorrhizal, 15 endomycorrhizal, and 4 carried infections of both types (Malloch and Malloch, 1981).

Nutrient availability does not seem to be a major factor limiting photosynthetic rates in lichens, at least it is not a limiting factor in *C. stellaris* at Schefferville. As opposed to boreal mosses, *C. alpestris* showed no enhancement of photosynthesis when fertilized with various nutrients, including nitrogen and phosphorus additions of up to 900 mg N m⁻² and 250 mg P m⁻².

4.10 Production

Biomass production is an integrative indicator of the functioning of the ecosystem. Addition of biomass, whether expressed on an area, weight, or per individual basis, reflects the net carbon gain minus the carbon diverted to storage or used in growth and maintenance respiration. Carbon fixation is a function of photosynthesis and reflects the environmental and physiological limits of photosynthesis. Carbon losses are also subject to environmental influences such as temperature, which affects respiration rates of all living plant tissue. The storage of carbon by the plant does not necessarily indicate an excess supply of photosynthate. Stored carbon may represent reserves that help insure survival of the individual. Some carbon is expended in seasonal growth, but other carbon reserves may be allocated to seed production in an annual or stored for next season's bud break and seed production in a perennial.

Calculating the carbon balance for a single plant is difficult; attempting to estimate the carbon balance for an ecosystem such as the taiga is rudimentary at best. Gas exchange of the foliage can be measured (Oechel and Lawrence, 1979; Vowinckel et al., 1975) to determine net carbon fixation by photosynthesis and losses to dark respiration of leaves and stems, but such measurements are difficult and require complex instrumentation. To calculate an aboveground carbon balance, gas exchange parameters must be characterized for all size classes and ages of foliage and stems and then combined with the calculated growth increment of each tissue type. The belowground components of gas exchange and growth are important as well, since roots, underground stems, and rhizomes can be 64–460% of the aboveground biomass (Larsen, 1980). Measurements of belowground growth are tedious and difficult, and measuring belowground gas exchange in the field is practically impossible. However, gas exchange of seedling roots can be measured in the laboratory (Lawrence and Oechel, 1983a, b; Limbach, et al., 1982).

The most common approach of assessing production in the taiga has been to measure aboveground biomass and increments of growth by the harvest method. Tree biomass ranges between 9590 and 163 360 kg ha^{-1} in spruce forests from Alaska to northern Quebec (Larsen, 1980). Tree biomass in an upland and a lowland spruce forest in Alaska averaged about 20 000 kg ha^{-1}; the average bio-

mass for Quebec forests at eight sites was over 52 000 kg ha^{-1}. Aspen sites in Canada which were between 6 and 52 years old had stem and branch biomasses of 21 500–51 200 kg ha^{-1}. Recent tree biomass estimates for taiga sites near Fairbanks, Alaska (Van Cleve et al., 1983) which were covered with muskeg, north-facing slope spruce, and birch forests were 16 597, 24 018 and 97 340 kg ha^{-1}, respectively.

Shrubs and non-vascular plants compose widely varying portions of the biomass at various sites in taiga. In the early successional birch forest of Alaska, the understory is poorly developed and contains only 95 kg ha^{-1} shrubs, which is about 0.1% of the tree biomass that is present. In black spruce forests the biomass of the heavy moss cover is about 4790 kg ha^{-1}, and the biomass of the scattered shrubs is 1490 kg ha^{-1}. Their combined biomass equals 26% of the tree biomass. In the muskeg where non-tree species comprise 52% of the living aboveground biomass, shrubs are a minor component with a biomass of 1160 kg ha^{-1}; non-vascular plants form the largest component of aboveground biomass, 7480 kg ha^{-1} (Van Cleve et al., 1983).

In the taiga yearly growth increments vary both with site and age of the stand. The factors of a site which influence production are water, nutrients, light, and temperature. In the taiga of central Alaska productivity and nutrient cycling are controlled by the temperatures of the forest floor and the mineral soil (Van Cleve et al., 1983). The effect of stand age is more subtle and is felt in the carbon losses involved in maintaining tissues that respire but do not photosynthesize, such as the stems and cambium of the bole. Aspen stands of 6, 15, and 52 years show an accumulation of stem and branch biomass from 21 500 to 91 800 kg ha^{-1}, but a decline in leaf biomass from 2600 to 1500 kg ha^{-1} (Pollard, 1972, cited in Larsen, 1980). The reduction in leaf area and the increased respiration of standing live wood have a negative feedback on production due to lower total net carbon availability.

Estimates of net primary production for forested regions, which were calculated as gross photosynthesis minus all respiration, ranged between 4800 and 12 800 kg ha^{-1} yr^{-1} for boreal conifer forests to about $14\,000\text{--}20\,000\,\mathrm{kg\,ha^{-1}\,yr^{-1}}$ for temperate deciduous forests (Olson, 1971). In these calculations the change in aboveground biomass was corrected for material shed during the growing season. Belowground growth was not included. Van Cleve *et al.* (1983) state that the average annual tree production in spruce, aspen, birch and poplar is about $460\,\mathrm{g\,m^{-2}}$ and that production in black spruce forests is only $110\,\mathrm{g\,m^{-2}}$.

Aboveground production from the taiga is roughly comparable to Olson's estimates. In the southern taiga production is $3719\,\mathrm{kg\,ha^{-1}\,yr^{-1}}$ in peatlands, $2840\text{--}9600\,\mathrm{kg\,ha^{-1}\,yr^{-1}}$ in aspens of various ages, and $5400\text{--}6400\,\mathrm{kg\,ha^{-1}}$ in alder (Larsen, 1980). In Alaska maximum production of $7040\,\mathrm{kg\,ha^{-1}\,yr^{-1}}$ was measured in 60 year old aspen, $4710\,\mathrm{kg\,ha^{-1}\,yr^{-1}}$ in birch, $3840\,\mathrm{kg\,ha^{-1}\,yr^{-1}}$ in white spruce, $2450\,\mathrm{kg\,ha^{-1}\,yr^{-1}}$ in black spruce, and $8390\,\mathrm{kg\,ha^{-1}\,yr^{-1}}$ in balsam poplar. In both the white and black spruce ecosystems the moss production is greater than that of either shrubs or trees (Van Cleve *et al.*, 1983).

4.11 Causes of timberline

Understanding the controls on northern and elevational treelines in the taiga is complicated by the movements of the position of treeline over the last 15 000 years (Ritchie, 1976; Whitehead, 1973). At about 8500 BP the forest treeline was 100 km further to the north on the Tuktoyaktuk Peninsula in the northwest district of Mackenzie than it is today (Ritchie and Hare, 1971). These forests were present until 5500 BP at Tuktoyaktuk and until 4800 BP in central Keewatin (Nichols, 1975). This northern extension of the taiga may have been due to higher mean summer temperatures of $+5\,^{\circ}\mathrm{C}$ near Tuktoyaktuk (Ritchie and Hare, 1971) and $+4\,^{\circ}\mathrm{C}\pm 1\,^{\circ}\mathrm{C}$ in central Keewatin (Nichols, 1975).

Owing to elevated temperatures and climatic change during the Hypsithermal period, the genus *Picea* probably extended 300 km north of its present position in Keewatin and reached the Beaufort Sea in the District of Mackenzie. Between 5500 BP and 4000 BP in the District of Mackenzie and after 4800 BP in Keewatin, treeline retreated to the south. By 2100 BP the treeline was 100 km south of its present position in the Keewatin. Treeline also advanced between 1200 and 1000 BP and then retreated to its present position in the Keewatin in about 600 BP. Extensive fires between 3500 and 600 BP also resulted in forest replacement by tundra (Nichols, 1975).

Changes in treeline have left outliers which are relic stands outside the current equilibrium distribution of spruce (Larsen, 1965; Nichols, 1975). Conversely, there are areas where the tundra appears to be invaded by forest (Griggs, 1934; Hansen and Chant, 1971; Marr, 1948). However, these areas may be confined to localized areas of disequilibrium.

Many factors influence the position of treeline in addition to past climatic changes and glaciation. The effect of fire must be considered as well as seed production, germination requirements, and physiology of seedlings and adult plants. However, poor photosynthetic performance does not appear to be a significant factor in explaining treeline in the subarctic (Auger, 1974; Vowinckel *et al.*, 1975). Photosynthetic rates may not limit carbon assimilation, but other factors such as nutrients, growing season length, or temperature may limit canopy development.

The $10\,^{\circ}\mathrm{C}$ isotherm for the warmest month of the year has been correlated with northern treeline (Löve, 1970). Hare (1950) used the 300 mm isopleth of annual potential evapotranspiration to indicate the limit of forest growth. Jeffree (1960) included both winter temperature and precipitation in his model to explain treeline. Bryson (1966) indicated the importance of the mean position of the Arctic frontal system during the summer in controlling tree distribution; he indicated that tree growth forms were found in areas dominated by temperate air masses in the summer. Since moisture is usually adequate in the north, temperature may be a controlling factor in tree distribution. The physiological impact of these correlations has yet to be proven.

Winter conditions can have marked effects on tissue loss and the carbon balance of trees. Winter desiccation, insolation, snow blasting, rime ice accumulation, branch movement, and abrasion can all markedly affect tissue mortality and reduce photosynthetic area, and therefore can affect plant carbon balance and growth rate. Werren (1978) found that net growth was markedly reduced in areas of greater exposure where tissue attrition was high. He concluded that biomass loss rather than reduced production restricts spruce growth, especially during the seedling establishment phase. Tree growth may also be limited by poor soil development, low available nitrogen, less stable soil moisture conditions, shorter growing periods, and increased soil cryogenic activity (Werren, 1978).

The inability of tree seeds to germinate or become established may also limit the northern distribution of trees. Temperatures below 15 °C limit seed germination (Black and Bliss, 1980; Fraser, 1970). Black and Bliss found a 4 °C temperature gradient over a 135 km transect south of the contemporary treeline. They felt that germination and establishment were limited by the temperature for a distance of about 40 km south of the present treeline. They also felt that regeneration could only occur during a 1–8 yr period following fire. Stand reproduction may be reduced by fire intervals of less than 100 yr or greater than 200 yr. Seed samples 1 yr old showed increasing germination of 0–7% over a north to south latitudinal range of about 1°.

4.12 Fire ecology

Since the summers are wet and the winters have snow cover for 4–8 mon, only the early spring and late fall have conditions favorable for fires, but in fact most fires in the taiga occur in midsummer (Heinselman, 1981). Across the taiga in North America, precipitation increases towards the south and east; the northwestern area of Canada and all of Alaska are drier and more prone to fire. The longer, sunny summer days in the northwest dry the crowns and understory of the forest and create conditions favorable to the occurrence of fire.

Fires in the taiga are either crown or surface fires (Fig. 4.12). Fire type and frequency are related to topography and community type (Rowe *et al.*, 1974). Upland and basin spruce burn with severe crown or surface fires; pine and mixed forest on level topography commonly have creeping surface fires;

and floodplain white spruce forests burn infrequently. Mean intervals between fires in the areas studied were 83–92 yr for uplands and basins, about 23 yr on more level terrain, and over 300 yr on floodplain sites. Only the severe crown or surface fires of the upland or basin spruce commonly result in the destruction of the organic and moss layers. Glacial deposits have yielded charcoal and the cones of jack pine and black spruce, suggesting that a similar fire regime may have existed many thousands of years ago (Heinselman, 1981).

Ignition has historically been by lightning, although human encroachment and development in the forest have resulted in more frequent man-caused fires, especially in the eastern forests where commerical exploitation of the forest is heavy. However, vast tracts of land are still so remote as to be practically free of man-made fires, and fire suppression efforts in developed areas has reduced the number of natural fires. The impact of man on the extent or frequency of burns is not clear. Of 6101 fires recorded in Alaska from 1950 to 1969, 70.5% were man-caused and 29.5% caused by lightning. Because man-caused fires usually occur close to fire suppression facilities, they burned only 22% of the total area destroyed (Barney, 1971).

The reproductive biology of many taiga tree species reflects their adaptation to frequent, yet patchy fire. Several conifers, particularly black spruce, jack pine, and lodgepole pine, have persistent, serotinous or semi-serotinous cones and early sexual maturity that favor their survival under fire frequency regimes as short as 20 yr. Aspen, balsam poplar, birch, and many herbs and shrubs survive fire well because of their capacity to resprout after a fire. In areas missed by a burn, surviving birch and aspen, and to a lesser extent larch and white spruce, are able to rapidly colonize nearby burned areas with light, wind dispersed seed (Zasada, 1971).

4.13 Establishment and reproduction

The complexity and variation in taiga plant communities and their associated environments makes generalization about reproductive biology and seedling establishment extremely difficult. However, certain patterns do emerge within the framework of

Figure 4.12 A controlled burn in the black spruce–moss forest of interior Alaska. The common spread of the fire into the tree crowns is demonstrated. This type of site commonly regenerates directly to the same forest type. Hardwood invasion only occurs on sites where the forest floor material is consumed by the fire and mineral soil is exposed.

Figure 4.13   An aerial view of a meandering river in central Alaska.

a few key communities. The disturbance cycle of erosion/deposition dominates the floodplains of interior Alaska as riverbeds meander, eroding existing vegetation on the outside of bends and depositing silt on the inside of river bends (Fig. 4.13). The pioneering species produce large numbers of short-lived, wind and/or water-dispersed seeds which are capable of colonizing the scattered sites of bare alluvium (Zasada and Viereck, 1975). Germination and establishment requirements of these pioneering species, predominantly alder (*Alnus tenuifolia*), balsam poplar (*Populus balsamifera*), and willow (*Salix* spp.), are minimal in terms of nutrients, but they demand an unshaded substrate and high soil temperatures. Pioneer species are also uniformly tolerant of flooding (Van Cleve and Viereck, 1981) and in the case of alder are capable of nitrogen fixation. There are limits to their tolerance; newly established stands are easily destroyed by deposition of silt

and/or disturbance by sheet ice carried in the floods of spring breakup. Only balsam poplar survives heavy silt deposition by producing adventitious roots which appear along the stem within the fresh layers of silt.

In the uplands of central Alaska, slopes and areas away from active water courses are not susceptible to extensive disturbance by water. The seedbed is usually organic litter of varied thickness rather than organic soil. The organic layer of the forest floor can be a thin layer of leaf litter or continuous layers of litter with living feather moss up to 30 cm deep.

In the earliest stages of succession, light-seeded hardwood species colonize open sites with mineral soil. Hardwood species are followed by white spruce. The spruce germinate in the shade and litter of the hardwood canopy and eventually as the spruce canopy develops they create even stronger shade. Mosses are represented discontinuously under the

Figure 4.14 A close view of the forest floor in an aspen-dominated forest. Note the low incidence of moss, being mostly restricted to the base of trees, semi-protected from the heavy litter fall of the deciduous trees.

hardwoods; they are usually clustered at the bases of tree trunks a few centimetres above the litter layer (Fig. 4.14). Mosses may be intolerant of litterfall (Grime, 1979; Viereck, 1970). A continuous but thin litter and moss layer forms under the spruce. Once a litter layer is developed, hardwood reproduction becomes predominantly vegetative by root sprouts, except when mineral soil is exposed by disturbance. The subshrubs and perennials present under a spruce canopy persist primarily by vegetative reproduction via underground shoots and rhizomes.

With time, most poorly drained floodplain and sloping sites become black spruce forests with a thick moss/litter carpet and permafrost dominated soils. Black spruce forests are most common on north-facing sites or those that receive limited amounts of solar irradiance. Under these conditions the potential seedbed consists of a continuous carpet of moss. Few seeds germinate on such a substrate, and most reproduction is by vegetative means. Even black spruce reproduce vegetatively and form new individuals by layering from lower branches buried in the moss. It is not clear when or if layered branches become physiologically distinct from the parent. Layering may only involve adventitious rooting.

Fire also plays an important role in species establishment. A fire may not completely clear the litter layer. A fire in a spruce forest or other forest type with a moss-dominated understory often only superficially burns the moss layer leaving it largely intact. The moisture regime prior to the fire is of great importance in determining the extent of the

Figure 4.15 The very commonly observed mosaic pattern of hardwood and spruce forests in interior Alaska. This pattern is fire mediated, the hardwoods invading continuous stands of conifers only when the forest floor is completely consumed by fire.

damage to the moss understory. If it is not completely removed, the moss/spruce canopy tends to reproduce itself with patches of hardwoods appearing only where the fire was hottest and a suitable seedbed was prepared (Fig. 4.15).

4.14 Succession

The size of the taiga with its many vegetation zones does not foster a singular successional pattern to a climax vegetation. Not only are there distinct successional patterns among the four vegetation zones, but within each zone there are also differences in succession on a regional basis (Larsen, 1980; Van Cleve and Viereck, 1981). The final stage of succession in each vegetation zone can be even further subdivided according to topography, based on drainage type and site water retention patterns.

Since disturbance which resets the succession sequence occurs frequently, especially as wildfire and less frequently as flooding, wind-throw, or insect damage, it has been suggested that a true climax may never be observed (Larsen, 1980).

Successional patterns in the taiga often consist of short cycles, with community types self replacing themselves after disturbance. In the black spruce–feather moss communities, the buildup of thick organic layers and cold, permafrost-dominated soils reduce the likelihood of the site recovering after disturbance to anything other than spruce–feather moss (Fig. 4.16). The thick organic and moss layers are seldom completely consumed in a fire due to their high moisture contents. If these layers are not completely removed, their insulating effect is retained and soils remain cold. Cold, organic soils are not favorable to the germination and

Figure 4.16   A winter view of a muskeg site in Alaska. Muskeg is typical of poorly drained, cold sites. In later stages of succession most of the spruce will die and *Sphagnum* will dominate the site.

establishment of most taiga plants. Black spruce and moss are able to reestablish rapidly after a fire. Even after a crown fire, the serotinous cones of black spruce remain to provide an ample seed source for stand replenishment. Moss propagules are widely disseminated, which also facilitates the black spruce–feather moss autosuccessional sequence.

Perhaps the longest successional series of species occurs on floodplains which are eroded by meandering rivers. Such cycles are common and have been documented especially in central Alaska (Van Cleve and Viereck, 1981; Viereck, 1970, 1975). Some areas are completely destroyed by rivers, while in other areas freshly deposited alluvium is available to pioneering vegetation in the classic sense of primary succession.

Van Cleve and Viereck (1981) document the pattern and process of succession on the floodplains of central Alaska. The first species to appear on exposed sandbars are pioneering shrubs, like willows and alder, *Equisetum*, and hardy perennials. After 2–5 yr, shrubs provide nearly 40% of the cover. Longer lived and potentially larger statured species, which later dominate the forest, also become established in the first five years. All the available mineral soil, which is necessary for seedling establishment, is exploited after 5 yr.

The earliest trees that invade a floodplain site are mostly balsam poplar (Fig. 4.17). They are tolerant of frequent flooding and gradually overtop the shrub and herbaceous layers. After initial establishment 20–30 yr are required for the balsam poplar canopy to close over the shrubs. When the canopy closes, shading is intensified, and the shrub layer is drastically altered. The pioneer shrubs begin to decline, and more shade-tolerant species appear. Shrub cover decreases to about 10% under the closed poplar canopy.

Continued, but more infrequent, flooding provides silt as a mineral seedbed on the floodplain site. White spruce can become established on this alluvium (Zasada and Gregory, 1969). Spruce persist only as long as flooding is mild; spruce cannot survive heavy siltation. The shift from balsam poplar to white spruce dominated forest takes up to 100 yr, and presupposes establishment of a white spruce understory. As white spruce mature, their dominance of the canopy is enhanced by the death of older, diseased balsam poplar. Patches of mineral soil produced by toppling of wind-thrown trees provides an ideal site for further white spruce establishment. The canopy at this stage is relatively open, but it soon closes as the white spruce crowns enlarge. Shrub distribution becomes patchy and the

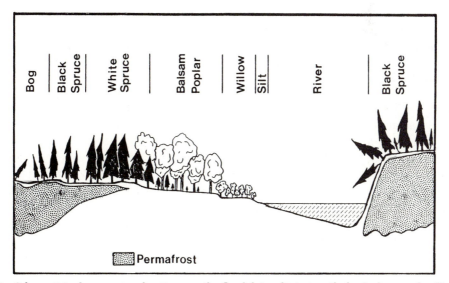

Figure 4.17 Schematicized successional pattern on the floodplain of interior Alaska (redrawn after Van Cleve and Viereck, 1981). See text for details of species and community replacement.

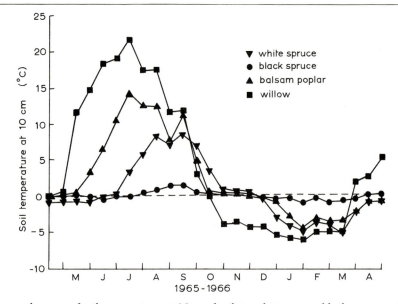

Figure 4.18   Seasonal course of soil temperature at 10 cm depth in white spruce, black spruce, willow, and balsam poplar stands in interior Alaska (data from Viereck, 1970). Data from 1965/66 season.

forest floor is eventually covered by feather mosses, particularly *Hylocomium* sp. and *Pleurozium* sp. After the initial deposition of a sandy substrate, 120–140 yr are required for a mature white spruce–moss community to develop. With continued moss growth, development of a deep soil organic layer, and decreasing soil temperatures (Fig. 4.18), white spruce are replaced by black spruce. The black spruce forest is one of the most nutrient impoverished and has the lowest productivity of the various vegetation types in the floodplain successional series.

4.15 Summary and research needs

The physiological ecology of the taiga is only beginning to be studied, much less understood. The widespread distribution of most species in the taiga and the natural variations in climatic and disturbance regimes add to the intrigue and challenge of studying this ecosystem. The recent and rapid reinvasion of previously glaciated areas and the occurrence of the same species as dwarf or krummholtz forms in permafrost-dominated bogs and at arctic treeline and as full statured trees with closed canopies in temperate forest, indicates the tremendous adaptability within the taiga species.

Little research has been done on the belowground component of the taiga. The belowground component of an ecosystem is very difficult to study, but the belowground activity, carbon costs of root systems, and nutrient acquiring capacities of species across soil temperature, successional, and disturbance gradients must be explored.

The floodplains of interior Alaska are ideal sites for future taiga research. Information is available on floodplain productivity and the successional sequence after disturbance (Van Cleve and Viereck, 1981). The existence of erosional disturbance, which leads to primary succession on alluvial sites, of fire disturbance with its shorter species replacement cycles, and of slower secondary successional patterns offer the opportunity for research on recovery and succession in a variety of sites. The mosaic pattern of the vegetation, which is the result of the patchiness of disturbance, offers the opportunity for research across various age classes and stages of community development which occur naturally on floodplains within the taiga.

Research is also needed in the more northern taiga systems and especially of the forest–tundra ecotone. Little information is available about the control of species distribution at extremes of their ranges.

References

Andrews, J.T. (1975) *Glacial systems. An approach to their environments*, Duxbury Press, North Scituate, MA, 191 pp.

Auger, S. (1974) *Growth and photosynthesis of Larix laricina (Du Roi) K. Koch in the subarctic at Schefferville. Quebec*. MSc Thesis, McGill University, Montreal, Quebec, Canada. 114 pp.

Barashkova, E.A. (1971) Photosynthesis in fruticose lichens *Cladonia alpestris* (L.) Rabh. and *C. rangiferina* (L.) Web. in the Taimyr Peninsula. (Fotosintez kustistykh lishaynikov *Cladonia alpestris* (L.) Rabh. i *C. rangiferina* (L.) Web. na Taymyre.) (Far Northern Agricultural Research Institute, Noril'sk, 1966.) IBP International Tundra Biome Transl. 4 (Translator: G. Belkov. National Research Council of Canada.) 10 pp.

Barney, R.J. (1971) Wildfires in Alaska – some historical and projected effects and aspects, in *Fire in the Northern Environment – A Symposium*, Pacific Northwest Forest and Range Expl. Station, USDA Forest Service.

Berry, J. and Bjorkman, D. (1980) Photosynthetic response and adaptation in temperature in higher plants, in *Annual review of plant physiology* (eds W.R. Briggs, P.B. Green, and R.L. Jones), Annual Review Inc. Palo Alto, CA, pp. 491–543.

Bigger, M. and Oechel, W.C. (1982) Nutrient effect on maximum photosynthesis in arctic plants. *Holarctic Ecology*, **5**, 158–63.

Black, R.A. and Bliss, L.C. (1980) Reproductive ecology of *Picea mariana* (Mill) B.S.P., at tree line near Inuvik, Northwest Territories, Canada, *Ecological Monographs*, **50**, 331–54.

Bliss, L.C. and Cantlon, J.E. (1957) Succession on river alluvium in northern Alaska. *American Midland Naturalist*, **58**, 452–69.

Brown, R.J.E. (1970) Permafrost as an ecological factor in the subarctic, in *Ecology of the Subarctic Regions*. Proceedings of the Helsinki Symposium, UNESCO, Paris, pp. 129–40.

Bryson, R.A. (1966) Airmasses, streamlines, and the boreal forest. *Geographical Bulletin*, **8**, 228–69.

Bryson, R.A. and Hare, F.K. (1974) Climates of North America, in *World Survey of Climatology*, Vol. 11, Elsevier Scientific, Amsterdam–London–New York, pp. 1–47.

Carstairs, A. and Oechel, W.C. (1978) Effects of several microclimate factors and nutrients on net carbon dioxide exchange in *Cladonia alpestris*. *Arctic Alpine Research*, **10**, 81–94.

Dyrness, C.T. and Grigal, D.F. (1979) Vegetation-soil

relationships along a spruce forest transect in interior Alaska. *Canadian Journal of Botany*, **57**, 2644–56.

Elliot, D.L. (1979) The current regenerative capacity of northern Canadian trees, Keewatin, NWT, Canada: some preliminary observations. *Arctic and Alpine Research*, **11**, 243–51.

Fraser, J.W. (1970) *Cardinal temperatures for germination of six provenances of white spruce seed*, Dept. Fisheries and Forestry, Canadian Forest Service, Ottawa, Canada, Publ. No. 1290.

Fraymouth, J. (1928) The moisture relations of terrestrial algae. III. The respiration of certain lower plants, including terrestrial algae, with special reference to the influence of drought. *Annals of Botany*, **42**, 75–100.

Griggs, R.F. (1934) The edge of the forest in Alaska and the reasons for its position. *Ecology*, **15**, 80–95.

Grime, J.P. (1979) *Plant strategies and vegetation processes*, Wiley, New York.

Hansen, R.I.C. and Chant, D.A. (1971) Changes in the northern limit of spruce at Dubawant Lake, Northwest Territories. *Arctic*, **24**, 233–4.

Hanson, H. (1953) Vegetation types in northwestern Alaska and comparisons with communities in other arctic regions. *Ecology*, **34**, 111–40.

Hare, F.K. (1950) Climate and zonal divisions of the boreal forest fromation in eastern Canada. *Geographical Review*, **40**, 615–35.

Hare, F.K. and Hay, J.E. (1974) The climate of Canada and Alaska, in *Climates of North America*, Vol. 11, in *World Survey of Climatology*, Elsevier Scientific, Amsterdam–London–New York, pp. 49–192.

Hare, F.K. and Ritchie, J.C. (1972) The boreal bioclimates. *Geographical Review*, **62**, 333–65.

Hay, J.E. (1969) Aspects of heat and moisture balance in Canada. PhD Thesis, University of London.

Heinselman, M.L. (1981) Fire intensity and frequency as factors in the distribution and structure of northern ecosystems, in *Fire Regimes and Ecosystem Properties*. Proceedings of the conference. USDA Forest Service General Tech. Rep. WO–26, pp. 7–57.

Hicklenton, P.R. and Oechel, W.C. (1976) Physiological aspects of the ecology of *Dicranum fuscens* in the subarctic. I. Acclimation and acclimation potential of CO_2 exchange in relation to habitat, light and temperature. *Canadian Journal of Botany*, **54**, 1104–19.

Hom, J.L. and Oechel, W.C. (1983) Photosynthetic capacity, nutrient content and nutrient use efficiency of different needle age classes of black spruce (*Picea mariana*) found in interior Alaska. *Canadian Journal of Forest Research*, **13**, 834–9.

Hopkins, D.M. (1959) Some characteristic of the climate in forest and tundra regions in Alaska. *Arctic*, **12**, 214–20.

Hosie, R.C. (1969) *Native trees of Canada*, Canadian Forestry Service, Department of Fisheries and Forestry, Ottawa, 380 pp.

Hultén, E. (1937) *Outline of the history of arctic and boreal biota during the quaternary period*, Bokforlags Aktiebolaget Thule, Stochkholm, 168 pp.

Hultén, E. (1968) *Flora of Alaska and neighboring territories. A manual of vascular plants*, Stanford University Press, Stanford, CA.

Hustich, I. (1949) Phytogeographical regions of Labrador. *Arctic*, **2**, 36–42.

Hustich, I. (1953) The boreal limits of conifers. *Arctic*, **6**, 149–62.

Jeffree, E.P. (1960) A climatic pattern between latitudes 40° and 70°N and its probable influence on biological distributions. *Proceedings of the Linnaean Society, London*, **171**, 89–121.

Kershaw, K.A. (1977) Physiological-environmental interactions in lichens, *New Phytologist*, **79**, 377–421.

Kershaw, K.A. and Field, G.F. (1975) Studies on lichen-dominated systems, XV. The temperature and humidity profiles in a *Cladonia alpestris* mat. *Canadian Journal of Botany*, **53**, 2614–20.

Kimball, S.L., Bennet, B.D. and Salisbury, F.B. (1973) The growth and development of montane species at near-freezing temperatures. *Ecology*, **54**, 168–73.

Landsberg, H.E. (1961) Solar radiation at the earth's surface. *Solar Energy*, **5**, 95–8.

Lange, O.L. and Kappen, L. (1972) Photosynthesis of lichens from Antarctica, in *Antarctic Terrestrial Biology* (ed. G.A. Llano), Antarctic Research Series, Vol. 20, American Geophysical Union, pp. 83–95.

La Roi, G.H. (1967) Ecological studies in the boreal spruce-fir forests of the North American taiga. I. Analysis of the vascular flora. *Ecological Monographs*, **37**, 229–53.

Larsen, J.A. (1965) The vegetation of the Ennadai Lake area, NTW: Studies in arctic and subarctic bioclimatology. *Ecological Monographs*, **35**, 37–59.

Larsen, J.A. (1980) *The boreal ecosystem*, Academic Press, New York, 500 pp.

Larsen, D.W. and Kershaw, K.A. (1975) Measurement of CO_2 exchange in lichens: a new method. *Canadian Journal of Botany*, **53**, 1535–41.

Lawrence, W.T. (1983) Soil temperature effects on carbon exchange in Taiga species of interior Alaska. PhD dissertation, San Diego State University and the University of California, Davis.

Lawrence, W.T., Cox, G.W. and Oechel, W.C. Photosynthesis of balsam poplar (*Populus balsamifera*) in the forest tundra. (In preparation).

Lawrence, W.T. and Oechel, W.C. Upward translocation of nutrients through feather moss understories. (In preparation).

Lawrence, W.T. and Oechel, W.C. Photosynthesis of interior Alaska taiga trees across soil temperature gradients among several community types. (In preparation).

Lawrence, W.T. and Oechel, W.C. (1983a) Effects of soil temperature on the carbon exchange of taiga seedlings: I. Root respiration. *Canadian Journal of Forest Research*, **13**, 840–9.

Lawrence, W.T. and Oechel, W.C. (1983b) Effects of soil temperature on the carbon exchange of taiga seedlings: II. Photosynthesis, respiration, and conductance. *Canadian Journal of Forest Research*, **13**, 850–9.

Lechowicz, M.J. (1978) Carbon dioxide exchange in *Cladina* lichens from subarctic and temperate habitats. *Oecologia*, **32**, 225–37.

Lechowicz, M.J. and Adams, M.S. (1974) Ecology of *Cladonia* lichens. II. Comparative physiological ecology of *C. mitis*, *C. rangiferina*, and *C. uncialis*. *Canadian Journal of Botany*, **52**, 411–22.

Limbach, W.E., Oechel, W.C. and Lowell, W. (1982) Photosynthetic and respiratory responses to temperature and light of three Alaskan tundra growth forms. *Holarctic Ecology*, **5**, 150–7.

Löve, D. (1970) Subarctic and subalpine: where and what? *Arctic and Alpine Research*, **2**, 63–73.

Malloch, D. and Malloch, B. (1981) The mycorrhizal status of boreal plants: species from northeastern Ontario. *Canadian Journal of Botany*, **59**, 2167.

Marr, J.W. (1948) Ecology of the forest tundra ecotone on the east coast of Hudson Bay. *Ecological Monographs*, **17**, 117–44.

Nichols, H. (1975) Palynological and paleoclimatic study of the Late Quaternary displacement of the boreal forest-tundra ecotone in Keewatin and Mackenzie, NTW, Canada. Occasional Paper No. 15. Institute of Arctic and Alpine Research, University of Colorado, Boulder, CO.

Oechel, W.C. (1976) Seasonal patterns of temperature response of CO_2 flux and acclimation in arctic mosses growing *in situ*. *Photosynthetica*, **10**, 447–56.

Oechel, W.C. and Collins, N.J. (1976) Comparative CO_2

exchange patterns in mosses from two tundra habitats at Barrow, Alaska. *Canadian Journal of Botany*, **54**, 1355–69.

Oechel, W.C. and Lawrence, W.T. (1979) Energy utilization and carbon metabolism in mediterranean scrub vegetation of Chile and California. I. Methods; A transportable cuvette field photosynthesis and data acquisition system and representative results for *Ceanothus greggii*. *Oecologia*, **39**, 321–35.

Oechel, W.C., Hicklenton, P., Sveinbjornsson, B., Miller. P.C. and Stoner, W.A. (1975) Temperature acclimation of photosynthesis in *D. fuscescens* growing *in situ* in the arctic and subarctic, in *Proceedings of the Circumpolar Conference on Northern Ecology*. 15–18 September 1975, Ottawa, Ontario, Canada, National Research Council of Canada, pp. I-131 to I-144.

Oechel, W.C., Van Cleve, K. and Lawrence, W.T. Control of forest metabolism and nutrient cycling by a minor community component. (In preparation).

Olson, J.L. (1971) Primary productivity: Temperate forests, especially American deciduous types, in *Productivity of Forest Ecosystems. Ecology and Conservation 4*, (ed. P. Duvigneaud) UNESCO, Paris, pp. 235–58.

Peterson, W.L. and Mayo, J.M. (1975) Moisture stress and its effect on photosynthesis in *Dicranum polysetum*. *Canadian Journal of Botany*, **53**, 2897–900.

Pollard, D.F. (1972) Above-ground dry matter production in three stands of trembling aspen. *Canadian Journal of Forest Research*, **2**, 27–33.

Porsild, A.E. (1957) *Illustrated flora of the Canadian Arctic Archipelago*, Bull. 146, National Museum of Canada, Ottawa, reprinted 1973, 209 pp.

Prudhomme, T.I. (1983) Carbon allocation to anti-herbivore compounds in a deciduous and an evergreen subarctic shrub species. *Oikos*, **40**, 344–56.

Pruit, W.O. (1978) *Boreal ecology*, Institute of Biology, Studies in Biology No. 91, Edward Arnold, London, 73 pp.

Pulling, H.E. (1918) Root habit and plant distribution in the far north. *Plant World*, **21**, 223–33.

Ritchie, J.C. (1976) The Late-Quaternary vegetational history of the western interior of Canada. *Canadian Journal of Botany*, **54**, 1793–818.

Ritchie, J.C. and Hare, F.K. (1971) Late-Quaternary vegetation and climate near the arctic tree line of northwestern North America. *Quaternary Research*, **1**, 331–42.

Rowe, J.S. Bergsteinsson, J.L., Padbury, B.A. and Hermesh, R. (1974) *Fire studies in the Mackenzie Valley*, Canadian Department of Indian Affairs and Northern Development INA Publ, QS-1567-000-EE-Al, ALUR. 73–74–61, 123 pp.

Sestak, Z., Catsky, J. and Jarvis, P.G. (1971) *Plant photosynthetic production. Manual of methods*, Junk, The Hague.

Skre, O. and Oechel, W.C. (1981) Moss functioning in different taiga ecosystems in interior Alaska. I. Seasonal, phenotypic, and drought effects on photosynthesis and response patterns. *Oecologia*, **48**, 50–9.

Skre, O., Oechel, W.C. and Miller, P.M. (1983) Moss leaf water content and solar radiation at the moss surface in a mature black spruce forest in central Alaska. *Canadian Journal of Forest Research*, **13**, 860–8.

Smith, E.M. and Hadley, E.B. (1974) Photosynthetic and respiratory acclimation to temperature in *Ledum groenlandicum* populations. *Arctic and Alpine Research*, **6**, 13–27.

Stålfelt, M.G. (1937) Der Gasaustausch der Moose. *Planta*, **27**, 30–60.

Tout, D.G. (1963) *The climate of Knob Lake*. MSc Thesis, McGill University, 242 pp.

Tout, D.G. (1964) The climate of Knob Lake. McGill Subarctic Station Paper No. 17, pp. 1–236.

Van Cleve, K., Dyrness, C.T., Viereck, L.A. Fox, J., Chapin, F.S. III, and Oechel, W.C. (1983) Taiga ecosytems in interior Alaska. *BioScience*, **33**, 39–44.

Van Cleve, K. and Viereck, L.A. (1981) Forest succession in relation to nutrient cycling in the boreal forest of Alaska, in *Forest succession concepts and application* (eds O.C. West, H.H. Shugart and D.B. Botkin), Springer-Verlag, New York, pp. 185–211.

van Zindern Bakker, E.M. Jr (1974) *An ecophysiological study of black spruce in Central Alberta*. PhD Dissertation, University of Alberta, Edmonton, Alberta. Canada.

Viereck, L.A. (1970) Forest succession and soil development adjacent to the Chena River in interior Alaska. *Arctic and Alpine Research*, **2**, 1–26.

Viereck, L.A. (1975) Forest ecology of the Alaska taiga, in *Proceedings of the circumpolar conference on northern ecology*. National Resource Council of Canada and Scientific Committee on Problems of the Environment, Ottawa, pp. I-1 to I-22.

Viereck, L.A. and Dyrness, C.T. (1980) *A preliminary classification system for vegetation of Alaska*, USDA Forest Service, Pacific NW Forest and Range Experimental Station, Gen. Tech. Rep. PNW-106, 38 pp.

Viereck, L.A. and Little, E.L. (1972) *Alaska trees and shrubs*, US Department of Agriculture, Agriculture Handbook 410, 265 pp.

Vowinckel, T., Oechel, W.C. and Boll, W.G. (1975) The

effect of climate on the photosynthesis of *Picea mariana* at the subarctic treeline. 1. Field measurements. *Canadian Journal of Botany*, **53**, 604–20.

Walter, H. (1979) *Vegetation of the earth and ecological systems of the geo-biosphere*, 2nd edn, Springer-Verlag, New York, 274 pp.

Walter, H., Harnickell, E. and Mueller-Dombois, D. (1975) *Climate diagram maps of the individual continents and the ecological regions of the earth*, Springer-Verlag, Berlin–Heidelberg–New York, 36 pp. + 9 maps.

Werren, G.L. (1978) *Winter stress in subarctic spruce associations: A Schefferville case study.*, MSc Thesis, McGill University, Department of Geology, Montreal, Quebec, Canada.

Whitehead, D.R. (1973) Late-Wisconson vegetational changes in unglaciated eastern North America.

Quaternary Research, **3**, 621–31.

Zasada, J.C. (1971) Natural regeneration of interior Alaska forests – seed, seedbed, and vegetative reproduction characteristics, in *Proceedings, Fire in the Northern Environment – A Symposium*, Pacific SW Forest and Range Experimental Station, USDA Forest Service, Fairbanks, AK, pp. 231–46.

Zasada, J.C. and Gregory, R.A. (1969) *Regeneration of white spruce with reference to interior Alaska: A literature review*. US Forestry Service Research paper PNW–79, 37 pp.

Zasada, J.C. and Viereck, L.A. (1975) The effect of temperature and stratification on germination in selected members of the Salicaceae in interior Alaska. *Canadian Journal of Forest Research*, **51**, 335–7.

5

Western montane forests

W.K. SMITH

5.1 Introduction

The Rocky Mountain, Sierra Nevada, Wasatch, and Cascade mountain ranges form the Cordilleran province (Gleason and Cronquist, 1964) and contain the Transition and Canadian zones of Merriam's early life zone classification (Merriam, 1898). In general, these temperate mountainous regions are dominated by evergreen coniferous forests that occur between elevations of about 2500–4200 m in the southwestern US and from 1700–3500 m near the US–Canada border. These forests are considered to be a southern extension of the transcontinental forests of Canada where lack of moisture and warmer temperatures has resulted in only relict forests on the tops of a few high desert mountains in the southwest USA. Although species diversity may be considerably greater in transition areas, nearly monospecific stands of conifers are often found over broad elevational limits.

A variety of habitats are distinctive within western montane forests. These include wet and dry meadows, rocky escarpments, lakes, ponds, and sphagnum bogs. A very limited amount of ecophysiological research has included plant species from these small-scale habitats (e.g. see Klikoff, 1965, for photosynthetic studies of meadow species of the Sierra Nevada mountains, USA). Moreover, these studies have dealt predominantly with boundaries and the physiological ecology of succession and reinvasion of tree species at the forest boundary. A

particularly high percentage of these studies has involved environmental–physiological explanations for the persistence of timberlines, especially the upper elevational limits of tree growth. This chapter separates the montane forest into understory and overstory species in order to recognize the environmental–physiological segregation between strata which is especially well developed in western coniferous forests. Within this overstory and understory categorization, individual species are discussed according to general elevational preference (Fig 5.1). Plant species considered in detail are those that have received the greatest attention in the literature. In most cases, these species are also those of greatest economic importance (such as many of the coniferous trees) even though the ecological or physiological significance of these species, in terms of specialized adaptations which they may express, may not necessarily warrant such an emphasis.

There are no summaries available concerning the physiological ecology or autecology of the western montane forest species. However, there is a recent autecological synthesis of primarily conifer species of the Pacific Northwest that provides a comparison of many of the species which are also abundant in the western montane forests (Minore, 1979). Table 5.1 is a ranking of autecological characteristics taken from Minore's synthesis for the tree species found in the northwest that also are abundant throughout the western montane forests. He concluded that much of the information compiled on

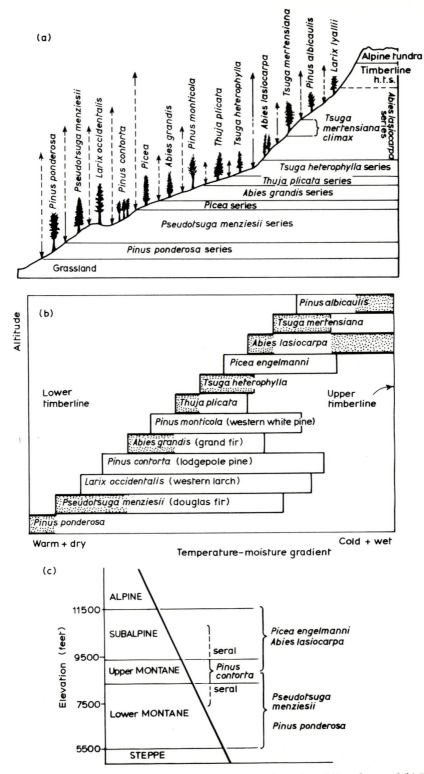

Figure 5.1 Representative vegetation zonation patterns for the northern (a and b) and central (b) Rocky Mountains, USA. Vertical bars in (a) designate elevational ranges (after Pfister *et al.*, 1977); hatched areas with dark borders in (b) represent hypothetical altitudinal range where reproduction is maintained under intense competition (after Daubenmire, 1956); (c) is for the mountains of Colorado and Wyoming (after Moir, 1969).

Table 5.1 Autecological rankings of several tree characteristics from greatest to least based on the synthesis of Minore (1979) for the Pacific Northwest of North America. Species listed are only those that are also abundant in the western montane forests.

Shade tolerance	Warm temperature tolerance	Frost tolerance	Drought tolerance	Moisture requirements	Excessive moisture tolerance
Tsuga heterophylla	Pinus ponderosa	Pinus contorta	Quercus kelloggii	Tsuga mertensiana	Pinus contorta
Taxus brevifolia	Quercus kelloggii	P. monticola	Pinus jeffreyi	Abies Procera	Thuja plicata
Thuja plicata	Pseudotsuga menziesii	Picea engelmannii	Pinus ponderosa	A. magnifica	Tsuga mertensiana
Tsuga mertensiana	Pinus lambertiana	Pinus ponderosa	P. contorta	A. concolor	Abies lasiocarpa
Abies lasiocarpa	Librocedrus decurrens	P. jeffreyi	Librocedrus decurrens	Populus trichocarpa	Sequoia sempervirens
A. grandis	Thuja plicata	Abies magnifica	Pseudotsuga menziesii	Abies grandis	Populus trichocarpa
A. concolor	Tsuga heterophylla	Tsuga mertensiana	Picea engelmannii	Thuja plicata	Umbellaria californica
A. magnifica	Abies grandis	Librocedrus decurrens	Abies grandis	Tsuga heterophylla	Pinus ponderosa
Pinus monticola	A. procera	Pinus lambertiana	Pinus lambertiana	Pseudotsuga menziesii	P. jeffreyi
Pseudotsuga menziesii	A. magnifica	Abies concolor	Larix occidentalis	Pinus ponderosa	Abies grandis
Pinus lambertiana	Tsuga mertensiana	A. lasiocarpa	Abies lasiocarpa	P. lambertiana	Tsuga heterophylla
Librocedrus decurrens	Abies lasiocarpa	A. grandis	Thuja plicata	Librocedrus decurrens	Pseudotsuga menziesii
Abies procera	Pinus albicaulis	Pseudotsuga menziesii	Pinus monticola	Quercus kelloggii	
Picea engelmannii	Pinus contorta	Thuja plicata	Abies concolor		
Pinus contorta		Tsuga heterophylla	Tsuga heterophylla		
Larix occidentalis			Abies magnifica		
Pinus ponderosa			Tsuga mertensiana		
Juniperus occidentalis					
Quercus kelloggii					

the northwest tree species is based on field observations and not field measurements or laboratory measurements under controlled experimental conditions. Although certain similarities may be expected among the physiological responses listed in Table 5.1 for the Pacific Northwest compared to the western montane species, care should be taken when making specific assumptions based on data from these very different geographical regions. In addition to Minore's autecological synthesis, a comprehensive bibliography of the effects of moisture, radiation, temperature and wind on forest trees prior to about 1972 is found in Anthony (1974).

5.2 Environment

Only one researcher (Baker, 1944) has attempted a comprehensive study of the climatic patterns found in the western United States. Winter snowfall in the western mountains originates in stormfronts that move in a southeasterly direction from the northwest Pacific states and Canada. Summer rainfall is more sporadic and often comes from southerly storms that originate in the Pacific Ocean (Sierra Nevada mountains) or the Gulf of Mexico (central and southern Rocky Mountains). Winter snow accumulation generally amounts to about 1–3 m water while summer rains add another 20–60 cm. Spring snowmelt begins in early May on southern exposures at low elevations and generally coincides with the onset of above-freezing nights, but may not begin until late July at higher elevations near timberline. Total frost-free nights range from less than 70 d to over 120 d depending primarily on elevation. Considerable soil drying may take place gradually through the summer resulting in surface soil water potential (10 cm depth) of less than -2.0 MPa in the more open stands such as lodgepole and ponderosa pine. Surface soils beneath the dense stands of spruce and fir at higher elevation may have much higher water potentials (> -0.5 MPa) throughout a given summer due partially to a prolonged snowpack and a lack of sunlight penetration to the forest floor (Smith, 1981). However, subsurface soil water may be strongly influenced by plant transpiration and potentially lead to dryer soils beneath a forested area than for adjacent open areas (Knapp and Smith, 1981). More detailed climatic information for the western mountains may be found in Baker (1944), especially temperature and precipitation patterns.

Although a substantial amount of research has characterized the physical–chemical environments within montane forests, the majority of this work was completed in areas outside the western forests of North America (see Geiger, 1965; Jarvis *et al.*, 1976; Reifsynder, 1955, 1967; Reifsynder and Lull, 1965; Rutter, 1968; Spurr and Barnes, 1980, for comprehensive reviews). As is the case for most montane forests, microclimate within the western forests is strongly influenced by changes associated with elevational increases and the substantial depletion of solar radiation and wind that occurs within the forest canopy. Only a few studies have attempted to comprehensively evaluate the quantitative changes in environmental parameters that are due solely to the general effects of elevational increase and which may also have important influences on plant ecophysiology (Gale, 1972, 1973; Smith and Geller, 1979). However, numerous studies have been concerned with specific microclimatic differences within the montane forests. As is the general case for the solar environments of other evergreen forests, a perennial leaf canopy results in a certain amount of seasonal constancy and the sunlight that penetrates to the forest understory may be predictably composed of relatively large and intense patches of sunlight, sunflecks of much shorter duration and lower irradiance, plus the diffuse and scattered radiation typical of shade light. Regardless, as substantial amounts of sunlight penetrate to the forest floor where wind speeds may be characteristically low, relatively large vertical and horizontal gradients in several microclimatic parameters can be anticipated according to canopy characteristics and tree spacing. For example, the open stands of ponderosa pine are an obvious contrast to the closed canopy of the 'doghair' stands of lodgepole pine or the dense stands of spruce and fir. Tree density within lodgepole pine forests may vary from less than 300 to more than 15 000 trees per hectare with the amount of leaf area per unit ground area varying from near 4 to over 14 (Moir and Francis, 1972). Despite the apparent differences in microclimate that occur in the western forests, very little information exists regarding microclimate variability and potential influences on community structure and function.

The high elevations associated with the western montane forests result in substantial increases in the amount of incident solar radiation. Montane forests between 2 and 4 km would be expected to receive from about 20% to 40% more incident solar irradiance, respectively, than at sea level on a clear day (Becker and Boyd, 1961). The solar spectrum at higher elevations also contains a larger proportion of the high energy ultraviolet wavelengths which could result in substantial effects on plant ecophysiology (Caldwell, 1981). Very little information exists regarding the effects of changes in the solar spectrum on the ecophysiology of western forest species.

A number of studies have considered the depletion of solar radiation within the montane forest canopy. Light penetration in a high elevation stand of firs and pines in the Sierra Nevada Mountains of California was found to be 18% of full sunlight for a crown density of 50% and only 6% for a full crown density (US Corps of Engineers, 1956). Wellner (1948) found that sunlight decreased to 10% of that in the open when basal tree area approached 400 ft^2 acre^{-1} in a western white pine (*Pinus monticola* Dougl.) stand. As has been found for other forests, radiation below the canopy is relatively enriched in red wavelengths (Atzet and Waring, 1970). Canopy gaps produce significant spatial and temporal patterning in the forest floor radiation environment which leads to variations in understory species composition. Young and Smith (1979, 1980, 1982) reported that the lodgepole pine and spruce–fir zones of SE Wyoming are characterized by the occurrence of relatively long-term sunpatches as well as the short-term sunflecks associated with wind generated trunk and branch movement. Sunpatches often lasted for nearly an hour with maximum solar irradiance approaching 80% of full sunlight. However, the most common sunpatch was generally 20–30 min in duration and about 40–60% of full sunlight irradiance. Also, cloud-cover during the summer months caused substantial reduction in total daily solar irradiance. Clouds were more frequent in the afternoon periods and resulted in an estimated 40% reduction in incident sunlight over the day. Cloudiness also resulted in greater diffuse sunlight levels within the understory compared to clear days (Young and Smith, 1983), including the photosynthetically active wavelengths (0.4–0.7 μm).

Air temperatures may decrease substantially with elevation in the western mountains due to the relatively dry air and, thus, high adiabatic lapse rates (Smith and Geller, 1979). Although air temperature may decrease more than 0.06 °C m^{-1} elevation increase, vertical profiles of air temperature at a given location may also be considerably steeper (near 3 °C m^{-1}), especially within the conifer canopy and near the forest floor (Bergen, 1972, 1974a; Fowells, 1948; Maguire, 1955). In summer, the canopy and understory areas may remain cooler than above-canopy air temperatures during early morning, while a reversed situation may exist during early evening. Virtually all periods with maximum temperature profiles are concomitant with periods of low windspeeds that may be less than 20 cm s^{-1} (Bergen, 1971, 1974a,b, 1977; Marston, 1956; Smith, 1979; Young and Smith, 1980). Windspeed may be five-fold lower inside the conifer canopy and often over ten times lower within the understory within a meter of the soil surface.

Ambient humidity and CO_2 may be low during both summer and winter with relative humidities often below 20% and vapor densities of greater than 15 g m^{-3} between the leaf and surrounding air (Baldwin and Barney, 1976; Knapp and Smith, 1981; Running, 1980). As reflected in the air-temperature profiles, absolute humidities in the understory and conifer canopy may remain substantially higher than above-canopy values in the early morning and evening when windspeeds are low. Understory plants exposed to large patches of direct sunlight often experienced vapor density deficits of only about 5 g m^{-3} in the early morning compared to over 20 g m^{-3} during midday (Smith, 1981; Young and Smith, 1979, 1980).

Although the amount of CO_2 in the ambient air is expected to decrease with elevation, only a few studies have considered the influence of variations in ambient CO_2 levels on the ecophysiology of the montane forest species. Mooney *et al.* (1966a) reported that reduced levels of ambient CO_2 did not appear to influence CO_2 uptake rates in *Oxyria* races at two different elevations and Gale (1972) provides evidence that the more rapid diffusion of CO_2 at low ambient pressures could compensate for the low concentrations at higher elevations. Also, considerable variation in CO_2 concentrations may exist for high elevation sites in the Rocky Mountains, with relatively high values (> 20% of ambient) within about 20 cm of the ground surface (Mooney *et al.*, 1966a; Smith and Young, unpublished data).

Very little information is available concerning variations in edaphic factors such as soil temperature, moisture, and nutrients in the montane forests. Soil surface temperature in summer may vary by over 30 °C during the midday period between fully shaded and sunpatch areas while subsurface temperatures are much more constant, especially under thick litter layers (Baldwin and Barney, 1976; Smith and Young, unpublished data). Subsurface soil appears to freeze only rarely within the forest due to a substantial snowpack early in the fall. Heat conducted from deeper in the soil generates surface temperatures that remain above freezing throughout much of the winter. However, exposed soils in the center of the larger wind-blown meadows could be expected to freeze to depths of over 20 cm.

Summer patterns of soil moisture within the forest are also coupled to the amount of winter snow deposition and other variables such as penetration and percolation, the depth of the litter layer, and the degree of canopy closure and sunlight penetration. Soil water potentials as shallow as 2 cm remained above − 0.2 MPa throughout the summer in a high elevation spruce–fir forest in SE Wyoming while an adjacent meadow area approached − 0.8 MPa at 2 cm by mid-July (Smith and Young, unpublished data). At a site about 1000 m lower within a relatively open lodgepole pine stand corresponding water potential values were − 1.7 MPa in the forest and − 2.6 MPa in a nearby open area. However, subsurface soils (> 5 cm depths) may be considerably wetter during summer in open areas compared to beneath a conifer overstory, possibly due to plant transpirational demands (Knapp and Smith, 1981).

Soil structure and nutrient content is extremely variable within the western montane forests with soil depths ranging from < 0.5 to > 3.0 m (Knight *et al.*, 1983). Also, root growth is often restricted to relatively shallow depths (< 40 cm) where considerable soil drying may occur during summer (Dahms, 1971; Johnston, 1975; Morgen, 1973). The abundance and availability of nitrogen is low in these coniferous forests and may serve as a limiting factor for tree growth (Fahey, 1977; Gosz, 1980), along with potassium and phosphorus (Fahey, 1983). Continued decomposer activity and mineralization beneath the snowpack during winter due to above-freezing soil temperatures may be important to the annual nutrient cycles of the western forests (Stark, 1972).

In general, the western montane forests occur at higher elevations and are relatively dry compared to more mesic coniferous forests along the western coasts of North America and throughout eastern North America. Smith and Geller (1979) provided a comprehensive evaluation of the influence of elevation on transpiration and predicted that transpirational demand would be similar between high elevation species of the central Rocky Mountains and desert species of the low, hot deserts of the southwest US. Moreover, the high transpiration rates predicted for high elevation species occurred at much lower air and leaf temperatures for the montane species (15–20 °C) compared to the desert species (30–37 °C). The higher transpirational demand for high elevation species was due to the greater leaf-to-air temperature difference and the more rapid diffusion of H_2O from the leaf at the lower ambient pressures of high elevations. This greater leaf-to-air temperature difference was due to the greater solar irradiance and lower air temperature with increasing elevation. Elevational effects on transpiration may be particularly pronounced for understory species with broad leaves compared to the narrow, needle-like leaves of the overstory conifers. The higher windspeeds at the overstory level could also tend to reduce evaporation demand of sunlit conifer trees via convective leaf cooling, while understory plants experience considerably less wind (Smith and Geller, 1980). Much less is known regarding the influence of elevational changes in microclimate on other important ecophysiological processes such as photosynthesis and respiration.

5.3 Ecophysiology

The majority of ecophysiological studies of montane plants in western North America has attempted to explain distribution patterns of the overstory conifers along elevational and geographic gradients, or their successional patterns. Numerous investigations have also dealt with the interpretation of upper and lower timberlines and the persistence of stable meadow and bog communities within the forest. These studies have focused primarily on the light, temperature, water, and nutrient requirements of photosynthesis and respiration that are necessary for growth and survival. The large majority of this work has involved either *Pinus ponderosa* Dougl. ex Laws. (ponderosa pine), *Pinus contorta* Dougl. ex Laud. var *latifolia* Engelm. (lodgepole pine), *P. engelmannii* Parry ex Engelm. (Engelmann spruce), or *Pseudotsuga menziesii* (Mirbel) Franco (Douglas fir). Owing to the abundance of Douglas fir in the Pacific Northwest, this species is not included in this chapter, but in Chapter 6 (Northwest Forests). For brevity, I have concentrated on summarizing studies which were conducted within the western montane forests of North America, even though some of the species have been studied in other portions of their distribution range in North America.

5.3.1 THE OVERSTORY

Overstory vegetation throughout western montane forests is composed predominantly of evergreen, coniferous trees with the exception of areas where *Populus tremuloides* Michx. (quaking, or trembling aspen) occurs in relatively large pure stands at low to middle elevations. Also, considerable variation in overstory species composition may occur along riparian systems where willows, poplars, alders, or birch may dominate and a general breakdown of the segregation between overstory and understory occurs (Preston, 1968). Details of the distributional and successional patterns of montane, coniferous

species of North America are contained in numerous publications (Cormack, 1953; Daubenmire, 1943b, 1952, 1956, 1968, 1978; Day, 1972; Larsen, 1930; Ostler and Harper, 1978; Pearson, 1920; Peterson and Peterson, 1975; Pfister *et al.*, 1977; Preston, 1968; Reed, 1976; Spurr and Barnes, 1980; Weaver, 1917; Whittaker and Niering, 1965). In general, the northern forests are characteristically more diverse in overstory conifer species compared to the southern forests (Fig. 5.1).

A summary of ecophysiological measurements related to the water and photosynthetic relations of overstory species is given in Tables 5.1 and 5.2, respectively. Ponderosa and lodgepole pine have received the majority of emphasis in the literature as being representative of low elevation, xeric conifers. Lodgepole pine also occurs as a predominant mid-elevation conifer, while Engelmann spruce has been included in a majority of studies dealing with high-elevation conifers. Quaking aspen has also received considerable attention because it is a deciduous overstory tree with photosynthetic stems and, thus, represents a relatively unique component of the western forests.

(a) *Pinus ponderosa*

Ponderosa pine occurs at the lowest elevational limits of the western montane forests and is often found in open stands on southerly slopes, often mixed with foothill shrubs and Douglas fir. Early, comprehensive research on the ecophysiology of ponderosa pine at its northeastern boundary in southwest North Dakota was provided by Hadley (1969). He found that transpiration and photosynthesis in ponderosa pine continued at soil water potentials of less than $-1.5\,\mathrm{MPa}$ with virtually complete recovery to maximum rates after prolonged periods of drought. Light compensation and saturation levels for photosynthesis were relatively high with a broad temperature range for maximum photosynthesis. Net photosynthesis was greatest early in the morning followed by a midday depression which occurred on clear but not cloudy days. A similar pattern in net photosynthesis for *P. ponderosa* was reported by Fritts (1966). Hadley (1969) also found that maximum rates of photosynthesis occur-

Table 5.2 Water relation parameters of western coniferous forest species. Values included are leaf conductance to water vapor diffusion (g^l), xylem pressure potential (ψ_p^{xylem}), osmotic potential (ψ_π), soil water potential (ψ^{soil}), leaf temperatures (T_L), leaf-to-air water vapor difference ($LAVD$), and transpiration (J_{wv}).

	Area	Material	g^l	Ψ_p^{xylem}	ψ_π	ψ^{soil}	T_L	$LAVD$	J_{WV}	Reference
Overstory										
Abies concolor	Washington and Oregon	greenhouse grown twigs. 3 yr old seedlings	5–8 psi infil. press. stomata open; 20.5 psi stomata closed	—	—	6.9% H₂O content at stomatal closure	—	—	4 g plant⁻¹ h⁻¹	Lopushinsky (1969)
Abies grandis	B.C. Canada	greenhouse grown 5 yr old seedlings	—	−0.9 MPa −1.2 −1.5 −1.8 −2.1	—	—	—	—	100% 80 60 40 20	Puritch (1973)
	E. Oregon	greenhouse grown 4 yr old seedlings	—	−0.1 MPa −0.3 −0.5 −0.9 −1.7	—	—	—	—	100% 80 60 40 20	Lopushinsky and Klock (1974)
	central Oregon (1350–1570 m, 120° and 350° aspect, 70° and 61° slope)	adult trees	1.0–2.8 bar infil. press. 0.07–0.10 mm s⁻¹	−0.5 to −3.0 MPa	—	—	—	—	—	Zobel (1974)
	E. Oregon	1–3 m tall trees, 0–3 yr needles	0.03–1.92 mm s⁻¹	−0.8 to −2.0 MPa	—	—	—	—	—	Running (1976)
Abies lasiocarpa	S.E. Wyoming 2990–3300 m (timberline)	mature trees, 0–3 yr old needles	—	lower site: −1.0 to −1.8 MPa upper site: −1.1 to −3.4	—	—	—	0–11.7 mm Hg	—	Lindsay (1971)
	B.C., Canada	greenhouse grown, 5 yr old seedlings	—	−1.0 MPa −1.1 −1.2 −1.4 −1.6 −2.4 −4.0	—	—	—	—	100% 80 60 40 20 10 0	Puritch (1973)
	S.E. Wyoming mixed conifer stand, 274 m	branch tips of 10–15 yr old trees, 1–4 yr old needles	3.7 mm s⁻¹	> −0.35 MPa	—	> −0.2 MPa	—	—	68 mg m⁻² s⁻¹	Smith (1979)
	S.E. Wyoming clear cut 3150 m understory 2840 m	branch tips 10–15 yr old trees, 1st yr needles	4.8–9.3 mm s⁻¹ 1.6–6.1	−0.3 to −1.7 MPa −0.4 to −1.9	—	—	4–26°C 7–26	5–22 g m⁻³ 4–17	2–82 mg m⁻² s⁻¹ 2–54	Knapp and Smith (1981)
	N. Colorado 2900 m	mixed, 2nd growth stand	0.2–1.5 mm s⁻¹	—	—	—	—	2–10 g m⁻³	—	Kaufmann (1982a)
Abies procera	W. Oregon, 200–1000 m open stand or clear cut	1–3 m trees	0.32–2.6 mm s⁻¹	—	—	—	—	—	—	Running (1976)
Abies magnifica	W. Oregon, open stand or clear cut 1000 m	1–3 m trees	0.07–4.7 mm s⁻¹	—	—	—	—	—	—	Running (1976)

Table 5.2 (*contd.*)

	Area	Material	g^l	ψ_p^{xylem}	ψ_π	ψ^{soil}	T_L	LAVD	J_{wv}	Reference
Picea engelmannii	Washington and Oregon	greenhouse grown twigs of 3 yr old seedlings	3–4 psi infil. press. stomata open, 13 psi stomata closed	—	—	7.1% H_2O content at stomatal closure	—	—	—	Lopushinsky (1969)
	S.E. Wyoming, 2990–3300 m (timberline)	mature trees. 0–3 yr old needles	—	lower site: − 1.0 to − 1.7 MPa upper site: − 1.1 to − 3.3	—	—	—	—	—	Lindsay (1971)
	N.E. Utah, Krummholtz	winter exposed needles above snowpack	—	< − 0.4 MPa in Jan. to < − 0.9 Mpa in spring	—	—	—	—	—	Hansen and Klikoff (1972)
	central Colorado 2770–3230 m	mature trees 15–50 yr old, 1.5–2 m	—	− 0.3 to − 1.1 MPa / − 3 to − 14 / − 7 to − 20	—	—	T_{soil} 0.1 − 15.0 °C / 5.1–10 / 0.1–5.0	—	2–15 mg m^{-2} s^{-1} / 2–15 / 2–10	Kaufmann (1975)
	central Colorado 2770–3230 m	mature trees 15–50 yr old 1.5–2 m	sun: 0.4–1.3 mm s^{-1} shade: 0.2–1.1	− 5 to − 20 MPa − 4 to − 10	—	—	7–24 °C	—	5–15 mg m^{-2} s^{-1} 2–11	Kaufmann (1976) Becwar et al. (1983)
	S.E. Wyoming clear cut 3150 m understory 2800 m	branch tips	1.2 mm s^{-1} 1.0	− 0.3 to − 1.9 MPa − 0.4 to − 1.8	—	—	4–26 °C 7–26	4–22 g m^{-3} 4–23	2–30 mg m^{-2} s^{-1} 1–20	Knapp and Smith (1981)
	N. Colorado 2900 m	mixed, 2nd growth stand	0.4–1.5 mm s^{-1}	—	—	—	—	2–10 g m^{-3}	—	Kaufmann (1982a)
Pinus contorta	Washington and Oregon	greenhouse twigs of 3 yr old seedlings	9–10 psi infil. press. stomata open, 30.4–32.8 psi stomata closed	− 1.6 MPa stomatal closure	—	—	—	—	—	Lopushinsky (1969)
	E. Oregon 3500–5000 ft	4 yr old potted plants grown in field	—	− 0.1 MPa − 0.3 − 0.4 − 0.6 − 0.8 − 2.0	—	—	—	—	100% 80 60 40 20 0	Lopushinsky and Klock (1974)
	central Colorado, 2535–2825 m south and north aspects, gentle to steep slopes	2–3 yr old greenhouse grown seedlings planted in field	—	north: − 0.9 to − 1.4 MPa south: − 1.0 to − 1.7	—	—	—	—	—	Baldwin and Barney (1976)
	S.E. Wyoming, 2800–2971 m dry site dry-mesic mesic	mature trees 1–3 yr old	0.17–6.7 mm s^{-1} 0.28–3.3 0.20–5.0	− 1.0 to − 1.9 MPa − 1.0 to − 1.6 − 0.9 to − 1.2	—	—	—	8–14 g m^{-3} 6–11 5–16 mg	20–22 mg m^{-2} s^{-1} 11–34 14–30	Fetcher (1976)

Table 5.2 (contd.)

	Area	Material	g^l	ψ_p^{xylem}	ψ_π	ψ^{soil}	T_L	LAVD	J_{wv}	Reference
	S.E. Wyoming, mixed conifer stand, 2740 m	branch tips of 10–15 yr old trees, 1–4 yr old needles	4.8 mm s^{-1}	> – 3.6 MPa	—	> – 0.2 MPa	—	—	7.4 mg m^{-2} s^{-1}	Smith (1979)
	S.E. Wyoming, 2900 m	sunlit twigs of 100 yr old trees, early spring	0.4–3.0 mm s^{-1} 0.1–1.1	—	—	wet soil dry soil	—	3.3–10.4 g m^{-3}	6.7–11.4 mg m^{-2} s^{-1}	Fahey (1979)
	central Colorado, level glacial outwash, 2700 m	branches of 10–60 yr old trees, current, 1, 2, 3 yr old needles	July: 0.4–3.0 mm s^{-1} Aug: 0.1–5.0	– 0.4 to – 1.3 MPa – 1.0 to – 1.5	—	—	—	2–20 g m^{-3} 1–15	—	Running (1980)
	S.E. Wyoming, clearcut 3150 m understory 2800 m	branch tips	0.4–4.0 mm s^{-1} 0.1–1.1	– 0.3 to – 1.6 MPa – 0.4 to – 1.5	—	—	4–25 °C 5–26	5–26 g m^{-3} 4–24	1–72 mg m^{-2} s^{-1} 1–21	Knapp and Smith (1981)
	N. Colorado 2900 m	mixed 2nd growth stand	0.4–1.8 mm s^{-1}	—	—	—	—	2–10 g m^{-3}	—	Kaufmann (1982a)
Pinus monticola	N. Idaho, south and north aspects	branches 30 m above ground	0.1–1.0 mm s^{-1}	– 1.0 to – 2.3 MPa	– 2.0 to – 2.5 MPa	—	—	—	—	Cline and Campbell (1976)
Pinus ponderosa	Washington and Oregon	greenhouse grown twigs of 3 yr old seedlings	25.7 psi infil. press.	– 1.7 MPa stomatal closure	—	7.6% H$_2$O content stomatal closure	—	—	—	Lopushinsky (1969)
	E. Oregon, 2500–5000 ft	4 yr old potted plants grown in field	—	– 0.1 MPa – 0.2 – 0.4 – 0.5 – 0.8 < – 2.0	—	—	—	—	100% 80 60 40 20 0	Lopushinsky and Klock (1974)
	central Colorado, 2535–2825 m south and north aspects, gentle to steep slopes	2–3 yr old greenhouse grown seedlings planted in field	—	north: – 1.2 to – 3.2 MPa south: – 1.5 to – 4.4	—	—	—	—	—	Baldwin and Barney (1976)
	W. Oregon, open stand or clear cut	1–3 m trees	0.05–3.23 mm s^{-1}	—	—	—	—	—	—	Running (1976)
Populus tremuloides	Ottawa, Canada, raised sphagnum plant bog	sunlit branches	3.1–5.0 mm s^{-1} 3.3–5.0	– 1.6 to – 2.1 MPa – 0.5 to – 1.8	—	—	—	—	—	Small (1972)
Tsuga heterophylla	B.C., Canada	sun and shade juvenile trees collected from field and potted in greenhouse	sun: 0.4–1.1 mm s^{-1} shade: 0.29–1.1 (0 to 43 klux)	—	—	—	—	—	—	Keller and Tregunna (1976)
	central Colorado 2750 m	potted 4 yr old seedlings grown in field	0.06–0.6 mm s^{-1}	– 1 to – 17 MPa	—	0 to – 19 MPa	—	10–30 g m^{-3}	0–9.0 mg^{-2} m^{-2} s^{-1}	Kaufmann (1979)

red on cool summer or mild autumn days when soil moisture was plentiful and nighttime respiratory losses were minimal. Finally, a combination of shallow soils and insufficient soil moisture due to excessive competition with grasses was believed to limit local establishment of *P. ponderosa* in certain areas. A greater abundance of trees was found on moist, eastern slopes compared with the dryer, southwest aspects which were dominated by grasses.

Helms (1970, 1971) studied the ecophysiology of ponderosa pine in the Sierra Nevada mountains of California and found that daily photosynthesis responded primarily to temperature and solar irradiance under apparently low stress conditions. However, under more stressful conditions of higher temperatures and lower soil moisture, photosynthesis decreased in response to increases in the ambient vapor-pressure deficit between the leaf and air. Helms (1970) also found that 2 year old needles were the most productive photosynthetically, as was the lower portion of the leaf canopy compared to the middle and upper sections. Other investigators have found that the growth of ponderosa pine may be especially poor at low temperatures (Cochran, 1972; Pharis *et al.*, 1970, 1972). Ponderosa pine seedlings were found to be more sensitive to low temperatures than seedlings of lodgepole pine in southcentral Oregon (Cochran, 1972), but were able to withstand lower temperatures than Douglas fir (Pharis *et al.*, 1970, 1972). Subfreezing or near-freezing nights led to substantial reductions in lodgepole pine photosynthesis and increased the time required to recover to pretreatment levels (Pharis *et al.*, 1972). This recovery was dependent upon the treatment temperature and the length of exposure and changes in apparent photosynthesis were correlated with changes in respiration rather than photosynthesis. Recent studies in the San Bernardino Mountains of southern California have shown substantial increases in field photosynthesis for ponderosa pine trees after enrichment with CO_2 (Green and Wright, 1977; Wright, 1974).

Information also exists regarding the water relations and stomatal behavior of *P. ponderosa*. Lopushinsky (1969) found that stomatal closure for potted ponderosa pine in Washington

and Oregon occurred at higher xylem pressure potentials of -1.4 to -1.7 MPa compared to values of -1.9 to -2.2 MPa for Douglas fire and near -2.5 MPa for grand fir (*Abies grandis*). Transpiration was reduced over 60% at a soil water potential of -0.5 MPa and about 88% at -1.0 MPa compared to smaller reductions for the two fir species. Running (1976) found that ponderosa pine had a wide range of leaf conductance values among six sites located throughout central and eastern Oregon. At the Metolius River site ponderosa pine had much higher leaf conductances than either grand or Douglas fir. Baldwin and Barney (1976) noted differences in xylem pressure potentials for seedlings of *P. ponderosa* from Idaho and Colorado grown on several different slopes and aspects. They attributed the higher xylem pressure potentials of the Idaho seedlings to their lower root to shoot biomass ratios and a greater abundance of fine roots. Recently, Jackson and Spomer (1979) reported that *P. ponderosa* had a greater leaf conductance when compared to Douglas, grand, and red fir, all growing within 100 m of one another. Ponderosa pine also had the lowest osmotic potentials at the point of incipient plasmolysis and a lower cell wall elasticity, but the greatest range of osmotic potentials during water stress compared to the three fir species. Along with these characteristics, the deep tap root of *P. ponderosa* and high leaf conductance may contribute to a greater adaptability to xeric conditions (Jackson and Spomer, 1979). Smith *et al.* (1984) also found that *P. ponderosa* and *P. contorta* had higher leaf conductances than subalpine fir, Engelmann spruce or limber pine (*Pinus flexilus*) growing naturally at the same location.

(b) *Pinus contorta*

Lodgepole pine is generally considered to be an early successional, pioneer conifer at low to middle elevations and may form relatively open stands to extremely dense 'doghair' stands throughout the western forests (Cormack, 1953; Day, 1972; Langenheim, 1962; Stahelin, 1943; Whipple and Dix, 1979). Lopushinsky (1975) reviewed much of the early work dealing with the ecophysiology of lodgepole pine including numerous references to other western forest conifers. Conflicting early reports showed that lodgepole pine transpiration was either less than other conifers such as ponderosa pine, Douglas fir, and Engelmann spruce (Gail and Long, 1935; Roeser, 1932; Sperry, 1936), or greater than these three species as well as limber pine (Bates, 1923). More recent work has shown that seedlings of six species of Rocky Mountain conifers had fairly similar transpiration rates, ranging from 1.84 to 4.27 g H_2O lost per g oven dry weight (Lopushinsky, 1969). Swanson (1975) reported that annual transpiration for a mixed stand of lodgepole pine and Engelmann spruce was near 150 mm for both, with only a 16 mm difference between the two species. However, whole-plant transpiration during the early spring was considerably greater in the spruce. Smith (1979) also measured similar transpiration rates per unit leaf area for understory saplings of lodgepole pine, subalpine fir, and common juniper and Running (1980) found relatively similar responses in transpiration to changes in vapor-pressure deficit in six conifer species, including lodgepole pine. Knapp and Smith (1981) concluded that the water relations of lodgepole pine in open, early successional sites contributed to its success as a pioneer species at low to middle elevations. Compared to Engelmann spruce and subalpine fir, lodgepole pine had the greatest leaf conductance values at the highest xylem pressure potentials as well as the largest reduction of leaf conductance under the shaded conditions of the understory. Kaufmann (1982a) provided seasonal comparisons of leaf conductance with photon flux density and the vapor difference between the leaf and air. At an irradiance level of near 400 $\mu E\,m^{-1}$, maximum leaf conductances were greatest for lodgepole pine (1.3 mm s^{-1}) followed by Engelmann spruce (0.6 mm s^{-1}) and subalpine fir (0.3 mm s^{-1}). Aspen had a considerably higher conductance (2.6 mm s^{-1}) than the conifers. Subalpine fir had the greatest reduction in leaf conductance (1.5 to 0.2 mm s^{-1}) for change in vapor-pressure deficit of from 2 to 10 g m^{-3} compared to lodgepole pine (1.8 to 0.4 mm s^{-1}). Engelmann spruce (1.5 to 0.4 mm s^{-1}), and aspen (4.5 to 1.2 mm s^{-1}). Also, stomatal closure in lodgepole pine and Engelmann spruce appeared less sensitive

to freezing nighttime temperatures than for subalpine fir. Smith *et al.* (1984) found that *P. contorta* had the highest leaf conductances compared to Engelmann spruce, subalpine fir, Douglas fir, ponderosa pine and limber pine growing at the same location.

The sapwood of lodgepole pine has been reported to be thicker than several other western conifers, with the exception of ponderosa pine and may contribute to more efficient stomatal behavior due to greater water reserves (Lassen and Okkonen, 1969). Reid (1961) found that the moisture content of the outer sapwood of lodgepole pine varied from 85 to 165% of oven dry weight compared to 16% for trees infested with pine bark beetles. Lodgepole pine was also found to have a much greater ratio of xylem cross-sectional area to total leaf area compared to Douglas fir or Engelmann spruce, but considerably less than ponderosa pine.

Several investigators have considered the influence of soil conditions on seed germination and growth of lodgepole pine. Early studies by Bates (1923) showed that lodgepole pine seedlings wilted at higher soil moisture levels and had poorer seed germination and seedling growth than either ponderosa pine, Douglas fir, or Engelmann spruce. Kaufmann and Eckard (1977) reported that seed germination and emergence decreased in both Engelmann spruce and lodgepole pine at low soil water potentials, but that considerably more time was required for lodgepole pine germination at the low temperature treatments. The authors concluded that this response may enable Engelmann spruce to become established in cooler environments compared to lodgepole pine. Colder soil temperatures (near freezing) may substantially reduce water uptake by roots of lodgepole pine (Running and Reid, 1980) as well as Engelmann spruce (Kaufmann, 1975). Brix (1979) found that lodgepole pine seeds were viable after drying to -11.0 MPa.

Lodgepole pine may also be able to establish on flooded soils. In contrast to earlier studies (Gail and Long, 1935), Cochran (1972) found that lodgepole pine seed germination in saturated soils was greater than for ponderosa pine seeds. Additional studies also have shown that root growth in lodgepole pine appears less sensitive to reduced O_2 levels (Boggie, 1974), possibly due to the increased capacity for O_2

transport in the root system (Coutts and Philipson, 1978a,b; Philipson and Coutts, 1978).

Substantial daily variation in stomatal behavior and transpiration in lodgepole pine appears sensitive to relatively short-term changes in environmental conditions. Nighttime air temperatures near freezing inhibited stomatal opening the following day (Fahey, 1979) and subsequent work by Kaufmann (1982a,b) and Smith *et al.* (1984) found a close correlation between the occurrence of subfreezing air temperatures at night and stomatal opening the following day in lodgepole pine, subalpine fir, and Engelmann spruce, as well as ponderosa pine, limber pine, and Douglas fir. Fetcher (1976) found that maximum leaf conductances to water-vapor loss were relatively high (2–3 mm s^{-1}) with little variation during the day in the early portion of the growth season in southeastern Wyoming. However, substantial decreases in leaf conductance occurred at midday in August when xylem water potentials decreased to less than -1.7 MPa. At xylem pressure potentials between -0.9 and -1.7 MPa, leaf conductances also decreased as the vapor-pressure deficit of the ambient air increased. Running (1980) found similar variations in daily and seasonal leaf conductance for lodgepole pine growing in north-central Colorado. A ten-fold decrease in leaf conductance to water vapor occurred between the maximum values in June and the minimum values of August and a high correlation existed between predawn leaf water potential (xylem pressure potential) and the maximum leaf conductance. Leaf conductance was also found to decrease at midday when ambient humidities approached the daily minimum. Knapp and Smith (1981) reported that *P. contorta* had greater stomatal closure at shaded, understory locations than for open sites, but had less of a response to the leaf-to-air water vapor difference when compared to Engelmann spruce.

Early studies on photosynthesis in lodgepole pine have been reviewed by Lopushinsky (1975). Photosynthetic rates of 44–107 nmol CO_2 g^{-1} s^{-1} (dry wt) were recorded at a light intensity of near 6000 ft c and at temperatures from 18–21 °C with 3–5 month old seedlings having considerably higher rates compared to 2 year old seedlings (Driessche and Wareing, 1966). Sweet and Wareing (1968) found considerable differences in net photo-

synthesis for 12–19 week old seedlings of lodgepole pine measured from four locations ranging from the California–Oregon coast to the Rocky Mountains in Alberta, Canada. Photosynthetic rates ranged from about 63 to 76 nmol CO_2 g^{-1} s^{-1} and light saturation occurred at nearly full sunlight.

Dykstra (1974) found that net photosynthesis for potted lodgepole pine seedlings reached maximum values of about 6.3 μmol CO_2 m^{-2} s^{-1} with light saturation occurring at near 400 W m^{-2} of PAR. The temperature optimum for photosynthesis for these greenhouse seedlings was near 20 °C with a four-fold reduction in photosynthesis occurring at about 2 °C. A near linear decrease in photosynthesis from maximum values down to zero occurred for a decrease in xylem pressure potential of from -0.3 MPa to -1.5 MPa. Finally, mesophyll conductance to CO_2 uptake was always less than the stomatal conductance, expecially under stress conditions, and there was a proportional increase in photosynthetic rate with increases in mesophyll and stomatal conductance to CO_2. Brix (1979) compared Douglas fir, white spruce, western hemlock, and lodgepole pine photosynthesis and found that lodgepole pine had relatively high photosynthesis expressed on a leaf area basis, but the lowest rate expressed on a unit weight basis. Also, photosynthesis in lodgepole pine showed the steepest decline with drying soil and began at a higher threshold soil water potential (< -1.0 MPa).

Smith (1979) measured net photosynthetic rates of near 8 mol m^{-2} s^{-1} for 12–18 year old lodgepole saplings in the understory of a mixed conifer stand in southeastern Wyoming and reported similar values for nearby subalpine fir saplings and common juniper (*Juniperus communis*) bushes. Also, aerodynamic resistances of individual shoots accounted for relatively small reductions in photosynthesis, but substantial increases in water-use efficiency, during relatively frequent periods of low wind speeds (< 20 cm s^{-1}) within the understory.

(c) *Picea engelmannii*

Although Engelmann spruce is an abundant climax conifer at high elevations extending to timberline, little information exists concerning environmental factors influencing the distribution in this species.

Oosting and Reed (1952) reviewed earlier investigations dealing with the autecology of Engelmann spruce, as well as subalpine fir, within the Rocky Mountains. Numerous studies have classified Engelmann spruce and subalpine fir as being either shade tolerant (Baker, 1944; Bates, 1917; Hodson and Foster, 1910; Kirkwood, 1922; Pearson, 1931; Roeser, 1924) or intolerant of full sun (Bethan, 1925; Korstian, 1925). Several early studies also found that spruce germination and survival were especially sensitive to high temperatures (Bates, 1923, 1924; Bates *et al.*, 1914; Kirkwood, 1922). Bates *et al.* (1914) stated that high soil temperatures were the most important factor inhibiting Engelmann spruce establishment at lower elevations. High temperatures caused by high insolation has also been reported to cause stem-girdling and death in Engelmann spruce seedlings (Korstian and Fetherolf, 1921). However, Kirkwood (1922) reported that both Engelmann spruce and subalpine fir seedlings had relatively high temperature tolerances.

More recent studies using improved instrumentation have also supported the importance of temperature to the germination and establishment of Engelmann spruce. Wardle (1968) stated that the survival of Engelmann spruce seedlings in open meadows in the Rocky Mountains of Colorado was positively correlated with lower minimum air temperatures during the night. Hellmers *et al.* (1970) also found that minimum night temperatures were the most important factor limiting the establishment and growth of Engelmann spruce seedlings, even when daytime temperatures were lower than nighttime values. The greatest seedling mortality occurred when low night temperatures were combined with high daytime temperatures. High day temperatures also had a detrimental effect on terminal bud formation. Kaufmann and Eckard (1977) found that seed germination and emergence in Engelmann spruce occurred more rapidly at low soil temperatures compared to lodgepole pine. Noble (1974) had previously reported that Engelmann spruce regeneration on cool north slopes was successful while virtually impossible on southern slopes without artificial shade. Noble (1972) and Kaufmann and Eckard (1977) found that differences in soil quality did not appear to influence seed germi-

nation, although subsequent survival and root growth of seedlings could be influenced by soil crusting and compaction.

Knapp and Smith (1982) found that the amount of incident sunlight at natural locations in the understory where *A. lasiocarpa* and *P. engelmannii* occurred was not different, although *A. lasiocarpa* seedlings were much more abundant on thicker litter layers. Also, Engelmann spruce seeds germinated more quickly and at lower temperatures than subalpine fir, but taproot development was greater in the fir. Thus, the greater abundance of *A. lasiocarpa* within mature spruce–fir forests of the central Rocky Mountains may be more related to substrate water relations than light requirements. In contrast, Patten (1963a) found that *P. engelmannii* seed germination was sensitive to light at low temperatures and hypothesized that germination was inhibited within the forest understory due to low light and cool temperatures. He observed that *P. engelmannii* was the main conifer to establish seedlings at the forest boundaries along grass meadow areas.

Other investigators have considered the importance of transpirational water relations on the survival and growth of Engelmann spruce. Kaufmann (1975) found that low soil temperatures of near $-5\,°C$ resulted in a marked reduction in transpiration in Engelmann spruce due to reduced root permeability to water uptake. Field values of the leaf conductance in Engelmann spruce were also found to decrease as the absolute humidity difference from the leaf to the air increased, especially under shade conditions (Kaufmann, 1976). Effects of air and soil temperature variations during this period did not appear to affect leaf conductance under natural conditions in the field. Hypothetically, stomatal closure at low light intensities when photosynthetic potential is low could lead to water conservation. However, a lack of stomatal closure in bright light, even under water stress, may partially explain the reduced growth and survival of planted Engelmann spruce seedlings in clear cuts (Ronco, 1970). Lopushinsky (1969) reported that transpiration rates in Engelmann spruce were much reduced at low soil water potentials when compared with lodgepole and ponderosa pine, but were greater when compared to Douglas and grand fir. Stomatal closure occurred at leaf water potentials of near -1.7 MPa, an intermediate value compared to the pines and firs. Significant decreases in xylem water potential occurred as the soil dried to about -0.8 MPa, followed by only slight decreases with further soil drying. Again, this response appeared intermediate to the pines and firs. More recently, Kaufmann (1979) reported that the leaf-to-air water vapor gradient had little effect on transpiration due to stomatal control for potted seedlings of Engelmann spruce. Sensitivity to water stress may contribute to the poor establishment of Engelmann spruce from seed on south aspects observed by Noble and Alexander (1977). Knapp and Smith (1981) reported that *P. engelmannii* had lower leaf conductances than either *P. contorta* or *A. lasiocarpa*, with much less variation between exposed and shaded understory trees. However, Engelmann spruce had the greatest increases in leaf conductance with decreasing vapor-pressure deficit of the air. As a result, transpiration in Engelmann spruce was less than half that of the pine and fir species.

Considerably less information exists concerning the photosynthetic performance of Engelmann spruce. Ronco (1970) reported that Engelmann spruce seedlings in open areas experienced a combination of high temperatures and insolation which led to decreased photosynthesis and high mortality. Optimal photosynthesis occurred under shaded conditions and soil moisture deficits of less than 10%. Photosynthesis approached zero at about 20% soil moisture deficits and light saturation of photosynthesis occurred at approximately one-third of full sunlight. The author concluded that a combination of water stress and excessive insolation was the primary cause of seedling mortality in Engelmann spruce at high elevations.

Knapp and Smith (1982) also found that light-saturated net photosynthesis was greater for 2–4 year old seedlings of *A. lasiocarpa* compared to *P. engelmannii* (1.6 compared to 1.2 μmol m^{-2} s^{-1}). Photosynthesis became saturated at 780 compared to 1060 μE m^{-2} s^{-1} of PAR, respectively, while light compensation points were similar (60–70 μE m^{-2} s^{-1}). The greater photosynthesis of subalpine fir at lower irradiance may contribute to the greater abundance of fir seedlings often observed in the understory of central montane forests (Knapp and Smith, 1982).

(d) *Populus tremuloides*

Quaking, or trembling, aspen is recognized as a shade-intolerant, early pioneer tree species of low to middle elevations and is the only deciduous tree found extensively throughout the central and northern montane forests in relatively pure stands. Although aspen is deciduous, it also has chlorophyll in the stems and trunk. Despite the obvious importance of the early successional status of this species, relatively little is known regarding the ecophysiology of quaking aspen. However, the ecological importance of vegetative reproduction by sprouting, as opposed to sexual reproduction and seed germination, in this species has been recognized for some time (Baker, 1918; Pearson, 1914). Sprouting by root suckers accounts for the great majority of establishment in open areas and seedlings are rarely found in nature (Baker, 1925; Ellison, 1943; Faust, 1936; Larsen, 1944; Moss, 1938).

Baker (1925) stated that aspen seeds were not plentiful and that germination was prevented by dry spring soils. However, a wide range of environmental conditions appear suitable for aspen seed germination as long as soil moisture is plentiful (Faust, 1936; McDonough, 1979; Moss, 1938). Because seedling establishment requires a wet surface layer of soil for at least a week due to the slow growth of the seedling root, conditions suitable for germination and survival of the young seedling under natural conditions may occur only during exceptional years (Moss, 1938). Litter layers may inhibit germination of aspen seeds due to less favorable physiochemical conditions as well as possible allelopathy (McDonough, 1979). Barnes (1966) concluded that under certain environmental conditions aspen may become a permanent rather than ephemeral pioneer species compared to other early successional species.

Considerable interest in aspen root suckering has led to several publications dealing with the growth dynamics of aspen root systems (Day, 1944; DeByle, 1964; Farmer, 1962; Sandberg, 1951; Zahner and

DeByle, 1965). These studies have revealed an extensive lateral root system that leads to prolific vegetative reproduction through sucker formation. Prolific suckering from roots in undisturbed forests may account for rapid invasion following disturbance (Maini, 1960). Individual suckers are supported by parent roots for up to 25 years when new roots begin to contribute to about half of the annual stem growth. Apical dominance as well as a host of environmental factors appear to be involved in the physiological initiation of root suckering (Farmer, 1962). Barnes (1966) and DeByle (1964) found root interconnections between aspen stems of up to 50 years of age, but also showed that several functionally independent root systems can exist within a stand due to death and decay of old root connections. A more recent study by Schier and Campbell (1978) showed that western aspen tended to have deeper and slightly larger parent roots compared to eastern aspen. They also found considerable variation in depth and size of lateral roots and the ability to initiate new sucker roots among individual clones of aspen. Increase in suckering has been attributed to better soil aeration (Weigle and Frothingham, 1911), warmer temperatures of darker soils (Maini, 1960; Maini and Horton, 1966; Shirley, 1932), and increased exposure to sunlight (Stoeckeler, 1960; Zehngraff, 1949). Schier and Campbell (1978) found that high burn intensities increased the depth at which suckers formed.

Only a few investigators have directly measured transpiration, photosynthesis, or respiration in *P. tremuloides*. Kaufmann (1982a) found that transpiration rates were from two to eight times more rapid per unit leaf area as compared to the co-occurring conifers lodgepole pine, Engelmann spruce, and subalpine fir at a solar irradiance of 400 μE m^{-2} s^{-1}. Photosynthetic studies have included measurements on leaves (Bourdeau, 1958a,b; Loach, 1967; Okafo and Hanover, 1978) as well as photosynthetic bark surfaces (Foote and Schaedle, 1976a,b, 1978; Perry, 1971; Schaedle, 1975; Strain and Johnson, 1963). A wide range of light-saturated photosynthesis rates (6.36–31.6 μmol m^{-2} s^{-1}) have been reported for quaking aspen (Bourdeau, 1958a,b; Foote and Schaedle, 1978; Loach, 1967). Okafo and Hanover (1978) reported that whole plant photosynthesis for *P. tremuloides* was approximately one-quarter that of single-leaf photosynthesis, respiration rates were about one-sixth to one-quarter the photosynthetic rate, and the CO_2 compensation point was near 52 ppm CO_2. Photosynthetic rates of stems and twigs were only one-quarter to one-seventh of leaf photosynthesis under laboratory and field conditions (Foote and Schaedle, 1978). However, bark photosynthesis may balance CO_2 losses due to stem respiration during leafless periods and, thus, provide an important adaptation for early spring growth.

(e) *Other overstory species*

Several species of conifers that often occur as major constituents of the overstory vegetation in western North America have received little attention in the literature. Of these studies, a major focus has been on water relations. Both grand fir (*Abies grandis*) and Douglas fir had the least reduction in transpiration at low soil water potentials as compared to pines and spruce (Lopushinsky and Klock, 1974). Also, much lower leaf water potentials occurred in grand fir (-2.5 MPa) than for Douglas fir (-1.9 MPa), Engelmann spruce (-1.6 MPa), or two pines (-1.4 to -1.7 MPa). The maximum leaf conductance value for grand fir was the lowest among seven other conifers measured (Jackson and Spomer, 1979; Running, 1976). Grand fir may have a relatively low acclimation potential for drought when compared to ponderosa pine, Douglas fir, and western red cedar (*Thuja plicata*), owing to a high osmotic pressure for incipient plasmolysis, a low water storage capacity, and a high leaf conductance (Jackson and Spomer, 1979). Seedlings of western hemlock (*Tsuga heterophylla*) were unable to survive water potentials lower than -6.0 MPa compared to substantially lower values for Douglas fir, white spruce (*Picea glauca*) and lodgepole pine (Brix, 1979). Cline and Campbell (1976) compared white pine (*Pinus monticola*) with three woody shrubs in northern Idaho and found that the leaf and osmotic water potentials and leaf diffusive conductance to water vapor of the pine were much less variable through the growth season. Also, the maximum and minimum values of these parameters were much lower than for the shrubs.

Germination of subalpine fir, Engelmann spruce, Douglas fir, and lodgepole pine was tested in various organic constituents (Daniel and Schmidt, 1972). Germination in subalpine fir was inhibited by its own O-horizon soil, but, this soil did not appear to inhibit germination of several other conifers. A similar situation occurred for Douglas fir, whereas the O-horizon from a lodgepole pine soil did not inhibit germination of lodgepole pine or any of the other conifer species. O-horizon soil from Engelmann spruce inhibited germination of Engelmann spruce and the other three conifer species. Knapp and Smith (1982) found that seedlings of *A. lasiocarpa* had greater light-saturated photosynthetic rates than *P. engelmannii* within the understory. *A. lasiocarpa* also had greater root penetration into litter and was found to occur more frequently on thicker litter layers in the field. These differences may partially account for the greater abundance of *A. lasiocarpa* seedlings in the understory where relatively thick litter layers undergo rapid drying in spring. Subalpine fir had the greatest leaf conductance and transpiration rates as compared to Englemann spruce and lodgepole pine in southeastern Wyoming (Knapp and Smith, 1981) for both exposed and understory locations. Fir also had the least reduction in leaf conductance in response to greater leaf-to-air vapor-pressure deficit.

Recent studies by Young and Hanover (1978) have shown that bud dormancy can be induced by nutrient and moisture stress in blue spruce (*Picea pungens*). Low levels of nitrogen, but not phosphorus and potassium, led to bud set under constant temperature regimes from 12 °C to 37 °C. Also, root to shoot ratios were decreased by low temperatures and increased by nutrient or water stress.

Photosynthesis in grand fir under natural conditions was found to respond to seasonal changes in leaf water potential primarily through changes in the mesophyll resistance to CO_2 uptake (Hodges, 1967). However, daytime variations in photosynthesis occurred in response to changes in leaf water potential and the leaf-to-air vapor-pressure deficit. Photosynthesis under sunny conditions was significantly less compared to shade conditions leading to a substantial midday depression in CO_2 fixation. Mooney *et al.* (1964) measured photosynthetic rates

for numerous species in the White Mountains of California. High elevation trees and shrubs had generally higher photosynthetic rates and photosynthesis/respiration ratios along with lower temperatures for peak photosynthesis. Bristlecone pine (*Pinus aristata*) had more constant photosynthesis and transpiration rates compared to adjacent big sagebrush (*Artemisia tridentata*) plants (Mooney *et al.*, 1966b).

5.3.2 THE UNDERSTORY

The physiological ecology of understory plants within the western coniferous forests of North America has been virtually neglected in the literature until recently. Little information is available regarding either the community or environmental characteristics of these understory habitats and even less is known about the physiological responses of understory species. This situation exists even for the economically important conifer seedlings and saplings that may be prevalent members of the understory community before regenerating the overstory.

One of the first studies to consider the ecophysiology of an understory species in the Rocky Mountains was the work of Janke (1968, 1970) on the abundant, deciduous subshrub *Vaccinium myrtillus* L. He found that plants growing in sunnier microhabitats within the understory had much lower stomatal resistances to water vapor diffusion, shorter aerial shoots, thicker stems, an upright branching habit, and more vertically oriented leaves than those growing in more shaded areas. Also, stomatal densities for leaves of sun plants were nearly twice those of shade plants. Knight *et al.* (1977) found that *Vaccinium scoparium* and *Carex geyerii* decreased in cover for areas with persistent snowbanks and exhibited accelerated phenological development. Persistent snowbanks appeared to significantly limit the growth season and possibly contributed to the patchy distribution patterns of understory plant species.

More recently, the small-scale distribution pattern of *Arnica cordifolia* and *A. latifolia* in the Medicine Bow Mountains of Wyoming has been associated with the temperature and water relations of these two species (Young and Smith, 1979). Individual populations of these two congeners were located in microhabitats that did not receive direct sunlight during midday, especially *A. latifolia*. Direct exposure to sunlight generally occurred during the early morning and late afternoon when air temperatures were lowest. These sunflecks were of a much lower intensity and had a shorter duration than the larger sunpatches that occurred during midday. Moreover, transpiration rates were exceptionally high during exposure to sunpatches due to elevated leaf temperatures and a significant increase in leaf conductance values. Leaf wilting was commonly observed during sunlight exposure along with substantial reductions in xylem pressure potentials (Young and Smith, 1979). The relatively tight coupling of the small-scale distribution patterns of *Arnica* with exposure to sunlight was further indicated by the occurrence of permanent wilting after sun exposure was experimentally prolonged in the field. The congener *A. cordifolia*, which occurs at lower elevations, was located in microsites that received more direct sunlight and withstood more prolonged exposure to direct sunlight during the day than *A. latifolia*.

Young and Smith (1980) further evaluated several physiological and morphological differences in populations of *A. cordifolia* growing in sun and shade microhabitats and found that considerable intraspecific variation was correlated with sun exposure. Plants growing in much sunnier microhabitats (greater frequency, intensities, and durations of sunpatches) had smaller, thicker leaves with larger stomata, and almost twice as many trichomes, in addition to having higher leaf temperatures, light saturation and temperature optima for photosynthesis, and transpiration rates compared to shade plants. Photosynthetic rates at light saturation for sun plants were over three times those measured for shade plants. Similar results have been reported for the water relations of seven other understory species (one shrub, one subshrub, and five herbaceous species) found sympatrically with *A.cordifolia* (Smith, 1981). Five other herbaceous understory species also exhibited midday wilting during exposure to sunpatches along with similar elevations in leaf temperature, high transpiration

rates, and low xylem pressure potentials. Stomatal closure did not occur under these conditions until plants were near permanent wilting. Morphologically, leaves of all seven species were hypostomous and bicolored with adaxial surfaces darker than abaxial surfaces. Also, stomatal opening in the early morning and closure in the evening were associated with the quantity of incident sunlight on their adaxial rather than abaxial leaf surfaces where the majority of stomata occurred. Stomata on the abaxial leaf surface closed in response to direct incident sunlight, indicating that hypostomy in these understory species may be a mechanism for avoiding the detrimental effects of direct sunlight on stomatal opening.

Recent simulation studies have shown a considerable influence of the phenotypic variation that is correlated with understory location on photosynthesis and water relations in *A. cordifolia* (Young and Smith, 1982). These simulations predicted that the most common leaf phenotype in its natural location would not have the greatest carbon gains compared to other leaf phenotypes but would have the greatest water-use efficiency (WUE). The greatest carbon gain was predicted for the smaller, sun-leaf phenotype that also had the lowest predicted WUE. Thus, phenotypic variations in leaf structure and physiology within the species appear to extend the small-scale distribution of *A. cordifolia* into shaded and sunnier microhabitats which could result in greater carbon gain in wet years along with greater water-use efficiency in dryer years. Also, the frequent occurrence of substantial cloudcover, especially during the afternoon period, resulted in a 37% increase in daily carbon gain and an 84% reduction in transpirational water loss in *A. latifolia* (Young and Smith, 1983). The resulting increase in WUE of over 700% indicates the potential importance of cloudcover to the daily photosynthesis and water relations of this understory species as well as other montane species.

5.3.3 TIMBERLINES

Western montane forests often form a mosaic of grass meadows bordered by forest which includes the low elevation transitions from foothill to forest and from forest to tundra at high elevations. The prevalence of these sometimes very abrupt ecotonal habitats has prompted numerous investigations of environmental and ecophysiological differences among the species involved. Several early investigations noted substantial differences between forest and meadow soils across timberline boundaries (Behan, 1958; Dunnewald, 1930; Thorp, 1931). More recent investigations have focused on the importance of fire and ensuing changes in microclimate and soil moisture on conifer seedling establishment (Billings, 1969; Daubenmire, 1968; Despain, 1973; McMinn, 1952; Olsen and Crockett, 1965; Patten, 1963a, 1969; Potter and Green, 1964; Root and Habeck, 1972; Strong and Parminter, 1980). Stahelin (1943) gave a detailed explanation of the potential successional stages following fire in the subalpine forests of the central Rocky Mountain. The author concludes that a relatively complex process is involved in the post-fire recovery to a climax stand of spruce–fir forest which depends upon the interaction of topography and prevailing environmental conditions. Severe fires may lead to a subalpine grassland or dry park that may persist indefinitely depending on fire frequency and the moisture conditions of the particular sites (Billings, 1969; Stahelin, 1943). In most investigations forest–meadow as well as elevational timberlines appear to be in a state of flux with conifers slowly reinvading the open fringe (Korterba and Habeck, 1971; Miles and Singleton, 1975; Patten, 1963b, 1969; Vale, 1978; White *et al.*, 1969). Miles and Singleton (1975) felt that a large dry park area in the Medicine Bow Mountains of Wyoming was gradually migrating eastward due to the prevailing westerly winds and the resulting destruction of trees on the west side due to snowdrift. However, Vale (1978) concluded, based on tree ages and seedling densities, that this same subalpine park was actually being reinvaded only on the west side. Numerous other investigators have associated the stability of forest–grassland boundaries with high soil temperatures and drought which have detrimental effects on conifer seedling establishment. However, the majority of these studies do not include measurements of plant water relations or other supporting ecophysiological data.

Specific investigations dealing with an ecophysiological interpretation of conifer tree boundaries in the western montane forests have focused extensively on lower and upper elevational limits of tree growth. Regarding lower timberlines, Daubenmire (1943a) found that the tolerance of several Rocky Mountain conifers to high soil surface temperatures and soil drought was greatest for the lower elevational species, although Douglas fir and ponderosa pine had similar tolerances despite the much lower distributional preferences of the pine. All of these species appeared to have equal tolerances for atmospheric drought, but the lowest elevation species also had the greatest root penetration rates in soil. Hadley (1969) studied the physiological ecology of ponderosa pine in North Dakota and found that competition for soil moisture with grasses severely retarded pine seedling growth. He concluded that this effect contributed to the observed stability of the ponderosa pine/grassland boundary. Wright (1966, 1968, 1970a,b) studied the CO_2 exchange rates of several low elevation pines in the San Bernardino mountains of southern California in an attempt to evaluate their lower elevational limits. Laboratory seedlings of knobcone pine (*Pinus attenuata*) had the greatest acclimation response in photosynthesis to low temperatures, followed by coulter pine (*P. coulteri*) and then sugar pine (*P. lambertiana*). This sequence was also the order of their elevational preferences from highest to lowest (Wright, 1970b). The author concluded that the ability to photosynthesize at low temperatures would be an adaptation for growth at dryer, lower elevations due to the simultaneous occurrence of low temperatures and the wetter portions of the year. Coulter and knobcone pines occurred at lower elevations within the chaparral and also had low photosynthetic efficiencies at low illumination levels as compared to coulter pine which occurred exclusively within the forest (Wright, 1970a). Recently, Bunce *et al.* (1979) found that the lower elevational limits of ponderosa pine and several shrubby species were correlated with the elevations at which their calculated annual carbon balances were zero.

The environmental and physiological factors influencing the formation and persistence of upper elevational timberlines has received considerable attention in the literature (Daubenmire, 1954; Griggs, 1938, 1946; Shaw, 1909; Tranquillini, 1979; Wardle, 1965, 1968, 1971). However, only a small portion of these studies have involved timberlines of the western montane forests. Griggs (1938, 1946) studied eleven timberline areas of the northern Rocky Mountains and concluded that North American timberlines appear to be static and are characterized by open areas created by a combination of fire and a slow rate of reinvasion by conifers. Daubenmire (1954) concluded that some environmental complex of variables was acting directly on the tree life form to prevent the annual heat requirements for respiration and foliage renewal. Wardle (1968) gave a detailed physiological explanation for the formation of timberline and the Krummholtz habit of Engelmann spruce in the Rocky Mountains of Colorado. He concluded that winter desiccation and the resulting death of needles was an immediate cause of the Krummholtz habit and, although the occurrence of timberline was correlated with minimum summer temperatures, other factors such as wind, snow accumulation, and insolation resulted in a complex explanation for the occurrence of the forest–tundra boundary. Lindsay (1971) measured the annual variation in leaf water potential of Engelmann spruce and subalpine fir at timberline in southern Wyoming and found that exposed timberline needles had much lower water potentials (-3.0 to -3.5 MPa) compared with needles from lower elevation forest sites and from within Krummholtz mats above timberline. Lindsay concluded that several factors contribute to winter desiccation at timberline including air temperature, wind, absolute humidity of the air, insolation, and the effects of soil temperature on soil moisture absorption. Of these, wind was believed to be the most important factor leading to needle desiccation damage (Lindsay, 1971). However, other factors such as the short growing season due to temperature and snow drifting were recognized as potential contributors to treeline formations. In fact, wind may actually act to decrease needle desiccation by reducing sunlight needle temperatures (Marchand and Chabot, 1978; Smith and Geller, 1979). Hansen and Klikoff (1972) found that Engelmann spruce

Krummholtz in the Wasatch Mountains of Utah developed leaf water potentials of below -9.0 MPa. Soils were found to freeze near Krummholtz mats by May 1, but not beneath normal spire-shaped trees of Engelmann spruce that had relatively deep snowpacks around their bases.

More recent studies have pointed to the possibility that microclimatic conditions during summer at timberline may also lead to less cuticle development and, thus, result in greater winter needle desiccation and mortality (Tranquillini, 1979; Wardle, 1981). However, Hadley and Smith (1983) found that needle mortality did not appear related to summer predisposition but to winter exposure and desiccation due to cuticle abrasion. These authors also concluded that numerous environmental–developmental factors may be interacting so that needle mortality could be caused by a variety of different factors during any given year.

A recent study found under-cooled tissue water during the winter in timberline species of Engelmann spruce and subalpine fir that did not become frozen until very near $-40\,°C$ (Becwar et al., 1981). Based on air temperature lapse rates, a common temperature of near $-40\,°C$ was also calculated to coincide with the elevations of a variety of timberlines in Colorado and other continental mountain regions of the western USA. Thus, the low temperature limits for subfreezing tissue water may be an important factor contributing to the elevational occurrence of upper timberlines.

Schulze et al. (1967) measured CO_2 exchange during the winter for bristlecone pine at timberline in the White Mountains of southern California. These authors calculated that at least half of the summer growing season was needed to replenish loss of CO_2 during the winter via respiration. Thus, annual carbon balance may certainly play an important role in the formation of upper timberlines, possibly via direct influence on such important factors as cuticle development.

5.4 Summary and perspectives

The majority of ecophysiological studies specifically involving plants within western montane forests have dealt primarily with overstory tree species.

Table 5.3 Photosynthetic parameters of western montane species. Values included are stomatal conductance (g^{st}), mesophyll conductance (g^m), and total leaf conductance (g^l) to CO_2 uptake: light saturation level (Lt. Sat.) and compensation level (Lt. Comp.) for photosynthesis; the CO_2 compensation point (CO_2 Comp.) and temperature optimum (Temp. Opt.) for photosynthesis, the xylem pressure (ψ_p^{xylem}) and soil water potential (ψ^{soil}); and photosynthesis (J_{CO_2})

	Area	Material	g^{st}	g^m	g^l	Lt. Sat.	Lt. Comp.	CO_2 Comp.	Temp. Opt.	ψ_p^{xylem}	ψ^{soil}	J_{CO_2}	Reference
Overstory													
Abies grandis	Douglas fir stand, LaGrand, Wash.	Planted seedlings, 2–4 inches of branch tip	5–6 psi (infil. press.)	—	—	4–600 ft.c.	—	—	20–25 °C	−14 to −20 bar	—	18.9–31.6 nmol g^{-1} s^{-1}	Hodges (1967)
	B.C., Canada	potted 5 yr old seedlings in greenhouse	—	—	—	—	—	—	—	−1.0 MPa −1.6 −2.0 −2.6 −3.0 −4.3 resp. −1.0 MPa −1.4 −3.8	—	100% 80 60 40 20 0 100% 80 75	Puritch (1973)
Abies lasiocarpa	mixed lodgepole pine/subalpine fir stand, S.E. Wyoming	4–10 in branch tips for understory saplings from 10–15 yr old	4.55 mm s^{-1}	1.35 mm s^{-1}	0.87 mm s^{-1}	—	—	—	—	> 0.3 MPa	—	8.2 μmol m^{-2} s^{-1}	Smith (1979)
	B.C., Canada	potted 5 yr old seedlings in greenhouse	—	—	—	—	—	—	—	−1.1 MPa −1.0 −1.4 −1.6 −2.9 resp. −0.7 MPa −1.4 −2.7	—	100% 80 60 40 20 0 100% 80 60	Puritch (1973)
	spruce–fir, S.E. Wyoming 3100 m	2–4 yr old field seedlings	—	—	—	780 μE m^{-2} s^{-1}	60–70 μE m^{-2} s^{-1}	—	—	—	—	1.8 μmol m^{-2} s^{-1}	Knapp and Smith (1982)
Abies procera	Douglas fir stand, La Grand, Wash.	planted seedlings, 2–4 inches of branch tip	5–8 psi (infil. press.)	—	—	—	—	—	—	−1.2 MPa	—	18.9–31.6 nmol g^{-1} s^{-1}	Hodges (1967)
Picea engelmannii	central Colorado	3 yr old potted seedlings in outdoor frames	—	—	—	—	5–7000 ft.c.	—	—	—	—	7.6–11.7 μmol m^{-1} s^{-1}	Ronco (1970)
	spruce–fir S.E. Wyoming, 3100 m	2–4 yr old field seedlings	—	—	—	1080 μE m^{-2} s^{-1}	60–70 μE m^{-2} s^{-1}	—	—	—	—	1.8 μmol m^{-2} s^{-1}	Knapp and Smith (1982)
Picea glauca	B.C., Canada	greenhouse grown, 16 cm high potted seedlings	—	—	—	—	—	—	—	−1.2 MPa −1.2 MPa −1.7 −2.8	—	9.8 μmol m^{-2} s^{-1} 57.5 nmol g^{-1} s^{-1} 100% of max 50% 0%	Brix (1979)
Picea pungens	B.C., Canada 2743 m	2–3 yr old potted field plants, current year foliage	—	—	—	0.82 × 10^5 ergs cm^{-2} s^{-1} (red light)	—	52 μl^{-1}	—	—	—	12.3 nmol g^{-1} s^{-1}	Clark and Lister (1975)

115

Table 5.3 (*contd.*)

Area	Material	g^{st}	g^m	g^l	Lt. Sat.	Lt. Comp.	CO_2 Comp.	Temp. Opt.	ψ_p^{xylem}	ψ^{soil}	J_{CO_2}	Reference
					0.73×10^5 erg cm^{-2} s^{-1} (blue light)	$83\,\mu l\,l^{-1}$					2.95 nmol g^{-1} s^{-1}	
Pinus aristata	S. Calif., 3230 m	field and green-house seedlings	—	—	—	—	—	—	20 °C 10–40 °C 10–40 °C	—	—	7.49 nmol g^{-1} s^{-1} resp. = 0.68–4.47	Mooney *et al.* (1964)
	S. Calif., 3094 m	field saplings, 1.5–2.5 cm diameter at breast height (d.b.h.)	—	—	—	—	—	—	10–15 °C 10–30 °C	—	—	5.90–8.40 nmol g^{-1} s^{-1}	Mooney *et al.* (1966b)
	S. Calif. 3094 m	2–3 m adult trees in field, 4–6.5 cm. d.b.h.	—	—	—	—	—	—	10–15 °C −5.30	—	—	7.26–7.94 nmol g^{-1} s^{-1} resp. = 1.36–0.41	Schultze *et al.* (1967)
Pinus attenuata	S. Calif.	greenhouse grown seeds to 3 yr old	—	—	—	—	—	—	15–20 °C	—	—	9.5–37.9 nmol g^{-1} s^{-1}	Wright (1970a)
	S. Calif. 1600 m	shoots of mature trees	—	—	—	—	—	—	—	—	—	8.17–31.6 nmol g^{-1} s^{-1}	Wright (1970b)
	S. Calif.	greehouse seedlings several cm tall	—	—	—	7200 ft.c.	—	—	—	—	—	37.9 nmol g^{-1} s^{-1}	Bryan and Wright (1976)
	S. Calif. seeds collected at 762, 1128, and 1372 m	greenhouse grown seedlings	—	—	—	—	—	175–350 ppm	—	−1.5 −3.0 MPa	8.6–14.3% soil moist		Hurt and Wright (1976)
Pinus contorta	B.C., Canada	Greenhouse grown 3 month old seedlings	—	—	—	3–5000 lumens ft^{-2}	—	—	—	—	—	113.5 nmol g^{-1} s^{-1}	Driessche and Wareing (1966)
	Pacific north west, 1219 m	12–19 wk old potted seedlings from 4 provences	—	—	—	20–30×10^4 erg cm^{-2} s^{-1}	0.32–0.39×10^4 erg cm^{-1} s^{-1}	—	—	—	—	8.85–11.4 nmol g^{-1} s^{-1}	Sweet and Wareing (1968)
	central Colorado	2 yr old seedlings grown in greenhouse at 2134 m	—	—	—	> 12,000 ft.c.	—	—	—	—	—	15.2 nmol m^{-3} s^{-1}	Ronco (1970)
	New Zealand timberline, 1800 m	20–30 cm tall seedlings collected and grown for 6 month in greenhouse	—	—	—	—	—	—	26–27 °C	—	—	20.9 nmol g^{-1} s^{-1}	Scott (1970)
	B.C., Canada	greenhouse grown, 25 cm 2 yr old	2.5 mm s^{-1}	1.3 mm s^{-1}	0.83 mm s^{-1}	380 W m^{-2}	—	—	20 °C	−0.3 MPa −0.7 −1.5	—	6.4 μmol m^{-2} s^{-1} 3.2 0.0	Dykstra (1974)

116

Table 5.3 (contd.)

	Area	Material	g^{st}	g^m	g^l	Lt. Sat.	Lt. Comp.	CO_2 Comp.	Temp. Opt.	ψ_p^{xylem}	ψ^{soil}	J_{CO_2}	Reference
	S.E. Wyoming, 2740 m	shoots from 10–15 yr old saplings in field	3.4 mm s^{-1}	1.4 mm s^{-1}	0.9 mm s^{-1}	—	—	—	—	—	—	9.3 μmol m^{-2}s^{-1}	Smith (1979)
Pinus coulteri	S. Calif.	greenhouse grown, 3 yr old, primary and secondary needles	—	—	—	1050 ft.c.	—	—	15–17 °C	—	—	10.7–19.5 g^{-1}s^{-1}	Wright (1970a)
	S. Calif. 1600 m, 40% S.W. slope	shoots from mature trees	—	—	—	—	—	—	—	—	—	June: 18.8–1.39 nmol g^{-1}s^{-1} Oct: 3.18–12.7	Wright (1970b)
Pinus flexilus	S. Calif. 3292 mm	branches of mature trees	—	—	—	—	—	—	20 °C	—	—	1.36 nmol g^{-1}s^{-1} (fresh weight, f.w.)	Mooney *et al.* (1964)
Pinus lambertii	S. Calif.	greenhouse grown seedlings to 3 yr old, primary and secondary needles	—	—	—	1050 ft. c.	—	—	24 °C	—	—	11.8–32.9 nmol g^{-1}s^{-1}	Wright (1970a)
	S. Calif., 1600 m	shoots of mature trees	—	—	—	—	—	—	—	—	—	June: 6, 36–14.5 nmol g^{-1}s^{-1} Oct: 0–18.8	Wright (1970b)
Pinus monophylla	S. Calif. 2195–2621 m	branches of mature trees	—	—	—	—	—	—	10–30 °C	—	—	1.8–4.3 nmol g^{-1}s^{-1} (f.w.) resp. 6.4–25.2	Mooney *et al.* (1964)
Pinus ponderosa	S.W. North Dakota, N.E. slope	field: > 6 cm d.b.h., 25–40 yr old lab: potted 10–15 yr old	—	—	—	7500 ft.c.	520 ft.c. (24 °C) 135 ft.c. (9 °C)	—	25–35 °C 7–41 °C	—	—	30.2 nmol g^{-1}s^{-1} resp. = 1.8–25.2 nmol g^{-1}s^{-1}	Hadley (1969)
	central Calif. 1300 m	4, 40, and 75 yr old trees up to 26 m ht; current, 2, 3, 4 yr old needles	—	—	—	—	—	—	—	—	—	12.7–25.2 nmol g^{-1}s^{-1}	Helms (1970)
	central Calif. 1300 m	50 year old trees	—	—	—	—	—	—	—	—	—	12.7–31.6 nmol g^{-1}s^{-1}	Helms (1971)
	S. Calif., 1725 m, open forest	mature trees	—	—	—	—	—	—	—	—	—	12.7–18.8 nmol g^{-1}s^{-1}	Green and Wright (1977)
Populus tremuloides	Massachusetts	potted root cutting grown in greenhouse	—	—	—	—	—	—	—	—	—	38.6–45.4 nmol g^{-1}s^{-1}(f.w.) resp. = 2.04–2.50	Bourdeau (1958a)
	S.E. Wyoming 2649 m	bark measurements, 40 yr old, 20 ft tall	—	—	—	—	—	—	—	—	—	2.27–3.18 nmol g^{-1}s^{-1}(f.w.) resp. = 2.50–3.41	Strain and Johnson (1963)
	New Haven, Connecticut	potted root cuttings	—	—	—	3–4000 ft.c. 44–58 W m^{-2}	120–300 ft.c.	—	—	—	—	20.2 μmol m^{-2}s^{-1} resp. = 1.59–2.95	Loach (1967)
	Syracuse, N.Y.	stem measurements of 3–5 yr old potted trees, outdoors	—	—	—	Summer 1800–2400 ft.c. 70–95 W m^{-2} Winter 9–1200 ft.c. 35–45 W m^{-2}	—	—	—	—	—	5.91 μmol m^{-2}s^{-1}	Foote and Schaedle (1976a)

Table 5.3   (*contd.*)

	Area	Material	g^{st}	g^m	g^l	Lt. Sat.	Lt. Comp.	CO_2 Comp.	Temp. Opt.	ψ_p^{xylem}	ψ^{soil}	J_{CO_2}	Reference
	Syracuse, N.Y.	stem measurements of 6–8 yr old potted plants in greenhouse and field	—	—	—	—	—	—	Summer 25 °C Winter 10–15 °C	—	—	6.36–31.6 μmol m^{-2} s^{-1} 0.023–0.295	Foote and Schaedle (1976b)
Tsuga heterophylla	B.C., Canada	2 yr old greenhouse grown	—	—	—	15–20 klux	—	—	—	—	—	12.7–31.6 nmol g^{-1} s^{-1}	Keller and Tregunna (1976)
		10–12 wk old seedlings grown in water culture	—	—	—	70 W m^{-2} (Photosynthetically Active Radiation, PAR)	—	—	—	—	—	8.17–86.3 nmol g^{-1} s^{-1} resp. = 10.7	Szaniawski and Wierzbicki (1978)

However, comprehensive physiological data on any given conifer species are not available (Table 5.2 and 5.3). Moreover, there is an obvious paucity of field data compared to laboratory and greenhouse measurements. Most of the general concepts related to successional changes of these montane forests following disturbance come from observational studies, with only meager physiological measurements to support these current ideas. A similar situation exists for virtually all of the explanations for both the large- and small-scale distribution patterns of plants within the western montane forests. Some of the most detailed physiological studies have focused on the formation and persistence of upper and lower treelines. Although recent research has been initiated within the forest understory, much less is known concerning the ecophysiological characteristics of these herbaceous and shrubby species. As a result, only a superficial summary of the potential ecophysiological factors influencing species interactions is possible.

The distribution and successional patterns of the western montane species are most obviously coupled to elevational effects such as the substantial decrease in air temperature that occurs with increasing elevation, along with numerous other biophysical factors. This biophysical coupling is reflected in the often sharp elevational demarcations between the dominant overstory conifers (Fig. 5.1). A strong influence by the physical environment, in combination with the occurrence of fire, results in a complex successional system. In many areas, disturbance appears to be followed by a relatively unchanging physiognomy whereas other areas show a more structured successional sequence of understory–overstory replacements. As an example, mature lodgepole pine of southeastern Wyoming may be found in monotypic stands over a wide range of elevations with various mixtures of lodgepole pine, spruce, or fir within the understory. There are even locations at relatively low elevations (2140 m) that have lodgepole pine, ponderosa pine, Douglas fir, Engelmann spruce, limber pine, and subalpine fir all growing naturally at the same site without any apparent differences in edaphic factors (Smith *et al.*, 1984). At higher elevations approaching timberline, Engelmann spruce and subalpine fir often occur together in about equal numbers as well as separately in almost pure stands. At middle elevations lodgepole pine appears to occur in seral stages both above and below a relatively narrow elevational band where climax stands occur (Fig. 5.1). Young stands of aspen may be observed with lodgepole, spruce, fir, or limber pine (*Pinus flexilis*) understories, plus numerous combinations. One may also find aspen stands that appear to represent a climax stage with trees greater than 300 years old. A much more detailed interpretation of habitat types and successional stages in the western forests can be found in Pfister *et al.* (1977).

According to the studies reviewed here, the growth and establishment of low elevation ponderosa pine may be particularly limited by the lower temperatures associated with higher elevations. Also, this species appears to have a capability for improved water status as evidenced by a high leaf conductance under drier soil conditions, possibly due to a more prolific root system and tolerance to cellular dehydration. Lodgepole pine appears to have similar adaptations to the warmer and drier conditions of lower elevations which include a high leaf conductance and xylem pressure potential under drying soil conditions, a large sapwood volume for water storage and conductance, and improved seed germination at high temperatures and greater soil dryness. Both lodgepole and ponderosa pine also close stomata with increasing vapor-pressure deficit from the leaf to the air.

The occurrence of Engelmann spruce and subalpine fir at higher elevations also appears related to cooler temperatures and wetter soils. Germination is higher at cooler temperatures, leaf conductance declines rapidly with soil drying or increases in the vapor-pressure deficit of the air, and mortality may be high for seddlings exposed to high insolation. However, it also appears as though spruce may require higher light levels than found in the understory environment as evidenced by the frequent occurrence of spruce at the exposed forest–meadow boundary. Another interpretation may involve the greater survival of spruce seedlings on a thinner litter layer or, ultimately, the wetter conditions associated with such substrates as mineral soils and fallen logs. The stomata of spruce also appear to be especially sensitive to changes in the leaf-to-air water vapor deficit. More information on photosynthetic and respiratory processes of western montane species is needed before attempting any explanation of distributional patterns based on physiological performance.

In addition to elevational effects the physiological ecology of plant succession in the western montane forests appears to be distinct in many respects from the temperate deciduous forests and tropical forests Bazzaz, 1979; Bazzaz and Pickett, 1980). The lower annual precipitation, a perennial overstory leaf canopy (except for aspen stands), and the more open nature of these conifer communities, results in an understory environment that is strongly influenced by a continuous mosaic of sunlight patterns. This situation is in contrast to the strong seasonality and more closed leaf canopy of the eastern deciduous and wet tropical forests. However, current interest in the dynamics of tropical 'gap phase' succession may be served by comparisons with similar studies of the much more persistent and predictable gaps of the western coniferous forests. There is a considerable lack of information on the physiological adaptations of gap species and the role of these species in forest community ecology.

Future ecophysiological studies in the western montane forests must include investigations on seed dispersal, germination, seedling growth and survival, and the growth and reproductive efforts of mature trees under natural conditions before the simplest of ecological questions concerning distribution, succession, or treeline boundaries can be comprehensively evaluated. The importance of the understory plant species to the vitality of the overstory tree species has not been elucidated, nor vice versa. This overstory–understory interaction may be altered drastically as clear-cutting policies give way to the more aesthetic, and potentially more productive, practice of selective thinning. Germination and seedling survival may play a potentially important role in the ecology of both overstory and understory species. The predominance of sexual reproduction in the overstory conifers compared to the asexual reproductive systems of many of the understory species throughout much of the Rocky Mountains presents an interesting comparison of these evolutionary 'strategies'. Mature, adult plants generally have low mortality (primarily due to fire or disease) and may not be as vulnerable to yearly fluctuations in the physical environment. However, the important influence of these mature plants on the observed species distribution patterns may be through the success of their reproductive efforts. An extremely small number of ecophysiological studies have attempted to correlate apparent physiological adaptations to an enhancement of reproductive effort beyond only intuitive speculations. As a simple illustration, a gain in seasonal carbon fixation or water-use efficiency does not automatically dictate that more seeds or more viable seeds will be produced, and, thus, result in enhanced reproductive fitness. This coupling between plant physiological measurements and corresponding changes in reproductive effort of adult plants has been a common omission in most ecophysiological studies.

An overriding concern in almost all evaluations of ecological fitness and physiological adaptation to any habitat is the potential influence that certain unusual years or even sequences of years may have on the evolution and ecology of these montane plant species. There is a distinct need for field studies conducted under natural conditions over long-term periods that will elucidate the effects of natural and man-made disturbances on individual species, physiology. The forests of the western mountains may be unique in many respects compared to other forests in North America and throughout the world. Much research is needed before the ecology of these areas will be understood from a historical as well as managerial viewpoint.

References

Anthony, J.B. (1974) *The effect of moisture, radiation, temperature, and wind upon forest trees*, Berkshire Community College Press, Pittsfield, Massachusetts, 155 pp.

Atzet, R. and Waring, R.H. (1970) Selective filtering of light by coniferous forests and minimum light energy requirements for regeneration. *Canadian Journal of Botany*, 48, 2163–7.

Baker, F.S. (1918) Aspen reproduction in relation to management. *Journal of Forestry*, 16, 389–98.

Baker, F.S. (1925) *Aspen in the central Rocky Mountain region*, United States Department of Agriculture Bulletin 1291, 46 pp.

Baker, F.S. (1944) Mountain climates of the western United States. *Ecological Monographs*, 14, 223–54.

Baldwin, V.C. and Barney, C.W. (1976) Leaf water potential in planted ponderosa and lodgepole pines. *Forest Science*, 22, 344–50.

Barnes, B.V. (1966) The clonal growth habit of American aspens. *Ecology*, 47, 439–97.

Bates, C.G. (1917) Forest succession in the central Rocky Mountains. *Journal of Forestry*, 15, 587–92.

Bates, C.G. (1923) Physiological requirements of Rocky Mountain trees. *Journal of Agriculture Research*, 24, 97–164.

Bates, C.G. (1924) *Forest types in the central Rocky Mountains as affected by climate and soil*, United States Department of Agriculture Bulletin 1233, 152 pp.

Bates, C.G., Notestein, N.B. and Keplinger, P. (1914) Climatic characteristics of forest types in the central Rocky Mountains. *Society of American Foresters Proceedings*, 9, 78–94.

Bazzaz, F.A. (1979) The physiological ecology of plant succession. *Annual Review of Ecology and Systematics*, 10, 351–71.

Bazzaz, F.A. and Pickett, S.I.A. (1980) Physiological ecology of tropical succession. A comparative review. *Annual Review of Ecology and Systematics*, 11, 287–310.

Becker, C.F. and Boyd, J.S. (1961) Availability of solar energy. *Agricultural Engineering*, 42, 302–5.

Becwar, M.R., Rajashekhar, C., Hansen-Bristow, K.J. and Bruke, M.J. (1981) Deep undercooling of tissue water and winter hardiness limitations in timberline flora. *Plant Physiology*, 68, 111–14.

Behan, M.J. (1958) *The vegetation and ecology of Dry Park in the Medicine Bow Mountains of Wyoming*, Msc thesis, University of Wyoming, Laramie, Wyoming.

Bergen, J.D. (1971) Vertical profiles of windspeed in a lodgepole pine stand. *Forest Science*, 17, 319–21.

Bergen, J.D. (1972) Topographic effects apparent in nocturnal temperature profiles in a conifer canopy. *Agricultural Meteorology*, 9, 39–50.

Bergen, J.D. (1974a) Spatial variation and scaling problems for vertical air temperature profiles in a pine stand. *Forest Science*, 20, 64–73.

Bergen, J.D. (1974b) *Variation of windspeed with canopy cover within a lodgepole pine stand*, United States Department of Agriculture, Forest Service Research Note, RM-252, 4 pp.

Bergen, J.D. (1977) Windspeed distribution in and near an isolated narrow forest clearing. *Agricultural Meteorology*, 17, 11–13.

Bethan, E. (1925) The conifers 'evergreens' of Colorado. *Colorado Magazine*, 2, 1–23.

Billings, W.D. (1969) Vegetational pattern near timberline as affected by fire–snowdrift interactions. *Vegetatio*, 19, 192–207.

Boggie, R. (1974) Response of seedlings of *Pinus contorta* and *Picea sitchensis* to oxygen concentration in culture solution. *New Phytology*, 73, 467.

Bourdeau, P.F. (1958a) Photosynthetic and respiratory rates in leaves of male and female quaking aspens. *Forest Science*, 4, 331–4.

Bourdeau, P.F. (1958b) Photosynthetic behavior of sun-

and shade-grown leaves in certain tolerant and intolerant tree species. *Abstracts in Bulletin Ecological Society America*, **39**, 84 pp.

Brix, H. (1979) Effects of plant water stress on photosynthesis and survival of four conifers. *Canadian Journal of Forestry Research*, **9**, 160–5.

Bryan, J. and Wright, R.D. (1976) The effect of enhanced CO_2 level and variable light intensities on net photosynthesis in competing mountain trees. *American Midland Naturalist*, **95**, 446–50.

Bunce, J.A., Chabot, B.F. and Miller, L.N. (1979) Role of annual leaf carbon balance in the distribution of plant species along an elevational gradient. *Botanical Gazette*, **140**, 288–94.

Caldwell, M.M. (1981) Plant response to solar ultraviolet radiation, *Physiological Plant Ecology I, Responses to the Physical Environment*, (eds O.L. Lange, P.S. Nobel, C.B. Osmond, H. Ziegler), Springer-Verlag, New York.

Clark, J.B. and Lister, G.R. (1975) Photosynthetic action spectra of trees, II. The relationship of cuticle structure to the visible and ultraviolet spectral properties of needles from four coniferous species. *Plant Physiology*, **55**, 401–6.

Cline, R.G. and Campbell, G.S. (1976) Seasonal and diurnal water relations of selected forest species. *Ecology*, **57**, 367–73.

Cochran, P.H. (1972) Temperature and soil fertility affect lodgepole and ponderosa pine seedling growth. *Forest Science*, **18**, 132–4.

Cormack, R.G.H. (1953) A survey of forest succession in the eastern Rockies. *Forestry Chronicles*, **29**, 218–32.

Coutts, M.P. and Philipson, J.J. (1978a) Tolerance of tree roots to waterlogging. I. Survival of sitka spruce and lodgepole pine. *New Phytology*, **80**, 63–9.

Coutts, M.P. and Philipson, J.J. (1978b) Tolerance of tree roots to waterlogging. II. Adaptation of sitka spruce and lodgepole pine to waterlogged soil. *New Phytology*, **80**, 71–7.

Dahms, W.G. (1971) *Growth and soil moisture in thinned lodgepole pine*, USDA Forest Service Research Paper PNW-127.

Daniel, T.W. and Schmidt, J. (1972) Lethal and nonlethal effects of the organic horizons of forested soils on germination of seeds from several associated-conifer species of the Rocky Mountains. *Canadian Journal of Forestry Research*, **2**, 179–84.

Daubenmire, R.F. (1943a) Soil temperature versus drought as a factor determining lower altitudinal limits of trees in the Rocky Mountains. *Botanical Gazette*, **105**, 1–13.

Daubenmire, R.F. (1943b) Vegetational zonation in the Rocky Mountains. *Botanical Review*, **9**, 326–93.

Daubenmire, R.F. (1952) Forest vegetation of northern Idaho and adjacent Washington and its bearing on concepts of vegetation classification. *Ecological Monographs*, **22**, 301–30.

Daubenmire, R.F. (1954) Alpine timberlines in the Americas and their interpretation. *Butler University Botanical Studies*, **11**, 119–36.

Daubenmire, R.F. (1956) Vegetation: identification of typal communities. *Science*, **151**, 291–8.

Daubenmire, R.F. (1968) *Plant Communities*, Harper and Row, New York, 300 pp.

Daubenmire, R.F. (1978) *Plant Geography*, Academic Press, New York, 338 pp.

Day, M.W. (1944) The root system of aspen. *American Midland Naturalist*, **32**, 502–7.

Day, R.J. (1972) Stand structure, succession, and use of southern Alberta's Rocky Mountain Forest. *Ecology*, **53**, 472–8.

DeByle, N.V. (1964) Detection of functional intraclonal aspen root connections by tracers and excavation. *Forest Science*, **10**, 386–96.

Despain, D.G. (1973) Vegetation of the Big Horn Mountains, Wyoming, in relation to substrate and climate. *Ecological Monographs*, **43**, 329–55.

Driessche, V.D. and Wareing, P.F. (1966) Dry-matter production and photosynthesis in pine seedlings. *Annals of Botany*, **30**, 673–82.

Dunnewald, T.J. (1930) Grass and timber soils distribution in the Big Horn Mountains. *Journals of the American Society of Agronomists*, **22**, 577–86.

Dykstra, G.F. (1974) Photosynthesis and carbon dioxide transfer resistance of lodgepole pine seedlings in relation to irradiance, temperature, and water potential. *Canadian Journal of Forestry Research*, **4**, 201–6.

Ellison, L. (1943) A natural seedling of western aspen. *Journal of Forestry*, **41**, 767–8.

Fahey, T.J. (1977) Changes in nutrient content of snow water during outflow from Rocky Mountain coniferous forests. *Oikos*, **32**, 422–8.

Fahey, T.J. (1979) The effect of night frost on the transpiration of *Pinus contorta* ssp. *latifolia*. *Oecologia Plantarum*, **19**, 483–90.

Fahey, T.J. (1983) Nutrient dynamics of above ground detritus in lodgepole pine (*Pinus contorta ssp. latifolia*) ecosystems, southeastern Wyoming. *Ecological Monographs*, **53**, 51–72.

Farmer, R.E. (1962) Aspen root sucker formation and apical dominance. *Forest Science*, **8**, 403–10.

Faust, M.E. (1936) Germination of *Populus grandidentata*

and *P. tremuloides* with particular reference to oxygen consumption. *Botanical Gazette*, 97, 808–21.

Fetcher, N. (1976) Patterns of leaf resistance to lodgepole pine transpiration in Wyoming. *Ecology*, 57, 339–45.

Foote, K.C. and Schaedle, M. (1976a) Physiological characteristics of photosynthesis and respiration in stems of *Populus tremuloides*. *Plant Physiology*, 58, 91–4.

Foote, K.C. and Schaedle, M. (1976b) Diurnal and seasonal patterns of photosynthesis and respiration in stems of *Populus tremuloides* Michx. *Plant Physiology*, 58, 651–5.

Foote, K.C. and Schaedle, M. (1978) The contribution of aspen bark photosynthesis to the energy balance of the stem. *Forest Science*, 24, 569–73.

Fowells, H.A. (1948) The temperature profile in a forest. *Journal of Forestry*, 46, 897–9.

Fritts, H.C. (1966) Growth rings of trees: their correlation with climate. *Science*, 154, 973–9.

Gail, F.W. and Long, E.M. (1935) A study of site, root development, and transpiration in relation to the distribution of *Pinus contorta*. *Ecology*, 16, 88–100.

Gale, J. (1972) Availability of carbon dioxide for photosynthesis at high altitudes: theoretical considerations. *Ecology*, 53, 494–7.

Gale, J. (1973) Experimental evidence for the effect of barometric pressure on photosynthesis and transpiration, in *Plant Response to Climate Factors*, UNESCO, Proceedings, Uppsala Symposium on Ecology and Conservation 5, pp. 289–94.

Geiger, R. (1965) *The climate near the ground*, Harvard University Press, Cambridge, Massachusetts, 249 pp.

Gleason, H.A. and Cronquist, A. (1964) *The natural geography of plants*, Columbia University Press, New York, 316 pp.

Gosz, J.R. (1980) Nitrogen cycling in coniferous ecosystems, in *Terrestrial Nitrogen Cycles* (eds R.S. Cark and T. Rosswall), Ecological Bulletin 33.

Green, K. and Wright, R. (1977) Field response of photosynthesis to CO_2 enhancement in ponderosa pine. *Ecology*, 58, 687–92.

Griggs, R.F. (1938) Timberlines in the northern Rocky Mountains. *Ecology*, 19, 548–64.

Griggs, R.F. (1946) The timberlines of northern America and their interpretation. *Ecology*, 27, 275–89.

Hadley, E.B. (1969) Physiological ecology of *Pinus ponderosa* in southwestern North Dakota. *American Midland Naturalist*, 81, 289–314.

Hadley, J.L. and Smith, W.K. (1983) Influence of wind exposure on needle desiccation and mortality for timberline conifers in Wyoming, USA. *Arctic and Alpine Research*, 15, 127–35.

Hansen, D.H. and Klikoff, L.G. (1972) Water stress in Krummholz, Wasatch Mountains, Utah, *Botanical Gazette*, 133, 392–4.

Hellmers, H., Genthe, M.K. and Ronco, F. (1970) Temperature affects growth and development of Engelmann spruce. *Forest Science*, 16, 447–52.

Helms, J.A. (1970) Summer net photosynthesis of ponderosa pine in its natural environment. *Photosynthetica*, 4, 243–53.

Helms, J.A. (1971) Environmental control of net photosynthesis in naturally growing *Pinus ponderosa* Laws. *Ecology*, 53, 92–101.

Hodges, J.D. (1967) Patterns of photosynthesis under natural environmental conditions. *Ecology*, 48, 234–42.

Hodson, E.R. and Foster, J.H. (1910) *Engelmann spruce in the Rocky Mountains*, United States Department of Agriculture, Forest Service Circular, 170.

Hurt, P. and Wright, R.D. (1976) CO_2 compensation point for photosynthesis: Effect of variable CO_2 and soil moisture levels. *American Midland Naturalist*, 95, 450–4.

Jackson, P.A. and Spomer, G.G. (1979) Biophysical adaptations of four western conifers to habitat water conditions. *Botanical Gazette*, 140, 428–32.

Janke, R.A. (1968) The ecology of *Vaccinium myrtillus*, using concepts of producibility, energy exchange, and transpiration resistance. PhD thesis, University of Colorado, Boulder, 149 pp.

Janke, R.A. (1970) Transpiration resistance in *Vaccinium myrtillus*. *American Journal of Botany*, 57, 1051–4.

Jarvis, P.G., James, G.B. and Landsberg, J.J. (1976) Coniferous forest, in *Vegetation and Atmosphere: Vol. 2 Case studies* (ed. J.L. Monteith), Academic Press, New York.

Johnston, R.S. (1975) *Soil water depletion by lodgepole pine on glacial till*, USDA Forest Research Note INT-199.

Kaufmann, M.R. (1975) Leaf water stress in Engelmann spruce: Influence of the root and shoot environments. *Plant Physiology*, 56, 841–4.

Kaufmann, M.R. (1976) Stomatal response of Engelmann spruce to humidity, light, and water stress. *Plant Physiology*, 57, 898–901.

Kaufmann, M.R. (1979) Stomatal control and the development of water deficit in Engelmann spruce seedlings during drought. *Canadian Journal of Forestry Research*, 9, 297–304.

Kaufmann, M.R. (1982a) Leaf conductance as a function of photosynthetic photon flux density and absolute humidity difference from leaf to air. *Plant Physiology*, 69, 1018–22.

Kaufmann, M.R. (1982b) Evaluation of season, temperature, and water stress effects on stomata using a leaf conductance model, *Plant Physiology*, **69**, 1023–6.

Kaufmann, M.R. and Eckard, A.N. (1977) Water potential and temperature effects on germination of Engelmann spruce and lodgepole pine seeds. *Forest Science*, **23**, 27–33.

Keller, R.A. and Tregunna, E.B. (1976) Effects of exposure on water relations and photosynthesis of western hemlock in habitat forms. *Canadian Journal of Forestry Research*, **6**, 40–8.

Kirkwood, J.E. (1922) *Forest Distribution in the Northern Rocky Mountains*, Montana State University Publication, Missoula.

Klikoff, L.G. (1965) Photosynthetic response to temperature and moisture stress in three timberline meadow species. *Ecology*, **46**, 516–17.

Knapp, A.K. and Smith, W.K. (1981) Water relations and succession in subalpine conifers in southeastern Wyoming. *Botanical Gazette*, **142**, 502–11.

Knapp, A.K. and Smith, W.K. (1982) Factors influencing understory seedling establishment of Engelmann spruce (*Picea engelmannii*) and subalpine fir (*Abies lasiocarpa*) in southeast Wyoming. *Canadian Journal of Botany*, **60**, 2753–61.

Knight, D.H., Rogers, B.S. and Kyte, C.R. (1977) Understory plant growth in relation to snow duration in Wyoming subalpine forest. *Bulletin Torrey Botanical Club*, **104**, 314–19.

Knight, D.H., Fahey, T.J. and Running, S.W. (1984) Water and nutrient outflow from contrasting lodgepole pine forests in Wyoming. *Ecological Monographs* (in press).

Korstian, C.F. (1925) Some ecological effects of shading coniferous nursery stock. *Ecology*, **6**, 48–51.

Korstian, C.F. and Fetherolf, N.J. (1921) Control of stem girdle of spruce transplants caused by excessive heat. *Phytopathology*, **11**, 985–90.

Korterba, W.D. and Habeck, J.R. (1971) Grasslands of the North Fork Valley, Glacier National Park, Montana. *Canadian Journal of Botany*, **49**, 1627–36.

Langenheim, J.H. (1962) Vegetation and environmental patterns in the Crested Butte Area, Gunnison County, Colorado. *Ecological Monographs*, **32**, 249–85.

Larsen, J.A. (1927) Relation of leaf structure of conifers to light and moisture. *Ecology*, **8**, 371–7.

Larsen, J.A. (1930) Forest types of the northern Rocky Mountains and their climatic controls. *Ecology*, **11**, 631–72.

Larson, G.C. (1944) More on seedlings of western aspen. *Journal of Forestry*, **42**, 452.

Lassen, L.E. and Okkonen, R. (1969) *Sapwood thickness of Douglas fir and five other western softwoods*, United States Department of Agriculture Forest Service Research Paper, FPL-124, 16 pp.

Lindsay, J.H. (1971) Annual cycle of leaf water potential in *Picea engelmannii* and *Abies lasiocarpa* at timberline in Wyoming. *Arctic and Alpine Research*, **3**, 131–8.

Loach, K. (1967) Shade tolerance in tree seedlings. I. Leaf photosynthesis and respiration in plants raised under artificial shade. *New Phytologist*, **66**, 607–21.

Lopushinsky, W. (1969) Stomatal closure in conifer seedlings in response to leaf moisture stress. *Botanical Gazette*, **130**, 258–63.

Lopushinsky, W. (1975) Water relations and photosynthesis in lodgepole pine, in *Management of lodgepole pine ecosystems, Vol. 1 Symposium: Proceedings of the Cooperative Extension Service College Agriculture* (ed. D.M.B. Baumgartner), Washington State University Press.

Lopushinsky, W. and Klock, G.O. (1974) Transpiration of conifer seedlings in relation to soil water potential. *Forest Science*, **20**, 181–6.

Maini, J.S. (1960) Invasion of grassland by *Populus tremuloides* in the northern Great Plains, PhD thesis, University of Saskatchewan.

Maini, J.S. and Horton, K.W. (1966) Vegetative propagation of *Populus* ssp. I. Influence of temperature on formation and initial growth of aspen suckers. *Canadian Journal of Botany*, **44**, 1183–9.

Maguire, W.P. (1955) Radiation, surface temperatures, and seedling survival. *Forest Science*, **1**, 277–85.

Marchand, P.J. and Chabot, B.F. (1978) Winter water relations of tree-line plant species on Mt. Washington, New Hampshire. *Arctic and Alpine Research*, **10**, 105–16.

Marston, R.B. (1956) Air movement under an aspen and on an adjacent opening. *Journal of Forestry*, **54**, 46.

McDonough, W.T. (1979) *Quaking aspen – seed germination and early seedling growth*, United States Department of Agriculture, Forest Service Research Paper INT-234, pp. 1–3.

McMinn, R.G. (1952) The role of soil drought in the distribution of vegetation in the northern Rocky Mountains. *Ecology*, **33**, 1–15.

Merriam, C.H. (1898) *Life zones and crop zones*, United States Department of Agriculture Division of Biological Surveys, Bulletin No. 10, 79 pp.

Miles, S.R. and Singleton, P.C. (1975) Vegetative history of Cinnabar Park in the Medicine Bow National Forest, Wyoming. *Soil Science Society of America Proceedings*, **39**, 965–7.

Minore, D. (1979) *Comparative autecological characteristics of north-western tree species*, United States Department of Agriculture Pacific Northwest Forest and Range Experiment Station Technical Report 87, 72 pp.

Moir, W.H. (1969) The lodgepole pine zone in Colorado. *American Midland Naturalist*, **81**, 87–98.

Moir, W.H. and Francis, R. (1972) Foliage biomass and surface area in three *Pinus contorta* plots in Colorado. *Forest Science*, **18**, 41–5.

Mooney, H.A., Wright, R.D. and Strain, B.R. (1964) The gas exchange capacity of plants in relation to vegetation zonation in the White Mountains of California. *American Midland Naturalist*, **72**, 281–97.

Mooney, H.A., Strain, B.R. and West, M. (1966a) Photosynthetic efficiency at reduced carbon dioxide tension. *Ecology*, **47**, 490–1.

Mooney, H.A., West, M. and Brayton, R. (1966b) Field measurements of the metalic responses of bristlecone pine and big sagebrush in the White Mountains of California. *Botanical Gazette*, **127**, 105–13.

Morgen, E.W. (1973) *Soil and rooting depths in lodgepole pine stands in northern Colorado and southern Wyoming*. College of Forestry and Natural Resources, Colorado State University, Research Note 19.

Moss, E.H. (1938) Longevity of seed and establishment of seedlings in species of *Populus*. *Botanical Gazette*, **99**, 529–42.

Noble, D.L. (1972) *Effects of soil type and watering on germination, survival and growth of Engelmann spruce; a greenhouse study*, United States Department of Agriculture Forest Service Research Note RM-216, 4 pp.

Noble, D.L. (1974) Natural regeneration of Engelmann spruce in clearcut openings in the central Rockies as affected by weather, aspect, seedbed and biotic factors. PhD thesis, Colorado State University, 187 pp.

Noble, D.L. and Alexander, R.R. (1977) Environmental factors affecting natural regeneration of Engelmann spruce in the central Rocky Mountains. *Forest Science*, **23**, 420–9.

Okafo, O.A. and Hanover, J.W. (1978) Comparative photosynthesis and respiration of trembling and big-tooth aspens in relation to growth and development. *Forest Science*, **24**, 103–9.

Olsen, O.C. and Crockett, D.H. (1965) Relation of two soil substrata to forest cover and non-forested openings in the Seven Devil Mountains, in *Forest Soil Relationships in North America* (ed. C.T. Youngberg), Oregon State University, Corvallis.

Oosting, J.H. and Reed, J.F. (1952) Virgin spruce fir forest in the Medicine Bow Mountains, Wyoming. *Ecological Monographs*, **22**, 69–91.

Ostler, W.K. and Harper, H.T. (1978) Floral ecology in relation to plant species diversity in the Wasatch Mountains of Utah and Idaho. *Ecology*, **59**, 848–61.

Patten, D.T. (1963a) Light and temperature influence on Engelmann spruce seed germination and subalpine forest advance. *Ecology*, **44**, 817–18.

Patten, D.T. (1963b) Vegetational pattern in relation to environments in the Madison Range, Montana. *Ecological Monographs*, **33**, 373–406.

Patten, D.T. (1969) Succession from sagebrush to mixed conifer forest in the northern Rocky Mountains. *American Midland Naturalist*, **82**, 229–40.

Pearson, G.A. (1914) The role of aspen in the reforestation of mountain burns in Arizona and New Mexico. *Plant World*, **17**, 249–60.

Pearson, G.A. (1920) Factors controlling distribution of forest types. *Ecology*, **1**, 289–308.

Pearson, G.A. (1931) *Forest types in the Southwest as determined by climate and soil*, United States Department of Agriculture. Technical Bulletin 247, 36 pp.

Perry, T.O. (1971) Winter season photosynthesis and respiration by twigs and seedlings of deciduous and evergreen trees. *Forest Science*, **17**, 41–3.

Peterson, P.V. and Peterson, Jr, P.V. (1975) *Native trees of the Sierra Nevada*, California Natural History Guide 36, University of California Press, Berkeley, 147 pp.

Pfister, R.D., Kovalchik, B.L., Arno, S.E. and Presby, P.C. (1977) *Forest habitat types of Montana*, United States Department of Agriculture, Forest Service General Technical Report, INT-34, Intermountain Forest and Range Experimental Station, Ogden, Utah, 174 pp.

Pharis, R.P., Hellmers, H. and Schuurmans, E. (1970) Effect of sub-freezing temperatures on photosynthesis of evergreen conifers under controlled environmental conditions. *Photosynthetica*, **4**, 273–9.

Pharis, R.P., Hellmers, H. and Schuurmans, E. (1972) The decline and recovery of photosynthesis of ponderosa pine seedlings subjected to low, but above freezing temperatures. *Canadian Journal of Botany*, **50**, 1965–70.

Philipson, J.J. and Coutts, M.P. (1978) Tolerance of tree roots to waterlogging. III. Oxygen transport in lodgepole pine and sitka spruce roots of primary structure. *New Phytology*, **80**, 341–9.

Potter, L.D. and Green, D.L. (1964) Ecology of ponderosa pine in western North Dakota. *Ecology*, **45**, 10–23.

Preston, R.J. (1968) *Rocky Mountain Trees*, 3rd edn, Dover Publications, New York.

Puritch, G.S. (1973) Effect of water stress on photosynthesis, respiration, and transpiration of four *Abies* species. *Canadian Journal of Forestry Research*, **3**, 293–8.

Reed, R.M. (1976) Coniferous forest habitat types of the Wind River Mountains, Wyoming. *American Midland Naturalist*, **95**, 159–73.

Reid, R.W. (1961) Moisture changes in lodgepole pine before and after attack by the mountain pine beetle. *Forestry Chronicles*, **37**, 368–75.

Reifsynder, W. and Lull, H.W. (1965) *Radiant energy in relation to forests*, United States Department of Agriculture, Technical Bulletin No. 1344, 111pp.

Roeser, J. (1924) A study of Douglas fir reproduction under various cutting methods. *Journal of Agriculture Research*, **28**, 1233–42.

Roeser, J. (1932) Transpiration capacity of coniferous seedlings and the problem of heat injury. *Journal of Forestry*, **30**, 381–95.

Ronco, F. (1970) Influence of high light intensity on survival of planted Engelmann spruce. *Forest Science*, **16**, 331–9.

Root, R.A. and Habeck, J.R. (1972) A study of high elevational grassland communities in western Montana. *American Midland Naturalist*, **87**, 109–21.

Running, S.W. (1976) Environmental control of leaf and water conductance in conifers. *Canadian Journal of Forestry Research*, **6**, 104–12.

Running, S.W. (1980) Environmental and physiological control of water flux through *Pinus contorta*. *Canadian Journal of Forestry Research*, **10**, 82–91.

Running, S.W. and Reid, C.P. (1980) Soil temperature influences on root resistance of *Pinus contorta* seedlings. *Plant Physiology*, **65**, 635–40.

Rutter, A.J. (1968) Water consumption by forests, in *Water Deficits and Plant Growth II* (ed. T.T. Kozlowski), Academic Press, New York.

Sandberg, D. (1951) The regeneration of quaking aspen by root suckers, Masters thesis, University of Minnesota (unpublished).

Schaedle, M. (1975) Tree photosynthesis. *Annual Review of Plant Physiology*, **26**, 101–15.

Schier, G.A. and Campbell, R.B. (1978) Aspen sucker regeneration following burning and clearcutting on two sites in the Rocky Mountains. *Forest Science*, **24**, 303–8.

Schulze, E.D., Mooney, H.A. and Dunn, E.L. (1967) Wintertime photosynthesis of bristlecone pine (*Pinus aristata*) in the White Mountains of California. *Ecology*, **48**, 1044–7.

Scott, D. (1970) Comparison between lodgepole pine and mountain beech in establishment and CO_2 exchange. *New Zealand Journal of Botany*, **8**, 357–60.

Shaw, C.H. (1909) The causes of timberline on mountains. *Plant World*, **12**, 169–81.

Shirley, H.L. (1932) II. Does light burning stimulate aspen suckers? *Journal of Forestry*, **30**, 419–20.

Small, E. (1972) Water relations of plants in raised *Sphagnum* peat bogs. *Ecology*, **53**, 726–8.

Smith, W.K. (1979) Importance of aerodynamic resistance to water use efficiency in three conifers under field conditions. *Plant Physiology*, **65**, 132–5.

Smith, W.K. (1981) Temperature and water relation patterns in subalpine understory plants. *Oecologia*, **48**, 353–9.

Smith W.K. and Geller, G.N. (1979) Plant transpiration at high elevations: theory, field measurements, and comparisons with desert plants. *Oecologia*, **41**, 109–22.

Smith, W.K. and Geller, G.N. (1980) Leaf and environmental parameters influencing transpiration: Theory and field measurements. *Oecologia*, **46**, 308–13.

Smith, W.K., Young, D.R., Carter, G.A., Hadley, J.L. and McNaughton, G.M. (1984) Autumn stomatal closure in six conifer species of central Rocky Mountains *Oecologia*, **63**, 237–42.

Sperry, O.E. (1936) A study of the growth, transpiration, and distribution of conifers of the Rocky Mountain National Park. *Bulletin of the Torrey Botanical Club*, **63**, 75–103.

Spurr, S.H. and Barnes, B.V. (1980) *Forest Ecology*, 3rd edn, John Wiley and Sons, New York, 687 pp.

Stahelin, R. (1943) Factors influencing natural restocking of high altitude burns by coniferous trees in the central Rocky Mountains. *Ecology*, **24**, 19–30.

Stark, N.J. (1972) Nutrient cycling pathways and litter fungi. *Bioscience*, **22**, 355–60.

Stoeckeler, J.H. (1960) *Soil factors affecting the growth of quaking aspen forests in the Lake States*, University of Minnesota Agriculture Experiment Station Technical Bulletin 233.

Strain, B.R. and Johnson, D.L. (1963) Cortical photosynthesis and growth in *Populus tremuloides*. *Ecology*, **44**, 581–4.

Strong, R.M. and Parminter, J.V. (1980) Conifer encroachment on the Chilcotin Grasslands of British Columbia. *The Forestry Chronicle*, **56**, 13–18.

Swanson, R.H. (1975) Water use by mature lodgepole pine, in *Management of lodgepole pine ecosystems, Vol. 1 Symposium Proceedings Cooperative Extension Service, College of Agriculture* (ed. D.M.B. Baumgartner), Washington State University Press.

Sweet, G.B. and Wareing, P.F. (1968) A comparison of the rates of growth and photosynthesis in first-year seedlings of four provences of *Pinus contorta* Dougl. *Annals of Botany*, **32**, 735–51.

Szaniawski, R.K. and Wierzbicki, A. (1978) Net photosynthetic rate of some coniferous species at diffuse high irradiance. *Photosynthetica*, **12**, 412–17.

Thorp, J. (1931) The effects of vegetation and climate upon soil profiles in northern and western Wyoming. *Soil Science*, **3**, 283–301.

Tranquillini, W. (1979) *Physiological ecology of the alpine timberline*, Springer-Verlag, Berlin.

US Corps of Engineers. (1956) *Snow hydrology*, Summary

report of snow investigations, North Pacific Division, Portland, Oregon, 437 pp.

Vale, T.R. (1978) Tree invasion of Cinnabar Park in Wyoming. *American Midland Naturalist*, **100**, 277–89.

Wardle, P. (1965) A comparison of alpine timberlines in New Zealand and North America. *New Zealand Journal of Botany*, **3**, 113–35.

Wardle, P. (1968) Engelmann spruce (*Picea engelmannii* Engel.) at its upper limits on the Front Range, Colorado. *Ecology*, **49**, 483–95.

Wardle, P. (1971) An explanation for alpine timberline. *New Zealand Journal of Botany*, **9**, 371–402.

Wardle, P. (1974) Alpine timberlines, in *Arctic and Alpine Environments* (ed. J.D. Ives and R.G. Barry), Methuen, London.

Wardle, P. (1981) Winter desiccation of conifer needles simulated by artificial freezing. *Arctic and Alpine Research*, **13**, 419–23.

Weaver, J.E. (1917) A study of the vegetation of south-eastern Washington and adjacent Idaho. *Nebraska University Studies*, **17**, 1–133.

Weigle, W.G. and Frothingham, E.H. (1911) *The aspens: their growth and management*, United States Department of Agriculture, Forest Service Bulletin, 93, pp. 1–34.

Wellner, C.A. (1948) Light intensity related to stand density in mature stands of western white pine type. *Journal of Forestry*, **46**, 16–19.

Whipple, S.A. and Dix, R.L. (1979) Age structure and successional dynamics of a Colorado subalpine forest. *American Midland Naturalist*, **101**, 142–58.

White, F.M., Johnson, J.R. and Nichols, J.T. (1969) Prairie-forest transition soils of the South Dakota Black Hills. *Soils Science Society of America Proceedings*, **33**, 932–6.

Whittaker, R.H. and Niering, W.A. (1965) Vegetation of the Santa Catalina Mountains. II. A gradient analysis of the south slope. *Ecology*, **46**, 429–52.

Wright, R.D. (1966) Lower elevational limits of montane trees. I. Vegetational and environmental survey in the San Bernardino Mountains of California. *Botanical Gazette*, **127**, 184–93.

Wright, R.D. (1968) Lower elevational limits of montane trees. II. Differential environment-keyed reponses of three conifer species. *Botanical Gazette*, **129**, 219–26.

Wright, R.D. (1970a) CO_2 exchange of seedling pines in the laboratory as related to lower elevational limits. *The American Midland Naturalist*, **83**, 321–9.

Wright, R.D. (1970b) Seasonal course of CO_2 exchange in the field as related to lower elevational limits of pines. *The American Midland Naturalist*, **83**, 291–300.

Wright, R.D. (1974) Rising atmospheric CO_2 and photosynthesis of San Bernardino Mountain plants. *The American Midland Naturalist*, **91**, 360–70.

Young, D.R. and Smith, W.K. (1979) Influence of sunflecks on the temperature and water relations of two subalpine understory congeners. *Oecologia*, **43**, 195–205.

Young, D.R. and Smith, W.K. (1980) Influence of sunlight on photosynthesis, water relations, and leaf structure in the understory species of *Arnica cordifolia*. *Ecology*, **48**, 353–9.

Young, D.R. and Smith, W.K. (1982) Simulation studies of the influence of understory location on transpiration and photosynthesis relations of *Arnica cordifolia* on clear days. *Ecology*, **63**, 1761–70.

Young, D.R. and Smith, W.K. (1983) Effect of cloudcover on photosynthesis and transpiration in the subalpine understory species, *Arnica latifolia*. *Ecology*, **64**, 681–7.

Young, E. and Hanover, J.W. (1978) Effect of temperature, nutrient, and moisture stresses on dormancy of blue spruce seedlings under continuous light. *Forest Science*, **24**, 458–66.

Zahner, R. and DeByle, N.V. (1965) Effect of pruning and parent root on growth of aspen suckers. *Ecology*, **46**, 373–5.

Zehngraff, P.J. (1949) Aspen as a forest crop in the Lake States. *Journal of Forestry*, **47**, 555–65.

Zobel, O.B. (1974) Local variation in intergrading *Abies grandis – A concolor* in the central Oregon Cascades. II Stomatal reaction to moisture stress. *Botanical Gazette*, **135**, 200–10.

6

Coniferous forests of the Pacific Northwest

JAMES P. LASSOIE, THOMAS M. HINCKLEY AND
CHARLES C. GRIER

6.1 Introduction

The coniferous forests described in this chapter are primarily those of Oregon, Washington, southern British Columbia, western Montana, and Idaho. This forest region is sharply bounded on the west by the Pacific Ocean, on the east by the crest of the Rocky Mountains, and ranges from central coastal California and southern Oregon to the southeast Alaskan coast. Altogether the region includes several major forest zones, their distributions mainly reflecting differences in annual temperature and moisture balances.

The purpose of this chapter is to describe the ecophysiology of major Northwest tree species in the context of their distribution and abundance in forests of the Pacific Coastal and western Rocky Mountain regions. Particular emphasis will be on the environmental factors and species adaptations influencing the unique productivities of these forests.

6.1.1 TOPOGRAPHY

The topography of the region is complex. A coastal mountain range borders the Pacific Ocean in northern California, Oregon, and southwestern Washington, and rises to become the Olympic Mountains of Washington's Olympic Peninsula. This range then continues northward into Canada and Vancouver Island. East of the coastal range is a pronounced geological depression. Its southern extreme is the Sacramento Valley of California, the Willamette Valley of Oregon occupies the middle region, while Puget Sound and the Straits of Georgia comprise the northern end of this depression. To the east lie the Sierra Nevada, Cascade, and British Columbia coastal ranges. This diverse topography produces a variety of forest habitats ranging from cool, moist coastal environments through dry, rain-shadows to cold, high mountain environments. Conifer forests range from sea level to about 1800 m in Oregon and to about 900 m in Alaska.

6.1.2 GEOLOGY AND SOILS

The geology, physiography, and soils of this steep, mountainous country are locally and regionally complex. Glaciation and vulcanism have shaped much of the landscape originally formed from volcanic, intrusive igneous, sedimentary, and metamorphic bedrocks. A chain of mostly dormant volcanos stretches from Mt Lassen in northern California to Mt Garabaldi in southern British Columbia. These volcanos began forming during the

Cenozoic Period and are still periodically active, as demonstrated by the recent dramatic eruption of Mt St Helens. Thus, volcanic ash can be an important component of forest soils throughout much of the region (Ugolini and Zasoski, 1979).

Superimposed on the complex terrain have been a series of recent glaciations, the last of which ended about 12 000 years ago. This glacial advance reached as far south as Olympia in Washington with alpine, valley, and piedmont glaciers covering much of the area north of this location. Except for Alaska and British Columbia, most present-day glaciers are restricted to high mountain areas. Forests commonly extend above the terminus of many of the active glaciers.

Soil types in the Pacific Northwest are primarily Inceptisols, Spodosols, and Ultisols. Past glaciation has influenced soil formation, especially in the northern part of this region and at upper elevations. Soils derived from the most recent glacial deposits tend to be immature and have low fertility for certain nutrient elements, particularly nitrogen. In contrast, soils of southwestern Washington and western Oregon are much older and reflect a more mature profile development.

6.1.3 CLIMATE

The maritime climate of the Pacific Northwest is characterized by high precipitation rates, low evaporative demands, and moderate temperatures (Franklin and Dyrness, 1973). Mean annual temperatures in the lowland areas range from 10 to 12 °C and from 5 to 7 °C in the mountains. Winters are generally mild. Even in the mountains, mean January temperatures are seldom less than − 5 °C in forested areas. The micro- and mesoclimates of these mountainous regions are strongly influenced by elevation, proximity to the Pacific Ocean, differences in climate on windward and leeward slopes, north–south oriented mountain ranges, and prevailing storm paths. The climate of this region is strongly seasonal with a preponderance of precipitation (usually more than 80%) occurring outside the growing season.

Generally, temperatures are higher in the summer and lower in the winter as one moves from the

Pacific Ocean to the interior, and from the windward to the leeward side of a mountain range. This gradient is most obvious from the west to the east side of the Cascade Range. Precipitation gradients are also locally strong in the Pacific Northwest; for example, the greatest precipitation gradient (about 9 cm precipitation km^{-1}) occurs between Mt Olympus (508.0 cm yr^{-1}) and Sequim (42.7 cm yr^{-1}) in Washington, a distance of only 52 km.

Elevation also plays a strong role in affecting the climate. For example, Seattle, Washington at an elevation of 38 m receives approximately 94 cm of rainfall and 35 cm of snowfall per year, while Paradise on the south side of Mt Rainier at 1821 m receives about 264 and 1362 cm yr^{-1}, respectively. A record 3600 cm of snow fell during the winter of 1973 at Paradise. This variable annual snowfall pattern obviously affects the length of the growing season at various locations (Fig. 6.1). Hawk *et al.* (1982) noted that as maximum snow depth increased from 50 to 750 cm, the date of snow melt changed from 19 March to 2 August.

6.1.4 VEGETATION ZONES

The topographic, climatic, and pedologic diversity of the Pacific Northwest has produced a number of recognizable forest zones. Moist forest dominants within the Pacific Coastal forest complex include *Tsuga heterophylla/Picea sitchensis* at low elevations and *Abies amabilis/Tsuga mertensiana* at high elevations. Dry forest dominants within the Rocky Mountain forest complex include low elevation *Juniperus occidentalis* and high elevation *Abies lasiocarpa*. Krajina (1965) and Franklin and Dyrness (1973) provide a more detailed discussion of the major forest zones and their specific forest habitats and successional sequences for the Pacific Northwest.

6.2 Community structure

The paleobotanical record indicates that coniferous forests have dominated the Pacific Northwest since the early Pleistocene; for some 1.5 million years (Waring and Franklin, 1979). There has been a major extinction of extensive deciduous hardwood

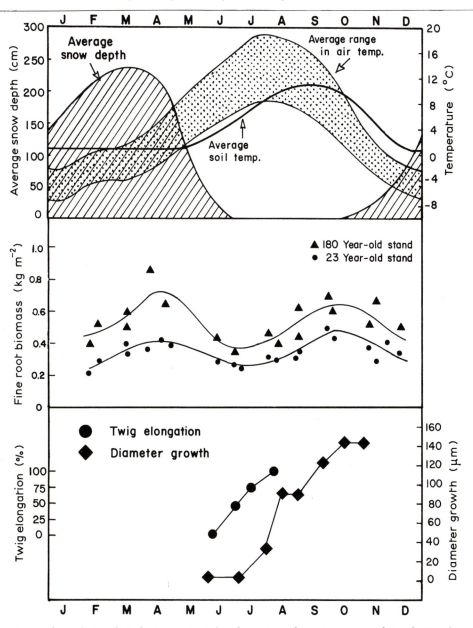

Figure 6.1 Seasonal trends in selected environmental and tree growth parameters typifying the Pacific Northwest. Environmental data are from USGS Weather Station records for Stampede Pass, Washington (1206 m); growth data are for *Abies amabilis* (after keyes and Grier, unpublished).

forests (e.g. *Carya* spp., *Platanus* spp., *Ulmus* spp.) and a massive expansion of disjunct coniferous forests (e.g. *Abies* spp., *Picea* spp., *Tsuga* spp.) since the early to middle Miocene (18–24 million years BP). The replacement has been nearly complete with the number of conifers now dominating hardwoods on a volume basis by over 1000 to 1 (Küchler, 1946), and with only a few hardwood species maintaining dominance either as pioneers where nitrogen deficiencies are severe (e.g. *Alnus rubra*) or as hardy species on water limited sites (e.g. *Arbutus menziessi, Quercus* spp.).

The abundance and wide distribution of conifers in the Pacific Northwest is atypical of northern temperate regions of the world (Waring and Franklin, 1979). The rapid loss of deciduous hardwood flora which occurred in the Pacific Northwest did not occur in the north temperate regions of Japan, China, Europe, and the eastern United States where hardwoods or hardwood–conifer mixes still dominate. The question then arises: why the abundance of conifers in the Pacific Northwest, and why are they so distributed across such a wide range of habitats? This question will now be addressed with specific reference to various ecophysiological characteristics discussed later in detail for various Northwest conifers.

6.2.1 CONIFER ABUNDANCE

Originally, it was thought that the loss of hardwoods from the Pacific Northwest resulted from low temperatures associated with Pleistocene glaciation. However, it now appears that this transition occurred before the abrupt climatic changes due to continental glaciation (Waring and Franklin, 1979). Three responses unique to conifers growing in the Northwest seem to contribute to their dominance (Franklin and Waring, 1979; Waring and Franklin, 1979): 1. the potential for active net photosynthesis outside the growing season; 2. water stress limitations on net photosynthesis during the growing season; and 3. high nutrient use efficiency (i.e. dry matter production per unit of nutrient).

Photosynthesis during the fall, winter, and early spring is only possible for evergreen plants. However, in northern latitudes or at high elevations such physiological responses are severely limited by subfreezing temperatures and/or light regimes (Tranquillini, 1979; Troeng and Linder, 1982). This is often not the case in most of the Pacific Northwest owing to the relatively mild temperatures even at higher elevations (Fig. 6.1). Moderate soil and air temperatures in conjunction with most conifers' broad temperature optima for photosynthesis mean that significant amounts of carbon can be fixed during periods of the year when deciduous trees lack

foliage (Emmingham and Waring, 1977; Mooney and Gulmon, 1982; Salo, 1974). Low respiration rates common at low temperatures further increase the amount of total daily net photosynthesis. In addition, net photosynthetic rates increase rapidly with increasing light levels and reach maxima at about one-third full sunlight (Krueger and Ruth, 1969). Stomatal opening occurs at much lower light intensities (Running, 1976). This provides net carbon uptake even in the lower canopy and on cloudy days; the latter being very characteristic of the Pacific Northwest. Increased light interception at low sun angles in winter also results from the long, conical crowns typical of conifers (Franklin and Waring, 1979) and from their distinctive branch, shoot, and needle arrangements (Leverenz and Jarvis, 1979). As a result, Emmingham and Waring (1977) have predicted that *Pseudotsuga menziesii* may assimilate over 50% of its annual carbon between October and May.

If conditions in this region are uniquely favorable for active photosynthesis during the nongrowing season, they are distinctly unfavorable during the summer. Summer precipitation is not adequate to recharge soil water to levels great enough to meet evapotranspirational demands. This, in association with the well-drained soils so common in the Northwest, can result in prolonged drought periods. Such conditions greatly limit stomatal conductances (Running, 1976) which reduces carbon dioxide uptake. Hence, Emmingham and Waring (1977) estimate that as much as 70% annual net photosynthesis may occur outside the summer months on very dry sites.

Even though water deficits affect hardwoods and conifers alike, the latter might have an advantage under drought conditions. Conifer needles exchange heat with the atmosphere better than broadleaves do thereby maintaining needle temperatures nearer to ambient which greatly reduces the needle to air vapor-pressure gradient. This is especially important during periods of drought-induced stomatal closure. The large foliar areas typical of many conifers may also collect condensation during summer. This may add water to the soil through fog drip (Stone, 1957) or delay early morning transpiration (Fritschen and Doraiswamy, 1973). In addition, conifers have very large internal water storage reservoirs in stems and foliage which can be utilized to moderate the effects of drought (Waring and Running, 1978). Lastly, the relatively small tracheids compared with vessels limit cavitation during periods of high water stress and aid in their later refilling, thereby maintaining water conduction continuity (Coutts, 1977, Tyree and Dixon; 1983).

Conifers in general have evolved a pattern of nutrient utilization different from most hardwood species. In northwestern ecosystems decomposition and nutrient mineralization occurs primarily outside the growing season since decomposition is limited during the warm, dry summers (Fogel and Cromack, 1977; Turner and Singer, 1976). In addition, substantial quantities of mineral nutrients can be immobilized in the large standing biomass and forest floor (Cole, 1981; Cole and Rapp, 1980).

Figure 6.2 Old-growth *Abies amabilis* stand near Mt St Helens, Gifford Pinchot National Forest, Washington (photo courtesy of J.F. Franklin).

Hence, tree species having high efficiencies of nutrient utilization appear to be favored in these environments. Conifers have less than half the annual nutrient requirement of most hardwoods, partly because they retain numerous age classes of foliage and thus have lower nutrient demands for foliage replacement. Furthermore, conifers have the capacity to internally redistribute a significant proportion of their annual nutrient requirement from older tissues. Although deciduous hardwoods also redistribute substantial nutrients from foliage prior to leaf fall, their total nutrient requirements are higher. For these reasons, conifers appear better adapted to the nutrient-poor regimes characteristic of much of the Northwest than are hardwood trees

(Meier, 1981; Waring and Franklin, 1979).

Northwest conifer species are generally much larger and longer lived than representatives of the same genera in other regions (Waring and Franklin, 1979; Fig. 6.2). The adaptive advantage of being large is probably associated with many of the characteristics of conifers previously discussed. As mentioned, large size means large storage areas for water and nutrients that can potentially buffer the adverse effects of environmental extremes. Furthermore, there are obvious competitive advantages of size, especially height. Species with the genetic potential to grow tall can overtop and outlive competitors, thereby becoming major components of forest ecosystems.

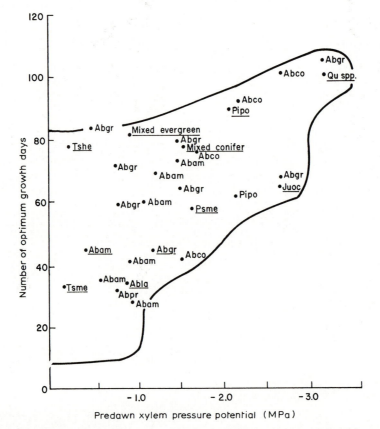

Figure 6.3 The distribution of Pacific Northwest tree species in relation to optimum growth days and predawn xylem pressure potential during the peak of a drought (after Waring *et al.*, 1972; Zobel *et al.*, 1976). The underlined names refer to the communities cited by Waring *et al.* (1972). Key: *Abies amabilis*, Abam; *A. concolor*, Abco; *A. grandis*, Abgr; *A. lasiocarpa*, Abla; *A. procera*, Abpr; *Pinus ponderosa*, Pipo; *Pseudotsuga menziesii*, Psme; *Tsuga mertensiana*, Tsme; *T. heterophylla*, Tshe; *Juniperus occidentalis*, Juoc; and *Quercus* spp., Qu spp. Shifts in distributions result from climatic differences associated with data being collected on different years. (from Hinckley *et al.*, 1983).

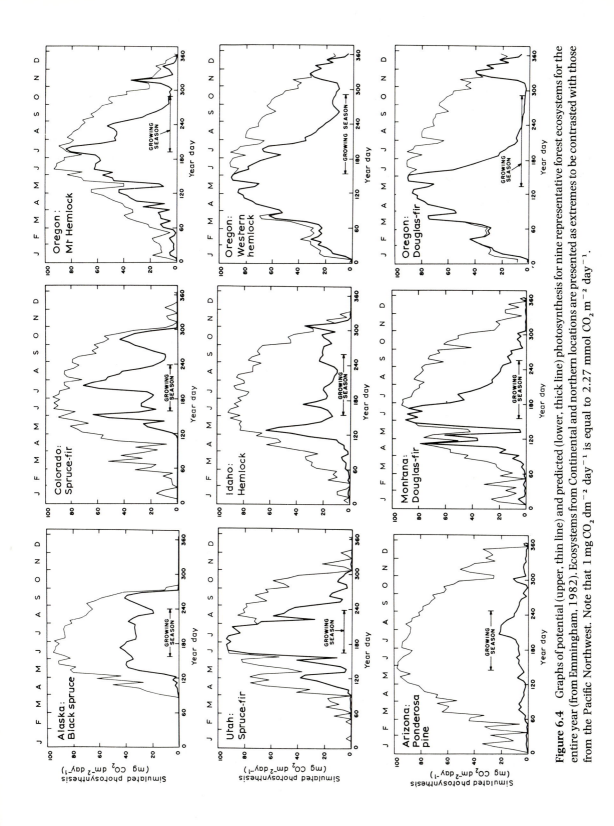

Figure 6.4 Graphs of potential (upper, thin line) and predicted (lower, thick line) photosynthesis for nine representative forest ecosystems for the entire year (from Emmingham, 1982). Ecosystems from Continental and northern locations are presented as extremes to be contrasted with those from the Pacific Northwest. Note that 1 mg CO_2 dm^{-2} day^{-1} is equal to 2.27 mmol CO_2 m^{-2} day^{-1}.

6.2.2 SPECIES DISTRIBUTIONS

The topography and climate of the Pacific Northwest result in moisture and temperature being the primary factors influencing species distribution (Franklin and Waring, 1979; Waring and Franklin, 1979). Hence, these factors, expressed as gradients, have been used to ordinate communities and to describe species distributions since the pioneering work by Waring (1969). Fig. 6.3 illustrates such a gradient for various species or community types common in the Pacific Northwest. As may be seen, observed distribution patterns (Franklin and Dyrness, 1973) range from cold, wet high elevation habitats (e.g. *Abies* spp., *Tsuga mertensiana*) to warm, dry habitats (*Quercus* spp., *Juniperus* spp.).

Emmingham and Waring (1977) and Emmingham (1982) have modelled annual patterns of potential and actual photosynthesis in different environments based on radiation, temperature, drought, preconditioning of foliage by frost, and low soil temperatures using data measured in nine forest stands (Fig. 6.4). Location and species composition of these stands ranged from the Alaskan interior (*Picea mariana*), to northern Arizona (*Pinus ponderosa*); and from coastal Oregon (*Pseudotsuga menziesii, Tsuga heterophylla*) to the Rocky Mountains of Montana (*P. menziesii*), Utah, and Colorado (*Picea engelmannii, Abies lasiocarpa*). For all of these locations, temperature and radiation did not appear to be limiting during the growing season, and potential photosynthesis ranged from 70 to 90 mg $CO_2 dm^{-2} d^{-1}$ (159.1–204 s mmol $CO_2 m^{-2} d^{-1}$) (Fig. 6.4). Instead, moisture deficits from either high evaporative demands or low soil moisture contents caused decreases in photosynthesis during the growing season. Even the coastal *Picea sitchensis* forests of Oregon with 240 cm of precipitation occurring per year achieved only 78% of their photosynthetic potential. The drier the environment, the greater the reduction in actual photosynthetic assimilation. For example, *Pseudotsuga menziesii* forests of the Willamette–Puget Forest achieved only 40% of their potential.

Large differences in photosynthetic potential were observed during the winter months for the nine locations (Fig. 6.4). At the coastal site, potential winter photosynthesis was near 40 mg $CO_2 dm^{-1} d^{-1}$ (90.9 mmol $CO_2 m^{-2} d^{-1}$) while in the interior of Alaska, it was zero. Emmingham (1982) also calculated that even alpine forests in the Cascades were capable of winter photosynthesis because of mild daytime air temperatures (see Fig. 6.1). Actual rates of photosynthesis dropped below potential whenever stomatal closure occurred. Stomatal closure was linked to soil and plant water deficits, to conditions of high evaporative demand, to low soil temperatures, and to a history of below freezing air temperatures.

6.2.3 FOREST PRODUCTIVITY

In Table 6.1 the net primary aboveground productivities of Pacific Northwest forests are compared with values reported for forest regions in other parts of North America. The average productivity of forests in the Pacific Northwest is about 1.74 kg m^{-2} yr^{-1} indicating that this is the most productive forest region in North America. However, Table 6.1 should be interpreted cautiously as it reports the results of numerous studies in a variety of unmanaged forest stands and not systematic surveys of productivity within specific forest zones.

In addition to high productivity, the forests of the Pacific Northwest are unique for the long life of individual species and their ability to accumulate large amounts of biomass (Table 6.2). The Pacific Coastal forest complex has the largest trees both in this part of North America and in the world (Shidei and Kira, 1977) (e.g. see Fig. 6.2). In addition to data given in Table 6.2, aboveground biomass values can range from a low of 15.0 kg m^{-2} in a 100 year old *Pinus ponderosa* forest (Grier *et al.*, in press) to 346.0 kg m^{-2} in an old-growth (over 1000 years old) *Sequoia sempervirens* stand (Fujimori, 1977).

The large aboveground biomass accumulations characteristic of coniferous forests in the Pacific Northwest are primarily the result of the very high proportion of support tissues (i.e. mostly stem xylem) compared to photosynthetic and meristematic tissues (Gholz, 1979). Hence, maintenance respiration is relatively low especially given the mild temperatures typical of the Pacific Northwest. Large

Table 6.1 Aboveground net primary production in major forested regions of North America and in major forest zones of the Pacific Northwest region. Values reported are for stands having closed canopies

| Region | Net primary production ($kg\,m^{-2}\,yr^{-1}$) | | | |
	Average*	Specific value	Reported range	Reference
Pacific Northwest[†]	1.74		0.31–3.22	Grier *et al.* (in press)
Pisi-Tshe	2.12		1.50–3.22	Grier *et al.* (in press)
26 yr old		3.22		Fujimori (1971)
120 yr old		2.00		Grier (1975)
Psme	1.34		0.80–2.41	Grier *et al.* (in press)
50 yr old, low site		0.73		Keyes and Grier (1981)
50 yr old, high site		1.37		Keyes and Grier (1981)
				Zavitkovski and
Alru[‡]		2.22		Stevens (1972)
				Keyes and Grier (1981)
Abam	0.97		0.19–1.66	Grier *et al.* (in press)
9 yr old		0.19		Keyes (1982)
23 yr old		0.59		Grier *et al.* (1981)
60 yr old		1.47		Keyes (1982)
180 yr old		0.45		Grier *et al.* (1981)
Abpr–Psme (130 yr old)[§]		1.66		Fujimori *et al.* (1976)
Pipo	0.42		0.20–1.25	Grier *et al.* (in press)
Rocky Mountain	0.98		0.19–1.75	Grier *et al.* (in press)
Southeast	0.89		0.30–1.55	Grier *et al.* (in press)
Central Hardwood	1.06		0.80–1.50	Grier *et al.* (in press)
Northeastern	1.03		0.41–1.26	Grier *et al.* (in press)

*Averages are not weighted by land area occupied by each forest type in a region.
[†]Major forest zones are given for the Pacific Northwest region: Pisi = *Picea sitchensis*; Tshe = *Tsuga heterophylla*; Psme = *Pseudotsuga menziesii*; Abam = *Abies amabilis*; Pipo = *Pinus ponderosa*.
[‡]Alru = *Alnus rubra*; a major seral species on disturbed forest lands within Psme Zone (Franklin and Dyrness, 1973).
[§]Abpr = *Abies procera*; a major seral species with Psme within Abam Zone (Franklin and Dyrness, 1973).

amounts of standing biomass, therefore, can be maintained for physical support and for storage and transport of water, nutrients, and carbohydrates with only a minimal expenditure of energy. Such an arrangement may be viewed as an adaptation for maximizing the amount of photosynthetic surface area that can be supported in a situation where light is limiting or where photosynthates are required in high amounts. Shade intolerant species (e.g. *Abies procera* and *Pseudotsuga menziesii*), shade tolerant climax species (e.g. *Tsuga heterophylla*), and extremely long-lived climax species (e.g. *Sequoia sempervirens*) typically amass very large aboveground biomasses (Table 6.2).

6.3 Physiological characteristics and responses

In spite of impressive net productivities and standing biomasses common to forests of the Pacific Northwest, annual photosynthate production, as in other regions, is limited by various edaphic and climatic variables; the most important being light, temperature, water, and nutrients (Emmingham, 1982; Franklin and Waring, 1979; Lassoie, 1982). These four limiting environmental factors will now be included in a discussion of ecophysiological responses of Northwest conifers to their environment. Emphasis will be on *Pseudotsuga menziesii* due to its dominance and economic importance throughout the region. However, responses of other common forest trees will be presented when possible.

6.3.1 LIGHT

It is well known that tree growth and development is influenced by light intensity, quality, and period. Of these, light intensity (particularly photosynthetic

Table 6.2   Above- and belowground biomass accumulation in major forest types of the Pacific Northwest (PNW) in comparison to other regions of the world (after Cannell, 1982).

Forest type*	Location	Age (yr)	Basal area ($m^2\,ha^{-1}$)	Biomass accumulation Above ($kg\,m^{-2}$)	Below ($kg\,m^{-2}$)
Psme–Tshe	PNW	450	62.1	53.6	8.2
Psme[†]	PNW	450	164.0	122.4[‡]	—
Tshe	PNW	150	84.4	86.5[‡]	—
Abpr	PNW	290	127.0	156.2[‡]	—
Abpr[†]	PNW	290	148.0	183.8[‡]	—
Pisi	PNW	130	111.2	149.2	—
Litu	E. USA	50	22.1	14.1	3.4
Fasa	W. Germany	122	28.3	27.4	3.7
Piab	W.Germany	80	57.4	32.2	—
Euob	Australia	51	63.3	31.2	7.5
Coniferous (subalpine)	Japan	290	53.0	19.3	6.4
Evergreen (wet tropical)	Manaus	—	31.0	73.4	25.5
Deciduous (dry tropical)	India	60	30.6	20.5	3.4

*Psme = *Pseudotsuga menziesii*, Tshe = *Tsuga heterophylla*, Abpr = *Abies procera*, Pisi = *Picea sitchensis*, Litu = *Liriodendron tuliperifera*, Fasa = *Fagus salvatica*, Piab = *Picea abies*, Euob = *Eucalyptus obliqua*.
[†]Best hectare.
[‡]Stemwood and bark only.

photon flux density) is the most important through its direct effects on photosynthetic phosphorylation, photorespiration, and stomatal activity, and its indirect effects on air and leaf temperatures. Thus, forest trees have evolved intra- and interspecific physiological and morphological characteristics which depend on the amount of light in their environment.

The stomata of most forest trees typically open rapidly with increasing light levels, and maximum stomatal conductances (minimum stomatal resistances) are usually reached at about $200\,\mu mol\,m^{-2}\,s^{-1}$ (Fig. 6.5; Jarvis and Morison, 1981; Leverenz, 1981a; Ludlow and Jarvis, 1971; Ng and Jarvis, 1980; Running, 1976: Tan *et al.*, 1977). It appears that the small increase in atmospheric vapor-pressure deficits commonly occurring in the early morning further enhances photoactive stomatal opening in *Pseudotsuga menziessii* (Meinzer, 1982a, c). In addition, mesophyll conductances to carbon dioxide increase with increasing light intensity (i.e. mesophyll resistances decrease; Fig. 6.5). Leaf conductance (generally assumed to be equal to stomatal conductance;

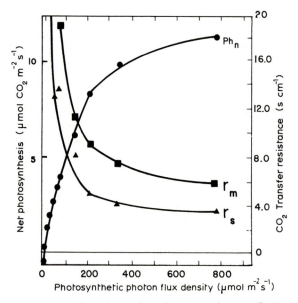

Figure 6.5 The relationships between photosynthetic photon flux density and net photosynthesis (Ph_n), stomatal (r_s) and mesophyll (r_m) resistances to carbon dioxide transfer in *Picea sitchensis* (from Ludlow and Jarvis, 1971). Note that conductance is the reciprocal of the resistance measurements given.

Hinckley *et al.*, 1978) maxima have been observed to differ between eight Northwest conifer species (*Abies grandis, A. magnifica* var. *shastensis, A. procera, Picea breweriana, P. sitchensis, Pinus ponderosa, Pseudotsuga menziesii*, and *Tsuga heterophylla*) composing forest communities at elevations between 200 and 1200m in Oregon (Running, 1976). However, most maxima ranged between about 0.2 and $0.4 \, \mathrm{cm \, s^{-1}}$; *P. menziesii* having a complete range (closed to open stomata) of 0.003–$0.33 \, \mathrm{cm \, s^{-1}}$ (Drew and Ferrell, 1979; Running, 1976; Salo, 1974).

Stomata are generally closed in the dark. However, at times of low evaporative demands and minimal tree water deficits during the late fall, winter, and early spring the stomata of *Picea sitchensis, Pseudotsuga menziesii*, and *Abies amabilis* have been observed to be open at night (Blake and Ferrell, 1977; Fry, 1965; Grace *et al.*, 1975; Hinckley and Ritchie, 1973; Leverenz, 1981a; Running, 1976). Blake and Ferrell (1977) suggested that very low levels of foliar abscisic acid are necessary for open stomata in *P. menziesii* at night. Such stomatal opening will result in transpirational water losses if nonsaturating vapor pressures develop during the night (see Section 6.3.3, Table 6.4).

Light saturation of net photosynthesis normally occurs between 600 and 800 μmol m^{-2} s^{-1} with little increase thereafter (Fig. 6.5; Larcher, 1969; Walker *et al.*, 1972). Because of mutual shading by needles on a given twig and among twigs, canopy net photosynthesis often tends to increase slowly as more needle surfaces become illuminated at light levels above saturation (Leverenz, 1981a; Leverenz and Jarvis, 1980a). Because of the three-dimensional nature of tree crowns and mutual shading among branches (Norman and Jarvis, 1974, 1975), complete light saturation of entire crowns is not possible (Jarvis and Leverenz, 1983).

Baker's (1934) hypothesis that there was a strong relationship between the degree of shade tolerance of a tree species and its photosynthetic response to light has been supported by more recent studies (e.g. Hodges and Scott, 1968; Krueger and Ruth, 1969). Net photosynthesis in shade-tolerant species exceeds that in intolerant species at low light intensities, but the reverse is true at high intensities.

The response of net photosynthesis to light intensity also varies greatly within and between individuals of a species. The saturating photosynthetic photon flux density reported in the literature for *Pseudotsuga menziesii* varies from about 270 to 780 μmol m^{-2} s^{-1} (Walker *et al.*, 1972). Since net photosynthetic rates are typically calculated on a leaf area basis, a portion of this intraspecific variation might be attributable to differences in dry weight and optical thickness in the needles resulting from the study trees being grown under different light environments (Krueger and Ruth, 1969; Leverenz and Jarvis, 1980b). Decreases in specific leaf weight (weight/area) typical of shade-acclimatized leaves (Drew and Ferrell, 1977; Krueger and Ruth, 1969; Lewandowska *et al.*, 1976; Magnussen and Peschl, 1981), will reduce calculated values of net photosynthesis expressed on a leaf area basis (Salo, 1974).

Also in response to growth in shade, total chlorophyll (a + b) concentration increases with a decrease in the chlorophyll a to b ratio (Lewandowska *et al.*, 1976; Magnussen and Peschl, 1981). This results from an absolute increase in the amount of chlorophyll b (Lewandowska *et al.*, 1976). Shade-tolerant trees appear better able to increase needle chlorophyll content when grown under increasing levels of shade than intolerant trees (e.g. *Picea sitchensis*, Lewandowska and Jarvis, 1977; *Abies grandis*, Magnussen and Peschl, 1981).

In similar fashion, needles throughout a conifer crown will be morphologically different depending on the light conditions occurring during their growth and development (Fig. 6.6). For example, shade needles in the lower canopy are shorter, narrower, and thinner, have lower total chlorophyll, and have a smaller biomass per unit leaf surface area than those growing in full sun (Aussenac, 1973; Leverenz and Jarvis, 1980b; Lewandowska and Jarvis, 1977; Lewandowska *et al.*, 1977; Norman and Jarvis, 1974; Phillips, 1967; Tucker and Emmingham, 1977). Stomatal densities also are higher in sun than shade needles (Aussenac, 1973; Phillips, 1967).

Leverenz (1974) observed higher net photosynthetic rates under field conditions in sun-acclimatized *Pseudotsuga menziesii* foliage than in

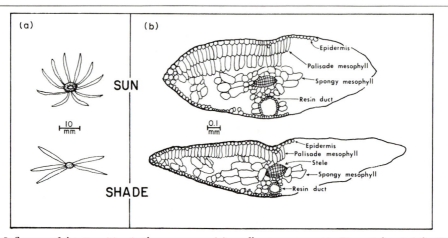

Figure 6.6 Influence of shoot position in the canopy on (a) needle arrangement in *Picea sitchensis* (after Leverenz and Jarvis, 1980a) and (b) needle morphology in *Tsuga heterophylla* (from Tucker and Emmingham, 1977).

foliage grown in the shade. The difference seemed associated with differences in needle conductances and specific leaf weights, as well as microenvironmental factors, especially the higher light intensity. More precise investigations under controlled environmental conditions using branches excised in the field from large crowns have revealed a 15% lower photosynthetic potential (i.e. maximum net photosynthesis at light saturation) for shade-acclimated *P. menziesii* foliage compared to foliage from more sunlit locations (Lassoie, 1982). In a similar study with *Picea sitchensis*, Leverenz and Jarvis (1979) found that sun needles had higher maximum net photosynthetic rates (11.4 compared with 8.6 μmol $CO_2\,m^{-2}\,h^{-1}$), stomatal conductances (0.26 compared with 0.18 cm s^{-1}), and mesophyll conductances (0.16 compared with 0.12 cm s^{-1}) than shade needles when measured under bilateral illumination at about 135 μmol m^{-2} s^{-1}. When needle surfaces were illuminated from the top only, bottom only, or bilaterally, these authors also found significant differences in maximum net photosynthetic rates for both sun (bilateral > top > bottom) and shade (bilateral = top > bottom) needles. These differences were primarily associated with different mesophyll conductances. Shoots at the top of the canopy, therefore, will have very different light-saturation curves than those growing in the lower canopy (Fig. 6.7).

The position of a conifer shoot on a branch also affects its photosynthetic capacity. At high light levels, terminal shade shoots of *P. menziesii* were found to have higher rates of net photosynthesis and greater stomatal conductances under field conditions than adjoining current-year lateral shoots (Leverenz, 1981b). On overcast days, daily average net photosynthetic rates did not differ with shoot order. The observed differences have been attributed to apical control of net photosynthesis in laterals by the terminal shoot (Leverenz, 1981b) since there were no between-shoot differences in microenvironment or water potentials (Hellkvist *et al.*, 1974). It is also possible that there were differences in shoot nutrition levels as the laterals were a lighter green than the terminals throughout the experiment. Leverenz and Jarvis (1980a) have reported similar differences in net photosynthesis and stomatal and mesophyll conductances for terminal and lateral shoots of *Picea sitchensis*. These differences existed in both sun and shade shoots.

With the exception of *Larix* spp., conifers in the Pacific Northwest retain their needles for as long as 22 years depending on species, microsite, and tree vigor (Long, 1982). In *Picea sitchensis*, the response of net photosynthetic rate to increasing photosynthetic photon flux densities has been observed to increase during the summer as current-year needles mature, but to decrease the following year (Watts *et al.*, 1976). Likewise, maximum net photosynthesis in *Pseudotsuga menziesii* foliage has been reported to

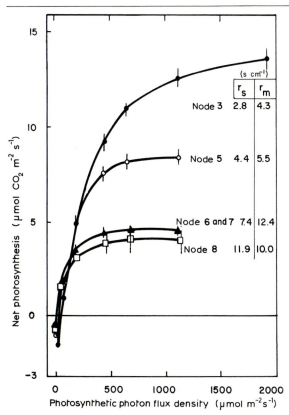

	r_s	r_m
Node 3	2.8	4.3
Node 5	4.4	5.5
Node 6 and 7	7.4	12.4
Node 8	11.9	10.0

Figure 6.7 The effect of developmental position in the canopy on the relationship between photosynthetic photon flux density and net photosynthesis in *Picea sitchensis* shoots (bars are two standard errors of five replicates). Also shown are minimal values of stomatal (r_s) and mesophyll (r_m) resistances to carbon dioxide transfer (from Jarvis *et al.*, 1976). Note that conductance is the reciprocal of the resistance measurements given.

decrease about 22% yr^{-1} for at least 4 years (Brix, 1969, 1971; Woodman, 1968) due to an age-related decrease in photoreactivity of stomata (Running, 1976; Watts *et al.*, 1976). However, Salo (1974) suggested that maximum net photosynthetic rates may not decrease greatly until *P. menziesii* needles are 3 years old. In *Abies amabilis*, a species that retains more age classes of foliage than *P. menziesii*, Teskey (1982) noted that 7 year old foliage had rates of net photosynthesis within 55% of those current or 1 year old foliage. In contrast, rates in 9 year old foliage were less than 6% of maximum.

Salo (1974) observed that the proportional contribution of each age class (current year, 1 and 2 years old) to total branch net photosynthesis closely approximated the proportional leaf area of each age class. Further work, using excised branches under optimal light and temperature conditions, has indicated that the photosynthetic capacities of second and third year needles, respectively, were only 5% and 25% lower than fully expanded, first year needles (Lassoie, 1982). Lassoie suggested that the variable effects that needle age seem to have on maximum net photosynthetic rates may be due to an age-related, atmospheric humidity–stomata interaction reported for *Pseudotsuga menziesii* by Running (1976) and to age-related and position-related changes in mesophyll conductances (Leverenz, 1981b).

While it seems that adjustments in photosynthetic efficiencies do occur in needles grown at different light levels and in needles of different ages, under field conditions differences in photosynthetic rates between foliage units at any moment are greatly affected by the immediate light, temperature, and water regimes (Lassoie, 1982). Hence, the possible adaptive/acclimatic value of environmentally induced morphological and biochemical changes in trees must be interpreted with respect to all factors in the microenvironment associated with their location.

6.3.2 TEMPERATURE

Temperature directly or indirectly affects all metabolic processes, especially those related to net photosynthesis (Lassoie, 1982) and the activity of meristematic tissues (Allen and Owens, 1972; Hinckley and Lassoie, 1981; Owens and Molder, 1973a,b, 1976). Temperature also influences the rate of mineralization of the forest floor which affects nutrient availability and uptake (Cole and Rapp, 1980). Therefore, the role of temperature in an elevationally diverse region like the Pacific Northwest is quite complex.

Temperature response curves for net photosynthesis in conifers generally are broad and have optima between 10 and 25 °C (Doehlert and Walker, 1981; Dykstra, 1974; Larcher, 1969; Neilson *et al.*,

1972; Salo, 1974; Walker *et al.*, 1972). When photosynthesis in *Pseudotsuga menziesii* was examined under controlled environmental conditions, the temperature response curve, at light levels in excess of 500 μmol m^{-2}s^{-1}, was quite flat between 2 and 25 °C with an optimum at 10 °C (Salo, 1974). Field studies by Dougherty and Morikawa (1980) and Leverenz (1981a,b), however, suggest that the optimum temperature range might be nearer to 15 °C. Variability appears to be associated with genetic diversity (Sorensen and Ferrell, 1973), possible seasonal adjustments, and the dependence of the temperature response on other environmental factors. For example, temperature optima are light (Brix, 1967; Lassoie, 1982; Webb, 1971) and perhaps vapor pressure (Leverenz, 1981b; Ng, 1978) dependent. The temperature optimum is higher in conifers growing in warm climates (e.g. *Abies concolor* and *Pinus ponderosa*) than conifers growing in cooler environments (e.g. *Abies amabilis* and *Pseudotsuga menziesii*) (Conard and Radosevich, 1981; Teskey, 1982).

In contrast to *Picea sitchensis* (Neilson *et al.*, 1972) and many other woody species, the temperature optimum for photosynthesis in *Pseudotsuga menziesii* and in *Abies amabilis* remains rather constant over the year (Dougherty and Morikawa, 1980; Leverenz, 1981a; Salo, 1974; Teskey, 1982). However, temperature preconditioning does have a large impact on leaf conductance (discussed later) and net photosynthesis in conifers such as *P. menziesii*. Salo (1974) and Lassoie *et al.* (1983) have reported negative net photosynthetic rates (i.e. net respiration) during daylight periods in the winter in response to prolonged, subfreezing temperatures. Low-temperature compensation points (i.e. net photosynthesis $= 0$) for conifers during the winter seem to range from -3 to -8 °C (Larcher, 1969; Ludlow and Jarvis, 1971; Neilson *et al.*, 1972), which is also in the expected freezing range for needles (Doehlert and Walker, 1981; Larcher, 1969). Such an ability to fix carbon at low needle temperatures has also been observed under field conditions in the winter (Lassoie *et al.*, 1983; Salo, 1974) and probably results in the accumulation of dry matter as has been observed in *Picea sitchensis* seedlings (Bradbury and Malcolm, 1978).

The effect of temperature on leaf conductance has been difficult to separate from other variables affecting stomatal activity (Lassoie, 1982). This is especially true since increasing temperatures cause concurrent changes in the leaf-to-air vapor-pressure gradient (Leverenz, 1981a,b; Neilson and Jarvis, 1975) which affects peristomatal transpiration (Lange *et al.*, 1971), and guard cell turgor. In *Pseudotsuga menziesii* and *Picea sitchensis* needle conductances are highest between 5 and 20 °C, but drop rapidly as temperatures increase (Lassoie, 1982; Neilson and Jarvis, 1975). Temperatures near freezing promote stomatal closure, or delay stomatal opening, in *Pseudotsuga menziesii* (Drew and Ferrell, 1979; Reed, 1968; Salo, 1974), *Pinus contorta* (Fahey, 1979; Kaufmann, 1982b), *Picea sitchensis* (Turner and Jarvis, 1975), *Picea engelmannii* (Kaufmann, 1976, 1982b), and *Abies lasiocarpa* (Kaufmann, 1982b). The low temperature effect may persist for several days even after environmental conditions have improved (Kaufmann, 1982b; Salo, 1974).

The factors discussed above are probably responsible for the observed changes in the temperature response curve for net photosynthesis with elevation reported by Dougherty and Morikawa (1980). With *Pseudotsuga menziesii* branches excised in the field during the winter, these authors observed a decrease in net photosynthesis at all needle temperatures between 5 and 35 °C under optimal light levels. Although the optimum temperature remained between 10 and 15 °C, the reduced photosynthetic capacity of foliage with elevation suggested a long-term preconditioning effect. In support, Neilson and Jarvis (1975) and Teskey (1982) have reported a similar temperature preconditioning effect acting directly on stomatal conductances in *Picea sitchensis* and *Abies amabilis*, respectively.

Net photosynthesis decreases significantly in *Pseudotsuga menziesii* at temperatures above 25 °C and becomes negligible for most conifers at 40 °C (Helms, 1965; Krueger and Ferrell, 1965; Teskey, 1982). Hence, daily net photosynthesis totals have been observed to be very low when average leaf temperatures are near 30 °C (Lassoie and Salo, 1981; Salo, 1974). This is due to an increase in respiration (Doehlert and Walker, 1981) and high temperature-induced stomatal closure (Leverenz, 1981b). The high temperature limit for positive net photosynthesis is determined by direct effects on photorespiration and membrane structure (Bauer *et al.*, 1975).

Since temperature directly affects all metabolic processes it has numerous interrelated influences on meristematic activity. As average daily temperatures increase in the spring, root growth is initiated, stem tissues rehydrate, buds swell, and aboveground growth commences (Hinckley and Lassoie, 1981) (see Fig. 6.1). Bark respiration is an exponential function of bark temperature during the period of active cambial growth (Hinckley and Lassoie, 1981). While internal water deficits at the cambium can stop or greatly reduce cell division as indicated by a decrease in respiration (Edwards and McLaughlin, 1978; Rook and Corson, 1978), the number of degree days (i.e. the duration of days above a certain threshold temperature) usually determines the amount of cambial growth (both growth rate and duration) (Dougherty and Morikawa, 1980; Lassoie, 1975).

Height growth in conifers is closely related to temperature, particularly soil temperature. Dougherty and Morikawa (1980) observed that the rate of height growth increased linearly in *Pseudotsuga menziesii* with soil temperatures above 8 °C. In contrast, height growth increased linearly, but at a slower rate in two middle to high elevation species (*Abies amabilis* and *A. procera*) at soil temperatures greater than 3 °C. These differences in threshold temperatures for growth are probably related to water absorption by roots. Low soil temperatures influence water absorption which affects water deficits throughout the xylem in *Picea engelmannii* (Kaufmann, 1975), *Pinus contorta* (Running and Reid, 1980), and *Abies amabilis, A. procera*, and *Pseudotsuga menziesii* (Teskey, 1982) (Fig. 6.8). Depending upon the root temperature regime a tree is adapted to, water absorption will be governed by metabolic, or membrane properties, by the physical properties of water, or by soil water potential. The transition temperature between physical (fast) and membrane-related (slow) water absorption is species dependent and usually decreases in species

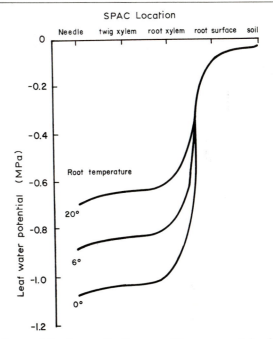

Figure 6.8 Composite diagram of the sources of development of leaf water potential in *Pinus contorta* seedlings at various root temperatures and at various locations along the soil–plant–atmosphere continuum (SPAC) (from Running and Reid, 1980).

adapted to higher elevations (Kaufmann, 1975; Running, 1980; Teskey, 1982).

6.3.3 WATER

The ecological importance of water is well illustrated by the stratified distribution and productivity of Northwest forest tree species along moisture gradients (Emmingham, 1982; Grier and Running, 1977; see Fig. 6.3). There is a strong relation between leaf area index of mature stands, a major determinant of community productivity, and precipitation (Gholz, 1979; Grier and Running, 1977). However, factors such as crown breakage due to wind or snow, soil nutrient status, and soil temperature will alter this relationship (Gholz, 1979). For stands at middle to low elevations, simple growing season water balances were strongly correlated with leaf area indices ($r^2 = 0.98$). Productivities at high elevations were primarily limited by soil and air temperatures (see Fig. 6.1).

The water relations of conifers common to the Pacific Northwest have received much attention from forest scientists (see reviews by Hinckley *et al.*, 1978; Lassoie, 1982; Whitehead and Jarvis, 1981). Work on water deficits due to drought has been emphasized as water excesses are limited to lowlands, bogs, thin tills over a compacted basal till, and artificially flooded areas and have limited effects on the distribution of major forest types throughout the region (Franklin and Dyrness, 1973). Moderate to severe water deficits can affect tree growth directly by reducing turgor pressures in meristematic tissues (Vaadia *et al.*, 1961) and indirectly by promoting stomatal closure which decreases carbon dioxide uptake and therefore photosynthesis (Beadle *et al.*, 1978, 1979). Severe water deficits also seem capable of directly disrupting the biochemistry of the net photosynthetic process (Beadle and Jarvis, 1977).

A drought begins when available soil water supplies and/or atmospheric evaporative demands cause prolonged periods of stomatal closure and/or reductions in growth processes. Most conifers in the Pacific Northwest are subjected to drought which may occur regularly or quite infrequently. Highly predictable or regular droughts are important in restricting the range of a species and reducing the likelihood of habitat invasion by less drought-resistant species. Such control is most likely exerted at the time of seed germination and seedling establishment. Because of the north-south orientation of the mountains in the Northwest, their proximity to the Pacific Ocean, and the Mediterranean pattern of rainfall which changes drastically with elevation and aspect, regular summer droughts are common throughout this region (Waring and Franklin, 1979).

Infrequent droughts may be important in allowing drought tolerant species to compete successfully in more mesic environments and in restricting the ranges of mesophytic species (Duhme, 1974). Although this pattern of drought in the Pacific Northwest is probably not as important as regular or predictable droughts, the exclusion of species from particular sites and the adjustment of stand leaf areas are probably determined by the infrequent drought.

Table 6.3 Comparative drought resistances of Pacific Northwest tree species (after Franklin and Dyrness, 1973; Lopushinsky, 1975; Minore, 1979; Wambolt, 1973). Observations are based on only responses to soil drought and not to winter desiccation

Resistance	Species
High	*Quercus garryana*
	Quercus kelloggii
	Pinus jeffreyi
	Pinus ponderosa
	Libocedrus decurrens, Arbutus menziesii
	Pseudotsuga menziesii, Pinus contorta
	Picea engelmannii
	Abies grandis
	Pinus lambertiana, Larix occidentalis
	Abies lasiocarpa, Thuja plicata, Pinus monticola
	Abies concolor, Picea breweriana
	Chamaecyparis lawsoniana, Abies procera
	Tsuga heterophylla, Picea sitchensis
	Abies amabilis
Low	*Abies magnifica, Tsuga mertensiana*

The morphological, physiological, and behavioral characteristics which affect the moisture interactions of forest trees can be grouped as to those related to either drought tolerance or avoidance (Levitt, 1980). Relative drought-resistance rankings for Northwest forest trees are based primarily on field observations of species occurrences and distributions (Table 6.3). Such a listing, however, fails to differentiate between the specific avoidance or tolerance characteristics which give a particular species its resistance ranking.

Most research on the drought tolerance of Northwest forest species has focused on stomatal behavior and osmotic adjustment. For example, osmotic potentials have been reported to vary with tissue age in coniferous species (Jackson and Spomer, 1979; Teskey, 1982), thus indicating a progressive adjustment with tissue development. For *Abies amabilis* such shifts occur seasonally in both current and 1 year old foliage (Fig. 6.9). These adjustments have significant effects on stomatal responses. When soil moisture was near field capacity in July and again in early March, predawn water potentials were -0.15 and $-0.74\,\mathrm{MPa}$, respectively; stomatal closure was observed when water potentials decreased to -1.54 and $-2.93\,\mathrm{MPa}$,

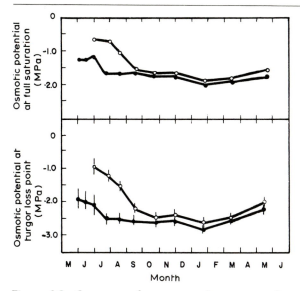

Figure 6.9 Summary of pressure–volume curves for current (open circles) and 1 year old (closed circles) foliage of *Abies amabilis*. Vertical lines on graph of osmotic potential at turgor loss point represent the 95% confidence interval for the estimated values (from Teskey, 1982).

respectively. Such shifts throughout the summer and into winter may be important in aiding trees to acquire water under dry or cold soil conditions.

The ability of a tree to maintain leaf water potentials above the turgor loss point can be used as a measure of its drought avoidance. This can involve active control or at least regulation of water loss through stomatal closure. In general, those species that close stomata at relatively high thresholds (i.e. lower xylem pressure potentials) would avoid stress-related physiological problems during periods of limited water availability. Hence, a possible ranking of increasing drought avoidance might be: *Abies* spp. > *Pinus* spp. = *Picea* spp. > *Pseudotsuga menziesii* (Table 6.4; Aussenac and Granier, 1978). Although most species seem to close their stomata abruptly, thereby preventing further water loss, some do not (e.g. *Abies grandis*; Table 6.4). Hence, *A. grandis* would appear to be more drought tolerant and, thus, able to actively photosynthesize at lower needle water potentials than other species in this genus (Puritch, 1973). Spomer (unpublished) compared the drought tolerance (estimated from their

osmotic potentials) of various conifer species with their successional positions (Fig. 6.10).

The specific control of stomatal activity by water deficits is a very important factor in determining the distribution of forest tree species since it arises as a 'normal' consequence of environmental conditions that promote low water uptake in relation to loss. Stomatal activity, however, is quite complex and is influenced by a variety of factors including light, leaf and soil water potentials, leaf to air vapor-pressure differences, leaf temperatures, abscisic acid levels, and internal carbon dioxide concentrations (Hinckley *et al.*, 1978; Jarvis, 1976; Jarvis and Morison, 1981; Lassoie, 1982; Tan *et al.*, 1977). In addition, environmental factors controlling stomata seldom act independently from one another. Although techniques are available for defining independent relations (Lassoie, 1982), the fact that there are reports of synergistic interactions occurring (e.g. Leverenz, 1981b; Meinzer, 1982a,b,c; Ng and Jarvis, 1980) may lead to reevaluation of the controlling influence of some environmental factors under field conditions (Lassoie, 1982).

During the day stomata will open photoactively and, barring low light or high vapor-pressure differences, remain open unless a threshold leaf water potential is reached (Table 6.4; Hinckley *et al.*, 1978; Lassoie, 1982). This threshold for *Pseudotsuga menziesii* saplings appears to be about -2.0 MPa as such deficits have been observed to trigger abrupt midday stomatal closure in the field (Running, 1976; Tan *et al.*, 1977). Waring and Running (1978) reported a threshold of -2.2 MPa in a 40 m, old-growth *P. menziesii* while stomatal closure seemed to be initiated near -1.7 MPa in seedlings (Waring and Running, 1978; Drew and Ferrell, 1979).

There seems to be marked seasonal adjustment in thresholds within some species. For example, during the summer Tan *et al.* (1977) observed an increase in the xylem pressure potential threshold for *Pseudotsuga menziesii* saplings in response to the first drought of the summer (Table 6.4). Hence, stomatal closure was initiated earlier but progressed more slowly during the second drought period, suggesting a change in drought resistance. Drew and Ferrell (1977) observed a similar response with *P. menziesii*

Table 6.4 Xylem pressure potential thresholds for stomatal closure, rates of transpiration at night, and rooting characteristics of common Pacific Northwest conifers

| Species | Condition and reference* | Thresholds (MPa) | | Night transpiration (% of day rate)[‡] |
		Initiate closure	Complete closure	
Abies amabilis	(a)	− 0.8	− 1.5	+ + + +
A. amabilis	(b)	—	− 1.8	
A. amabilis	(c₁)	—	− 1.6	
A. amabilis	(c₂)	—	− 2.9	
A. balsamea	(a)	− 1.0	− 1.3	
A. lasiocarpa	(a)	− 0.9	− 1.9	
A. grandis	(a)	− 0.9	− 2.4	+ + + +
A. grandis	(d)	—	− 2.5	
Picea engelmannii	(d)	—	1.6	+
P. sitchensis	(e₁)	—	1.6(− 1.1)[†]	+
P. sitchensis	(e₂)	—	2.1(− 1.4)[†]	
P. sitchensis	(e₃)	—	3.4(− 2.1)[†]	
Pinus contorta	(d)	—	− 1.5 to − 1.6	+
P. ponderosa	(d)	—	− 1.6 to − 1.7	+
P. ponderosa	(f)	− 1.8	—	
Pseudotsuga menziesii	(d)	—	− 1.9 to − 2.2	+ + +
P. menziesii	(f)	− 2.0	—	
P. menziesii	(g₁)	− 1.8	− 2.0	
P. menziesii	(g₂)	− 1.5	− 2.2	
P. menziesii	(h₁)	− 1.0	− 2.2	
P. menziesii	(h₂)	− 1.5	− 1.9	
Tsuga heterophylla	(b)	—	− 1.5	
T. heterophylla	(f)	—	− 2.1	

*(a) potted 3–5 year old seedlings, autumn, based on controlled environment measurements of transpiration (Puritch, 1973); (b) stand, 2–4 m trees, summer (Kotar, 1978); (c) stand, 2–4 m trees for (c₁) summer and (c₂) winter (Teskey, 1982); (d) potted 2 year old seedlings, winter, based on controlled environment measurements of transpiration (Lopushinsky, 1969); (e) stand, 11 m trees, based on field measurements of needle conductances for (e₁) winter, (e₂) spring, and (e₃) summer (Hellkvist et al., 1974; Richards, 1973); (f) stand, 1–3 m trees, summer, field measurements of needle conductances (Running, 1976); (g) stand, 7–10 m trees, summer based on field measurements of needle conductances for (g₁) first dry period and (g₂) second dry period of 1974 (Tan et al., 1977); (h) potted 1 year old seedlings, greenhouse, based on greenhouse measurements of needle conductance for (h₁) summer and (h₂) autumn (Drew and Ferrell, 1979).
[†]Water potential value at zero turgor is given; saturated osmotic potential is in parenthesis.
[‡]Night transpiration (Hinckley, 1971; Hinckley and Ritchie, 1973; Hodges, 1967; Lopushinsky, 1975): + = 10%, + + = 10–20%, + + + = 20–30%, + + + + = 30%.

seedlings between the summer and autumn. During the active summer growth period the higher threshold and more gradual rate of stomatal closure indicated an advantage to maintaining high turgor pressures necessary for the growth processes while also maintaining significant net photosynthetic rates. Some of this change probably reflected seasonal osmotic adjustments needed to maintain needle conductances at maximum levels during periods of low water supply (discussed earlier). In addition, Drew and Ferrell (1977) reported that seedlings grown at low light levels had lower thresholds for stomatal closure, thus making them less drought resistant than seedlings grown under full sunlight. Similarly, shade-intolerant pioneer species appear to close stomata at higher water potentials than more shade-tolerant species (Table 6.4).

Stomata also are sensitive to high rates of water loss due to atmospheric conditions. A ranking of species from most to least sensitive to high evaporative demand (i.e. low atmospheric humidity) would be: *Picea engelmannii = P. sitchensis = Pinus contorta > P. ponderosa = Pseudotsuga men-*

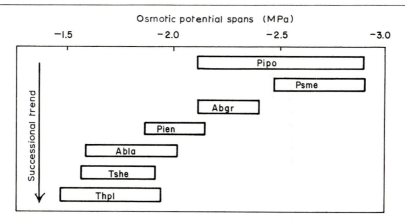

Figure 6.10 Representation of osmotic potential spans between osmotic potential at full saturation and osmotic potential at the turgor loss point for *Pinus ponderosa* (Pipo), *Pseudotsuga menziesii* (Psme), *Abies grandis* (Abgr), *A. lasiocarpa* (Abla), *Picea engelmannii* (Pien), *Tsuga heterophylla* (Tshe), and *Thuja plicata* (Thpl) (after Jackson and Spomer, 1979; Spomer, unpublished).

ziesii = Tsuga heterophylla > Abies concolor (Barker, 1973; Running, 1976, 1980; Rutter, 1977).

The exact mechanism by which atmospheric humidity controls stomatal response has been described by Farquahar (1978). Rapid decreases in leaf conductances have been shown as needle to air vapor-pressure differences increase from about 0.2 kPa in *Picea sitchensis* (Grace *et al.*, 1975) and from 0.5 kPa in *Pseudotsuga menziesii* (Leverenz, 1981a; Running, 1976). Similar results have been observed in all Northwest conifers studied to date (Franklin and Waring, 1979; Kaufmann, 1976, 1982a; Neilson and Jarvis, 1975; Running, 1976; Watts *et al.*, 1976). Obviously, vapor-pressure differences will have significant indirect effects on the response of net photosynthesis to light intensity due to its direct influence on stomatal conductance (Jarvis, 1981). As mentioned earlier, low atmospheric vapor-pressure differences appear to be important in maintaining open stomata in the dark. Atmospheric humidity and leaf temperature interact synergistically in their effect on stomata (Hall *et al.*, 1976, Leverenz, 1981b) which masks the controlling influence of the latter at high temperatures (Lassoie, 1982).

Maximum stomatal conductance during the day is closely correlated with predawn xylem pressure potential (Running, 1976) and/or vapor-pressure differences developed during the day between the

needles and the air. Tan *et al.* (1977) observed that vapor-pressure deficit clearly influenced the relationship between stomatal conductance and leaf

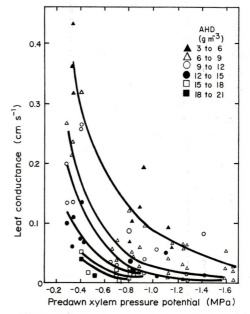

Figure 6.11 Relation between leaf conductance and predawn xylem pressure potential in various *Pseudotsuga menziesii* seedlings (from Hallgren, 1978). The data were collected on 20 days and then separated into six ranges of absolute humidity deficit (AHD). Each data point represents the mean of 7–15 measurements.

water potential. Therefore, *Pseudotsuga menziesii* stomata will remain relatively closed on days when atmospheric evaporative demands are high even though soil moisture levels are high (i.e. high predawn xylem pressure potentials) (Fig. 6.11). Such 'fine tuning' to atmospheric evaporative demand by stomata possibly may be an adaptation to prevent low water potentials in shoots when water flux across the root is limited.

Running (1976) has commented that daily thresholds for stomatal closure are not directly affected by the progressive decrease in predawn xylem pressure potentials during a drought period; however, the length of time available for carbon dioxide uptake to occur, unaffected by stomatal closure, would progressively decrease (Lassoie and Salo, 1981). In fact, during severe droughts when predawn xylem pressure potentials are very low (e.g. -1.6 MPa in large, field-grown *Pseudotsuga menziesii*; Lassoie and Salo, 1981), stomata are essentially closed all day as hydraulic factors almost completely override the photoactive stomatal response illustrated in Fig. 6.5 (Hinckley *et al.*, 1978; Lassoie, 1982).

There appears to be a considerable amount of both inter- and intraspecific variation in the control of stomata by water potential (Table 6.4). One might expect that a portion of this variation has resulted from the confounding effects of other environmental variables, such as light and humidity, and from the fact that the actual controlling parameter, turgor pressure in the stomatal guard cells and surrounding epidermal cells, must be inferred from indirect measurements (Lassoie, 1982). However, some amount of this variation is ecologically significant as it indicates important adaptations and acclimatizations to drought.

6.3.4 NUTRIENTS

Franklin and Waring (1979) consider nutrients to rank third, behind temperature and moisture, in their effects on the evolution of Pacific Northwest forests. Even so, the relatively young age of Northwest soils and their glacial origin, in association with seasonal climatic patterns which are not conducive to rapid decomposition rates (Fogel and Cromack, 1977; Turner and Singer, 1976), suggest that nutrient budgets may be an important factor affecting forest productivity within the region. Nitrogen appears to be the major growth-limiting nutrient owing to its biological origin (Gessel *et al.*, 1973). Thus, a significant growth response has been observed in over 80% of the *Pseudotsuga menziesii* stands fertilized with nitrogen (Gessel *et al.*, 1969).

Nutrient status, in combination with water, appear to limit the ability of a site to accumulate foliar biomass (Daucet *et al.*, 1976; Grier and Running, 1977) which obviously influences stemwood production. For example, irrigation of young *Pseudotsuga menziesii* stands with nutrient-rich waste water has been found to increase aboveground biomass increment by over 380% compared to stands irrigated with identical amounts of river water (1.64 compared to 0.49 kg m^{-2} yr^{-1}; Schiess and Cole, 1981).

A considerable amount of research has been conducted to delineate nutrient budgets and cycling dynamics for forests within the Pacific Northwest (Abee and Lavender, 1972; Cole *et al.*, 1968; Cole *et al.*, 1977; Fredriksen, 1972; Grier and Cole, 1972; Johnson *et al.*, 1982; Sollins *et al.*, 1980; Turner and Singer, 1976). Although this information has yielded a better understanding and appreciation for the internal workings of coniferous ecosystems (Edmonds, 1982b), it has not provided an understanding of the possible controlling influence that nutrients might exert on forest tree productivities and distributions. Nutrient manipulations through the addition of fertilizer, however, have provided some insight into this area. The addition of nitrogen has been known to enhance coniferous tree growth and stand development since work in the 1960s investigated the feasibility of using fertilizer as a silvicultural tool in the Pacific Northwest (Gessel *et al.*, 1969). However, it was not until the late 1970s that an understanding of the specific physiological processes involved emerged. Even now, knowledge seems to be limited primarily to one species, *Pseudotsuga menziessi*, due to its economic importance in the region.

The addition of nitrogen is known to improve site quality, reduce the time for canopy closure to occur, and improve aboveground productivities of forest stands (perhaps by shifting belowground production to aboveground components) (Johnson *et al.*, 1982;

Keyes and Grier, 1981). In spite of these observations, there are only limited data relating nutrient additions and total stand responses, that is, those including both above- and belowground production. Hence, the most comprehensive view for nutrients as a limiting environmental factor involves the influence of nitrogen on aboveground processes in *Pseudotsuga menziesii*.

If the stimulatory influence of nitrogen fertilization can be viewed as an indicator of nutrient control, then two major effects may be identified. One involves morphological and growth changes associated with the various structural components of the tree and the other concerns physiological changes related to net photosynthesis of the foliage.

The addition of nitrogen fertilizer (ammonium nitrate) in the spring has been found to significantly increase stem, branch, and needle growth in stand-grown *Pseudotsuga menziesii* on nitrogen-poor sites (Brix, 1971, 1972, 1981a,b; Brix and Ebell 1969; Brix and Mitchell, 1980). Individual needles increase in dry weight (Brix, 1981a) and in both length and width during the spring and summer immediately following fertilization, thereby producing new needles with larger surface areas (averaging 0.78 cm² compared to 0.65 cm² per needle; Brix

and Ebell, 1969). In contrast, the number of needles per shoot does not increase until the second year, since *P. menziesii* has performed buds with the number of needle primordia being fixed the year before emergence (Allen and Owens, 1972). For a similar reason, lateral shoot production does not increase until the second year (Brix, 1981a; Brix and Ebell, 1969). Needle retention does not seem to be affected by nitrogen fertilization (Brix, 1981a; Brix and Ebell, 1969) except in very nitrogen-deficient soils where only a few age classes of needles are retained under normal conditions (Brix, 1981a; Turner and Olson, 1976). Hence, nitrogen additions enhance the total needle surface area of individual shoots, the number of shoots per branch, and in severe situations the number of foliage age classes retained. Although yearly needle production tends to peak 2/3 years after fertilization (Brix, 1981a), this will mean larger crowns and large photosynthetic surface areas in the years following fertilization (Table 6.5).

Concurrent increases in stem basal area growth following fertilization indicate a close relationship between changes in foliar surface area and cambial activity. The effect is to increase the rate of tracheid cell production and to delay the transition from

Table 6.5 Main branch (whorl 9) characteristics 5 yr after fertilization and tree characteristics 7 yr after fertilization for 24 year old *Pseudotsuga menziesii* (after Brix, 1981a)

	Treatment	
	Control	*+448 kg N ha⁻¹*
Branch characteristics:		
Total number of needles	21 000	60 100
Number of shoots	269	572
Needles per shoot	78	105
Average shoot length (cm)	4.58	5.76
Needle density (no. cm⁻¹)	16.3	17.1
Total needle weight (g)	82.8	247.3
1000 needle weight (g)	3.95	4.11
Tree characteristics:		
Needle surface area (m²)	25.19	48.65*
Crown projection area (m²)	3.56	5.02
Leaf area index for trees	7.08	9.80*
Crown closure (%)	84	90
Leaf area index for stand	5.92	8.79*

*Significantly different ($p = 0.05$) from control.

early- to latewood (Brix, 1972; Brix and Mitchell, 1980). For example, Brix and Mitchell (1980) with 24 year old *Pseudotsuga menziesii* found that the addition of 448 kg N ha^{-1} increased the maximum radial production of tracheids from 0.8 cells per day to 1.53; neither the size of individual cells nor the seasonal duration of cell production was found to change significantly. These authors also reported that cambial growth was controlled primarily by temperature during the spring and by soil water during the summer, while the cessation of growth in the autumn was determined by the shortness of the photoperiod.

The addition of nitrogen enhances height growth as well as stem diameter growth. For example, Brix and Ebell (1969) reported that height growth in *Pseudotsuga menziesii* increased about 30% during the summer following fertilization, and about 50% during each of the following 2 years: no significant differences were observed the fourth year. Similar results have been reported by Brix (1981a) where the stimulatory effect of nitrogen fertilization on height growth was observed to last no longer than 2 years following an application of 448 kg N ha^{-1}.

Internal changes also have been noted in *P. menziesii* needles immediately following fertilization. For example, Brix (1971) has recorded significant increases in chlorophyll (a + b) and in needle nitrogen levels during the summer following a springtime fertilization with 448 kg N ha^{-1}, an observation made earlier by Brix and Ebell (1969). However, the effect seems to be less significant the following year except for a continued higher chlorophyll content in the 1 year foliage developed immediately after fertilizing. Brix (1981b) noted no significant increases in foliar nitrogen at lower fertilization rates (112 and 224 kg N ha^{-1}) even though stem growth was enhanced.

In addition to changes in needle and shoot characteristics, one should expect that nitrogen fertilization should also increase foliar photosynthetic capacities as nitrogen is important to this process. In fact, Brix (1981b) has reported a convexed, curvilinear relationship between foliar per cent nitrogen and the per cent of maximum net photosynthesis (at 30 000 k lumen and 20 °C) for current year *Pseudotsuga menziesii* foliage which was op-

timum at 1.74% nitrogen (dry weight basis). This value is remarkably similar to the 1.75% optimum reported by Gessel *et al.* (1969) for maximum stem growth in the species.

Brix (1971) has reported that fertilization with 448 kg N ha^{-1} greatly increased the photosynthetic capacity of new shoots of *P. menziesii* in the summer immediately following fertilization. However, in 1 year old shoots, no significant effect was observed in spite of the fact that both chlorophyll and nitrogen concentrations increased considerably and in mid-July reached levels comparable to those in current year shoots. The year after fertilization, increased net photosynthetic rates were recorded in June and July only with 1 year old foliage as no differences were recorded in the spring or autumn. Increased photosynthetic rates of new shoots expanded during the second year following the application of nitrogen occurred only in June. Throughout the 2–3 year experiment, fertilization increased dark respiration rates in both current and 1 year old foliage.

Brix (1981b) attributes the increase in photosynthetic capacity with the addition of nitrogen to increased chlorophyll contents, increased needle conductances for carbon dioxide (Natr, 1975), enhanced activity of carboxylating enzymes (Mooney and Gulmon, 1982; Natr, 1975), and improved utilization of carbohydrates (Sweet and Wareing, 1966). Brix (1972) has observed improved tree water balances (i.e. lower tree water deficits) during the day in response to nitrogen fertilizer, which could have direct effects on stomatal conductances for carbon dioxide. Hence, he observed a 59% increase in diameter growth when springtime fertilization (448 kg N ha^{-1}) was combined with summer irrigation (25 mm water weekly); fertilization and irrigation alone increased growth 16% and 15%, respectively. Such enhancement suggests better root growth and/or less suberization of new roots and, thus, an increase in the number of fine roots in response to nitrogen. Although fertilization has been reported to increase conifer rooting densities primarily by stimulating the growth of short roots in the uppermost 10–20 cm of soil (Hermann, 1977), the possible importance of this effect has yet to be examined in detail.

In summary, it appears that nitrogen fertilization

increases crown surface areas of individual trees and shoot net photosynthetic rates. At the stand level, however, increases in stand leaf area due to increased tree leaf areas may be offset considerably by an average increase in tree mortality following fertilization (about 18%; Grier, unpublished). The effect is further moderated by the distribution and arrangement of foliage which become very important if they fail to optimize available light. For example, Brix (1971) did not find a nitrogen-induced increase in net photosynthesis below about $400\,\mu\text{mol}\,\text{m}^{-2}\,\text{s}^{-1}$ in *Pseudotsuga menziesii*, but the influence of improved nitrogen levels became more important at higher light levels (e.g. a 34% increase at about $500\,\mu\text{mol}\,\text{m}^{-2}\,\text{s}^{-1}$ and a 78% increase at about $1000\,\mu\text{mol}\,\text{m}^{-2}\,\text{s}^{-1}$). Hence, factors affecting light regimes (e.g. stand density, crown depths and shapes, and foliage distributions) are very important in determining the influence of nitrogen on the photosynthetic process.

6.4 Tree structure

Knowledge of tree structure and leaf area distribution is a prerequisite for detailed ecosystem-level studies related to productivity, microclimate, carbon and nutrient cycling, and competition (Long, 1982), and for ecophysiological studies of photosynthesis, respiration, water status, and growth (Hinckley and Lassoie, 1981; Lassoie, 1982). For example, estimates of foliar distribution are among the parameters necessary for extrapolation of shoot-level gas exchange data to a whole tree or a stand basis (Kinerson and Fritschen, 1971; Leverenz, 1981b; Monsi *et al.*, 1973; Salo, 1974; Woodman, 1971) or for estimations of canopy resistance to water loss (Leverenz *et al.*, 1982; Tan and Black, 1976; Tan *et al.*, 1978). Crown structure varies considerably with environment. For example, Gholz (1979) determined biomass and leaf area along a transect from the Pacific Ocean to the interior of Oregon in stands past the age of canopy closure (Fig. 6.12). Live biomass densities along this transect ranged from $2.7\,\text{kg}\,\text{m}^{-3}$ in a 55 m tall *Picea sitchensis* stand located near the Ocean to $0.3\,\text{kg}\,\text{m}^{-3}$ in an 8 m tall *Juniperus occidentalis* stand in interior Oregon. The *P. sitchensis* stand carried $3.18\,\text{kg}\,\text{m}^{-2}$ of foliage with a leaf area (all sides) of $44.4\,\text{m}^2\,\text{m}^{-2}$. In contrast, the *J. occidentalis* stand had $0.46\,\text{kg}\,\text{m}^{-2}$ with a leaf area (all sides) of $2.0\,\text{m}^2\,\text{m}^{-2}$. These large differences in the physical structure of forests were primarily associated with differences in growing season site water balances (Grier and Running, 1977). These balances were, in turn, related to geographic and pedologic effects on soil water storage and evaporative demand.

6.4.1 LEAF AREA DEVELOPMENT

A detailed treatment of crown dynamics is dependent on an understanding of patterns of individual

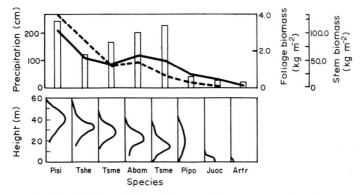

Figure 6.12 Relative vertical distribution of foliage biomass of overstory woody plants, total annual precipitation (bars), total foliage biomass (solid line), and total stem biomass (dashed line) for eight sites which ranged from the coast to the interior of Oregon (after Gholz, 1979), Key: *Abies amabilis* (1500 m), Abam; *Artimesia tridentata* (1200 m), Artr; *Juniperus occidentalis* (1356 m), Juoc; *Picea sitchensis* (200 m), Pisi; *Pinus ponderosa* (870 m), Pipo; *Tsuga heterophylla* (365 m), Tshe; *T. mertensiana* (1590 m), Tsme.

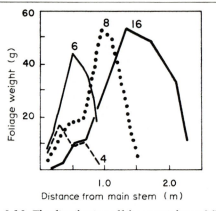

Figure 6.13 The distribution of foliage weight on different branches within the crown of a codominant *Pseudotsuga menziesii* (after Jensen, 1976). The numbers next to each line refer to the whorl number from the top of the tree.

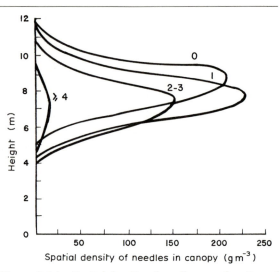

Figure 6.14 Spatial density of needles as a function of height in a *Picea sitchensis* canopy. Spatial density is the dry weight of needles per unit volume of canopy. The numbers on the curves are the ages of the needles in years; 0 indicates current (1971) year's needles (from Norman and Jarvis, 1974).

branch growth; on needle arrangement, morphology, and retention; and on the ratio of photosynthesis to respiration for different aged and positioned needles. It is also dependent on an understanding of the competitive interaction among individual trees in a stand.

In a detailed investigation of branch growth, Jensen (1976) found that the extension of primary branch internodes in *Pseudotsuga menziesii* was highly dependent on the position of the branch within the crown (Fig. 6.13; Jensen and Long, 1983). As branches became older and lower in the crown, the center of foliar mass moved towards the apical end of the branch apparently to minimize self-shading and maximize photosynthetic efficiency (Monsi *et al.*, 1973; Norman and Jarvis, 1975). There is also a shift in the distribution of different age classes of foliage with height (Fig. 6.14). Depending upon the density of the forest and the height of surrounding vegetation, a point is reached where photosynthetic production of a branch decreases below its growth and maintenance requirements. At this point, the branch no longer contributes to the carbohydrate balance of the tree (i.e. it becomes nonfunctional); continued tissue respiration depletes the branch carbohydrate reserves and eventually it dies (Hinckley and Lassoie, 1981; Woodman, 1971).

By either using the distribution of foliage to branch position relationship (Figs. 6.13 and 6.14),

or the known linear relationships between branch or stem sapwood area and foliage biomass (discussed later; Grier and Waring, 1974; Long *et al.*, 1981) it is possible to determine the overall distribution of foliage within the crown of a tree. This information can then be related to known patterns of branch growth (Jensen, 1976; Jensen and Long, 1983), stem growth (Dobbs and Scott, 1971), and net photosynthesis (Kinerson and Fritschen, 1971; Woodman, 1971) (Fig. 6.15). It has been suggested that the position of maximum radial increment may correspond to the height of the maximum 'effective' (i.e. highest net photosynthesis rates) foliar biomass (Assmann, 1970; Larson, 1963). However, if one compares stem cross-sectional increment to ring width along a tree stem, another pattern emerges (Fig. 6.15). The greatest increment is added considerably below the point of maximum 'effective' foliar biomass.

Foliar biomass of individual trees may continue to increase throughout most of the life of the tree (e.g. 200–1000 years in conifers). (Gholz *et al.*, 1976; Tadaki, 1977). In contrast, stands or aggregates of trees reach maximum foliar accumulation at a

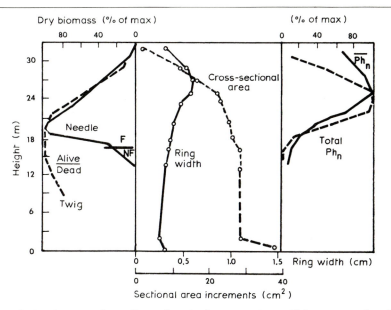

Figure 6.15 Vertical distribution of needle and twig biomass ring width, cross-sectional area increments ($= [r_x^2 - r_{x-1}^2]\pi$; where r = radius and x = year), average daily net photosynthesis ($\overline{Ph_n}$), total net photosynthesis ($=$ [needle biomass] $[\overline{Ph_n}]$) in a dominant *Pseudotsuga menziesii*. Branches which have the same number of rings as the whorl to which they are attached are defined as functional (F), those having less are defined as nonfunctional (NF) (from Hinckley and Lassoie, 1981).

much earlier age; that is, from 15 to 80 years depending on site quality and initial stand density (Mar:Moller, 1947; Tadaki *et al.*, 1970; Turner and Long, 1975). This means that on a stand basis, foliar area and productivity reach an early equilibrium while respiring biomass continues to accumulate. The ratio of support biomass (stem and branch) to production biomass (foliage) (see Fig. 6.12) should decrease as environmental stresses increase. Calculated ratios ranged from 45.9 for the coastal *Picea sitchensis* stand to 3.7 for the interior *Juniperus occidentalis* stand, and to 17.8 for the moist, but high elevation stand to *Tsuga mertensiana*. As environmental stresses such as moisture and temperature increase, respiration either remains relatively constant or increases as photosynthesis decreases (Emmingham and Waring, 1977). Therefore in stressed environments, there should be less support tissue compared to production tissue.

6.4.2 STEM FORM AND FUNCTION

A consequence of transpirational water loss at the foliage–atmosphere interface is a localized decrease in foliar water content and water potential which eventually is transmitted to tissues throughout the tree. Contrary to strict adherence to the caternary theory of water transport, water required for transpiration can come from a number of internal sources in addition to the soil reservoir (Lassoie, 1982). There are two categories of internal sources of water in forest trees: 1. elastic tissues which undergo dimensional changes when water is exchanged with the transpirational stream (e.g. foliage, fruits, buds, and phloem-cambium-immature xylem); and 2. inelastic tissues that do not (e.g. mature lignified sapwood).

The majority of literature on stored water has dealt with examinations of elastic phloem–cambium–immature xylem tissue owing to the ease at which dimensional fluctuations may be monitored in tree stems (Lassoie, 1975, 1982). In general, diurnal fluctuations result from the development of daily tree water deficits while seasonal fluctuations are indicative of seasonal changes in soil moisture (Lassoie, 1973, 1979). The magnitudes of both diurnal and seasonal shrinkages are closely related to the size and morphological condition of the storage zone involved, which vary with the season, tree vigor, and height along the stem; as well as with internal water relations of the trees (Lassoie, 1979). The actual amount of water stored near the cambium of a large *Pseudotsuga menziesii* has been shown to account for only a minimal amount of the water lost during the summer, but at least 5% of the water transpired daily (Lassoie, 1979).

Both the conducting sapwood and the non-conducting heartwood appear to be significant storage areas (Jarvis, 1975; Running, 1980; Waring and Running, 1978). Water content changes in the sapwood and heartwood (Hinckley *et al.*, 1978) and the cavitation of water column in the sapwood have been reported (Waring and Running, 1978; Tyree and Dixon, 1983). Richards (1973) estimated that between sunrise and solar noon as much as 80% of the water lost from *Picea sitchensis* trees came from stored reserves and not the soil. Waring and Running (1978) estimated that the total storage capacity of an old-growth *Pseudotsuga menziessi* forest was $267\,\mathrm{m^3}$ water ha^{-1} (26.7 mm), 75% of which was stored in the stemwood. In a study by Running (1980), readily available water from internal storage was calculated to be capable of providing 0.6 h maximum transpiration in *Pinus contorta*. Needle water content contributed only 4% of the total storage.

Water storage serves principally as a buffer during periods of high transpiration or low soil moisture. Food storage (carbohydrate reserves) is also an important buffer in woody plants because of the seasonality of growth in temperate trees. Cycles of rapid growth and intense utilization of photosynthates are matched with cycles of no growth and storage. Reserves are critical for deciduous species to meet respiration and growth needs when there are no leaves present (McLaughlin *et al.*, 1980). However, reserves are also important to evergreen species (Emmingham and Waring, 1977; Ross, 1972; Webb, 1980). In addition to providing a path of lateral water and food transport, ray parenchyma cells of the phloem and xylem are a major site for carbohydrate storage in the roots and stems of forest trees. High concentrations of reserve materials are also found in needles, roots, and phloem, and in evergreen species the foliage may be the most important location of carbohydrate reserves (Ericsson *et al.*, 1980).

As a consequence of transport and storage properties of the xylem, one might anticipate a high degree of correlation between the functional xylem and a tree's foliage area. Morikawa (1974) and Kline *et al.* (1976) have noted that the absolute amount of water transported in a conifer is linearly related to sapwood cross-sectional area. More recent work with *Pinus sylvestris* using radioactive phosphorus to estimate xylem water movement has shown a 1:1 relationship on a daily basis between water transport in the sapwood and water loss from the needles (Waring *et al.*, 1979). When water transport in the xylem was compared to water loss on a subdaily basis, a hysteresis loop developed due to the contribution of stored water to the transpiration stream (Hinckley *et al.*, 1978; Morikawa, 1974; Running, 1980; Swanson, 1972; Waring *et al.*, 1980b).

A sensitive linear relationship between sapwood cross-sectional area and either foliage biomass or leaf area of individual trees has been observed for all Pacific Northwest conifer species examined to date (Grier and Waring, 1974; Kaufmann and Troendle, 1981; Snell and Brown, 1978; Waring *et al.*, 1982). Although the linear relationships did not change with stand density or age, a slight bias was introduced in leaf–sapwood area relations due to site influences on leaf areas. Individual tree leaf areas versus sapwood areas were not biased by changes in site quality (Grier and Running, 1977). Whitehead (1978) reported that the relationship between leaf area and sapwood area in *Pinus sylvestris* was independent of tree spacing, but was dependent on site quality.

Recent investigations have built upon the 'pipe-theory' (Shinozaki *et al.*, 1964) and the work of Grier and Waring (1974). Since water is confined to the outer annual ring of ring-porous trees, Rogers and Hinckley (1979) found a better relationship between the area of the current annual ring and foliar biomass in *Quercus alba* and *Q. velutina* growing in mid-Missouri than when either stem diameter at breast height (1.3 m) or sapwood cross-sectional area was used. Similar results have been observed in mid-New York with *Q. bicolor* (Lassoie and Malecki, unpublished). Whether this relationship holds for *Quercus* species common to the Northwest (e.g. *Q. garryana* and *Q. kelloggii*) has yet to be verified.

Long *et al.* (1981) postulated that sapwood cross-sectional area at any location along a conifer stem should predict the foliar biomass above that point. They expanded on the 'pipe-model' analysis of tree form originally used by Shinozaki *et al.* (1964) to differentiate structural and water conduction features of the stem (Fig. 6.16). Sapwood cross-sectional area was shown to be linearly related to foliage biomass above the point examined anywhere along the stem. This relationship has been extended recently to include hardwoods (Lassoie and Malecki, unpublished). In addition, Long *et al.* (1981) observed that individual branch sapwood cross-sectional area was also linearly related to the foliage on that branch. In large trees, an adequate structure is provided by the total stem for support of the crown; however, the entire cross-sectional area of the stem is not needed for water conduction. Therefore, heartwood formation occurs as transport functions decrease and xylary elements become nonfunctional.

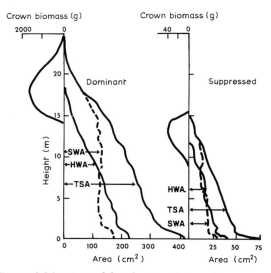

Figure 6.16 Vertical distribution of crown biomass and heartwood (HWA), sapwood (SWA) and total cross-sectional (TSA) areas for a dominant and suppressed *Pseudotsuga menziesii* (after Long *et al.*, 1981).

Waring and co-workers have proposed a measure of the photosynthetic efficiency, or vigor, of a tree's crown (Waring and Pitman 1980; Waring *et al.*, 1980a). They reasoned that if sapwood cross-sectional area provides an estimate of the size of a tree's crown, then the area of the last annual growth increment should be related to the efficiency of that crown. A tree vigor index was then defined as being equal to basal area increment divided by sapwood cross-sectional area. Individual tree vigor indices can be related to insect and pathogen susceptibilities (Larsson *et al.*, 1983; Mitchell *et al.*, 1983; Waring and Pitman, 1980) and have been used to determine mean tree vigor and its relation to stand growth (Waring *et al.*, 1980a).

6.5 Conclusion

This chapter has discussed ecophysiology of coniferous forest tree species common to the Pacific Northwest. It has provided some physiological support for the unique abundance and wide distribution of large, long-lived conifers throughout this northern temperate region. As with plant species in other biomes presented in this text, these conifers seem well adapted to their environment; specifically to wet mild winters, dry warm summers, and nutrient-poor soils. In addition, the foliage of Northwest conifers appears capable of various morphological and physiological adjustments in response to spatial and temporal variations in light, temperature, and water regimes. These adaptations and adjustments assure the production of adequate amounts of photosynthates for growth, development, reproduction, and maintenance while preventing damaging water losses.

Research efforts during the US International Biological Program (Edmonds, 1982a) and afterwards have provided a wealth of information concerning the biotic and abiotic control of photosynthesis in numerous conifer species and the ecological structure and functioning of components of the coniferous forest biome. Such work, however, has yielded only a qualitative understanding of the connection between the photosynthetic process and the productivity and distribution of trees and forests (Lassoie, 1982). Given the current knowledge base outlined in this chapter, the next decade will most likely be dedicated to providing the connection specifically in relation to intra- and interspecific differences, above- and belowground processes, and site variabilities. Such information will provide the scientific community with a more complete picture of the physiological basis for the distribution and abundance of conifers within the Pacific Northwest.

References

Abee, A. and Lavender, D.P. (1972) Nutrient cycling in throughfall and litterfall in 450 year old Douglas-fir stands, in *Proceedings – Research on Coniferous Forest Ecosystems – A Symposium* (ed. by J.F. Franklin, L.J. Dempster and R.H. Waring), United States Department of Agriculture Forest Service, Portland, Oregon.

Allen, G.S. and Owens, J.N. (1972) *The Life History of Douglas-fir*, Environment Canada, Forest Service, Information Canada, Ottawa.

Assmann, E. (1970) *The Principles of Forest Yield Study: Studies in the Organic Production, Structure Increment, and Yield of Forest Stands*, Pergamon Press, Oxford, England.

Aussenac, G. (1973) Effets de conditions microclimatiques différentes sur la morphologie et la structure anatomique des aiguilles de quelques résineux. *Annales des Sciences Forestières*, **30**, 375–92.

Aussenac, G. and Granier, A. (1978) Quelques résultats de cinétique journalière du potentiel de seve chez les arbres forestiers. *Annales des Sciences and Forestières*, **35**, 19–32.

Baker, F.S. (1934) *Theory and Practice of Silviculture*, McGraw-Hill, New York, New York.

Barker, J.E. (1973) Diurnal patterns of water potential in *Abies concolor* and *Pinus ponderosa*. *Canadian Journal of Forest Research*, **3**, 967–84.

Bauer, H., Larcher, W. and Walker, R.B. (1975) Influence of temperature stress on CO_2-gas exchange, in *Photosynthesis and Productivity in Different Environments* (ed. J.P. Cooper), Cambridge University Press, New York, New York.

Beadle, C.L. and Jarvis, P.G. (1977) The effects of shoot water status on some photosynthetic partial processes in Sitka spruce. *Physiologia Plantarum*, **41**, 7–13.

Beadle, C.L., Turner, N.C. and Jarvis, P.G. (1978) Critical water potential for stomatal closure in Sitka spruce. *Physiologia Plantarum*, **43**, 160–5.

Beadle, C.L., Jarvis, P.G. and Neilson, R.E. (1979) Leaf conductance as related to xylem water potential and carbon dioxide concentration in Sitka spruce. *Physiologia Plantarum*, **45**, 158–66.

Blake, J. and Ferrell, W.K. (1977) The association between soil and xylem water potential, leaf resistance and abscisic acid content in droughted seedlings of Douglas-fir (*Pseudotsuga menziesii*). *Physiologia Plantarum*, **39**, 106–9.

Bradbury, I.K. and Malcolm, D.C. (1978) Dry matter accumulation by *Picea sitchensis* seedlings during winter. *Canadian Journal of Forest Research*, **8**, 207–13.

Brix, H. (1967) An analysis of dry matter production in Douglas-fir seedlings in relation to temperature and light intensity. *Canadian Journal of Botany*, **45**, 2063–72.

Brix, H. (1969) Effect of temperature on dry matter production of Douglas-fir seedlings during bud dormancy. *Canadian Journal of Botany*, **47**, 1143–6.

Brix, H. (1971) Effects of nitrogen fertilization on photosynthesis and respiration in Douglas-fir. *Forest Science*, **17**, 407–14.

Brix, H. (1972) Nitrogen fertilization and water effects on photosynthesis and earlywood-latewood production in Douglas-fir. *Canadian Journal of Forest Research*, **2**, 467–78.

Brix, H. (1981a) Effects of thinning and nitrogen fertilization on branch and foliage production in Douglas-fir. *Canadian Journal of Forest Research*, **11**, 502–11.

Brix, H. (1981b) Effects on nitrogen fertilizer source and application rates on foliar nitrogen concentration, photosynthesis, and growth of Douglas-fir. *Canadian Journal of Forest Research*, **11**, 775–80.

Brix, H. and Ebell, L.F. (1969) Effects of nitrogen fertilization on growth, leaf area, and photosynthesis rate in Douglas-fir. *Forest Science*, **15**, 189–95.

Brix, H and Mitchell, A.K. (1980) Effects of thinning and nitrogen fertilization on xylem development in Douglas-fir. *Canadian Journal of Forest Research*, **10**, 121–8.

Cannell, M.G.R. (1982) *World Forest Biomass and Primary Production Data*, Academic Press, New York, New York.

Cole, D.W. (1981) Nutrient cycling in world forests, in *Proceedings – XVII International Union of Forestry Research Organizations World Congress (Division 1)*, Japanese International Union of Forestry Research Organizations Committee, Tsukuba Norin Kenkyudanchi – nai Ibaraki 305, Japan.

Cole, D.W. and Rapp, M. (1980) Elemental cycling in forest ecosystems, in *Dynamic Properties of Forest Ecosystems* (ed. D.E. Reichle), Cambridge University Press, Cambridge, Massachusetts.

Cole, D.W., Gessel, S.P. and Dice, S.F. (1968) Distribution

and cycling of nitrogen, phosphorus, potassium and calcium in a second growth Douglas-fir ecosystem, in *Symposium Primary Productivity and Mineral Cycling in Natural Ecosystems* (ed. H.E. Young). University of Maine Press, Orono, Maine.

Cole, D.W., Turner, J. and Bledsoe, C. (1977) Requirement. and uptake of mineral nutrients in coniferous ecosystems, in *The Belowground Ecosystem: A Synthesis of Plant-Associated Processes* (ed. J.K. Marshall), Range Science Department, Science Series No. 26, Colorado State University, Fort Collins, Colorado.

Conard, S.G. and Radosevich, S.R. (1981) Photosynthesis, xylem pressure potential, and leaf conductance of three Montane chaparral species in California. *Forest Science*, **27**, 627–39.

Coutts, M.P. (1977) The formation of dry zones in the sapwood of conifers. II. The role of living cells in the release of water. *European Journal of Forest Pathology*, **7**, 6–12.

Daucet, R., Berglund, J.V. and Farnsworth, C.E. (1976) Dry matter production in 40-year-old *Pinus banksiana* stands in Quebec. *Canadian Journal of Forest Research*, **6**, 357–67.

Dobbs, R.C. and Scott, D.R.M. (1971) Distribution of diurnal fluctuations in stem circumference of Douglas-fir. *Canadian Journal of Forest Research*, **1**, 80–3.

Doehlert, D.C. and Walker, R.B. (1981) Photosynthesis and photorespiration in Douglas-fir as influenced by irradiance, CO_2 concentration, and temperature. *Forest Science*, **27**, 641–50.

Dougherty, P.M. and Morikawa, Y. (1980) Influence of elevation on the growth, photosynthesis, and water relations of *Pseudotsuga menziesii*. *Plant Physiology*, **65**, (Supplement), 155.

Drew, A.P. and Ferrell, W.K. (1977) Morphological acclimation to light intensity in Douglas-fir seedlings. *Canadian Journal of Botany*, **55**, 2033–42.

Drew, A.P. and Ferrell, W.K. (1979) Seasonal changes in the water balance of Douglas-fir (*Pseudotsuga menziesii*) seedlings grown under different light intensities. *Canadian Journal of Botany*, **57**, 666–74.

Duhme, F. (1974) Die Kennzeichnung der ökologischen Konsitution von Gehölzen im Hinblick auf der Wasserhaushalt. *Dissertationes Botanicae*, **28**, 1–115.

Dykstra, G.F. (1974) Photosynthesis and carbon dioxide transfer resistance of lodgepole pine seedlings in relation to irradiance, temperature, and water potential. *Canadian Journal of Forest Research*, **4**, 201–6.

Edmonds, R.L. (ed.) (1982a) *Analysis of Coniferous Forest Ecosystems in the Western United States. US/IBP Synthesis Series 14*, Hutchinson Ross Publishing Company, Stroudsburg, Pennsylvania.

Edmonds, R.L. (1982b) Introduction, in *Analysis of Coniferous Forest Ecosystems in the Western United States, US/IBP Synthesis Series 14*, (ed. R.L. Edmonds), Hutchinson Ross Publishing Company, Stroudsburg, Pennsylvania.

Edwards, N.T. and McLaughlin, S.B. (1978) Temperature-independent diel variations of respiration rates in *Quercus alba* and *Liriodendron tulipifera*. *Oikos*, **31**, 200–6.

Emmingham, W.H. (1982) Ecological indexes as a means of evaluating climate, species distribution, and primary production, in *Analysis of Coniferous Forest Ecosystems in the Western United States, US/IBP Synthesis Series 14* (ed. R.L. Edmonds), Hutchinson Ross Publishing Company, Stroudsburg, Pennsylvania.

Emmingham, W.H. and Waring, R.H. (1977) An index of photosynthesis for comparing forest sites in western Oregon. *Canadian Journal of Forest Research*, **7**, 165–74.

Ericson, A., Larrson, S. and Tenow, O. (1980) Effects of early and late season defoliation on growth and carbohydrate dynamics in Scots Pine. *Journal of Applied Ecology*, **17**, 747–69.

Fahey, T.J. (1979) The effect of night frost on the transpiration of *Pinus contorta* spp. *latifolia*. *Oecologia Plantarum*, **14**, 483–90.

Farquahar, G.D. (1978) Feedforward responses of stomata to humidity. *Australian Journal of Plant Physiology*, **5**, 787–800.

Fogel, R. and Cromack, K. Jr (1977) Effects of habitat and substrate quality on Douglas-fir litter decomposition in Western Oregon. *Canadian Journal of Botany*, **55**, 1632–40.

Franklin, J.F. and Dyrness, C.T. (1973) *Natural Vegetation of Washington and Oregon*. United States Department of Agriculture Forest Service General Technical Report PNW-8.

Franklin, J.F. and Waring, R.H. (1979) Distinctive features of the northwestern coniferous forest: Development, structure, and function, in *Forests: Fresh Perspectives from Ecosystem Analysis* (ed. R.H. Waring), Oregon State University Press, Corvallis, Oregon.

Fredriksen, R.L. (1972) Nutrient budget for a Douglas-fir forest on an experimental watershed in western Oregon, in *Proceedings – Research on Coniferous Forest Ecosystems – A Symposium* (ed. J.F. Franklin, L.J. Dempster and R.H. Waring), United States Department of Agriculture Forest Service, Portland, Oregon.

Fritschen, L.J. and Doraiswamy. P. (1973) Dew: An addition to the hydraulic balance of Douglas-fir. *Water Resources Research*, **9**, 891–4.

Fry, K.E. (1965) A Study of Transpiration and Photosynthesis in Relation to the Stomatal Resistance and Internal Water Potential in Douglas-fir, PhD thesis, University of Washington, Seattle, Washington.

Fujimori, T. (1971) *Primary Productivity of a Young Tsuga heterophylla Stand and Some Speculations about Biomass of Forest Communities on the Oregon Coast*, United States Department of Agriculture Forest Service Research Paper PNW-123.

Fujimori, T. (1977), Stem biomass and structure of a mature *Sequoia sempervirens* stand on the Pacific Coast of northern California. *Journal of the Japanese Forestry Society*, **59**, 435–41.

Fujimori, T., Kawanabe, S., Saito, H., Grier, C.C. and Shidei, T. (1976) Biomass and primary production in forests of three major vegetation zones of the northwestern United States. *Journal of the Japanese Forestry Society*, **58**, 360–73.

Gessel, S.P., Cole, D.W. and Steinbrenner, E.C. (1973) Nitrogen balances in forest ecosystems of the Pacific Northwest. *Soil Biology and Biochemistry*, **5**, 19–34.

Gessel, S.P., Stoate, T.N. and Turnbull, K.J. (1969) *The Growth and Behavior of Douglas-fir with Nitrogenous Fertilizer in Western Washington*, Institute of Forest Products Contribution No. 7, University of Washington, Seattle, Washington.

Gholz, H.L. (1979) Limits on Aboveground Net Primary Production, Leaf Area, and Biomass in Vegetational Zones of the Pacific Northwest, PhD thesis, Oregon State University, Corvallis, Oregon.

Gholz, H.L., Fitz, F.K. and Waring, R.H. (1976) Leaf area differences associated with old-growth forest communities in the western Oregon Cascades. *Canadian Journal of Forest Research*, **6**, 49–57.

Grace, J., Malcolm, D.C. and Bradbury, I.K. (1975) The effect of wind and humidity on leaf: Diffusive resistance in Sitka spruce seedlings. *Journal of Applied Ecology*, **12**, 931–40.

Grier, C.C. (1975) Wildfire effects on nutrient distribution and leaching in a coniferous ecosystem. *Canadian Journal of Forest Research*, **5**, 599–607.

Grier, C.C. and Cole, D.W. (1972) Elemental transport changes occurring during development of a second-growth Douglas-fir ecosystem, in *Proceedings – Research in Coniferous Forest Ecosystems – A Symposium* (eds J.L. Franklin, L.J. Dempster and R.H. Waring), United States Department of Agriculture Forest Service, Portland, Oregon.

Grier, C.C. and Running, S.W. (1977) Leaf area of mature northwestern coniferous forests: Relation to site water balance. *Ecology*, **58**, 893–9.

Grier, C.C and Waring, R.H. (1974) Conifer foliage mass related to sapwood area. *Forest Science*, **20**, 205–6.

Grier, C.C., Lee, K.M., Nadkarni, N.M. and Klock, G.O. (in press) *Productivity of forests of the United States, its relation to site factors and management practices: A review*, United States Department of Agriculture Forest Service, Pacific Northwest Forest and Range Experiment Station, Portland, Oregon.

Grier, C.C., Vogt, K.A., Keyes, M.R. and Edmonds, R.L. (1981) Biomass distribution and above- and below-ground production in young and mature *Abies amabilis* zone ecosystems of the Washington Cascades. *Canadian Journal of Forest Research*, **11**, 155–67.

Hall, A.E., Schulze, E.D. and Lange, O.L. (1976) Current perspectives of steady-state stomatal responses to environment, in *Water and Plant Life – Problems and Modern Approaches, Ecological Studies*, Vol. 19 (eds O.L. Lange, L, Kappen, and E.D. Schulze), Springer-Verlag, New York, New York.

Hallgren, S.W. (1978) Plant Water Relations in Douglas-fir Seedlings and Screening Selected Families for Drought Resistance, MS thesis, Oregon State University, Corvallis, Oregon.

Hawk, G.M., Long, J.N. and Franklin, J.F. (1982) Relations between vegetation and environment, in *Analysis of Coniferous Forest Ecosystems in the Western United States, US/IBP Synthesis Series 14* (ed. R.L. Edmonds), Hutchinson Ross Publishing Company, Stroudsburg, Pennsylvania.

Hellkvist, J., Richards, G.P. and Jarvis, P.G. (1974) Vertical gradients of water potential and tissue water relations in Sitka spruce with the pressure chamber. *Journal of Applied Ecology*, **11**, 637–68.

Helms, J.A. (1965) Diurnal and seasonal patterns of net assimilation in Douglas-fir, *Pseudotsuga menziesii* (Mirb.) Franco. *Ecology*, **46**, 698–708.

Hermann, R.K. (1977) Growth and production of tree roots: A review, in *The Belowground Ecosystem: A Synthesis of Plant-Associated Processes* (ed. J.K. Marshall), Range Science Department Science Series 26, Colorado State University, Fort Collins, Colorado.

Hinckley, T.M. (1971) Plant Water Stress and its Effect on Ecological Patterns of Behavior in Several Pacific Northwest Conifer Species, PhD thesis, University of Washington, Seattle, Washington.

Hinckley, T.M. and Lassoie, J.P. (1981) Radial growth in conifers and deciduous trees: A comparison. *Mitteilungen der Forstlichen Bundesversuchsanstalt, Wien*, **142**, 17–56.

Hinckley, T.M. and Ritchie, G.A. (1973) A theoretical model for calculation of xylem sap pressures from climatological data. *American Midland Naturalist*, **90**, 56–69.

Hinckley, T.M., Lassoie, J.P. and Running, S.W. (1978) Temporal and spatial variations in the water status of forest trees. *Forest Science Monograph*, No. 20.

Hinckley, T.M., Teskey, R.O., Waring, R.H. and Morikawa, Y. (1983) The water relations of true firs, in *True Fir Symposium Proceedings* (eds C.D. Oliver and R.M. Kenady), Institute of Forest Resources Contribution No. 45, University of Washington, Seattle, Washington.

Hodges, J.D. (1967) Patterns of photosynthesis under natural environmental conditions. *Ecology*, **48**, 234–42.

Hodges, J.D. and Scott, D.R.M. (1968) Photosynthesis in seedlings of six conifer species under natural environmental conditions. *Ecology*, **49**, 973–81.

Jackson, J.D. and Spomer, G.C. (1979) Biophysical adaptations of four western conifers to habitat water conditions. *Botanical Gazette*, **140**, 428–32.

Jarvis, P.G. (1975) Water transfer in plants, in *Heat and Mass Transfer in the Environment of Vegetation* (eds D.A. deVries and N.K. vanAlfen), Scripta Book Co, Washington, D.C.

Jarvis, P.G. (1976) The interpretation of the variations in leaf water potential and stomatal conductance found in canopies in the field. *Philosophical Transactions of the Royal Society of London, Series B (Biological Sciences)*, **273**, 593–610.

Jarvis, P.G. (1981) Production efficiency of coniferous forests in the UK, in *Physiological Processes Limiting Plant Productivity* (ed. C.B. Johnson), Butterworths, London, UK.

Jarvis, P.G. and Leverenz, J.W. (1983) Productivity of temperate deciduous and evergreen forests, in *Encyclopedia of Plant Physiology, Vol. 12D, Physiological Plant Ecology: Productivity and Ecosystem Processes*, (ed O.L. Lange, P.S. Nobel, C.B. Osmond and H. Ziegler) Springer-Verlag, New York, New York.

Jarvis, P.G. and Morison, J.J.L. (1981) The control of transpiration and photosynthesis by stomata, in *Stomatal Physiology* (eds P.G. Jarvis and T.A. Mansfield), Cambridge University Press, New York, New York.

Jarvis, P.G., James, G.B. and Landsberg, J.J. (1976) Coniferous forest, in *Vegetation and the Atmosphere*, Vol. 2, Case Studies (ed. J.L. Monteith), Copyrighted by Academic Press, New York, New York.

Jensen, E.C. (1976) The Crown Structure of a Single Codominant Douglas-fir, MS thesis, University of Washington, Seattle, Washington.

Jensen, E.C. and Long, J.N. (1983) Crown structure of a codominant Douglas-fir. *Canadian Journal of Forest Research*, **13**, 264–9.

Johnson, D.W., Cole, D.W., Bledsoe, C.S., Cromack, K., Edmonds, R.L., Gessel, S.P., Grier, C.C., Richards, B.N. and Vogt, K.A. (1982) Nutrient cycling in forests of the Pacific Northwest, in *Analysis of Coniferous Forest Ecosystems in the Western United States, US/IBP Synthesis Series 14* (ed. R.L. Edmonds), Hutchinson Ross Publishing Company, Stroudsburg, Pennsylvania.

Kaufmann, M.R. (1975) Leaf water stress in Engelmann spruce: Influence of the root and shoot environments. *Plant Physiology*, **56**, 841–4.

Kaufmann, M.R. (1976) Stomatal response of Engelmann spruce to humidity, light, and water stress. *Plant Physiology*, **57**, 898–901.

Kaufmann, M.R. (1982a) Leaf conductance as a function of photosynthetic photon flux density and absolute humidity difference from leaf to air. *Plant Physiology*, **69**, 1018–22.

Kaufmann, M.R. (1982b) Leaf conductance during the final season of a senescing aspen branch. *Plant Physiology*, **70**, 655–7.

Kaufmann, M.R. and Troendle, C.A. (1981) The relationship of leaf area and foliage biomass to sapwood conducting area in four subalpine forest tree species. *Forest Science*, **27**, 477–82.

Keyes, M.R. (1982) Ecosystem Development in *Abies amabilis* Stands of the Washington Cascades: Root Growth and its Role in Net Primary Production, PhD thesis, University of Washington, Seattle, Washington.

Keyes, M.R. and Grier, C.C. (1981) Above-ground and below-ground net production in 40-year-old Douglas-fir stands on low and high productivity sites. *Canadian Journal of Forest Research*, **11**, 599–605.

Kinerson, R.S. and Fritschen, L.J. (1971) Modeling a coniferous forest canopy. *Agricultural Meteorology*, **8**, 439–45.

Kline, J.R., Reed, K.L., Waring, R.H. and Steward, M.L. (1976) Field measurements of transpiration in Douglas-fir. *Journal of Applied Ecology*, **13**, 273–83.

Kotar, J. (1978) Relationships of early seedling development to altitudinal distribution of *Abies amabilis*. *Bulletin of the Torrey Botanical Club*, **105**, 289–95.

Krajina, V.J. (1965) Biogeoclimatic zones and classification of British Columbia. University of British Columbia Department of Botany, Vancouver, British Columbia, Canada. *Ecology of Western North America*, **2**, 1–17.

Krueger, K.W. and Ferrell, W.K. (1965) Comparative photosynthesis and respiratory response to temperature and light by *Pseudotsuga menziesii* var. *menziesii* and var. *glauca* seedlings. *Ecology*, **46**, 794–801.

Krueger, K.W. and Ruth, R.H. (1969) Comparative photosynthesis of red alder, Douglas-fir, Sitka spruce and western hemlock seedlings. *Canadian Journal of Botany*, **47**, 519–27.

Küchler, A.W. (1946) The broadleaf deciduous forests of the Pacific Northwest. *Annuals of the Association of American Geographer*, **36**, 122–47.

Lange, O.L., Lösch, R., Schulze, E.D. and Kappen, L. (1971) Responses of stomata to changes in humidity, *Planta*, **100**, 76–86.

Larcher, W. (1969) The effect of environmental and physiological variables on the carbon dioxide gas exchange of trees. *Photosynthetica*, **3**, 167–98.

Larson, P.R. (1963) Stem form development of forest trees. *Forest Science Monograph*, No. 5.

Larsson, S., Oren, R., Waring, R.H. and Barrett, J.W. (1983) Attacks of mountain pine beetle as related to tree vigor of ponderosa pine. *Forest Science*, **29**, 395–402.

Lassoie, J.P. (1973) Diurnal dimensional fluctuations in a Douglas-fir stem in response to tree water status. *Forest Science*, **19**, 251–5.

Lassoie, J.P. (1975) Diurnal and Seasonal Basal Area Fluctuations in Douglas-fir Tree Stems of Different Crown Classes in Response to Tree Water Status, PhD thesis, University of Washington, Seattle, Washington.

Lassoie, J.P. (1979) Stem dimensional fluctuations in Douglas-fir of different crown classes. *Forest Science*, **25**, 132–44.

Lassoie, J.P. (1982) Physiological activity in Douglas-fir, in *Analysis of Coniferous Forest Ecosystems in the Western United States*, US/IBP Synthesis Series 14 (ed. R.L. Edmonds), Hutchinson Ross Publishing Company, Stroudsburg, Pennsylvania.

Lassoie, J.P. and Salo, D.J. (1981) Physiological response of large Douglas-fir to natural and induced soil water deficits. *Canadian Journal of Forest Research*, **11**, 139–44.

Lassoie, J.P., Dougherty, P.M., Reich, P.B., Hinckley, T.M., Metcalf, C.M. and Dina, S.D. (1983). Ecophysiological investigations of understory eastern redcedar in central Missouri. *Ecology*, **64**, 1355–66.

Leverenz, J.W. (1974) Net Photosynthesis as Related to Shoot Hierarchy in a Large Dominant Douglas-fir, MS thesis, University of Washington, Seattle, Washington.

Leverenz, J.W. (1981a) Photosynthesis and transpiration in large forest-grown Douglas-fir: Diurnal variation. *Canadian Journal of Botany*, **59**, 349–56.

Leverenz, J.W. (1981b) Photosynthesis and transpiration in large forest-grown Douglas-fir: Interactions with apical control. *Canadian Journal of Botany*, **59**, 2568–76.

Leverenz, J.W. and Jarvis, P.G. (1979) Photosynthesis in Sitka spruce (*Picea sitchensis* [Bong.] Carr.). VIII: The effects of light flux density and direction on the rate of net photosynthesis and stomatal conductance of needles. *Journal of Applied Ecology*, **16**, 919–32.

Leverenz, J.W. and Jarvis, P.G. (1980a) Photosynthesis in Sitka spruce (*Picea sitchensis* [Bong.] Carr.). IX: The relative contribution made by needles at various points on the shoot. *Journal of Applied Ecology*, **17**, 59–68.

Leverenz, J.W. and Jarvis, P.G. (1980b) Photosynthesis in Sitka spruce (*Picea sitchensis* [Bong.] Carr.). X: Acclimation to quantum flux density within and between trees. *Journal of Applied Ecology*, **17**, 697–708.

Leverenz, J., Deans, J.D., Ford, E.D., Jarvis, P.G., Milne, R. and Whitehead, D. (1982) Systematic spatial variation of stomatal conductance in a Sitka spruce plantation. *Journal of Applied Ecology*, **19**, 835–51.

Levitt, J. (1980) *Response of Plants to Environmental Stresses*, Vol. II, *Water, Radiation, Salt, and Other Stresses*, Academic Press, New York, New York.

Lewandowska, M. and Jarvis, P.G. (1977) Changes in chlorophyll and carotenoid content, specific leaf area and dry weight fractions in Sitka spruce, in response to shading and season. *New Phytologist*, **67**, 31–8.

Lewandowska, M., Hart, J.W. and Jarvis, P.G. (1976) Photosynthetic electron transport in plants of Sitka spruce subjected to different light environments during growth. *Physiologia Plantarum*, **37**, 269–74.

Lewandowska, M., Hart, J.W. and Jarvis, P.G. (1977) Photosynthetic electron transport in shoots of Sitka spruce from different levels in a forest canopy. *Physiologia Plantarum*, **41**, 124–8.

Long, J.N. (1982) Productivity of western coniferous forests, in *Analysis of Coniferous Forest Ecosystems in the Western United States*, US/IBP Synthesis Series 14 (ed. R.L. Edmonds), Hutchinson Ross Publishing Company, Stroudsburg, Pennsylvania.

Long, J.N., Smith, F.W. and Scott, D.R.M. (1981) The role of stem sapwood and heartwood in the mechanical and physiological support of Douglas-fir crowns. *Canadian Journal of Forest Research*, **11**, 459–64.

Lopushinsky, W. (1969) Stomatal closure in conifer seedlings in response to leaf moisture stress. *Botanical Gazette*, **130**, 250–63.

Lopushinsky, W. (1975) Water relations and photosynthesis in lodgepole pine, in *Management of Lodgepole Pine Ecosystems Symposium Proceedings* (ed. D.M. Baumgartner), Washington State University, Cooperative Extension Service, Pullman, Washington.

Ludlow, M.M. and Jarvis, P.G. (1971) Photosynthesis in

Sitka spruce (*Picea sitchensis* [Bong.] Carr.). *Journal of Applied Ecology*, **8**, 925–53.

Magnussen, S. and Peschl, A. (1981) Die Einwirkung verschiedener Beschattungsgrade auf die Photosynthese und die Transpiration junger Weiss- und Küstentannen. *Allgemeine Forst und Jagdzeitung*, **152**, 82–93.

Mar: Moller, C. (1947) The effect of thinning, age, and site on foliage increment, and loss of dry matter. *Journal of Forestry*, **45**, 393–404.

McLaughlin, S.B., McConathy, R.K., Barnes, R.L. and Edwards, N.T. (1980) Seasonal changes in energy allocation by white oak (*Quercus alba*). *Canadian Journal of Forest Research*, **10**, 379–88.

Meier, C.E. (1981) *The Role of Fine Roots in Nitrogen and Phosphorus Budgets of Young and Mature Abies amabilis Ecosystems*, PhD thesis, University of Washington, Seattle, Washington.

Meinzer, F.C. (1982a) The effect of vapor pressure on stomatal control of gas exchange in Douglas-fir (*Pseudotsuga menziesii*) saplings. *Oecologia*, **54**, 236–42.

Meinzer, F.C. (1982b) The effect of light on stomatal control of gas exchange in Douglas-fir (*Pseudotsuga menziesii*) saplings. *Oecologia*, **54**, 270–4.

Meinzer, F.C. (1982c) Models of steady-state and dynamic gas exchange responses to vapor pressure and light in Douglas-fir (*Pseudotsuga menziesii*) saplings. *Oecologia*, **55**, 403–8.

Minore, D. (1979) *Comparative Autecological Characteristics of Northwestern Tree Species: A Literature Review*, United States Department of Agriculture Forest Service General Technical Report PNW-87.

Mitchell, R.G., Waring, R.H. and Pitman, G.B. (1983) Thinning lodgepole pine increases tree vigor and resistance to mountain pine beetle. *Forest Science*, **29**, 204–11.

Monsi, M., Uchijima, Z. and Oikawa, T. (1973) Structure of foliage canopies and photosynthesis. *Annual Reveiw of Ecology and Systematics*, **4**, 301–27.

Mooney, H.A. and Gulmon, S.L. (1982) Constraints on leaf structure and function in reference to herbivory. *BioScience*, **32**, 198–206.

Morikawa, Y. (1974) Sap flow in *Chamaecyparis obtusa* in relation to water economy of woody plants. *Bulletin of the Tokyo University Forests*, **66**, 252–96.

Natr, L. (1975) Influence of mineral nutrition on photosynthesis and the use of assimilates, in *Photosynthesis and Productivity in Different Environments, International Biological Program, Programme 3* (ed. J.P. Cooper), Cambridge University Press, London, UK.

Neilson, R.E. and Jarvis, P.G. (1975) Photosynthesis in Sitka spruce (*Picea sitchensis* [Bong.] Carr.). VI. Response of stomata to temperature. *Journal of Applied Ecology*, **12**, 879–91.

Neilson, R.E., Ludlow, M.M and Jarvis, P.G. (1972) Photosynthesis in Sitka spruce (*Picea sitchensis* [Bong.] Carr.). II. Response to temperature. *Journal of Applied Ecology*, **9**, 721–45.

Ng. P.A.P. (1978) *Response of Stomata to Environmental Variables in Pinus sylvestris L.* PhD thesis, University of Edinburgh, Edinburgh, Scotland, UK.

Ng, P.A.P and Jarvis, P.G. (1980) Hysteresis in the response of stomatal conductance in *Pinus sylvestris* L. needles to light: Observations and a hypothesis. *Plant, Cell and Environment*, **3**, 207–16.

Norman, J.M. and Jarvis, P.G. (1974) Photosynthesis in Sitka spruce (*Picea sitchensis* [Bong.] Carr.). III. Measurement of canopy structure and interception of radiation. *Journal of Applied Ecology*, **11**, 375–98.

Norman, J.M. and Jarvis, P.G. (1975) Photosynthesis in Sitka spruce (*Picea sitchensis* [Bong.] Carr.). V. Radiation penetration theory and a test case. *Journal of Applied Ecology*, **12**, 839–78.

Owens, J.N. and Molder, M. (1973a) A study of DNA and mitotic activity in the vegetative apex of Douglas-fir during the annual growth cycle. *Canadian Journal of Botany*, **51**, 1395–409.

Owens, J.N. and Molder, M. (1973b) Bud development in western hemlock. I. Annual growth cycle of vegetative buds. *Canadian Journal of Botany*, **51**, 2223–31.

Owens, J.N. and Molder, M. (1976) Bud development in Sitka spruce. I. Annual growth cycle of vegetative buds and shoots. *Canadian Journal of Botany*, **54**, 313–25.

Phillips, R.A. (1967) *Stomatal Characteristics Through a Tree Crown*, MS thesis, University of Washington, Seattle, Washington.

Puritch, G.S. (1973) Effects of water stress on photosynthesis, respiration, and transpiration of four *Abies* species. *Canadian Journal of Forest Research*, **3**, 293–8.

Reed, K.L. (1968) *The Effects of Sub-zero Temperatures on the Stomata of Douglas-fir*, MS thesis, University of Washington, Seattle, Washington.

Richards, G.P. (1973) *Some Aspects of the Water Relations of Sitka Spruce*, PhD thesis, University of Aberdeen, Aberdeen, Scotland, UK.

Rogers, R. and Hinckley, T.M. (1979) Foliar mass and area related to current sapwood area in oak. *Forest Science*, **25**, 298–303.

Rook, D.A. and Corson, M.J. (1978) Temperature and irradiance and the total daily photosynthetic production of a *Pinus radiata* tree. *Oecologia*, **36**, 371–82.

Ross, S.D. (1972) *The Seasonal and Diurnal Source-sink Relationships for Photoassimilated ¹⁴C in the Douglas-*

fir Branch, PhD thesis, University of Washington, Seattle, Washington.

Running, S.W. (1976) Environmental control of leaf water conductance in conifers. *Canadian Journal of Forest Research*, **6**, 104–12.

Running, S.W. (1980) Relating plant capacitance to the water relations of *Pinus contorta*. *Forest Ecology and Management*, **2**, 237–52.

Running, S.W. and Reid, C.P.P. (1980) Soil temperature influences in root resistance of *Pinus contorta* seedlings. *Plant Physiology*, **65**, 635–40.

Rutter, M.R. (1977) An Eco-physiological Field Study of Three Sierra Conifers, PhD thesis, University of California, Berkeley, California.

Salo. D.J. (1974) Factors Affecting Photosynthesis in Douglas-fir, PhD thesis, University of Washington, Seattle, Washington.

Schiess, P. and Cole, D.W. (1981) Renovation of wastewater by forest stands, in *Municipal Sludge Application to Pacific Northwest Forest Lands* (ed. C.S. Bledsoe), Institute of Forest Resources Contribution No. 41. University of Washington, Seattle, Washington.

Shidei, T. and Kira, T. (eds) (1977) *Primary Productivity of Japanese Forests, Japanese International Biological Program Synthesis*, Vol. 16, University of Tokyo Press, Tokyo, Japan.

Shinozaki, K. Yoda, K., Hozumi, K. and Kira, T. (1964) A quantitative analysis of plant form: The pipe model theory. 1. Basic analysis. *Japanese Journal of Ecology*, **14**, 97–105.

Snell, J.A.K. and Brown, J.K. (1978) Comparison of tree biomass estimators – DBH and sapwood area. *Forest Science*, **24**, 455–7.

Sollins, P., Grier, C.C., McCorison, F.M., Jr, Cromack, K., Fogel, R. and Fredriksen, R.L. (1980) The internal element cycles of an old-grown Douglas-fir ecosystem in Western Oregon. *Ecological Monographs*, **50**, 261–85.

Sorensen, F.C. and Ferrell, W.K. (1973) Photosynthesis and growth of Douglas-fir seedlings when grown in different environments. *Canadian Journal of Botany*, **51**, 1689–98.

Stone, E.C. (1957) Dew as an ecological factor. I. A review of the literature. *Ecology*, **38**, 407–13.

Swanson, R.H. (1972) Water transpired by trees is indicated by heat pulse velocity. *Agricultural Meteorology*, **10**, 277–81.

Sweet, G.B. and Wareing, P.F. (1966) The role of plant growth in regulating photosynthesis. *Nature (London)*, **210**, 77–9.

Tadaki, Y. (1977) Leaf biomass, in *Primary Productivity of Japanese Forests, Japanese International Biological Pro-*

gram Synthesis, Vol. 16 (eds T. Shidei and T. Kira), University of Tokyo Press, Tokyo, Japan.

Tadaki, Y., Hatiya, K., Tochiaki, K., Miyauchi, H. and Matsuda, U. (1970) Studies on the production structure of forest. 16. Primary productivity of *Abies veitchii* forest in the subalpine zone of Mt. Fuji. *Bulletin of the Government Forest Experiment Station, Tokyo*, **229**, 1–20.

Tan, C.S. and Black, T.A. (1976) Factors affecting the canopy resistance of a Douglas-fir forest. *Boundary-Layer Meteorology*, **10**, 475–88.

Tan, C.S., Black, T.A. and Nnyamah, J.U. (1977) Characteristics of stomatal diffusion resistance in a Douglas-fir forest exposed to soil water deficits. *Canadian Journal of Forest Research*, **7**, 595–604

Tan, C.S., Black, T.A. and Nnyamah, J.U. (1978) A simple diffusion model of transpiration applied to a thinned Douglas-fir stand. *Ecology*, **59**, 1221–9.

Teskey, R.O. (1982) Acclimation of *Abies amabilis* to Water and Temperature Stress in a Natural Environment, PhD thesis, University of Washington, Seattle, Washington.

Tranquillini, W. (1979) *Physiological Ecology of the Alpine Timberline*, Springer-Verlag, New York, New York.

Troeng, E. and Linder, S. (1982) Gas exchange of a 20-year-old stand of Scots pine. 1. Net photosynthesis of current and one-year-old shoots within and between seasons. *Physiologia Plantarum*, **54**, 7–14.

Tucker, C.S. and Emmingham, W.H. (1977) Morphology changes in leaves of residual western hemlock after clear and shelterwood cutting. *Forest Science*, **23**, 12, 561–76.

Turner, N.C. and Jarvis, P.G. (1975) Photosynthesis in Sitka spruce (*Picea sitchensis* [Bong.]. Carr.). IV. Response to soil temperature. *Journal of Applied Ecology*, 561–76.

Turner, J. and Long, J.N. (1975) Accumulation or organic matter in a series of Douglas-fir stands. *Canadian Journal of Forest Research*, **5**, 681–90.

Turner, J. and Olson, P.R. (1976) Nitrogen relations in a Douglas-fir plantation. *Annuals of Botany*, **40**, 1185–93.

Turner, J. and Singer, M.J. (1976) Nutrient distribution and cycling in a subalpine coniferous forest ecosystem. *Journal of Applied Ecology*, **13**, 295–301.

Tyree, M.T. and Dixon, M.A. (1983) Cavitation events in *Thuja occidentalis* L.: Ultrasonic accoustic emissions from the sapwood can be measured. *Plant Physiology*, **72**, 1094–99.

Ugolini, F.C. and Zasoski, R.J. (1979) Soils derived from tephra, in *Volcanism and Human Habitation* (eds P.D.

Sheets and D.K. Grayson), Academic Press, New York, New York.

Vaadia, Y., Raney, F.C. and Hagan, R.M. (1961) Plant water deficits and physiological processes. *Annual Review of Plant Physiology*, **12**, 265–92.

Walker, R.B., Scott, D.R.M., Salo, D.J. and Reed, K.L. (1972) Terrestrial process studies in conifers: A review, in *Proceedings – Research on Coniferous Forest Ecosystems – A Symposium* (eds J.F. Franklin, L.J. Dempster, and R.H. Waring), United States Department of Agriculture Forest Service, Portland, Oregon.

Wambolt, C.L. (1973) Conifer water potential as influenced by stand density and environmental factors. *Canadian Journal of Botany*, **51**, 2333–7.

Waring, R.H. (1969) Forest plants of the Siskiyous: Their environmental and vegetational distribution. *Northwest Science*, **43**, 1–17.

Waring, R.H. and Franklin, J.F. (1979) Evergreen coniferous forests of the Pacific Northwest. *Science*, **204**, 1380–86.

Waring, R.H. and Pitman, G.B. (1980) *A Simple Model of Host Resistance to Bark Beetles*. Forest Research Laboratory Note **65**, Oregon State University, Corvallis, Oregon.

Waring, R.H. and Running, S.W. (1978) Sapwood water storage: Its contribution to transpiration and effect upon water conductance through the stems of old-growth Douglas-fir. *Plant, Cell and Environment*, **1**, 131–40.

Waring, R.H., Reed, K.L. and Emmingham, W.H. (1972) An environmental grid for classifying coniferous forest ecosystems. In *Proceedings – Research on Coniferous Forest Ecosystems – A Symposium* (eds J.F. Franklin, L.J. Dempster, and R.H. Waring). United States Department of Agriculture Forest Service, Portland, Oregon.

Waring, R.H., Schroeder, P.E., and Oren, R. (1982) Application of the pipe model theory to predict canopy leaf area. *Canadian Journal of Forest Research*, **12**, 556–60.

Waring, R.H., Thies, W.G. and Muscato, D. (1980a) Stem growth per unit area: A measure of tree vigor. *Forest Science*, **26**, 112–17.

Waring, R.H., Whitehead, D. and Jarvis, P.G. (1979) The contribution of stored water to transpiration in Scots pine. *Plant, Cell and Environment*, **2**, 309–17.

Waring, R.H., Whitehead, D. and Jarvis, P.G. (1980b) Comparison of an isotropic method and the Penman-Monteith equation for estimating transpiration from Scots pine. *Canadian Journal of Forest Research*, **10**, 555–58.

Watts, W.R., Neilson, R.E. and Jarvis, P.G. (1976) Photosynthesis in Sitka spruce (*Picea sitchensis* [Bong.] Carr.) VII. Measurement of stomatal conductance and $^{14}CO_2$ uptake in a forest canopy. *Journal of Applied Ecology*, **13**, 623–39.

Webb, W.L. (1971) Photosynthetic Models for a Terrestrial Plant Community, PhD thesis, Oregon State University, Corvallis, Oregon.

Webb, W.L. (1980) Starch content of conifers defoliated by the Douglas-fir tussock moth. *Canadian Journal of Forest Research*, **10**, 535–40.

Whitehead, D. (1978) The estimation of foliage area from sapwood basal area in Scots pine. *Forestry*, **51**, 137–49.

Whitehead, D. and Jarvis, P.G. (1981) Coniferous forests, in *Water Deficits and Plant Growth*, Vol. 6, *Woody Plant Communities*, (ed. T.T. Kozlowski), Academic Press, New York, New York.

Woodman, J.N. (1968) The Relationship of Net Photosynthesis to Environment within the Crown of a Large Douglas-fir Tree, PhD thesis, University of Washington, Seattle, Washington.

Woodman, J.N. (1971) Variation of net photosynthesis within the crown of a large forest-grown conifer. *Photosynthetica*, **5**, 50–4.

Zavitkovski, J. and Stevens, R.D. (1972) Primary productivity of red alder ecosystems. *Ecology*, **53**, 235–42.

Zobel, D.B., McKee, W.A., Hawk, G.M. and Dyrness, C.T. (1976) Relationships of environment to composition, structure, and diversity of forest communities of the central Western Cascades of Oregon. *Ecological Monographs*, **46**, 135–56.

7

Annuals and perennials of warm deserts

JAMES EHLERINGER

7.1 Introduction

Warm deserts have held a certain attraction and fascination for plant ecologists and physiologists for many decades. The reasons for this interest lie in the different types of plant assemblages, the large diversity of life forms, and the remarkable plant adaptations which have arisen in response to the environmental extremes of high air and soil temperatures, high solar radiation levels, low relative humidities, low precipitation levels, and extended drought periods. Substantial progress has been made in understanding the ecology of some of the dominant plant species, including the relationships between form and function, adaptation at morphological, physiological, and biochemical levels, and factors affecting distributional ranges.

This chapter deals with the physiological ecology of both perennials and annuals of warm deserts, but excludes cacti and other succulents which are discussed in Chapter 8. The Sonoran Desert and its various sub-divisions and the low elevation regions of the Mohave Desert such as Death Valley are included in this discussion, but the Chihuahuan Desert of northern Mexico is not. Together these regions comprise the warm deserts; the higher elevation cold deserts will be covered in Chapter 9. The location of warm and cold deserts is indicated in Fig. 8.1.

7.2. The physical environment

Before discussing the ways in which plants have adapted to warm deserts, it is essential to have an

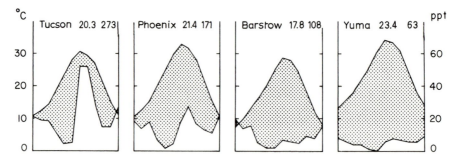

Figure 7.1 Climate diagrams for representative sites within warm deserts. Sites were chosen to represent a range of precipitation levels, both in amount and seasonality (from Sellers and Hill, 1974, and US Weather Bureau Records). Left hand axis is mean monthly temperature and right hand axis is mean monthly precipitation

understanding of the physical environment and the constraints it imposes on plant activity. High temperatures and drought are two factors which severely limit plant growth, and are evident in representative climate diagrams of desert sites (Fig. 7.1). Precipitation varies between sites in both amount and seasonality, and all locations have drought periods ranging from 5 to 12 months. Both the length and magnitude of the drought (difference between precipitation and temperature curves) are greatest at the driest sites. These data suggest that at the drier sites, potential evapotranspiration always exceeds precipitation and that plants are continuously exposed to drought conditions.

Although the climate diagrams indicate high mean monthly temperatures throughout the year, they underestimate the actual potential for stress during growth periods in winter and early spring. Mean maximum daily air temperatures range from 15 to 20 °C in winter, from 25 to 35 °C in spring, and 38 to 43 °C in the summer, depending on location

(Ives, 1949; Sellers and Hill, 1974; US Weather Bureau records). Thus, daily air temperature fluctuations may be quite high and plants can be exposed to temperatures in excess of 30 °C for most of the growth periods. Since these are weather screen data (usually measured at ~ 1.5 m), plants growing near the surface (0–30 cm) will be exposed to much higher temperatures. Soil surface temperatures greater than 60 °C are common during summer months (Ehleringer *et al.*, unpublished; Sellers and Hill, 1974; Terjung *et al.*, 1970). This would imply that plants growing close to the surface in the summer would be exposed to air temperatures of 45–55 °C.

Skies are often clear and levels of solar radiation are generally higher than in other ecosystems (Sellers and Hill, 1974; Thekaekara,1976). The greater incident solar radiation should enhance plant productivity during periods of available soil moisture. However, clear nighttime skies and low relative humidities result in high net radiation loss

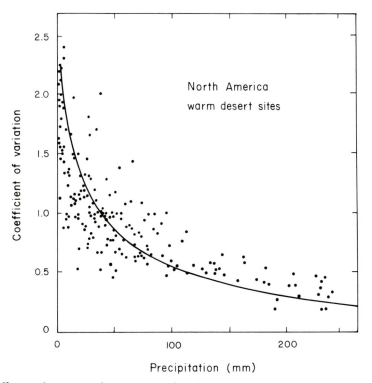

Figure 7.2 The coefficient of variation of precipitation plotted as a function of mean precipitation for warm desert sites (modified from Ehleringer and Mooney, 1983).

and large daily fluctuations in air temperatures.

Equally important as the range of mean annual precipitation is the variability or predictability of that rain. Between one season and the next, precipitation is often highly predictable (Pianka, 1967). However, from year to year the coefficient of variation increases asymptotically as mean precipitation decreases (Fig. 7.2). This is also the case for precipitation falling within individual seasons. Consequently, precipitation is highly unpredictable at low precipitation sites. In fact, the mean annual precipitation exceeds the mode in desert locations so that most years are drier than the average (McDonald, 1956).

An important but neglected component in considerations of the desert physical enviornment is the soil. Typical desert soils, termed aridisols, are clearly distinguishable from soils of other regions by their lack of development (Steila, 1976). Low precipitation results in reduced weathering, shallow soil depths, low concentration of organic matter and low cation exchange capacity. Soils on slopes and bajadas (alluvial fans) tend to be rocky, while those on flat plains are high in clay particles and often have caliche layers 1–2 m from the surface. Desert pavement, a surface of finely packed rocks in a swelling clay matrix, has a very low infiltration rate, and is common in many of the drier areas. Unlike most ecosystems, not all drainage basins empty into the oceans. Those that do not, create highly alkaline regions at their termini, many of which are ancient lake beds. This combination of soil characteristics tends to accentuate the arid conditions already imposed on the vegetation by the climate.

7.3 Phenology and life cycle adaptations

7.3.1 OPPORTUNISTIC RESPONSE

The feature which most generally characterizes warm desert plants is their opportunistic response to water availability. Water is the cue for plants to commence vegetative growth (Beatley, 1974; Ehleringer and Björkman, 1978a). Most perennial shrubs, irrespective of whether they have deciduous or evergreen leaves, are capable of vegetative growth and reproductive activity in either winter (November–April) or summer (July–October) seasons. A critical minimum amount of rainfall, usually in the form of a large storm, is necessary to trigger vegetative growth in perennials. This minimum precipitation is thought to be approximately 25 mm (Beatley, 1974). Because of the pronounced geographical gradient in seasonality of precipitation, most of the annual growth in the northern and western portions of warm deserts occurs in winter and spring. Only in the southern and eastern portions is vegetative growth predominantly in summer.

Germination of annuals is also triggered by a critical minimal rainfall (Beatley, 1974; Juhren *et al.*, 1956; Went 1948, 1949). For winter annuals, 10–15 mm is necessary to induce germination. Higher germination percentages and more vegetative growth occur with greater rainfall amounts (Beatley, 1969; Went and Westergaard, 1949). Specific temperature requirements for germination for winter and summer annual floras prevent germination during the wrong season (Koller, 1972).

The timing of the initial heavy rainstorms is variable. Based on weather records it appears that late November to early December is the most likely period for the start of the winter precipitation season in northern portions of the warm deserts. About one year in four, these rains will not come until January or later (Beatley, 1974). Summer rainfall is just as variable (McDonald, 1956), and usually begins in late July. The effects of precipitation timing and amounts on the germination of particular annual or perennial species are not known.

7.3.2 LIFE FORMS

One result of the large spatial and temporal variability in the physical environment of warm deserts is a high diversity of plant life forms (Fig. 7.3). Warm deserts stand out from adjacent vegetation types (chaparral, cold desert, pinyon – juniper woodland, thorn forest and grassland) by the differences in physiognomy of its dominant plants. Shreve (1951) describes 25 major life forms for the Sonoran Desert. These include the dominants of the Raunkiaer life form system, but also forms varying in leaf duration, character of stem, degree of succulence, and seasonality.

Figure 7.3 The variety of plant life forms present in warm deserts is evident in this photograph of the Anzo-Borrego desert in Southern California.

The distribution of life forms is not uniform, suggesting adaptive and competitive limitations of specific life forms. Some desert regions are characterized by complex communities, whereas other have near monospecific stands. Evergreen and deciduous-leaved shrubs dominate bajadas as well as the plains. Trees are for the most part phreatophytic or restricted to water courses. Succulents of various types are much more common on well-drained rocky slopes than the silty plains. As a class winter ephemerals do not appear to be restricted to certain microsites, although particular species may be (Forseth *et al.*, 1983). Summer ephemerals appear most abundant along water courses and plains, and facultative perennials are most common on disturbed sites.

While annuals occur in virtually all warm desert sites, their abundance as an adaptive life form is very much affected by the variability in precipitation. Schaffer and Gadgil (1975) have shown that as the variability in precipitation from year to year increases, so does the percentage of annuals that comprise the flora. It appears that in low precipitation sites, the probability of surviving extended drought periods becomes so low that perennial life forms be they shrubs, succulents or root perennials, are selected against. It should be noted, however, that while the diversity of perennial forms declines with precipitation levels, certain perennials such as *Larrea divaricata* may predominate even at the driest locations.

7.3.3 VEGETATIVE AND REPRODUCTIVE GROWTH

Several contrasting adaptations occur in the vegetative and reproductive features of annuals and perennials. In perennials, vegetative growth occurs rapidly following the 'triggering' rain storm. Presumably carbohydrate reserves from the previous year are the source of carbon. Following this initial development, vegetative growth occurs at a reduced rate and ceases at flowering in zoophilous species. In anemophilous perennials, vegetative growth may continue after flowering, since flowering occurs so early in the growing season. More data on specific dominant perennials are needed to verify the generality of this pattern. In contrast, winter annuals exhibit slow vegetative growth until the onset of warmer spring temperatures (Beatley, 1974). Also unlike perennials, these same winter annuals will add vegetative growth following late season precipitation (Beatley,1974; Clark and Burk, 1980). Death in annuals appears to be caused by a shortage of water. In years with above normal and late season precipitation, many annuals are able to persist and become facultative perennials (Beatley, 1967, 1969, 1970).

Sexual reproduction occurs by either zoophily or anemophily in perennials. Flowering appears cued by photoperiod in anemophilous species and occurs in late winter. Developing fruits often have large green (presumably photosynthetic) bracts and mature over the remainder of the growing season (~ 4 mon). In contrast, zoophilous perennials flower late in the growing season, and reproduction is presumably induced by water stress or high carbohydrate levels. Fruits of these species are also green, have photosynthetic activity (Werk and Ehleringer, 1983), and mature in a relatively short period (~ 1 mon). For the most part, vegetative growth and reproductive activity are distinct in these perennials. This is not the case for most winter and summer annuals.

In most annuals reproductive activity begins after plants have attained some minimal size, and then both vegetative growth and reproductive activity progress simultaneously until the plant dies. This continued flowering can be either determinate (as in the axillary flowering of *Mohavea*) or indeterminate (as in terminal spikes of *Lupinus*). In indeterminate flowering species, new vegetative growth is axillary. Early and continuous flowering of annuals is presumably an adaptive feature in desert habitats, where the rainfall and the end of the growing season are both unpredictable. Reproductive output (number of seeds produced) appears to be directly related to both plant size and carbon balance (Ehleringer *et al.*, unpublished data). Therefore, small differences in water availability and slope geometry affecting solar radiation levels between microsites will have a large impact on reproductive output.

Environmental gradients also affect aspects of the reproductive biology of perennials. Dioecy is common in many desert plants (Freeman *et al.*, 1976; Simpson, 1977), including representatives of the genera *Atriplex*, *Ephedra*, *Jatropha*, and *Simmondsia*. As sexes within a single population segregate along microenvironmental gradients (Freeman *et al.*, 1976), dioecy can be considered as an adaptation to stressful conditions. Male plants tend to be more common in microsites where high salinity, high radiation, or low soil moisture increase drought stress, while female plants are overrepresented on mesic microsites. Since there is a differential cost to the production of male versus female reproductive structures, the plant carbon balance is likely to be a key factor in these sex ratio clines. The observed distributions are consistent with the notion that female reproduction is more costly in terms of carbon and should therefore be more common on microsites where the carbon gaining capacity of the plant is greater. Also consistent with this notion is the observation that sex expression in *Atriplex canescens* is a function of annual environmental conditions (McArthur, 1977). In mild wet years, the sex ratio is female biased, while in cold dry years it is male biased.

7.3.4 COMPETITION

When nondesert plant communities are compared, one common feature is the high degree of life form similarity in the dominant species, irrespective of differences in phylogenetic relationships. In these situations it might be concluded that strong competition has produced this uniformity. In warm deserts though, the distances between plants are great, and under only the rarest of situations do canopies of neighboring plants overlap. As a consequence, Shreve (1951) concludes... 'there has been much less competition. The greatest "struggle" of the plants has not been with one another, but with the environment. Therefore the conditions tending toward the elimination of certain types and the survival and dominance of a relatively uniform one have not been operative'. For perennial and annual warm desert plant species this is probably the case for aboveground competitive interactions, with the possible exceptions of the establishment phase in some species and high density situations in micro-depressions. However, there is evidence for below-ground competition in several perennial species.

Vast areas of the desert consist of a two species community, *Larrea divaricata* (= *L. tridentata*) and *Ambrosia dumosa* (= *Franseria dumosa*). The horizontal spatial dispersion patterns of the long-lived evergreen shrub *Larrea* are often regular and for the short-lived drought deciduous shrub *Ambrosia* are contagious (Fonteyn and Maball 1978; Woodell *et al.*, 1969). In an elegant experiment, Fonteyn and Mahall (1978) followed the development of leaf water stress of single individuals in which there had been differential and/or total removal of the other neighboring plants (Table 7.1). Their results imply that:

1. competition is not occurring between *Larrea* individuals;

2. competition is occurring between *Ambrosia* individuals; and

3. competition is occurring between *Larrea* and *Ambrosia* individuals. Presumably the distribution of

Table 7.1 Effects of differentially removing plant species surrounding a central individual in a *Larrea divaricata–Ambrosia dumosa* community on its leaf water potential (from Fonteyn and Mahall, 1978). Values are means ± 1 s.d. Units are MPa

Treatment	*Larrea*	*Ambrosia*
Control	− 3.90 ± 0.13	− 4.42 ± 0.32
Total removal	− 3.37 ± 0.16	− 3.05 ± 0.34
Larrea removal	− 3.93 ± 0.23	− 4.46 ± 0.39
Ambrosia removal	− 3.46 ± 0.27	− 3.65 ± 0.28

Larrea was originally contagious, but as a result of competition for water and the resultant elimination of individuals, the plants now occupy a regular distribution in which competition no longer occurs. Competition occurs among *Ambrosia* individuals, because they are short lived and have higher recruitment rates.

7.4 Leaf and canopy adaptations

7.4.1 LEAF ENERGY BALANCE

One result of the large spacing between plants is that leaves are potentially exposed to high radiation loads in this high temperature environment. The consequence can be seen in the broad array of adaptations found in leaves of warm desert plants. The relevance of different morphological features is made quantitatively clear through an analysis of leaf energy budgets as developed by Gates (1980). As a review of this topic will not be provided here, the reader is referred to the work by Gates for an introduction and the specific equations. The leaf energy budget is a balance of energy inputs (absorbed solar and infrared radiation) and losses (reradiation, convection, and transpiration) which determines the physiologically important parameters of tissue temperature and water loss.

The preponderance of small, highly dissected, or compound leaves as in *Acacia, Ambrosia, Larrea, Pectis* and *Prosopis* has been considered an adaptation to warm desert environments (Schimper, 1903; Warming, 1909). The smaller leaf size (1–10 mm width) results in a higher heat transfer coefficient, and consequently leaf temperature will remain close to or slightly above air temperatures (Gates, 1980; Smith and Geller, 1980). This will be of particular value to plants experiencing high solar radiation loads and high air temperatures, especially when wind speed and transpiration rates may be low.

Not all desert plants are characterized by small leaf size, however. A number of drought-deciduous perennials such as *Datura, Encelia,* and *Hyptis* as well as annuals such as *Amaranthus, Atrichoseris,* and *Proboscidea* have leaf widths in the 30–200 mm range. Smith (1978) and Smith and Geller (1980) have pointed out that large leaf size can be beneficial and result in significant leaf undertemperatures (leaf temperature lower than air temperature) if sufficient soil moisture is present to allow the maintenance of high transpiration rates. Additionally, a low leaf absorptance to solar radiation will enhance this advantage. Many of these large-leaved species are restricted to microhabitats having greater soil moisture contents. In response to decreases in water availability or to increases in air temperatures, large-leaved species show substantial decreases in leaf size (Cunningham and Strain, 1969; Mooney *et al.*, 1977b; Smith and Nobel, 1977).

Two principal adaptations exist to reduce the amount of solar radiation absorbed by the leaf under stress conditions. First is a steep leaf inclination, and second, a decrease in leaf absorptance to solar irradiance. Leaves with inclinations greater than 70° occur in a number of evergreen perennial shrubs, including *Atriplex hymenelytra* and *Simmondsia chinensis* (Mooney *et al*, 1977b). Steep leaf angles result in decreased energy loads, and thus lowered leaf temperatures during stressful midday conditions of the late spring and summer months. During winter and early spring, steep leaf angles result in a more constant solar radiation load incident on the leaf than is the case with more horizontal orientations.

Leaf absorptance may be reduced by increasing surface reflectance through epidermal modifications including hairs (*Brickelia, Encelia*), salt glands (*Atriplex*), and waxes and spines (*Opuntia*) (Ehleringer, 1981) (Fig. 7.4). Over the 400–700 nm waveband (photosynthetically useful wavelengths), green leaves such as those of *Prosopis juliflora* typically absorb 85% of the incident solar radiation. In contrast, species with well-developed epidermal modifications may have leaf absorptances as low as 30–40% (Ehleringer and Björkman, 1978a; Ehleringer *et al.*, 1976; Mooney *et al.*, 1977b). Changes in leaf absorptance are reversible within the same leaf in species utilizing salt glands to reflect light (Mooney *et al.*, 1977b) and probably also in species using waxes. However, reversible changes in the same leaf do not occur in species utilizing hairs (Ehleringer, 1982; Ehleringer and Björkman, 1978a). In those situations, older leaves are ab-

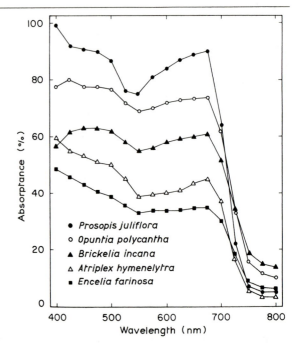

Figure 7.4 Leaf absorptances over the 400–800 nm waveband for a number of warm desert species (from Ehleringer, 1981).

scised and new leaves with different spectral characteristics are produced.

Leaf temperature and energy balance depend on the leaf absorptance over the entire solar spectrum, not just over the 400–700 nm waveband. However, as one might expect, the two are very tightly correlated (Ehleringer, 1981). For example, leaf absorptances in the 400–700 nm band of 85% (green leaf) and 40% (white leaf) correspond to 50% and 17% absorptances over the entire solar band, respectively. As a consequence of this three-fold difference in leaf absorptance to total solar radiation, there are substantial decreases in both leaf temperature and transpiration (a result of decreased leaf water-vapor pressure). Decreases in leaf temperature of 5–10 °C as a result of decreased leaf absorptance are common (Ehleringer and Mooney, 1978; Smith, 1978; Smith and Geller, 1980). Just exactly what this means to the plant in terms of competition or extending growth activity into the drought period is not known.

Reduction of leaf absorptance as an adaptation to hot, arid conditions is best developed in shrubs,

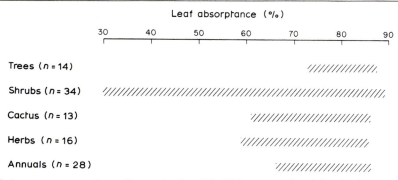

Figure 7.5 Leaf absorptances to solar radiation in the 400–700 nm waveband for the dominant life forms in the Mohave and Sonoran Deserts (from Ehleringer, 1981).

occurs to a lesser extent in cacti and perennial herbs, and is rare or absent in trees and annuals (Fig. 7.5). Within the shrub life form, several dominant species are capable of varying leaf absorptance from 50% to 85% (400–700 nm) depending on environmental conditions, and a few such as *Encelia farinosa* produce leaves whose absorptance will vary from 30% to 85% (Ehleringer, 1981; Ehleringer and Björkman, 1978a; Mooney *et al.*, 1977b). As a rule, herbaceous and shrub species with reduced leaf absorptances occur on drier, exposed bajadas and rocky slopes and not in ravine bottoms or along water courses where glabrate-leaved species predominate. Within the trees and annuals, most of the variation in leaf absorptances are the result of differences in leaf thickness (transmittance) and not surface reflectance. The absence of reduced leaf absorptances in these two groups is not surprising, since trees occur primarily along water courses and annuals have an ephemeral life history. One notable exception to this trend is *Dicoria canescens*, which is a pubescent leaved, spring and summer active annual of sand dune habitats, and has a late season leaf absorptance of 66% (Ehleringer, 1981).

7.4.2 CANOPY MORPHOLOGY

The air temperatures, humidities, wind speeds, and consequently leaf temperatures to which a plant canopy is exposed are very much dependent on where the canopy is within the microclimatic profile. Soil surfaces are heated quickly because of high incident solar radiation and low soil evapora-tion. Microclimatic profiles are steep (Fig. 7.6), and there may be significant temperature (and thus metabolic) differences between leaves at different heights.

During the cool winter and early spring periods, annuals take advantage of this microclimatic profile by growing as rosettes just above the soil surface (Ehleringer *et al.*, 1979; Mulroy and Rundel, 1977). Winter annuals often have highly dissected leaves (Mulroy and Rundel, 1977). Since wind speeds are low near the soil surface, leaf dissection may be of advantage in keeping leaf boundary layer small. Notable exceptions include *Atrichoseris platyphylla*, a species with orbicular leaves up to 8 cm diameter and closely appressed to the ground surface.

As air temperatures increase rapidly in the spring, many winter ephemerals simply lengthen inter-nodes (e.g. *Geraea canescens*, *Lupinus arizonicus*, *Malvastrum rotundifolium*, and *Phacelia calthafolia*) so that leaves are raised above the hot surface into a cooler part of the microclimatic profile. When temperatures get hot, other plants, including *Atri-choseris platyphylla* and *Eriogonum rexfordii*, bolt and allow the basal leaves to wither away. Still other prostrate annuals such as *Astragalus lentiginosus*, *Coldenia nuttalii*, and *Polygonum aviculare* raise and lower their branches diurnally apparently to avoid high soil surface temperatures (Ehleringer *et al.*, unpublished).

Leaf canopies of perennial herbs and shrubs fit either of two qualitatively distinct morphologies. They are tightly packed and hemispherical (*Ambrosia dumosa*, *Encelia farinosa* and *Hymenoclea*

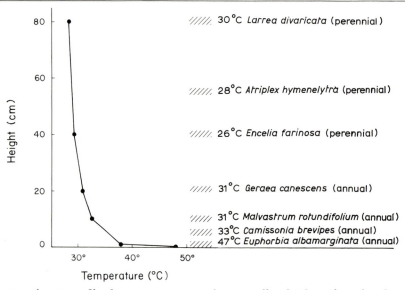

Figure 7.6 Left: microclimatic profile of air temperature as a function of height above the soil surface at midday in late March on a bajada in Death Valley, California. Right: representative leaf temperatures measured on the dominant plant species at approximately mid-canopy height.

salsola) or loosely arranged and amorphous (*Hyptis emoryi*, *Larrea divaricata*, and *Simmondsia chinensis*). The functional significance of either of the two canopy types is not understood.

7.5 Photosynthetic and water relations adaptations

7.5.1 ADAPTATION MODEL

Throughout this discussion, the assumption is made that natural selection favors plants whose form (physiology, morphology, physiognomy) maximizes net carbon gain, since such plants should have the greatest resources for (a) reproduction, (b) survival during drought periods (perennials only), and (c) competition with other plants for additional water and nutrients.

A model which relates net carbon gain to increases in environmental stress in desert plants is given in Fig. 7.7. Available data for photosynthetic and water relations adaptations in warm desert plants are consistent with this model. In this model a genotype is thought to be capable

of expressing multiple phenotypes in response to changes in the physical environment. For each plant species, and specifically each genotype, there is a combination of environmental parameters that is optimal in the sense of producing the highest net carbon gain. Deviations in the environment from these optimal conditions (e.g. increases or decreases in solar radiation, temperature, humidity, available soil moisture) will decrease net carbon gain. As the environmental stress increases, the plant may acclimate to express a new phenotype, a different combination of physiological, morphological, and physiognomic characteristics, which enhances net carbon gain. However, this form of adaptation has an associated cost. A measure of this cost is that while net carbon gain of the acclimated phenotype under the environmental stress will be greater than for the previous phenotype, it will be less should the environment return to its previous state. The transition in expression of one phenotype to another is gradual and continuous as the environmental stress increases, but may be abrupt when the environmental stress is removed (such as when rainfall removes long-term drought).

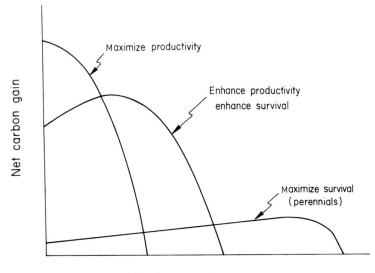

Figure 7.7 Adaptation model relating net carbon gain of a plant with environmental stress. Each curve represents a specific and different phenotypic expression (combination of physiology, morphology, and physiognomy) by a single genotype. See text for further details.

7.5.2 MECHANISMS TO MAXIMIZE NET CARBON GAIN

How net carbon gain is maximized depends on a balance of the photosynthetic capacities of leaves and other structures, on allocation patterns and phenology, and on carbon losses (maintenance respiration, predator defense, and leaching by the various plant parts). As yet there have been no complete and detailed studies of carbon balance in any warm desert species. Essentially all of the information available relates to the photosynthetic characteristics of different plants. While these data are by no means complete, distinct patterns are emerging.

Both C_3 and C_4 photosynthetic pathways are present in the annual and perennial floras (Eickmeier, 1978; Mooney *et al.*, 1974; Mulroy and Rundel, 1977; Stowe and Teeri, 1978; Syvertsen *et al.*, 1976; Teeri and Stowe, 1976). Annuals and grasses of the winter flora utilize the C_3 pathway exclusively, whereas C_4 is predominant in the summer flora. This trend is consistent with the notion that C_4 photosynthesis is a necessary component for achieving high photosynthetic rates in hot temperature environments. Within the perennials, the C_4 pathway is rare except for

1. halophytic species, where it is common, and
2. short-lived summer active perennials, where it occurs infrequently. The latter category includes *Boerhaavia coccinea* and *Tidestromia oblongifolia*, and in both cases all closely related species are annuals. In northern warm desert sites, the low percentage of C_4 perennials is perhaps explained by the fact that most of the rains and the growth of perennials occurs during the cooler winter – spring periods. However, this does not explain the absence of C_4 photosynthesis among perennials of southern warm desert locations, where precipitation occurs primarily during the summer.

The C_4 photosynthetic pathway is generally considered to be more productive, but is not a consistent predictor of maximum photosynthetic capacity in warm desert plants. Whereas C_4 photosynthetic capacities of 82 μmol m^{-2} s^{-1} for *Amaranthus palmeri* (Ehleringer, 1983b), 67 μmol m^{-2} s^{-1} for *Hilaria rigida* (Nobel, 1980), and 56 μmol m^{-2} s^{-1} for *Tidestromia oblongifolia* (Björkman *et al.*, 1980) have been measured, other species such as *Atriplex hymenelytra* and *A. lentiformis* have maximum rates of 30 μmol m^{-2} s^{-1} or less (Mooney *et al.*, 1977b; Pearcy, 1977). Several C$_3$ species including *Camissonia claviformis* (Mooney *et al.*, 1976), *Encelia californica* (Ehleringer and Björkman, 1978b), and *Malvastrum rotundifolium* (Forseth and Ehleringer, 1982; Mooney and Ehleringer, 1978) have photosynthetic capacities equivalent to the higher rates in C$_4$ species.

Rates of photosynthesis in desert plants are higher than those in plants from other ecosystems. Annuals have greater photosynthetic capacities than do perennial species, and among perennials, rates are higher in deciduous-leaved species than in evergreen-leaved species (Ehleringer and Mooney, 1983). Taken as a whole, there appears to be a strong negative correlation between maximum leaf photosynthetic rate and life expectancy.

One characteristic typical of desert plants is that the leaf photosynthetic rates are not saturated by midday irradiances (Fig. 7.8). High stomatal conductances, high enzyme contents, and high meso-phyll surface volume ratios are necessary to achieve these high photosynthetic rates (Armond and Mooney, 1978; Ehleringer, 1983b; Ehleringer and Björkman, 1978b; Longstreth *et al.*, 1980; Mooney *et al.*, 1976, 1977a). The quantum yield of desert plants appears not to be different from that of other species (Ehleringer and Björkman, 1977) and, thus, high photosynthetic rates occur because leaves are able to utilize the higher irradiances.

In many warm desert species, a leaf adaptation called diaheliotropism or solar tracking occurs, potentially resulting in high photosynthetic rates throughout the day (Ehleringer and Forseth, 1980). Solar-tracking leaves move during the day so that they remain perpendicular to the sun's direct rays at all times. This results in a 38% or higher daily quantum flux on solar tracking leaves over those with fixed orientations (Fig. 7.9). Given the capacity to utilize high irradiances, species with solar tracking leaves should have greater rates of daily net carbon gain and be at a competitive advantage.

Solar tracking leaves occur in both winter and

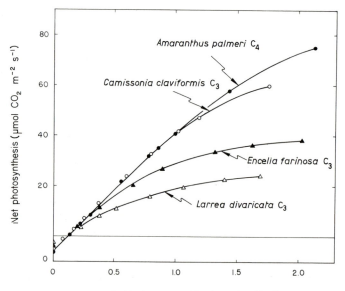

Figure 7.8 Response of photosynthesis to changes in quantum flux (400–700 nm) for *Amaranthus palmeri* (a summer annual), *Camissonia claviformis* (a winter annual), *Encelia farinosa* (a drought deciduous perennial), and *Larrea divaricata* (an evergreen perennial). Measurements were made under normal atmospheric conditions and at a leaf temperature of 30 °C, except for *Amaranthus palmeri* which was measured at 40 °C (from Ehleringer, 1983b; Ehleringer and Björkman, 1978b; Mooney, unpublished data; Mooney *et al.*, 1976).

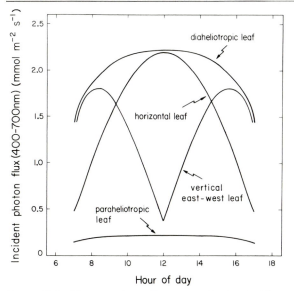

Figure 7.9 Photosynthetically active radiation incident on three leaf types over the course of a midsummer day: a diaheliotropic leaf (cosine of incidence = 1.0); a fixed leaf angle of 0°, the horizontal leaf, and a paraheliotropic leaf (cosine of incidence = 0.1) (from Ehleringer and Forseth, 1980).

summer annuals as well as in many perennials (Table 7.2). The ability to solar track is independent of plant family and photosynthetic pathway. The percentage of the flora that has solar trackers increases as the length of the growing season decreases (Ehleringer and Forseth, 1980). Many solar trackers also exhibit paraheliotropic or leaf cupping movements. This results in leaf orientation parallel to the sun's rays and reduces incident solar radiation (Fig. 7.8); this will be considered later as an adaptation to drought.

7.5.3 PHOTOSYNTHETIC ADAPTATION TO TEMPERATURE

Three possibilities exist for adapting to the seasonally fluctuating thermal environments of warm deserts (Fig. 7.10). These are a short leaf duration, metabolic adjustment, and morphological adjustment. All three possibilities have been found in warm desert species, and the adaptation employed by a particular species appears independent of

Table 7.2 Photosynthetic pathway and families of common solar tracking species in warm deserts (from Ehleringer and Forseth, 1980)

Species	Photosynthetic pathway	Family
Winter ephemerals		
Abronia villosa	C_3	Nyctaginaceae
Astragalus lentiginosus	C_3	Fabaceae
Coldenia nuttallii	C_3	Boraginaceae
Lotus saluginosus	C_3	Fabaceae
Lupinus arizonicus	C_3	Fabaceae
Malvastrun rotundifolium	C_3	Malvaceae
Oxystylis lutea	C_3	Capparidaceae
Palafoxia linearis	C_3	Asteracae
Sphaeralcea coulteri	C_3	Malvaceae
Summer ephemerals		
Allionia incarnata	C_4	Nyctaginaceae
Amaranthus palmeri	C_4	Amaranthaceae
Boerhaavia wrightii	C_4	Nyctaginaceae
Cleome subulata	C_3	Capparidaceae
Dicoria canescens	C_3	Asteraceae
Eriogonum deflexum	C_3	Polygonaceae
Euphorbia abramsiana	C_4	Euphorbiaceae
Helianthus annuus	C_3	Asteraceae
Hymenothrix wislizenii	C_3	Asteraceae
Kallstroemia grandiflora	C_4	Zygophyllaceae
Portulaca oleracea	C_4	Portulaceae
Proboscidea parviflora	C_3	Martyniaceae
Solanum rostratum	C_3	Solanaceae
Tidestromia lanuginosa	C_4	Amaranthaceae
Trianthema portulacastrum	C_4	Aizoaceae
Tribulus terrestris	C_4	Zygophyllaceae
Perennials		
Abutilon parvulum	C_3	Malvaceae
Acacia angustissima	C_3	Mimosaceae
Boerhaavia annulata	C_3	Nyctaginaceae
Boerhaavia coccinea	C_4	Nyctaginaceae
Cassia bauhinioides	C_3	Caesalpiniaceae
Cercidium microphyllum	C_3	Caesalpiniaceae
Croton california	C_3	Euphorbiaceae
Dalea emoryi	C_3	Fabaceae
Marina divaricata	C_3	Fabaceae
Prosopis juliflora	C_3	Mimosaceae
Sida lepidota	C_3	Malvaceae
Sphaeralcea abbigua	C_3	Malvaceae
Stylosanthes viscosa	C_3	Fabaceae

taxonomic association and photosynthetic pathway.

In plants characterized by short leaf duration, leaves are present only during a narrow air temper-

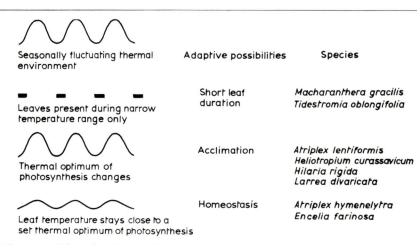

Seasonally fluctuating thermal
environment

Leaves present during narrow
temperature range only

Thermal optimum of
photosynthesis changes

Leaf temperature stays close to a
set thermal optimum of photosynthesis

Adaptive possibilities

Short leaf
duration

Acclimation

Homeostasis

Species

Macharanthera gracilis
Tidestromia oblongifolia

Atriplex lentiformis
Heliotropium curassavicum
Hilaria rigida
Larrea divaricata

Atriplex hymenelytra
Encelia farinosa

Figure 7.10   Three possibilities for photosynthetic adaptation to a seasonally fluctuating thermal environment and examples of species utilizing each possibility (after Mooney, 1980a).

ature range, that includes either hot summer or cool winter temperatures. Ephemerals and seasonal, short-lived herbaceous perennials appear to use this adaptive possibility. The temperature response curves for photosynthesis are vastly different for annuals depending on the season in which the species normally grow (Fig. 7.11). The temperature

optima for photosynthesis of winter active *Camissonia claviformis* and summer active *Amaranthus palmeri* are 23 °C and 42 °C, respectively, reflecting the differences in leaf temperatures found during these seasons. Studies of thermal acclimation potential in the annual, *Machaeranthera gracilis*, have shown that acclimation does not occur (Monson

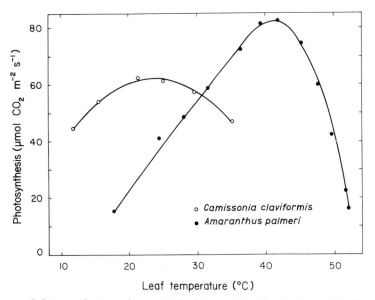

Figure 7.11   Response of photosynthesis to changes in leaf temperature for *Camissonia claviformis*, a winter annual, and *Amaranthus palmeri*, a summer annual. Measurements were made on outdoor grown plants under normal atmospheric conditions and midday irradiance levels (from Ehleringer, 1983b and Ehleringer *et al.*, 1979).

and Szarek, 1979). Thermal acclimation also does not occur in the summer active herbaceous perennial *Tidestromia oblongifolia* (Björkman *et al.*, 1980). While data for other annuals and herbaceous perennials are unavailable, Seemann *et al.* (1979) have reported as much as 8 °C shifts in the fluorescence–temperature curves of several winter annuals. These shifts imply increased thermal stability of the photosynthetic apparatus when plants are exposed to higher temperatures, although there may be no change in the photosynthetic temperature optimum.

Adaptations involving metabolic adjustment entail biochemical and physiological changes within a leaf (often an evergreen leaf) so that the temperature optimum for photosynthesis as well as the thermal stability at both ends of the curve changes in concert with changes in air temperature (Björkman *et al.*, 1980; Raison *et al.*, 1980). Temperature acclima-

tion results in a higher photosynthetic rate under the new environmental conditions, but the maximum rate is often less than that observed under optimal temperatures (see *Larrea* in Fig. 7.12), suggesting a cost associated with acclimation to stress conditions. Species which adjust metabolically to changes in the thermal environment tend to be perennials capable of growing during both winter and summer seasons. These encompass a diversity of forms including evergreen shrubs such as *Atriplex lentiformis* (Pearcy, 1977) and *Larrea divaricata* (Mooney *et al.*, 1977a), the prostrate herb *Heliotropium curassavicum* (Mooney, 1980b), the grass *Hilaria rigida* (Nobel, 1980), the fern *Notholaena parryi* (Nobel, 1978), and the tree *Chilopsis linearis* (Strain and Chase, 1966).

The third adaptive possibility, a homeostatic response, involves changes in leaf spectral characteristics and/or stomatal conductances so that midday leaf temperatures remain fairly constant through the season, even though there may be large fluctuations in air temperatures. Species utilizing this adaptation are similar to those characterized by short leaf duration in that the photosynthetic temperature response is fixed and does not show acclimation (see *Encelia farinosa* in Fig. 7.12). To maintain constant leaf temperatures, leaf absorptance may fluctuate from 30% to 85% (400–700 nm)(Ehleringer, 1981; Ehleringer and Björkman, 1978a; Mooney *et al.*, 1977b). Species in this category would include shrubs such as *Atriplex hymenelytra* and *Encelia farinosa*. Shrub species that use transpiration to regulate temperature are restricted to washes and include *Encelia frutescens* (Ehleringer and Cook, unpublished) and *Hymenoclea salsola* (Strain and Chase, 1966).

That physiological adjustment and homeostatic response are of adaptive value has been shown by calculations of the net carbon gain under different environmental conditions. Mooney (1980a) has shown for *Larrea divaricata* (physiological adjustment) and Ehleringer (1980) has shown for *Encelia farinosa* (homeostatic response) that the daily net carbon gain is greater under suboptimal environmental conditions when the species adjust (express new phenotype) than if no adjustment were to occur.

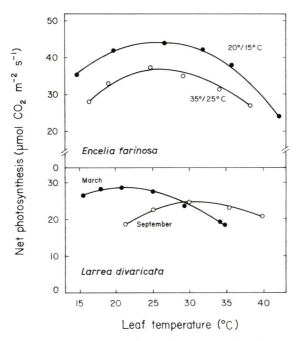

Figure 7.12 Response of photosynthesis to changes in leaf temperature for *Encelia farinosa* and *Larrea divaricata* under contrasting growth conditions. Measurements were made under normal atmospheric conditions and midday irradiance levels (from Ehleringer and Björkman, 1978b; Mooney *et al.*, 1977a).

7.5.4 ADAPTATION TO DROUGHT

Warm deserts are well known for their extended droughts. Plant adaptations to drought fall into two catergories: mechanisms to survive through extended drought periods and mechanisms to enhance productivity under drought conditions. Along warm desert gradients of increasing drought, several trends are evident within the vegetation: a decrease in both leaf and canopy size, an increase in drought-deciduousness, and an increase in drought and desiccation tolerance. Additionally, within a site shrub density is correlated with soil water potential and shrub height is negatively correlated with soil water potential (Balding and Cunningham, 1974).

While there has been substantial progress in understanding how plants grow and cope with increasing drought or temperature stress under desert conditions, very little information is available on the mechanisms of adaptation to long periods of inactivity (Levitt, 1980). The picture emerging suggests that most perennial plants are capable of withstanding desiccation to leaf water potentials far below those water potentials at which stomates close and growth ceases.

In addition to the temperature adaptations, that is, the ability for leaf tolerance characteristics to change as the environment changes (discussed in previous section), osmotic adjustment is a principal means of enhancing photosynthetic activity into drought periods. Osmotic adjustment involves an active increase in solute concentrations (osmotic component of water potential) as leaf water potentials decline so that positive turgor pressure is maintained, and therefore stomates can remain open. The ranges of osmotic adjustment that occur in warm desert plants are largely unknown. While osmotic pressure data from Walter and Stadelmann (1974) suggest that osmotic adjustment is high in summer ephemerals, shrubs, and trees, and low in winter ephemerals, ferns, and rain phanerophytes, our knowledge is sufficiently sparse as to be inconclusive. From their data, it is not possible to distinguish between the ranges over which a plant can remain photosynthetically active, and the ranges of water potentials over which the plant is simply able to survive.

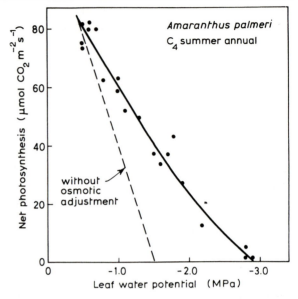

Figure 7.13 Net photosynthesis as a function of leaf water potential in *Amaranthus palmeri*. Measurements were made under normal atmospheric conditions, midday irradiances, and 40 °C leaf temperature. The dashed line is based on pressure–volume water relations curves and indicates the expected response if no osmotic adjustment were to occur in leaves of this species (from Ehleringer, 1983b).

In terms of net carbon gain, the advantages of osmotic adjustment are evident in the net photosynthesis – leaf water potential response curve for *Amaranthus palmeri* (Fig. 7.13). Without osmotic adjustment, leaves would wilt, and net photosynthesis would reach zero at leaf water potentials below -1.55 MPa (Ehleringer, 1983b). However, because of the osmotic adjustment that occurs, *A. palmeri* is able to grow with enhanced productivity under mild drought conditions and have positive photosynthetic rates down to leaf water potentials of -3 MPa. Although osmotic adjustment results in a higher rate of net photosynthesis than would have occurred without the adjustment the photosynthetic rate is never as high as it was under high water potentials, again suggesting a cost associated with the new phenotypic expression under stress conditions.

Photosynthetic rates decline in response to decreased leaf water potentials (Ehleringer 1983b; Forseth and Ehleringer, 1982; Mooney, 1980a; Odening *et al.*, 1974). The decline in photosynthesis in a long-term drought cycle is associated with decreases in both leaf conductance and intrinsic photosynthetic capacity. As a result, the stomatal diffusion limitations to photosynthesis are constant or increase only slightly as leaf water potentials decrease. In those species in which stomatal diffusion limitation does not increase with decreasing leaf water potential (e.g. *Amaranthus palmeri* and *Malvastrum rotundifolium*, water use efficiency (photosynthesis/transpiration ratio under a constant vapor-pressure deficit gradient) also remains constant. In contrast, the stomatal diffusion limitations on photosynthesis increase in *Lupinus arizonicus* as leaf water potentials decrease, resulting in a higher water use efficiency under low leaf water potentials. However, the adaptive value of an increased water use efficiency is as yet not demonstrated for desert plants.

As the drought period develops, some species can switch from leaf to stem photosynthesis. In a number of trees such as *Cercidium microphyllum* and in rain phanerophytes such as *Fouqueria splendens* and *Idria columnaris*, leaves abscise at relatively high leaf water potentials, and the plants rely on bark photosynthesis. For those cases studied, photosynthetic activity of stems can result in substantial carbon gain to the plant (Adams and Strain, 1968; Mooney and Strain, 1964; Szarek and Woodhouse, 1978). Presumably, stem photosynthesis also gives the plant a higher water use efficiency during drought periods, although the overall photosynthetic rate is reduced.

For those species which use leaf reflectance as a morphological adjustment in photosynthetic adaptation, the percentage reflectance by the leaf depends on the extent of drought development. Ehleringer *et al.* (1976) found that leaf reflectance decreased linearly with increases in precipitation in *Encelia farinosa*. Similar patterns can be seen for other species with pubescent leaves (Ehleringer, 1983a). The drought stress cue to which *E. farinosa* responds by producing differentially pubescent leaves (thus varying in leaf absorptance) is leaf water potential (Ehleringer, 1982).

A number of solar tracking annuals with compound leaves show an interesting adaptation to brief drought periods. This involves paraheliotropism or leaf cupping and serves to reduce the amount of solar radiation striking the leaf by reducing the cosine of the angle of incidence (Forseth and Ehleringer, 1982). Consequently, paraheliotropic leaves have reduced temperatures and rates of water loss when compared to strictly diaheliotropic leaves with similar leaf conductances, and can therefore survive brief drought periods without developing low leaf water potentials (Forseth and Ehleringer, 1980). In rainy periods after an intermittent drought, paraheliotropic leaves resume diaheliotropic leaf movements.

7.6 Summary

Water is the environmental parameter most affecting plant life in warm deserts. Growth, productivity, and phenological activity are tightly linked to the brief periods in which adequate soil moisture is available. In response to these selective pressures, warm desert plants have evolved a diversity of physiological, morphological and life cycle adaptations that allow plants to maximize net carbon gain during periods of high soil moisture availability, to enhance carbon gain during early drought, and to maximize survival through extended drought periods.

References

Adams, M.S. and Strain, B.R. (1968) Photosynthesis in stems and leaves of *Cercidium floridium*: spring and summer diurnal field response and relation to temperature. *Oecologia Plantarum*, **3**, 285–97.

Armond, P.A. and Mooney, H.A. (1978) Correlation of photosynthetic unit size and density with photosynthe-

tic capacity. *Carnegie Institution Washington Yearbook*, **77**, 234–7.

Balding, F.R. and Cunningham, G.L. (1974) The influence of soil water potential on the perennial vegetation of a desert arroyo. *Southwest Naturalist* **19**, 241–8.

Beatley, J.C. (1967) Survival of winter annuals in the

northern Mojave Desert. *Ecology*, **48**, 745–50.

Beatley, J.C. (1969) Biomass of desert winter annual plant populations in southern Nevada. *Oikos*, **20**, 261–73.

Beatley, J.C. (1970) Perennation in *Astragalus lentiginosus* and *Tridens pulchellus* in relation to rainfall. *Madroño*, **20**, 326–32.

Beatley, J.C. (1974) Phenological events and their environmental triggers in Mohave Desert ecosystems. *Ecology*, **55**, 856–63.

Björkman, O., Badger, M.R. and Armond, P.A. (1980) Response and adaptation of photosynthesis to high temperatures, in *Adaptations of Plants to Water and High Temperature Stress* (eds N.C. Turner and P.J. Kramer), Wiley–Interscience, New York, pp. 233–49.

Clark, D.D. and Burk, J.H. (1980) Resource allocation patterns of two California Sonoran Desert ephemerals. *Oecologia*, **46**, 86–91.

Cunningham, G.L. and Strain, B.R. (1969) Ecological significance of seasonal leaf variability in a desert shrub. *Ecology*, **50**, 400–8.

Ehleringer, J. (1980) Leaf morphology and reflectance in relation to water and temperature stress, in *Adaptations of Plants to Water and High Temperature Stress* (eds N.C. Turner and P.J. Kramer), Wiley–Interscience, New York, pp. 295–308.

Ehleringer, J. (1981) Leaf absorptances and Mohave and Sonoran Desert plants. *Oecologia*, **49**, 366–70.

Ehleringer, J. (1982) The influence of water stress and temperature on leaf pubescence development in *Encelia farinosa*. *American Journal of Botany*, **69**, 670–5.

Ehleringer, J. (1983a) Ecology and ecophysiology of leaf pubescence in North American desert plants, in *Biology and Chemistry of Plant Trichomes* (eds E. Rodriguez, P. Heley and I. Mehta), Plenum Press, New York, pp. 113–132.

Ehleringer, J. (1983b) Ecophysiology of *Amaranthus palmeri*, a Sonoran Desert summer ephemeral. *Oecologia*, **57**, 107–12.

Ehleringer, J. and Björkman, O. (1977) Quantum yields for CO_2 uptake in C_3 and C_4 plants: dependence on temperature, CO_2, and O_2 concentration. *Plant Physiology* **59**, 86–90.

Ehleringer, J. and Björkman, O. (1978a) Pubescence and leaf spectral characteristics in a desert shrub, *Encelia farinosa*. *Oecologia*, **36**, 151–62.

Ehleringer, J. and Björkman, O. (1978b) A comparison of photosynthetic characteristics of *Encelia* species possessing glabrous and pubscent leaves. *Plant Physiology*, **62**, 185–90.

Ehleringer, J. and Forseth, I. (1980) Solar tracking by plants. *Science*, **210**, 1094–8.

Ehleringer, J. and Mooney, H.A. (1978) Leaf hairs: effects on physiological activity and adaptive value to a desert shrub. *Oecologia*, **37**, 183–200.

Ehleringer, J. and Mooney, H.A. (1983) Photosynthesis and productivity of desert and mediterranean-climate plants, *Encyclopedia of Plant Physiology*, Vol. 12D, Springer-Verlag, New York, pp. 205–31.

Ehleringer, J., Björkman O. and Mooney, H.A. (1976) Leaf pubescence: effects on absorptance and photosynthesis in a desert shrub. *Science*, **192**, 376–7.

Ehleringer, J., Mooney, H.A. and Berry, J.A. (1979) Photosynthesis and microclimate of a desert winter annual. *Ecology*, **60**, 280–6.

Eickmeier, W.G. (1978) Photosynthetic pathway distributions along an aridity gradient in Big Bend National Park, and implications to enhanced resource partitioning. *Photosynthetica*, **12**, 290–7.

Fonteyn, P.J. and Mahall, B.E. (1978) Competition among desert perennials. *Nature*, **275**, 544–5.

Forseth, I. and Ehleringer, J. (1980) Solar tracking response to drought in a desert annual. *Oecologia*, **44**, 159–63.

Forseth, I.N. and Ehleringer, J. (1982) Ecophysiology of two solar tracking desert water annuals. II. Leaf movements, water relations, and microclimate. *Oecologia*, **54**, 41–9.

Forseth, I.N., Ehleringer, J., Werk, K.S. and Cook, C.S. (1983) Field water relations of Sonoran Desert annuals. *Ecology*, **65**, 1436–45.

Freeman, D.C., Klikoff, L.G. and Harper, K.T. (1976) Differential resource utilization by the sexes of dioecious plants. *Science*, **193**, 597–9.

Gates, D.M. (1980) *Biophysical Ecology*, Springer-Verlag, New York.

Ives, R.L. (1949) Climate of the Sonoran Desert region. *Annals Association of American Geographers*, **39**, 143–87.

Juhren, M., Went, F.W. and Phillips, E. (1956) Ecology of desert plants. IV. Combined field and laboratory work on germination of annuals in the Joshua Tree National Monument, California. *Ecology*, **37**, 318–30.

Koller, D. (1972) Environmental control of seed germination, in *Seed Biology*, Vol. 2 (ed. T.T. Kozlowski) Academic Press, New York, pp. 1–101.

Levitt, J. (1980) *Responses of Plants to Environmental Stresses*, Academic Press, New York.

Longstreth, D.J., Hartsock, T.L. and Nobel, P.S. (1980) Mesophyll cell properties for some C_3 and C_4 species with high photosynthetic rates. *Physiologica Plantarum*, **48**, 494–8.

McArthur, E.D. (1977) Environmentally induced changes of sex expression in *Atriplex canescens*. *Heredity*, **38**, 97–103.

McDonald, J.E. (1956) *Variability of precipitation in an arid region: a survey of characteristics for Arizona*, University of Arizona Institute of Atmospheric Physics Technical Report No. 1.

Monson, R.K. and Szarek, S.R. (1979) Ecophysiological studies of Sonoran Desert plants. V. Photosynthetic adaptions of *Machaeranthera gracilis*, a winter annual. *Oecologia*, **41**, 317–27.

Mooney, H.A. (1980a) Seasonality and gradients in the study of stress adaptations, in *Adaptations of Plants to Water and High Temperature Stress* (eds N.C. Turner and P.J. Kramer) Wiley–Interscience, New York, pp. 279–94.

Mooney, H.A. (1980b) Photosynthetic plasticity of populations of *Heliotropium curassavicum* L. originating from differing thermal regimes. *Oecologia*, **45**, 372–6.

Mooney, H.A. and Ehleringer, J. (1978) The carbon gain benefits of solar tracking in a desert annual. *Plant, Cell, Environment*, **1**, 307–11.

Mooney, H.A. and Strain, B.R. (1964) Bark photosynthesis in ocotillo, *Madroño*, **17**, 230–3.

Mooney, H.J., Troughton, J.H. and Berry, J.A. (1974) Arid climates and photosynthetic systems. *Carnegie Institution Washington Yearbook.*, **73**, 793–805.

Mooney, H.A., Ehleringer, J. and Berry, J.A. (1976) High photosynthetic capacity of a winter desert annual in Death Valley. *Science*, **194**, 322–4.

Mooney, H.A., Björkman, O. and Collatz, G.J. (1977a) Photosynthetic acclimation to temperature in the desert shrub, *Larrea divaricata*. I. Carbon dioxide exchange characteristics of intact leaves. *Plant Physiology*, **61**, 406–10.

Mooney, H.A., Ehleringer, J. and Björkman, O. (1977b) The energy balance of leaves of the evergreen desert shrub *Atriplex hymenelytra*. *Oecologia*, **29**, 301–10.

Mulroy, T.W. and Rundel, P.W. (1977) Annual plants: adaptations to desert environments. *Bioscience*, **27**, 109–14.

Nobel, P.S. (1978) Microhabitat, water relations and photosynthesis of a desert fern, *Notholaena parryi*, *Oecologia*, **31**, 293–309.

Nobel, P.S. (1980) Water vapor conductance and CO_2 uptake for leaves of a C_4 desert grass, *Hilaria rigida*. *Ecology*, **61**, 252–8.

Odening, W.R., Strain, B.R. and Oechel, W.C. (1974) The effect of decreasing water potential on net CO_2 exchange of intact desert shrubs. *Ecology*, **55**, 1086–95.

Pearcy, R.W. (1977) Acclimation of photosynthetic and respiratory carbon dioxide exchange to growth temperature in *Atriplex lentiformis* (Torr.) Wats. *Plant Physiology*, **59**, 795–9.

Pianka, E.R. (1967) On lizard species diversity: North American flatland deserts. *Ecology*, **48**, 333–51.

Raison, J.K., Berry, J.A., Armond, P.A. and Pike, C.S. (1980) Membrane properties in relation to the adaptation of plants to temperature stress, in *Adaptations of Plants to Water and High Temperature Stress* (eds N.C. Turner and P.J. Kramer) Wiley–Interscience, New York, pp. 261–73.

Schaffer, W.M. and Gadgil, M.D. (1975) Selection for optimal life histories in plants, in *Ecology and Evolution of Communities* (eds M.L. Cody and J.M. Diamond), Belknap Press, Cambridge, MA, pp. 142–57.

Schimper, A.F.W. (1903) *Plant Geography Upon a Physiological Basis*, Clarendon Press, Oxford.

Seemann, J.R., Downton, W.J.S. and Berry, J.A. (1979) Field studies of acclimation to high temperature: winter emphemerals in Death Valley. *Carnegie Inst. Wash. Yb.*, **78**, 157–62.

Sellers, W.D. and Hill, R. (1974) *Arizona Climate 1931–1972*, University of Arizona Press, Tucson.

Shreve, F. (1951) *Vegetation of the Sonoran Desert*, Carnegie Institute Washington Publication No. 591.

Simpson, B.B. (1977) Breeding systems of dominant perennial plants of two disjunct warm desert ecosystems. *Oecologia*, **27**, 203–26.

Smith, W.K. (1978) Temperatures of desert plants: another perspective on the adaptability of leaf size. *Science*, **201**, 614–16.

Smith, W.K. and Geller, G.N. (1980) Leaf and environmental parameters influencing transpiration: theory and field measurements. *Oecologia*, **46**, 308–13.

Smith, W.K. and Nobel, P.S. (1977) Influences of seasonal changes in leaf morphology on water-use efficiency for three desert broadleaf shrubs. *Ecology*, **58**, 1033–43.

Steila, D. (1976) *The Geography of Soils*, Prentice Hall, Englewood Cliffs, New Jersey.

Stowe, L.G. and Teeri, J.A. (1978) The geographic distribution of C_4 species of the Dicotyledonae in relation to climate. *Amer. Nat.*, **112**, 609–23.

Strain, B.R. and Chase, V.C. (1966) Effect of past and prevailing temperatures on the carbon dioxide exchange capacities of some woody desert perennials. *Ecology*, **47**, 1043–5.

Syvertsen, J.P., Nickell, G.L., Spellenberg, R.W. and Cunningham, G.L. (1976) Carbon reduction pathways and standing crop in three Chihuahuan Desert plant communities. *Southwest. Nat.*, **21**, 311–20.

Szarek, S.R. and Woodhouse, R.M. (1978) Ecophysiological studies of Sonoran Desert plants. IV. Seasonal photosynthetic capacities of *Acacia greggii* and *Cercidium microphyllum*. *Oecologia*, **37**, 221–9.

Teeri, J.A. and Stowe, L.G. (1976) Climatic patterns and the distribution of C_4 grasses in North America. *Oecologia*, **23**, 1–12.

Terjung, W.H., Ojo, S.O. and Swarts, S.W. (1970) A nighttime energy and moisture budget in Death Valley, California, in mid-August. *Geog. Ann.*, **52A**, 160–73.

Thekaekara, M.P. (1976) Solar radiation measurement: techniques and instrumentation. *Solar Energy*, **18**, 309–25.

Walter, H. and Stadelmann, E. (1974) A new approach to the water relations of desert plants, in *Desert Biology*, Vol. 2 (ed. G. Brown), Academic Press, New York, pp. 213–310.

Warming, E. (1909) *Oecology of Plants: An Introduction to the Study of Plant Communities*, Oxford University Press, London, UK.

Went, F.W. (1948) Ecology of desert plants. I. Observations on germination in the Joshua Tree National Monument, California. *Ecology*, **29**, 242–53.

Went, F.W. (1949) Ecology of desert plants. II. The effect of rain and temperature on germination and growth. *Ecology*, **30**, 1–13.

Went, F.W. and Westergaard, M. (1949) Ecology of desert plants. III. Development of plants in the Death Valley National Monument, California. *Ecology*, **30**, 26–38.

Werk, K.S. and Ehleringer, J. (1983) Photosynthesis by flowers of two shrubs, *Encelia farinosa* and *Encelia californica*. *Oecologia*, **57**, 311–15.

Woodell, S.R.J., Mooney, H.A. and Hill, A.J. (1969) The behavior of *Larrea divaricata* (creosote bush) in response to rainfall in California. *Journal of Ecology*, **57**, 37–44.

8

Desert succulents

PARK S. NOBEL

8.1 Introduction

Succulents such as agaves and cacti comprise one of the most interesting and characteristic groups of plants coping with the low and sporadic rainfall and often high temperatures of deserts. For the North

Figure 8.1 Deserts of North America. Adapted from Hastings and Turner (1965) and McGinnies *et al.* (1968).

American deserts – Chihuahuan, Great Basin, Mojave, and Sonoran (Fig. 8.1) – rainfall is generally less than 250 mm yr^{-1} with an average of about 170 mm. The Chihuahuan Desert (500 000 km²) is characterized by primarily summer rainfall and 17 °C average annual temperature, the Great Basin Desert (500 000 km²) by primarily winter rainfall and 10 °C (including considerable freezing weather and frequent snowfall), the Mojave Desert (40 000 km²) also by primarily winter rainfall but a higher average annual temperature of 18 °C, and the Sonoran Desert (310 000 km²) by approximately equal summer and winter rainfall and 22 °C average annual temperature (Bacheller, 1980; Barbour *et al.*, 1980; Hastings and Turner, 1965). Mean daily maximum temperatures in July are quite similar for the four deserts, generally 34–40 °C. Although succulent plants occur in all four deserts, they are more common in the warmer and drier regions, being rather infrequent in the Great Basin Desert.

In addition to the obvious water storage, succulent plants also exhibit Crassulacean acid metabolism (CAM), characterized by diurnal fluctuations in tissue acidity and nighttime stomatal opening (Kluge and Ting, 1978; Osmond, 1978; Ting and Gibbs, 1982). Stomatal opening at night minimizes transpirational water loss, since the surface temperatures and hence the water vapor concentration drop from the tissue to the air are then much lower than during the daytime. We can illustrate this critical aspect of the ecological utility of CAM by a simple calculation. Suppose that the air has a water vapor content of $5\,\mathrm{g\,m^{-3}}$, which is fairly representative for the Sonoran Desert, and that the intercellular air spaces in the chlorenchyma are saturated with water vapor. Then the water vapor concentration drop from the tissue to the air would be $25\,\mathrm{g\,m^{-3}}$ for a daytime tissue surface temperature of $30\,^{\circ}\mathrm{C}$, but only $4\,\mathrm{g\,m^{-3}}$ for a nighttime tissue surface temperature of $10\,^{\circ}\mathrm{C}$. For the same stomatal opening, the transpiration rate would therefore be six-fold higher in the daytime compared to the nighttime.

Although tissue succulence has obvious water storage attributes, the key to CAM is actually the 'succulence' of the cells, brought about by a large central vacuole (Kluge and Ting, 1978). Nearly all CO_2 uptake by most desert CAM plants occurs at night, the CO_2 being incorporated into organic acids, such as malic acid, which are stored in the large vacuoles of the chlorenchyma cells. During the daytime, the organic acids released from the vacuoles are decarboxylated, which leads to a decrease in tissue acidity. The CO_2 released in the chlorenchyma during the daytime is prevented from leaving the plant by the closing of the stomates, and it is then fixed into carbohydrates and other photosynthetic products by the conventional C_3 pathway of photosynthesis.

Because of the two-step strategy for CO_2 fixation, the low maximal CO_2 uptake rates, and the relatively low surface area for CO_2 uptake relative to total volume, CAM plants tend to grow rather slowly. This has important implications for seedling establishment and reproduction. Also, the massiveness of the photosynthetic tissue has special influences on the thermal properties. These topics plus other aspects of the water relations and gas exchange will be considered below, paying particular attention to *Agave deserti* and cacti of the Sonoran Desert.

8.2 Water relations

Since the key environmental factor affecting the ecophysiology of desert succulents is the availability of water, we will begin by considering their response to rainfall.

8.2.1 RAINFALL, SOIL WATER POTENTIAL, AND STOMATAL OPENING

Desert succulents tend to have shallow roots, leading to rapid responses to the rather infrequent rains. The mean rooting depths of a common desert agave, *A. deserti*, and a barrel cactus, *Ferocactus acanthodes*, are about 8 cm (Nobel, 1976, 1977a). Both species respond within 24 h to a 19 mm rainfall on dry soil and nearly full stomatal opening occurs in 48 h. Stomatal opening and tissue acidity changes for the sympatric *Opuntia basilaris* take place in about 30 h (Szarek and Ting, 1975a; Szarek *et al.* (1973), as does a measurable swelling of stems of *O. discata* (Cannon, 1911). New roots are formed by *Opuntia puberula* a few hours after watering (Kausch, 1965), indicating that the response is fairly immediate.

The minimum amount of rainfall on dry soil leading to detectable changes in stomatal activity of desert succulents is about 6 mm (Nobel, 1976; Szarek and Ting, 1975b). For rainfalls greater than about 6 mm on dry sandy soil, the soil water potential near the roots surpasses $-0.4\,\mathrm{MPa}$ (-4 bar), which is generally greater than the water potential of the plants, e.g. representative tissue osmotic potentials for cacti and agaves are about $-0.5\,\mathrm{MPa}$ (Nobel, 1976; 1977a; Soule and Lowe, 1970; Walter and Stadelmann, 1974). Water can then enter the plants, leading to nocturnal stomatal opening. The maximum water vapor conductance for both *A. deserti* and *F. acanthodes* is about $2\,\mathrm{mm\,s^{-1}}$ (Fig. 8.2), which is much lower than is generally observed for mesic plants, suggesting that succulent plants do not have high transpiration rates even when stomates are fully open. Indeed, stomatal frequencies for succulent plants are fairly low, generally $20-50\,\mathrm{mm^{-2}}$, and the guard cells are of average size (Gentry and Sauck, 1978; Kluge and Ting, 1978). The minimum water vapor conductance of $0.01-0.02\,\mathrm{mm\,s^{-1}}$ observed for *A. deserti* (Nobel, 1976), *F. acanthodes* (Nobel, 1977a), and three species of *Opuntia* (Ting and Szarek, 1975) indicates that cuticular transpiration can be quite low for desert succulents. Such observations are consistent with their thick waxy cuticles, which helps lead to excellent water conservation.

Another facet of the water relations of succulent

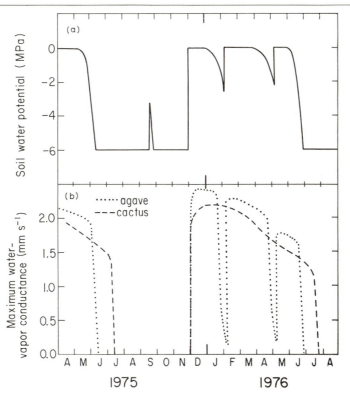

Figure 8.2 Seasonal variation in soil water potential (a) and stomatal opening (b) for *Agave deserti* (....) and *Ferocactus acanthodes* (- - -). Soil water potential was measured in the root zone (soil thermocouple psychrometers were 10 cm below the soil surface). The maximum water vapor conductance generally occurred between 3 and 5 a.m. Data were obtained in the University of California Philip L. Boyd Deep Canyon Desert Research Center at 33° 38′ N, 116° 24′ W, 850 m (Nobel, 1976, 1977a).

plants is the ability to store water and hence to have nocturnal stomatal opening even when the plants are not taking up water from the soil. Fig. 8.2 indicates that *F. acanthodes* can have appreciable nocturnal stomatal opening for about 40 d after the soil water potential goes below − 0.5 MPa and *A. deserti* for about 10 d. This reflects the greater volume-to-surface ratio for the cactus, namely, V/S for its stem was 10 900 cm³/4300 cm² or 2.5 cm, while that of an agave leaf was 348 cm³/380 cm² or 0.92 cm (V/S indicates the average depth of tissue supplying water to the surface). By comparison, V/S for a typical leaf of a C_3 or C_4 plant is only 0.01 to 0.02 cm.

Once the plants can no longer obtain water from the soil, the continuation of cuticular transpiration, however slight, would lead to a loss of tissue water

and hence the osmotic potential of the cells would be expected to decrease. Indeed, the osmotic potential of *F. acanthodes* went from − 0.31 to − 0.64 MPa after a 7 month drought (Nobel, 1977a). For a drought of 5 months, the osmotic potential of leaf cells of *A. deserti* went from − 0.60 to − 1.34 MPa (Nobel, 1976). The somewhat faster rate of dehydration for the agave reflects its lower V/S, even though it had a slightly lower water vapor conductance than did the cactus.

8.2.2 SEEDLING ESTABLISHMENT

Water relations play a critical role in the survival of desert succulents during the seedling stage. To survive drought, the seedlings must develop suffi- cient water storage tissue during the previous wet season, despite growing slowly. For example, globular seedlings of *Carnegiea gigantea* in the field grew only 3.3 mm in height during the first year (Steenbergh and Lowe, 1969). Direct solar irradiation can raise the temperature of desert soil to 80 °C (Hadley, 1970) which can prove lethal to seedlings and which will also increase the water vapor concentration drop to the air, causing the soil to dry rapidly. Suitable microhabitats are thus necessary for survival, and these can be provided by 'nurse plants' or other obstructions to solar irradiation (Steenbergh and Lowe, 1977; Turner *et al.*, 1966, 1969).

As the seedlings grow, their volume-to-surface ratio increases (Fig. 8.3a). Thus, the amount of tissue water that can be called upon in the case of drought increases. This morphological aspect together with changes in the water vapor conductance causes the length of drought that can be tolerated to increase rapidly with age, e.g. a 1 mon old seedling of *A. deserti* or *F. acanthodes* can barely tolerate a 5 day drought, while a 6 mon old seedling can endure over 3 months of drought (Fig. 8.3b). This has important implications for predicting the locations where seedling establishment can occur, since the periods favorable for growth can be compared with the length of ensuing drought. Using local measurements of soil water potential and rainfall, only the

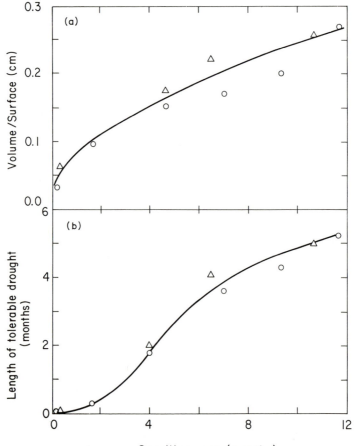

Figure 8.3 Influence of age on the volume-to-surface ratio (a) and the length of drought that can be tolerated (b) for seedlings of *Agave deserti* (△) and *Ferocactus acanthodes* (○). Plants were grown in the laboratory under wet conditions and tolerable drought length was calculated from seedling morphology, measured water vapor conductances, and maximum tolerable desiccation (Jordon and Nobel, 1979, 1981).

year 1967 out of a stretch of 17 years was predicted to be suitable for the establishment of *A. deserti* at a specific site in the northwestern Sonoran Desert (Jordan and Nobel, 1979). This agrees with the estimated mean year of germination for the six smallest agaves found at that site, all of which were protected by nurse plants or rocks. At a moister site nearby, nearly half of the years were suitable for the establishment of seedlings of *F. acanthodes* (Jordan and Nobel, 1981). The infrequent pulses of establishment of *C. gigantea* in certain regions can be correlated with heavier than average rainfall (Brum, 1973; Jordan and Nobel, 1982). The seedlings germinate in response to summer rains, but are often killed by the late spring drought of the following year (Steenbergh and Lowe, 1969, 1977). This may explain why *C. gigantea* is absent in the more arid parts of the Sonoran Desert in California (Shreve, 1911; Steenbergh and Lowe, 1977).

8.2.3 ANNUAL TRANSPIRATION

Nocturnal stomatal opening and a low maximum water vapor conductance cause transpiration by succulent plants to be relatively low. When averaged over the ground area occupied by the roots, *A. deserti* transpired 35% of the rainfall incident in a year (27 out of 78 mm, Nobel, 1976), and *F. acanthodes* transpired 22% (34 out of 154 mm, Nobel, 1977a). This represents a fairly high use of the incident rainfall, since about half of it in each year considered fell in two rather heavy rainstorms with considerable runoff, and some water would evaporate from the soil or percolate beyond the rooting depths of these succulents.

The water cost of reproduction is another interesting aspect of the ecophysiology of desert succulents. Flowers of *F. acanthodes* have a large surface area and water loss (about $3.6\,g\,d^{-1}$ per flower for the 7 days that each flower is present). The total water required for reproduction (from the bud stage through to the production of viable seed) represented 4% of the water annually taken up from the soil (Nobel, 1977a). For the monocarpic *A. deserti* during the 5 month period from inflorescence emergence to when the seeds were viable, 3.1 kg water was stored in the large central inflorescence (up to 4 m tall), 4.3 kg was transpired from its surface, and 10.8 kg was lost from its lateral flower branches, mostly from the flowers and fruit (Nobel, 1977b). Essentially all of this 18 kg water came from the leaves, which decreased in thickness from 4.1 to 1.4 cm at midleaf over the 5 month period. Thus, the leaves serve as a water reservoir for this monocarpic plant, as well as supplying the dry matter needed for the growth of the inflorescence.

8.3 CO_2 uptake and acidity changes

A prerequisite for growth is a net CO_2 gain. For CAM plants this is generally accomplished by periods of nocturnal stomatal opening and CO_2 uptake when water is available. During extended dry periods the stomates remain closed and there is very little gas exchange with the environment.

8.3.1 DIURNAL CAM PATTERNS

When succulent plants are operating in the 'full-CAM' mode (Neales, 1975), stomatal opening and CO_2 uptake occur only at night (Fig. 8.4). Daytime CO_2 uptake is often greater for young plants, e.g. for a 22 day old seedling of *A. deserti* under fairly wet conditions, 80% of the net CO_2 uptake occurred during the daytime, the value was 65% for a 119 day old seedling (Jordan and Nobel, 1979), but generally less than 10% for adults. However, plentiful water can cause appreciable gas exchange to occur at the beginning and end of the daytime (Kluge and Ting, 1978). For the leaf succulent *A. deserti* under laboratory conditions, excessive watering can even cause a nearly complete shift from the CAM to the C_3 mode, leading to daytime stomatal opening and CO_2 uptake (Fig. 8.4). Daytime CO_2 uptake also increases as water becomes more available for *Agave americana* (Neales *et al.*, 1968) and *Dudleya farinosa* (Bartholomew, 1973). Strictly adhering to the water-conserving CAM mode may not be necessary when water is readily available, although nocturnal gas exchange is the usual pattern observed in the field for desert succulents, especially cacti (Eickmeier and Bender, 1976; Hanscom and Ting, 1978; Mooney *et al.*, 1974; Szarek and Troughton, 1976).

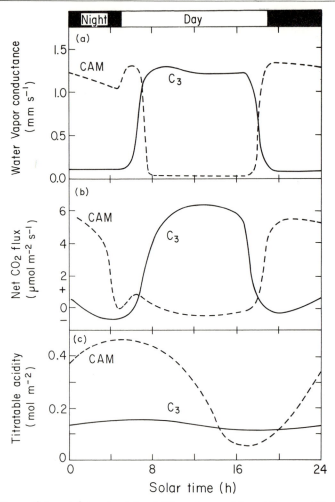

Figure 8.4 Diurnal patterns of stomatal opening (a), net CO_2 exchange (b), and tissue acidity (c) for *Agave deserti* in the normal CAM mode (- - -) and after excessive watering for 12 weeks, which converted it to the C_3 mode (—). Adapted from Hartsock and Nobel (1976).

As CO_2 is taken up at night and organic acids are formed, the tissue acidity gradually rises, generally reaching a peak near dawn (Fig. 8.4c). During the day, the stomates are closed in the full-CAM mode, the organic acids move out of the vacuoles and become decarboxylated, and the tissue acidity decreases to its minimum value near dusk. Such diurnal patterns of tissue acidity are not only diagnostic for CAM plants, but also facilitate field studies of some of their physiological responses.

8.3.2 RESPONSES TO PHOTOSYNTHETICALLY ACTIVE RADIATION

Although the desert is characteristically a high light environment, CO_2 uptake and acidity changes by succulents are often limited by the amount of photosynthetically active radiation (PAR) reaching the chlorenchyma. For instance, the PAR for 90% saturation of nocturnal acidity increases is 20–25 mol of photons m^{-2} day^{-1} for many desert

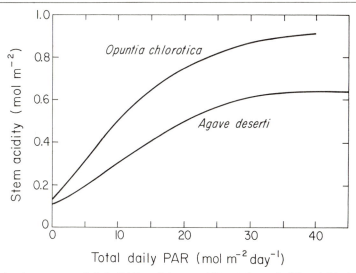

Figure 8.5 Relationship between total daily PAR and tissue acidity at the end of the night. Adapted from Nobel and Hartsock (1978) for *Agave deserti* and Nobel (1980a) for *Opuntia chlorotica*.

succulents (Fig. 8.5, Nobel, 1977a, 1980a, 1982a). Although this is far below the maximum PAR on a horizontal surface (e.g. 66 mol m^{-2} day^{-1} on the summer solstice at 34 °N), the PAR incident on the surfaces of succulents is often considerably less. At 34 °N in the Sonoran Desert, the total daily PAR on leaves of *A. deserti* averaged about 10 mol m^{-2} day^{-1} for a winter day and 20 mol m^{-2} day^{-1} for a summer day (Woodhouse *et al.*, 1980). At the same latitude, the maximal daily PAR on clear days varies seasonally and directionally from about 7 to 33 mol m^{-2} day^{-1} for vertical unshaded surfaces of cacti (down to about 3 mol m^{-2} day^{-1} for ribbed surfaces), with the annual average PAR being about 22 mol m^{-2} day^{-1} (Nobel, 1980a). Thus, any appreciable shading can lead to a suboptimal PAR and hence tend to reduce the nocturnal CO_2 uptake and acidity changes for vertically oriented surfaces of cacti and other desert succulents.

Since PAR can be limiting for desert succulents, the height and orientation of the photosynthetic surfaces can be critical (Nobel, 1982a). *Stenocereus gummosus*, a ceroid cactus which occurs over the entire 1300 km extent of Baja California, Mexico, is short (mean height of 0.7 m) on coastal bluffs in the northern part of its range where the surrounding

vegetation is sparse and mainly composed of short species. It is tall (mean height of 4.1 m) in the fairly tall subtropical forests near the southern part of its range where the PAR near the ground is greatly reduced (Nobel, 1980a). The increased stem height in the subtropical forest ensures interception of adequate PAR there.

East–west facing cladodes (flattened stems of a platyopuntia) of *Opuntia amyclaea* in central Mexico received more PAR and had greater dry matter accumulation in the fall than did north–south facing ones (Rodriquez *et al.*, 1976), emphasizing the importance of orientation. Cladodes of *O. chlorotica* tended to face north–south in a canyon site in the Mojave Desert, which maximized PAR absorption during the winter when water was most available (Nobel, 1980a). Orientations of this species at three different sites in the Sonoran Desert are shown in Fig. 8.6 At the western Sonoran site (33° 35' N, 116° 26' W), most of the rainfall occurs in the winter May–August rainfall/November–February rainfall = 0.48, US Weather Bureau, 1964) when facing north–south would lead to the most PAR absorption (Nobel, 1980a), consistent with the observation that approximately twice as many cladodes faced within 15° of north–south compared to within 15°

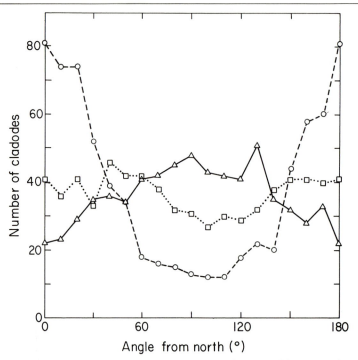

Figure 8.6 Orientation of horizontal axis of erect terminal cladodes of *Opuntia chlorotica* in the Sonoran Desert. Angle classes are 10° wide centered on the indicated angles, which were measured clockwise from true north. A total of 660 cladodes for which the sun's rays were not obstructed were measured at each site. Data were obtained at 33° 35′N, 116° 26′W, 1190 m on 2 July 1980 (△); at 31° 48′N, 110° 46′W, 1540 m on 4 August 1980 (○); and at 34° 30′N, 113° 23′W, 870 m on 7 August 1980 (□).

of east–west at this site (the tendency to face north–south was highly significant, $P < 0.001$). At the northern Sonoran site (34° 30′ N, 113° 23′ W), the rainfall patterns become more equal (May–August/November–February = 0.73), but the more favorable summer temperatures presumably lead to more growth then, when facing east–west would lead to more PAR absorption by *O. chlorotica* (the tendency to face east–west was significant, $0.025 < P < 0.05$). At the eastern Sonoran site (31° 48′ N, 110° 46′ W), the rainfall is mainly in the summer (May–August/November–Februray = 2.10) and the temperatures are also conducive to growth then; this helps explain the highly significant tendency to face east–west ($P < 0.001$), e.g. nearly six times more cladodes faced within 15° of east–west than within 15° of north–south there (Fig. 8.6), leading to maximization of PAR absorption at times of the

year when growth is favored (Nobel, 1980a). Indeed, cladodes on 23 species of platyopuntias have been shown to orient preferentially in the direction maximizing intercepting of PAR, taking into consideration the effects of topographical blockage of incoming radiation as well as seasonality of rainfall (Nobel, 1982b). Observed phototropic responses, the higher productivity of favorably oriented cladodes, and the tendency of cladodes to orient similarly to the underlying cladode presumably account for the overall orientation patterns found (Nobel, 1982c).

8.3.3 TEMPERATURE EFFECTS

Nocturnal CO_2 uptake by desert CAM plants has a fairly low temperature optimum, generally 10–15 °C (Dinger and Patten, 1974; Nisbet and Patten, 1974; Nobel 1977a, 1978; Nobel and

Hartsock, 1978; Patten and Dinger, 1969). This apparently reflects both stomatal properties as well as those of the chlorenchyma. For example, the water vapor conductance approximately halves for each 10 °C increase in tissue temperature for *A. deserti* (Nobel, 1976; Nobel and Hartsock, 1979) and *F. acanthodes* (Nobel, 1977a). The mesophyll CO_2 conductance more than halves for each 10 °C change from the optimal temperature for *A. deserti* (Nobel and Hartsock, 1978). Daytime temperatures from 15 °C to 40 °C have relatively little influence on the nocturnal CO_2 uptake by *A. deserti*. Also, daytime temperatures are not as important as nighttime ones in influencing nocturnal CO_2 exchange by *A. lecheguilla* (Eickmeier and Adams, 1978) and by cacti (Patten and Dinger, 1969).

As has been shown for other species (e.g. Berry and Björkman, 1980; Mooney and West, 1964), desert succulents can exhibit temperature acclimation for CO_2 uptake in response to changing environmental temperatures. For example, the high temperature CO_2 compensation point varies seasonally for *O. phaeacantha* var. *discata* (Nisbet and Patten, 1974) and *O. basilaris* showed acclimation one day after shifting from day/night air temperatures 20 °C/10 °C to 40 °C/30 °C (Gulmon and Bloom, 1979). Temperature acclimation was investigated in more detail for *Coryphantha vivipara* (Nobel and Hartsock, 1981), a cactus occurring in the Great Basin, Mojave, and Sonoran Deserts. The temperature optimum for nocturnal CO_2 uptake was normally near 10 °C, the nighttime air temperature used here (Fig. 8.7). This optimum was uninfluenced by a 20 °C increase in daytime temperature, but it increased 13 °C in response to a 20 °C increase in nighttime temperature. The half-time for the increase was 8 d and that for the reversion to 10 °C upon lowering the nighttime temperature to the original value was 4 d (Fig. 8.7). Indeed, raising the day/night air temperatures from 10 °C/10 °C to 30 °C/30 °C

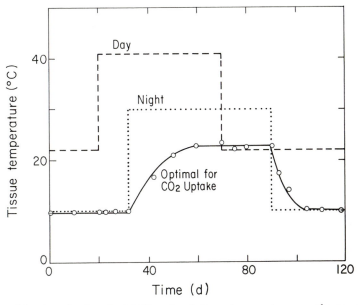

Figure 8.7 Influence of daytime (- - -) and nighttime (....) tissue temperatures on the temperature optimum for nocturnal CO_2 uptake (—) by *Coryphantha vivipara*. Plants were maintained in a growth chamber with 12 h days at $350\,\mu\text{mol m}^{-2}\text{s}^{-1}$ PAR. Daytime air temperature was increased from 10 °C to 30 °C on day 20 and reduced back to 10 °C on day 70. Nighttime air temperature was increased from 10 °C to 30 °C on day 33 and reduced back to 10 °C on day 90. Tissue temperatures were measured 0.5 mm beneath the stem surface with copper–constantan thermocouples 0.13 mm in diameter. CO_2 uptake was measured for a series of tissue temperatures and the optimal one determined graphically (see Nobel and Hartsock, 1981).

caused the optimal temperature of nocturnal CO_2 uptake by six species of cacti and three species of agave to shift from an average of 12 °C to an average of 20 °C, with halftime of generally only 1–2 d (Nobel and Hartsock, 1981).

8.3.4 ANNUAL GROWTH

Since desert succulents tend to grow quite slowly, plants of small to moderate size can be relatively old. For *Carnegiea gigantea* (Fig. 8.8) in the field, flower bearing begins for plants about 2 m tall, which are about 32 yr old; the plants live approximately 200 yr with a height up to about 10 m (Steenbergh and Lowe, 1976, 1977). The annual increment in height for a 2 m tall *C. gigantea* is about 11 cm, that for *Ferocactus acanthodes* 34 cm tall is about 1.4 cm (Nobel, 1977a), and for slightly taller *F. wislizenii* it is 2 cm (MacDougal and Spalding, 1910). Barrel cacti rarely exceed 2 m in height and presumably reach 100 yr in age, while most species of *Coryphantha* and *Mammillaria* do not exceed 20 cm tall and may be equally as old at senescence. *Agave deserti* is commonly known as the century plant, although the age when its inflorescence emerges in the last year of its life may be somewhat less than 100 yr.

Measurement of gas exchange has enabled the net CO_2 uptake and water use efficiency (mass CO_2 fixed/mass H_2O transpired) to be estimated on an annual basis for desert succulents. For a leaf of *A. deserti* 30 cm in length, the net CO_2 uptake (calculated as carbohydrate) was 40% of the leaf dry weight, and the water use efficiency was 40 g CO_2/kg H_2O (Nobel, 1976). For an *F. acanthodes* 34 cm tall, the annual carbohydrate gain represented 15% of the stem dry weight and the water use efficiency was 14 g CO_2/kg H_2O (Nobel, 1977a). Such water use efficiencies of CAM plants are much higher than observed for C_3 and C_4 plants, whose annual values are about 1–3 g CO_2/kg H_2O (Neales, 1975; Neales *et al.*, 1968; Szarek and Ting, 1975b; Ting and Szarek, 1975). On an annual basis, about 6% of the total water entering the stem in the transpirational stream was stored by *F. acanthodes*, which is a far higher water storage than for most C_3 and C_4 plants. Additionally, *F. acanthodes* does not shed its photosynthetic tissue, and so dry weight is

also advantageously conserved compared to deciduous species.

Although productivity of desert succulents has been little studied, a few comments can still be made. Using the fractional uptake of incident rainfall and the water use efficiency presented above, *A. deserti* could annually produce 740 g carbohydrate per m² of ground occupied by the roots and *F. acanthodes* could produce 320 g m⁻² yr⁻¹ (Nobel, 1976, 1977a), which are rather high values. They are also consistent with measured productivities for cacti under agronomic conditions, which range from 220 to 2400 g carbohydrate m⁻² yr⁻¹ for *Opuntia* (assuming that carbohydrate was 10% of wet weight; Griffiths, 1915; Le Houerou, 1970; Metal, 1965).

8.4 Thermal relations

Tissue temperatures markedly affect physiological processes and ultimately the ecology of desert

Figure 8.8 *Carnegiea gigantea* from western Arizona.

succulents. The influence of temperature on the water vapor concentration drop, stomatal opening, and the biochemical reactions underlying CO_2 uptake including its seasonal shifts have already been discussed. Succulent tissues can also be inactivated by exposure to temperature extremes. Cactus stems can rise up to 15 °C or more above air temperature in direct sunlight (Gates *et al.*, 1968; Gibbs and Patten, 1970; Mozingo and Comanor, 1975; Patten and Smith, 1975; Smith, 1978), leading to tissue temperatures up to about 60 °C during the daytime. Low nighttime temperatures have long been recognized as influencing the northern and altitudinal distribution of *C. gigantea* (Shreve, 1911; Fig. 8.8).

8.4.1 MODELLING, MORPHOLOGY, AND CACTUS DISTRIBUTION

A simulation model has been developed to predict the surface temperatures of cacti as a function of morphology (e.g. height, diameter, spine coverage, apical pubescence) and environmental conditions

(Lewis and Nobel, 1977). In addition to the usual quantities in an energy budget analysis of leaves, heat conduction within the stem and to the ground, as well as heat storage in various subvolumes of the stem, are also included in the model. Use of the simulation model has given insights into various environmental relationships of succulents, including distribution patterns as well as the effect of transpiration. Specifically, transpiration can reduce the surface temperatures, avoiding high temperature damage. However, stomates of CAM plants are generally not open during the daytime. Moreover, the maximum difference in simulated surface temperature during rapid transpiration (maximum water vapor conductance of $2.6 \, \text{mm s}^{-1}$ at night) compared to one without any transpiration was only about 2 °C for *C. gigantea* (Fig. 8.9). The approximate halving of water vapor conductance for each 10 °C increase in surface temperature discussed above ensures that transpiration does not increase appreciably with temperature. In fact, incorporating this temperature response of stomatal

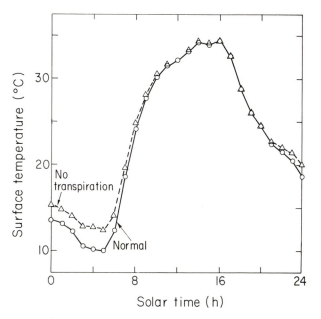

Figure 8.9 Predicted average surface temperature for a *Carnegiea gigantea* 5.64 m tall with measured stomatal opening and hence normal transpiration (—) and the simulated absence of transpiration (---). Morphological and microclimatic measurements were made at the Kofa Game Range in Arizona at 33° 32′N, 114° 10′W, 410 m on 20 April 1977 (Nobel, 1978). Simulations were performed using an energy budget model that predicted surface temperatures to within 1 °C of measured values (Lewis and Nobel, 1977; Nobel, 1978).

opening into the simulation model, the water loss for *F. acanthodes* was found to be approximately the same for winter and summer days (Lewis and Nobel, 1977).

The lowest wintertime temperature can be the most important factor influencing the distribution of desert succulents (Shreve, 1914). Of the 65 species of arborescent ceroid cacti in the Sonoran Desert, only three species [*C. gigantea*, *Stenocereus thurberi* (= *Lemaireocereus thurberi*), and *Lophocereus schottii*] occur further north than the frost line, and frost damage is common for all three at the northernmost part of their ranges in Arizona (Shreve, 1911; Steenbergh and Lowe, 1977; Turnage and Hinckley, 1938). Freezing damage and the lowest surface temperatures generally occur in the vicinity of the stem apex. In this regard, the simulation model predicted that the minimum apical temperatures under the same environmental conditions would be 7.7, 5.9, and 3.9 °C for *C. gigantea*, *S. thurberi*, and *L. schottii*, respectively (Nobel, 1980b), which is the same relative order as their northernmost limits (34° 56′ N, 32° 38′ N, and 31° 55′ N, respectively).

Various morphological features affect the predicted interspecific differences in minimum temperature. Niering *et al.* (1963) indicated that *C. gigantea* increases in diameter from the southern to the northern part of its range; from 30°25′ N to 34°50′ N, the mean diameter increases from 33 to 44 cm, which the model indicated would cause the minimum apical temperature to increase 1.7 °C (Nobel, 1980b). A dense layer of apical pubescence 10 mm thick, which occurred for *C. gigantea*, would cause the minimum apical temperature to increase by 2.4 °C (apical pubescence was lacking for *S. thurberi* and *L. schottii*). Finally, the spine shading of the apex increases from about 15% for *L. schottii* to 70% for *S. thurberi*, which presumably helps extend the range of the latter species to higher latitudes.

A similar computer analysis has been performed for the four species of *Ferocactus* that occur in the southwestern United States (Nobel, 1980b). *F. acanthodes* had the highest minimum apical temperatures under a given set of environmental conditions and hence was predicted to range to the coldest sites, *F. wislizenii* next, then *F. covillei*, and finally *F. viridescens*. Indeed, *F. acanthodes* ranges to

36°52′ N, 115°16′ W, 1590 m in Nevada; *F. wislizenii* and *F. covillei* range to central Arizona; while *F. viridescens* is restricted to southern California fairly near the coast. (In addition to the sites reported in Nobel, 1980d, *F. wislizenii* was observed at 33° 46′ N, 112° 58′ W, 600 m and 31° 48′ W, 1550 m – where frost damage was prevalent – and *F. covillei* at 32° 48′ N, 112° 6′ W, 640 m; apical spine shading and pubescence were within the standard deviations found at the other sites.)

The simulation model indicated that the increase in diameter accompanying stem growth raised the minimum apical temperature more than 3 °C for *C. gigantea*, indicating that the shortest and therefore thinnest stems would be most vulnerable to frost damage (Nobel, 1980c). Indeed, two catastrophic freezes at Saguaro National Monument (1962 and 1971) killed 15% of natural populations of *C. gigantea* under 0.5 m tall and only 2% that were 0.5 to 3.8 m tall (Steenbergh and Lowe, 1976). Nurse plants raised the effective environmental temperature for infrared radiation at the colder sites by 10 °C, which raised the minimum stem surface temperature just over 1 °C for plants under 50 cm tall. This could help extend the range of *C. gigantea* northward by offering some protection against freezing (Nobel, 1980c). In addition to protecting young plants from low temperatures, nurse plants have also been proposed to protect *C. gigantea* from high temperatures (Despain, 1974; Shreve, 1931; Turner *et al.*, 1966), to reduce water loss (Turner *et al.*, 1966), and to help avoid consumption by rodents or lagomorphs (Niering *et al.*, 1963; Steenbergh and Lowe, 1969).

8.4.2 TEMPERATURE EXTREMES

For those cacti that can survive freezing, the temperature tolerance varies considerably. Shreve (1911) suggested that *C. gigantea* was restricted to those regions where freezing temperatures did not last 24 h, a hypothesis that has received considerable support (Niering *et al.*, 1963). However, *Echinocereus polyacanthus* and *Opuntia versicolor* can withstand 66 h continuous freezing and *O. missouriensis* can withstand 375 h (Shreve, 1911, 1914). Death occurs at − 8 °C for *O. ficus-indica*,

-10°C for *O. fusicaulis*, -17°C for *O. castillae*, and -18°C for *O. ellisiana* (Uphof, 1916).

Coryphantha vivipara, which ranges from northern Mexico all the way to southern Canada, can occur in habitats subjected to considerable freezing. When its stems were cooled at a rate similar to that occurring in the field, supercooling to about -6°C occurred followed by an exothermic reaction that presumably reflected the freezing of extracellular water (Nobel, 1981). As cooling resumed, intracellular water apparently left the cells and crystallized on to the extracellular ice that had already formed. Such intracellular dehydration accounted for the shrunken appearance of the protoplasts as freezing progressed, and it was apparently ultimately responsible for the low temperature damage. *Coryphantha vivipara*, *O. Polyacantha*, and *Pediocactus simpsonii*, all of which range to over 3000 m elevation in southern Wyoming, can tolerate tissue temperatures of -20°C, while *O. bigelovii* and *O. ramosissima*, which are restricted to much warmer habitats, could not tolerate -10°C (Nobel, 1982b). Such studies of the comparative cold sensitivity of cacti have supplemented the information from the modelling efforts discussed above. For instance, *Lophocereus schottii* was found to be slightly more sensitive to low temperatures than the two other columnar cacti considered and *Ferocactus covillei* was more sensitive than other barrel cacti (Nobel, 1982d). Thus, both the morphological factors affecting stem temperature and the actual tissue sensitivity to low temperatures must be taken into account in predicting the ranges. Moreover, the tissue tolerance of subzero temperatures has been shown to acclimate to decreasing environmental temperatures for ten species of cacti (Nobel, 1982d), which can have a major influence on northern range boundaries.

Certain desert succulents can withstand very high temperatures. For instance, *Opuntia* spp. can grow at 58°C and survive a tissue temperature of 62°C (MacDougal and Working, 1921), and *O. ficus-indica* can survive 63°C (Konis, 1950). Studies with *O. bigelovii* growing at day/night air temperatures of 30°C/20°C indicated that various activities which depend on membrane integrity were decreased by 50% for a 1 h treatment at 51–60°C;

nocturnal acid accumulation, which depends on stomatal opening and enzymatic reactions as well as membrane properties, was half-inactivated at a lower temperature, 46°C (Didden-Zopfy and Nobel, 1982). However, heat acclimation occurred. As the air temperature was raised 10°C, the high temperature tolerance increased about 3°C for the membrane properties and fully 6°C for the nocturnal acid accumulation. Such heat acclimation, which has also been observed for leafy desert perennials (Downton *et al.*, 1980), is apparently necessary for survival of *O. bigelovii* in the field.

8.5 Conclusions

Emphasis in this chapter has been on certain important physical factors, e.g. soil water, PAR, and air temperature, that affect the physiological ecology of desert succulents. Owing to their opacity and rigidity, orientation with respect to solar irradiation plays an especially crucial role for the photosynthetic tissues of these CAM plants. In this regard, Fig. 8.10 portrays the seasonal changes in PAR available for a vertical surface facing east or south. Although the optimal temperature for nocturnal CO_2 uptake by both cacti and agaves can acclimate to the prevailing environmental temperature, seasonal changes in temperature do affect the net carbon gain. However, the overriding factor for desert succulents is rainfall, which can be quite unpredictable for deserts. Taking into consideration both the PAR and temperature responses of *F. acanthodes*, Fig. 8.10 shows the estimated maximum net CO_2 uptake for the east and south sides of this barrel cactus when the nocturnal stomatal opening is not limited by soil water anytime during the year. For a year with late summer rainfall and winter/early spring rainfall, periods when the average monthly rainfall is highest (Fig. 8.10), the productivity of the south-facing side would be about the same as the east-facing side in the late summer but about twice as great in the winter/early spring. That is, the high PAR for the east-facing surface comes at a time when rainfall tends to be low, while the higher rainfall and higher PAR are better synchronized for a south-facing surface (Fig. 8.10). However, the average monthly rainfall is not a

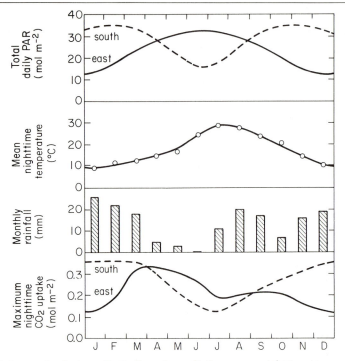

Figure 8.10   Seasonal changes in photosynthetically active radiation, mean nighttime temperature, monthly rainfall, and maximum nighttime CO_2 uptake for a barrel cactus. PAR data are for unshaded vertical surfaces facing east or south on clear days at 34 °N. Temperatures and rainfall are based on 20 yr of weather records appropriate for the desert site indicated in Fig. 8.2. Maximum nighttime CO_2 uptake assumes that water is not limiting as well as the observed temperature and PAR responses for CO_2 uptake of *Ferocactus acanthodes* (Nobel, 1976; Nobel and Hartsock, 1981).

reliable index of productivity due to the episodic nature of the desert storms. In fact, no appreciable net CO_2 uptake may occur during the summer in one year, while in the next year the summer growing season might be longer than the winter one for the Sonoran Desert site chosen here.

The interplay between environment, gas exchange, morphology, and distribution is clearly becoming better understood for desert succulents. For instance, ribs of a cactus have only secondary influences on the thermal relations and their obvious increase of the surface area may not enhance CO_2 uptake, since PAR is generally limiting and the increased area has a lower PAR per unit area (Nobel, 1980a). Ribbing can facilitate stem swelling and the accompanying water storage by providing structural flexibility (Nobel, 1977a). The observation that PAR may be limiting for succulents even in the high radiation environment of a desert has

aided interpretation of the orientation of cladodes of platyopuntias, the intersite height variation of ceroid cacti (Nobel, 1980a, 1982a), and the leaf arrangement for the basal rosette of agaves (Woodhouse *et al.*, 1980). In addition to their traditional role of protecting stems against herbivory, spines also influence the distribution of cacti by moderating diurnal temperature extremes (Nobel, 1978, 1980b). However, spines reduce the PAR incident on the stem surface, which can reduce productivity, since PAR is generally limiting for these CAM plants (Nobel, 1983).

Increased understanding of the biochemistry of CAM metoblism is now available (Kluge and Ting, 1978; Osmond, 1978; Ting and Gibbs, 1982). The role of the central vacuole for storage of organic acids has been demonstrated, but the control mechanisms are still unclear. The ability of succulent plants to utilize daytime stomatal opening and C_3

metabolism with an accompanying greater daily CO_2 uptake either early in life or when water subsequently is plentiful has obvious ecological implications. However, the overriding water conservation attributes of nocturnal stomatal opening are clearly adaptive in the desert, as is the high water use efficiency of CAM plants. This is nowhere in greater evidence than in the seedling stage, where the development of a sufficiently high volume-to-surface ratio facilitates endurance of the inevitable droughts that characterize deserts.

Acknowledgements

Helpful comments on the manuscript were provided by B. Didden-Zopfy, G.N. Geller, V.T. Ishihara, P.W. Jordan, and R.M. Woodhouse; T.L. Hartsock technically assisted with Fig. 8.7. Financial support was provided by Department of Energy contract DE-AM03-76-SF00012.

References

Bacheller, M.A. (ed.) (1980) *The Hammond Almanac*, Hammond Almanac, Maplewood, New Jersey.

Barbour, M.G., Burk, J.H. and Pitts, W.D. (1980) *Terrestrial Plant Ecology*, Benjamin/Cummings, Menlo Park, California.

Bartholomew, B. (1973) Drought response in the gas exchange of *Dudleya farinosa* (Crassulaceae) grown under natural conditions. *Photosynthetica*, **7**, 114–20.

Berry, J. and Björkman, O. (1980) Photosynthetic response and adaptation to temperature in higher plants. *Annual Review of Plant Physiology*, **31**, 491–543.

Brum, G.D. (1973) Ecology of the saguaro (*Carnegiea gigantea*): Phenology and establishment in marginal populations. *Madroño*, **22**, 195–204.

Cannon, W.A. (1911) *The Root Habits of Desert Plants*, Carnegie Institution of Washington, Washington, D.C.

Despain, D.G. (1974) The survival of saguaro (*Carnegiea gigantea*) seedlings on soils of differing albedo and cover. *Journal of the Arizona Academy of Sciences*, **9**, 102–7.

Didden-Zopfy, B. and Nobel, P.S. (1982) High temperature tolerance and heat acclimation of *Opuntia bigelovii*. *Oecologia*, **52**, 176–80.

Dinger, B.E. and Patten, D.T. (1974) Carbon dioxide exchange and transpiration in species of *Echinocereus* (Cactaceae), as related to their distribution within the Pinaleno Mountains, Arizona. *Oecologia*, **14**, 389–411.

Downton, W.J.S., Seemann, J.R. and Berry, J.A. (1980) Thermal stability of photosynthesis in desert plants. *Carnegie Institution of Washington Year Book*, **79**, 143–5.

Eickmeier, W.G. and Adams, M.S. (1978) Gas exchange in *Agave lecheguilla* Torr. (Agavaceae) and its ecological implications. *Southwestern Naturalist*, **23**, 473–86.

Eickmeier, W.G. and Bender, M.M. (1976) Carbon isotope ratios of Crassulacean acid metabolism species in relation to climate and phytosociology. *Oecologia*, **25**, 341–7.

Gates, D.M., Alderfer, R. and Taylor, E. (1968) Leaf temperature of desert plants. *Science*, **159**, 994–5.

Gentry, H.S. and Sauck, J.R. (1978) The stomatal complex in *Agave*: groups Deserticolae, Campaniflorae, and Umbelliflorae. *Proceedings of the California Academy of Sciences, 4th series*, **41**, 371–87.

Gibbs, J.G. and Patten, D.T. (1970) Plant temperature and heat flux in a Sonoran desert ecosystem. *Oecologia*, **5**, 165–84.

Griffiths, D. (1915) *Yields on Native Prickly Pear in Southern Texas*, US Department of Agriculture Bulletin No. 208, pp. 1–11.

Gulmon, S.L. and Bloom, A.J. (1979) C_3 photosynthesis and high temperature acclimation of CAM in *Opuntia basilaris* Engelm. and Bigel. *Oecologia*, **38**, 217–22.

Hadley, N.F. (1970) Micrometeorology and energy exchanged in two desert arthropods. *Ecology*, **51**, 434–44.

Hanscom, Z. and Ting, I.P. (1978) Irrigation magnifies CAM-photosynthesis in *Opuntia basilaris* (Cactaceae). *Oecologia*, **33**, 1–15.

Hartsock, T.L. and Nobel, P.S. (1976) Watering converts a CAM plant to daytime CO_2 uptake. *Nature*, **262**, 574–6.

Hastings, J.R. and Turner, R.M. (1965) *The Changing Mile. An Ecological Study of Vegetation Change with Time in the Lower Mile of an Arid and Semiarid Region*, The University of Arizona Press, Tucson.

Jordan, P.W. and Nobel, P.S. (1979) Infrequent establishment of seedlings of *Agave deserti* (Agavaceae) in the northwestern Sonoran Desert. *American Journal of Botany*, **66**, 1079–84.

Jordan, P.W. and Nobel, P.S. (1981) Seedling establishment of *Ferocactus acanthodes* in relation to drought. *Ecology*, **62**, 901–6.

Jordan, P.W. and Nobel, P.S. (1982) Height distribution of two species of cacti in relation to rainfall, seedling establishment, and growth. *Botanical Gazette*, **143**, 511–7.

Kausch, W. (1965) Beziehungen zwischen Wurzelwachstum, Transpiration und CO_2-Gaswechsel bei einigen Kakteen. *Planta (Berlin)*, **66**, 229–38.

Kluge, M. and Ting, I.P. (1978) *Crassulacean Acid Metabolism. Analysis of an Ecological Adaptation*, Ecological Studies Series, Vol. 30, Springer-Verlag, Berlin.

Konis, E. (1950) On the temperature of *Opuntia* joints. *Palestine Journal of Botany, Jerusalem Series*, **5**, 46–55.

Le Houerou, H.N. (1970) North Africa: past, present, future, in *Arid Lands in Transition* (ed. H.E. Dregne), publication No. 90, American Association for the Advancement of Science, Washington, D.C., pp. 227–78.

Lewis, D.A. and Nobel, P.S. (1977) Thermal energy exchange model and water loss of a barrel cactus, *Ferocactus acanthodes*. *Plant Physiology*, **60**, 609–16.

MacDougal, D.T. and Spalding, E.S. (1910) *The water-balance of succulent plants*, Carnegie Institution of Washington, Washington, D.C.

MacDougal, D.T. and Working, E.B. (1921) A new high-temperature record for growth. *Carnegie Institution of Washington Year Book*, **20**, 47–8.

McGinnies, W.G., Goldman, B.J. and Paylore, P. (eds) (1968) *Deserts of the World. An Appraisal of Research into their Physical and Biological Environments*, University of Arizona Press, Tucson.

Metral, J.J. (1965) Les cactus fourragers dans le Nord-Est du Brazil plus particulièrement dans l'état du Ceara. *Agronomia Tropical (Maracay)*, **20**, 248–61.

Mooney, H.A. and West, M. (1964) Photosynthetic acclimation of plants of diverse origin. *American Journal of Botany*, **51**, 825–7.

Mooney, H., Troughton, J.H. and Berry, J.A. (1974) Arid climates and photosynthetic systems. *Carnegie Institution of Washington Year Book*, **73**, 793–805.

Mozingo, H.N. and Comanor, P.L. (1975) Implications of the thermal response of *Ferocactus acanthodes*. *Cactus and Succulent Journal (US)*, Supplement, **47**, 22–8.

Neales, T.F. (1975) The gas exchange patterns of CAM plants, in *Environmental and Biological Control of Photosynthesis* (ed. R. Marcelle), W. Junk, The Hague, pp. 299–310.

Neales, T.F., Patterson, A.A. and Hartney, V.J. (1968) Physiological adaptation to drought in the carbon assimilation and water loss of xerophytes. *Nature*, **219**, 469–72.

Niering, W.A., Whittaker, R.H. and Lowe, C.H. (1963) The saguaro: a population in relation to environment. *Science*, **142**, 15–23.

Nisbet, R.A. and Patten, D.T. (1974) Seasonal temperature acclimation of a prickly-pear cactus in southcentral Arizona. *Oecologia*, **15**, 345–52.

Nobel, P.S. (1976) Water relations and photosynthesis of a desert CAM plant, *Agave deserti*. *Plant Physiology*, **58**, 576–82.

Nobel, P.S. (1977a) Water relations and photosynthesis of a barrel cactus, *Ferocactus acanthodes*, in the Colorado Desert. *Oecologia*, **27**, 117–33.

Nobel, P.S. (1977b) Water relations of flowering of *Agave deserti*. *Botanical Gazette*, **138**, 1–6.

Nobel, P.S. (1978) Surface temperatures of cacti – influences of environmental and morphological factors. *Ecology*, **59**, 986–96.

Nobel, P.S. (1980a) Interception of photosynthetically active radiation by cacti of different morphology. *Oecologia*, **45**, 160–6.

Nobel, P.S. (1980b) Morphology, surface temperatures, and northern limits of columnar cacti in the Sonoran Desert. *Ecology*, **61**, 1–7.

Nobel, P.S. (1980c) Morphology, nurse plants, and minimum apical temperatures for young *Carnegiea gigantea*. *Botanical Gazette*, **141**, 188–91.

Nobel, P.S. (1980d) Influences of minimum stem temperatures on ranges of cacti in southwestern United States and central Chile. *Oecologia*, **47**, 10–15.

Nobel, P.S. (1981) Influence of freezing temperatures on a cactus, *Coryphantha vivipara*. *Oecologia*, **48**, 194–8.

Nobel, P.S. (1982a) Interaction between morphology, PAR interception, and nocturnal acid accumulation in cacti, in *Crassulacean Acid Metabolism* (eds I.P. Ting and M. Gibbs), American Society of Plant Physiologists, Rockville, Maryland, pp. 260–77.

Nobel, P.S. (1982b) Orientation of terminal cladodes of platyopuntias. *Botanical Gazette*, **143**, 219–24.

Nobel, P.S. (1982c) Orientation, PAR interception, and nocturnal acidity increases for terminal cladodes of a widely cultivated cactus, *Opuntia ficus-indica*. *American Journal of Botany*, **69**, 1462–9.

Nobel, P.S. (1982d) Low-temperature tolerance and cold hardening of cacti. *Ecology*, **63**, 1650–6.

Nobel, P.S. (1983) Spine influences on PAR interception, stem temperature, and nocturnal acid accumulation by cacti. *Plant, Cell and Environment*, **6**, 153–9.

Nobel, P.S. and Hartsock, T.L. (1978) Resistance analysis of nocturnal carbon dioxide uptake by a Crassulacean acid metabolism succulent, *Agave deserti*. *Plant Physiology*, **61**, 510–14.

Nobel, P.S. and Hartsock, T.L. (1979) Environmental

influences on open stomates of a Crassulacean acid metabolism plant, *Agave deserti. Plant Physiology,* **63**, 63–6.

Nobel, P.S. and Hartsock, T.L. (1981) Shifts in the optimal temperature for nocturnal CO_2 uptake caused by changes in growth temperature for cacti and agaves. *Physiologia Plantarum,* **53**, 523–7.

Osmond, C.B. (1978) Crassulacean acid metabolism: a curiosity in context. *Annual Review of Plant Physiology,* **29**, 379–414.

Patten, D.T. and Dinger, B.E. (1969) Carbon dioxide exchange patterns of cacti from different environments. *Ecology,* **50**, 686–8.

Patten, D.T. and Smith, E.M. (1975) Heat flux and the thermal regime of desert plants, in *Environmental Physiology of Desert Organisms* (ed. N.F. Hadley), Dowden, Hutchinson, and Ross, Stroudsburg, Pennsylvania, pp. 1–19.

Rodriquez, S.B., Perez, F.B. and Montenegro, D.D. (1976) Eficiencia fotosintetica del nopal (*Opuntia* spp.) in relación con la orientación de sus cladodios. *Agrociencia,* **24**, 67–77.

Shreve, F. (1911) The influence of low temperature on the distribution of the giant cactus. *Plant World,* **14**, 136–46.

Shreve, F. (1914) The role of winter temperatures in determining the distribution of plants. *American Journal of Botany,* **1**, 194–202.

Shreve, F. (1931) Physical conditions in sun and shade. *Ecology,* **12**, 96–104.

Smith, W.K. (1978) Temperatures of desert plants: another perspective on the adaptability of leaf size. *Science,* **201**, 614–16.

Soule, O.H. and Lowe, C.H. (1970) Osmotic characteristics of tissue fluids in the sahuaro giant cactus (*Cereus giganteus*). *Annals of the Missouri Botanical Garden,* **57**, 265–351.

Steenbergh, W.F. and Lowe, C.H. (1969) Critical factors during the first year of life of the saguaro (*Cereus giganteus*) at Saguaro National Monument, Arizona. *Ecology,* **50**, 825–34.

Steenbergh, W.F. and Lowe, C.H. (1976) Ecology of the saguaro: I. The role of freezing weather in a warm-desert plant population, in *Research in the Parks, National Park Service Symposium Series, No. 1,* US Government Printing Office, Washington, D.C., pp. 49–92.

Steenbergh, W.F. and Lowe, C.H. (1977) *Ecology of the Saguaro: II. Reproduction, germination, establishment, growth, and survival of the young plant. National Park Service Scientific Monograph Series, No. 8,* US Government Printing Office, Washington, D.C.

Szarek, S.R. and Ting, I.P. (1975a) Physiological responses to rainfall in *Opuntia basilaris* (Cactaceae). *American Journal of Botany,* **62**, 602–9.

Szarek, S.R. and Ting, I.P. (1975b) Photosynthetic efficiency of CAM plants in relation to C_3 and C_4 plants, in *Environmental and Biological Control of Photosynthesis* (ed. R. Marcelle), W. Junk, The Hague, pp. 289–97.

Szarek, S.R. and Troughton, J.H. (1976) Carbon isotope ratios in Crassulacean acid metabolism plants. Seasonal patterns from plants in natural stands. *Plant Physiology,* **58**, 367–70.

Szarek, S.R., Johnson, H.B. and Ting, I.P. (1973) Drought adaptation in *Opuntia basilaris. Plant Physiology,* **52**, 539–41.

Ting, I.P. and Gibbs, M. (eds) (1982) *Crassulacean Acid Metabolism,* American Society of Plant Physiologists, Rockville, Maryland.

Ting, I.P. and Szarek, S.R. (1975) Drought adaptation in Crassulacean acid metabolism plants, in *Environmental Physiology of Desert Organisms* (ed. N.F. Hadley), Dowden, Hutchinson, and Ross, Stroudsburg, Pennsylvania, pp. 152–67.

Turnage, W.V. and Hinckley, A.L. (1938) Freezing weather in relation to plant distribution in the Sonoran Desert. *Ecological Monographs,* **8**, 529–50.

Turner, R.M., Alcorn, S.M., Olin, G. and Booth, J.A. (1966) The influence of shade, soil, and water on saguaro seedling establishment. *Botanical Gazette,* **127**, 95–102.

Turner, R.M., Alcorn, S.M. and Olin, G. (1969) Mortality of transplanted saguaro seedlings. *Ecology,* **50**, 835–44.

Uphof, J.C. Th. (1916) *Cold-resistance in spineless cacti,* University of Arizona Agricultural Experiment Station Bulletin No. 79, pp. 119–44.

US Weather Bureau (1964) *Climatic Summary of the United States – Supplement for 1951 through 1960,* US Government Printing Office, Washington, D.C.

Walter, H. and Stadelmann, E. (1974) A new approach to the water relations of desert plants, in *Desert Biology,* Vol. II (ed. G.W. Brown, Jr), Academic Press, New York, pp. 213–310.

Woodhouse, R.M., Williams, J.G. and Nobel, P.S. (1980) Leaf orientation, radiation interception, and nocturnal acidity increases by the CAM plant *Agave deserti* (Agavaceae). *American Journal of Botany,* **67**, 1179–85.

9

Cold desert

M. CALDWELL

9.1 Introduction

Uplifting of the Sierra Nevada and the Cascades in the late Pliocene provided the massive rain shadow that causes the Great Basin desert climate. The extensive shrub-dominated cold desert of North America, centred in the Great Basin, is the focus of this chapter. In a broad sense, the cold desert occupies much of the Intermountain West between the Sierra Nevada and Cascades to the west and the Rocky Mountains to the east. The northern extent of the Intermountain Region occurs somewhat north of the US – Canadian border where increasing moisture permits the transition to forested vegetation. The shrub component of the vegetation is much reduced on the Columbian Plateau and, therefore, this northern part is not always included as part of the Great Basin Desert (MacMahon, 1979). With decreasing elevation to the south, the transition to the warm deserts occurs gradually in southern Nevada and Utah (West, 1983).

Average precipitation and temperature of the Great Plains shortgrass prairie, east of the Rocky Mountains, is similar to the Great Basin. However, much of the annual precipitation of the western Great Plains comes from convectional storms originating in the Gulf of Mexico during the summer season, while in the Great Basin, summer precipitation is much less important. Dominance of the shrub life form in the Great Basin is thought to be largely due to the paucity of summer precipitation (West, 1983).

Compared to warm deserts of North America, the cold desert is a rather monotonous expanse dominated by a few shrub species, principally of the Asteraceae and Chenopodiaceae. In marked contrast to the warm deserts, succulents and native annuals are insignificant components of the Great Basin vegetation. A few perennial grasses (principally *Agropyron, Oryzopsis, Poa* and *Sitanion*) provide the only other life form of significance in the cold desert (MacMahon, 1979).

Two major vegetation zones of the Great Basin are generally recognized following the original descriptions of Billings (1949). The major zone is dominated by sagebrush, *Artemisia tridentata* (Fig. 9.1). Drier and warmer areas, principally west central Nevada and Utah and the southern areas of these two states, are dominated by shadscale, *Atriplex confertifolia*. Drainage of much of the Great Basin, especially in Nevada and Utah, is internal and, thus, extensive areas of halomorphic soils occur in both vegetation zones. Where these saline soils occur, *Atriplex confertifolia* and other halophytes dominate. Soil salinity in these areas contributes significantly to site aridity. As much as -3.5 MPa of the total soil water potential can be due to the osmotic component (Moore and Caldwell, 1972).

The Great Basin climate is distinctly continental with warm summers and prolonged, cold winters. Simplified diagrams in Fig. 9.2 depict climate at three sites in the Great Basin with a comparison to locations in each of the three major warm deserts. About 60% of the annual precipitation occurs

Figure 9.1 View of a sagebrush community on the west slope of the Desatyoa range, Churchill County, Nevada.

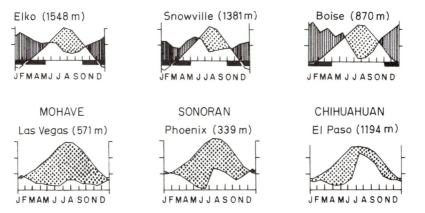

Figure 9.2 Simplified climate diagrams for three sites in the Great Basin (top) and a representative site from each of the major warm deserts of North America (bottom), following the format of Walter and Lieth (1960). The abscissa represents the twelve months of the year beginning in January and the blackened bar along the abscissa indicates months of year when the mean daily minimum is below 0 °C. One division on the ordinate represents either 10 °C or 20 mm precipitation. The site elevation is indicated in parentheses for each location.

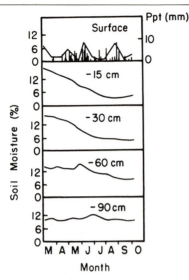

Figure 9.3 Soil moisture content at several depths in the soil profile beneath a mixed community of *Atriplex confertifolia* and *Ceratoides lanata*. Vertical bars in the top frame represent individual precipitation events. From Caldwell *et al.* (1977b).

during the period from October to April and much of this is in the form of snow. Melting snow and spring rains result in the primary annual recharge of the soil profile. Spring and summer precipitation is generally effective in only wetting the surface layers of the profile except in years of unusually high summer rainfall (Fig. 9.3; Moore and Caldwell, 1972).

Although it is one of the most extensive vegetation provinces in the continental US, the cold desert has received comparatively little study. Land has been used primarily for grazing in the sagebrush zone during spring and fall and in the shadscale zone during the winter. More attention is now focusing on the cold desert as the combined pressures of water and energy development, mining, recreation, grazing and, most recently, military missile basing systems, compete for the same ecosystem resources.

This chapter summarizes the limited information on the physiological ecology of Great Basin species with an emphasis on the shrubs that dominate this landscape. The implications of physiological ecology for community structure and function will be emphasized.

9.2 Photosynthesis in the cold desert environment

The primary, and usually only, recharge of the soil profile occurs in the spring of each year in the cold desert. Effective competition for this moisture resource for most species is coupled with active photosynthesis and growth during this period (Caldwell *et al.*, 1977b; DePuit and Caldwell, 1973, 1975). Photosynthetic activity on representative days during the year is shown for *Atriplex confertifolia* and the co-occuring *Ceratoides lanata* (Fig. 9.4). Other cool desert species, including *Artemisia tridentata* and perennial *Agropyron* bunchgrasses, also exhibit most of their photosynthetic activity in the spring and early summer (DePuit and Caldwell, 1973, 1975; Caldwell *et al.*, 1981). Despite the prevailing cool spring temperatures (Fig. 9.2) species possessing C_4 photosynthesis also are active in growth and photosynthetic activity during this period (West and Wein, 1971; Caldwell *et al.*, 1977b).

Atriplex confertifolia possesses all of the physiological characteristics of C_4 photosynthesis (of the NAD malic enzyme type) including Kranz anatomy, initial carbon products, and low carbon dioxide compensation points, even when grown at low temperatures (Caldwell *et al.*, 1977a; Welkie and Caldwell, 1970). Under natural conditions *Atriplex* exhibits temperature optima for photosynthesis around 16 °C in the spring which corresponds with mean daily maximum temperatures that range between 14 and 21 °C at this time of year (Caldwell *et al.*, 1977a; White, 1976). The remarkably similar timing of photosynthetic activity during the year for C_4 and C_3 species is quite in contrast to ephemeral and perennial herbs of the Sonoran Desert and grasses of the Great Plains. In these latter communities, C_3 species are active during the cooler seasons and C_4 species delay growth and photosynthetic activity until warm seasons of the year (Boutton *et al.*, 1980; Johnson, 1975; Tieszen, 1970). Presumably, *Atriplex* has evolved the capacity for photosynthesis at low temperatures in order to compete with cool season species for the moisture resource available in the spring.

Atriplex confertifolia has been a successful element of the Great Basin flora in warmer and drier periods

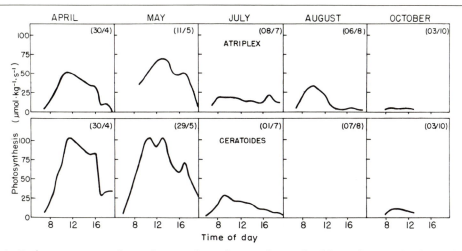

Figure 9.4 Daily progression of net photosynthesis for *Atriplex confertifolia* and *Ceratoides lanata* during several nearly cloudless days for a 6 month period. Exact dates are given in parentheses for the year 1970 in each frame. From Caldwell *et al.* (1977b).

of the Pleistocene and also in cooler pluvial conditions. Stable carbon isotope ratios of subfossil deposits of this species from pack rat middens in a Nevada cave indicate this species to have employed C_4 photosynthesis some 10 000 years ago in the Pleistocene and also some 40 000 years previous in pluvial times (Troughton *et al.*, 1974).

The evolutionary accommodation of photosynthetic potential at low temperatures by *Atriplex confertifolia* appears to be at the sacrifice of the high photosynthetic rates characteristic of C_4 species. Maximum rates measured on this species in the field or under optimum controlled conditions in the laboratory have not exceeded $14\,\mu\text{mol}\,\text{m}^{-2}\,\text{s}^{-1}$. Maximum rates of the few C_3 shrubs that have been studied in the Great Basin are somewhat greater, but still not in excess of $19\,\mu\text{mol}\,\text{m}^{-2}\,\text{s}^{-1}$ for *Ceratoides lanata* and *Artemisia tridentata* (Caldwell *et al.*, 1977b; Depuit and Caldwell, 1975). Rates of C_3 grasses in the Great Basin can be somewhat higher (Caldwell *et al.*, 1981). Although photosynthetic rates are not impressive in Great Basin species, photosynthesis/transpiration (*P/T*) ratios are high when compared with nonagricultural species from other arid habitats. In Fig. 9.5, *P/T* ratios for *Atriplex* and *Ceratoides* are shown during the course of the year for plants under field conditions. Each datum point represents the average *P/T* ratio for a day, or a

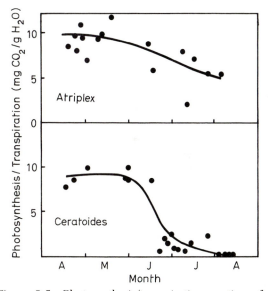

Figure 9.5 Photosynthesis/transpiration ratios for *Atriplex confertifolia* and *Ceratoides lanata*. Each point represents the average daily value for this ratio. From Caldwell *et al.* (1977b).

major portion thereof. In the spring, *P/T* ratios for both species were greater than daily ratios reported for several aridland species in Western Australia (Hellmuth, 1971) and in the Negev Desert of Israel, with the exception of *Artemisia herba-alba*, which displayed daily ratios equal to those of these two cool desert species (Lange *et al.*, 1969). High P/T ratios of the Great Basin species are attributable in large part to the cool environmental temperatures prevailing during the spring period of the year when transpiration, but not photosynthesis, is limited. Later in the season P/T ratios of both species declined. This decline was particularly abrupt for *Ceratoides*. Nevertheless, the greater ratios exhibited by *Atriplex* in July and August are of marginal benefit since most of the carbon gain of both species takes place before the hot summer.

Because C_4 species can still effectively conduct photosynthesis at low intercellular carbon dioxide concentrations, stomates can be closed partially, limiting water loss without limiting photosynthesis. Thus, one should expect greater stomatal diffusion resistances to generally occur in C_4 species (Björkman, 1975). The average diffusion resistance during the middle 7 h of the day when plants are most active in photosynthesis is plotted for several days throughout the season for *Atriplex confertifolia* and *Ceratoides lanata* (Fig. 9.6). Leaf diffusion resistance increased during the growing season as water became limiting. However, there was no difference in diffusion resistance between the two species. Therefore, it is apparent that the C_4 species was not

exhibiting this predicted property under field conditions.

Cold desert species generally have not been found to exhibit striking temperature acclimation during the season of photosynthetic activity. Net photosynthesis of *Artemisia tridentata* is shown as a function of leaf temperature at constant irradiance ($1150 \, \mu\text{mol m}^{-2}\text{s}^{-1}$) in Fig. 9.7. These experiments were conducted with plants in their natural soil environment. Although photosynthetic capacity declined markedly during the season due to combined effects of leaf age and decreasing soil moisture, the temperature optimum for photosynthesis changed by only 6–7 °C. This was also found to be the case for *Ceratoides lanata*, *Agropyron spicatum*, and *Xanthocephalum sarothrae*; *Atriplex confertifolia* did exhibit slightly more acclimation as expressed in the photosynthetic temperature optimum (DePuit

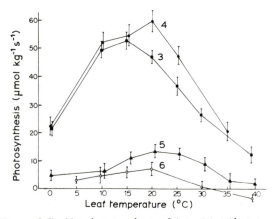

Figure 9.7　Net photosynthesis of *Artemisia tridentata* as a function of leaf temperature at four different phenological stages. Stage 3 represents the time of emergent large new leaves on vegetative branches (usually mid-May). Phenological Stage 4 represents the time of rapid new vegetative shoot and leaf growth and the initiation of reproductive shoots (late May to mid-June). Stage 5 (early July to mid-August) represents the period of reduced vegetative growth, a period of reproductive shoot and bud growth and growth of ephemeral leaves on reproductive shoots while leaves formed in phenological Stage 3 are being shed. Stage 6 (late August to mid-September) represents the period of final growth of reproductive shoots, flower bud development and a period of little vegetative growth. Adapted from DePuit and Caldwell (1973).

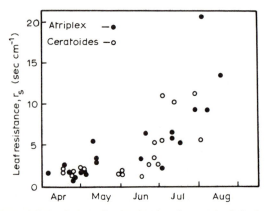

Figure 9.6　Seasonal progression of average daily leaf diffusive resistance for *Atriplex confertifolia* and *Ceratoides lanata*. Each point represents the average resistance for the middle 7 h portion of the day when the plants were most active in photosynthesis. From Caldwell *et al.* (1977b).

and Caldwell, 1975; White, 1976). West and Mooney (1972) have reported somewhat greater acclimation in high elevation species of *Artemisia*. At lower elevations a substantial degree of photosynthetic acclimation to temperature might be of limited value, since most of the carbon gain for these Great Basin species takes place in a comparatively short period of time during the spring and early summer.

In late summer when soil moisture is limited, average leaf age is well advanced and leaf temperatures are high, negative net photosynthetic rates can occur as indicated in Fig. 9.7. This phenomenon also has been observed occasionally for *Atriplex confertifolia* and *Ceratoides lanata* although not as frequently as for *Artemisia* (Caldwell *et al.*, 1977b; DePuit and Caldwell, 1975). In experiments using irrigation to reduce the late season water stress of *Artemisia*, negative net photosynthetic rates at high leaf temperatures were relieved (DePuit and Caldwell, 1975).

The significance of the evergreen habit of shrubs in the cold desert is not altogether clear since seasonal photosynthetic activity is largely limited to spring and early summer. Several prominent shrubs of the Great Basin possess leaves throughout the year. For example, overwintering leaves on *Artemisia tridentata* and *Atriplex confertifolia* are formed in mid to late summer and persist until spring the following year, when they are gradually replaced by summer leaves. Apart from the spring period of rapid photosynthesis, periods of limited carbon dioxide uptake have been measured under field conditions as late as October, yet rates are generally so limited that this does not constitute a significant portion of the annual carbon gain. When costs of leaf maintenance are considered, the value of the evergreen habit becomes questionable. During cold winter periods, low leaf temperatures and frozen soils usually preclude significant photosynthetic activity. The value of overwintering foliage may lie, then, chiefly in the availability of foliage for photosynthesis during favorable early spring periods which occur unpredictably from year to year in the Great Basin. ^{14}C-labelling of overwintering foliage indicates these leaves to be quite photosynthetically active when environmental conditions are favorable early

in the spring (Caldwell and Camp, unpublished data).

9.3 The moisture constraint

Although most of the annual carbon gain and transpiration of Great Basin shrubs occurs in the spring and early summer (Fig. 9.8), these shrubs are able to extract moisture from soils of very low water potential during late summer. *Atriplex confertifolia* and *Ceratoides lanata* transpired and extracted water from soils with water potentials of less than − 7 MPa and when plant xylem pressure potentials were − 11.5 MPa (Moore *et al.*, 1972b). Campbell and Harris (1977) reported that *Artemisia tridentata* could continue to draw some moisture from a soil profile with water potentials of − 6 to − 7 MPa. Capacity to conduct photosynthesis at plant water potentials more negative than − 5 MPa has also been verified in several Great Basin shrubs (Detling and Klikoff, 1973; White, 1976).

Root growth is initiated in upper layers of the soil profile in early spring before visible evidence of vegetative growth and persists as late as November, some two to three months after completion of fruit development (Fig. 9.9). Progression of root growth to increasing depths in the soil profile during the season is apparent in the three shrubs that have been studied. Root extension in soils with water potentials of − 6 to − 7 MPa occurred for all three species, but was most evident for *Atriplex confertifolia* (Fernandez and Caldwell, 1975).

The capacity to utilize soil moisture when soil water potentials are low, which usually occurs at the time of year when P/T ratios are also low, may seem to be of marginal value for annual carbon gain. For a species such as *Atriplex* there is sufficient photosynthesis during periods of limited soil moisture in the late summer and fall to account for 18% of the annual carbon gain, but for *Ceratoides* this late season activity accounts for only 4% of the yearly total. These estimates are based on simulations for years of above- and below-average precipitation (Fig. 9.8). Effective depletion of soil moisture by some species may be of greater consequence for excluding competitors rather than for the small contribution to annual carbon gain.

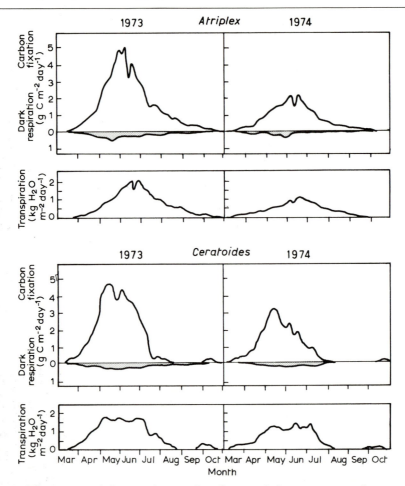

Figure 9.8   Seasonal progression of photosynthetic carbon fixation, dark respiration, and transpiration per square meter ground area of monospecific communities of *Atriplex confertifolia* and *Ceratoides lanata* for years of above-average (1973) and below-average (1974) precipitation. These estimates were derived from plant gas exchange measurements and assessments of community foliage during the season. Adapted from Caldwell *et al.* (1977b).

At the community level, the quantity of water that moves through the transpiration stream of vegetation appears to be limited to about half of the annual precipitation. Estimates of annual community transpiration as well as independent estimates of water extraction by root systems in the shrub communities over a period of several years are shown in Fig. 9.10. Since there is essentially no overland flow or subsurface drainage in these communities, evapotranspiration should be equivalent to annual precipitation with an allowance for a storage term. Thus, the half of the annual precipita-

tion which does not pass through the transpiration stream of the vegetation presumably is lost by evaporation from the soil surface or sublimation from the snowpack. While loss of half of the annual precipitation directly to evaporation substantially reduces moisture available to cold desert vegetation, the evaporation component in warm deserts is likely to be larger still. Estimates from a lysimeter study using a single shrub of *Larrea tridentata* indicated that only 7% of the annual precipitation moves through the transpiration stream (Sammis and Gay, 1979).

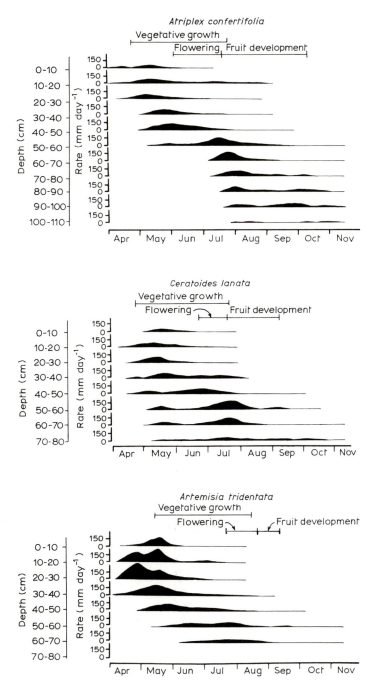

Figure 9.9 Rates of daily root elongation for three cold desert shrubs during the growing season in 10 cm depth intervals in the soil. Rates are expressed as average millimeters of root growth per day for the visible root system in an observation window of root observation chambers in the field. The periods of active shoot elongation, denoted as vegetative growth, and the principal periods of flowering and fruit development are indicated for each species for the year of measurement (1973). From Fernandez and Caldwell (1975).

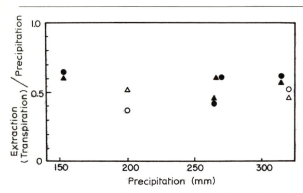

Figure 9.10 Annual water extraction from the soil profile in monospecific communities of *Atriplex confertifolia* (represented by solid triangles) and *Ceratoides lanata* (represented by dots) calculated from water content data of 4 yr and represented as the fraction of annual precipitation for the water year (October to September). Annual transpirational water loss as derived from Fig. 9.7 is also represented as a fraction of annual precipitation during the year. This is denoted by open symbols. From Caldwell *et al.* (1977b).

9.4 The salinity constraint

Extensive areas of halomorphic soils occur in the cold desert. In much of the northern Great Basin, distribution of sagebrush and halophytic shrubs is determined by the degree of soil salinity. *Artemisia tridentata* is a moderately salt-tolerant xerophyte that can exclude salt in the process of water uptake by the root system (Breckle, 1975). However, sagebrush cannot compete with the more salt-tolerant species of *Atriplex* and other halophytes in saline areas. A schematic representation of the distribution of prominent Great Basin shrubs in relation to habitat aridity and salinity is given in Fig. 9.11. Greasewood, *Sarcobatus vermiculatus*, and the pickle weeds such as *Allenrolfea occidentalis*, though extremely salt-tolerant, are limited to moist habitats or areas with accessible ground water. Species such as *Ceratoides lanata* and species of *Atriplex* can tolerate the combined stresses of low soil moisture and salinity.

The manner in which these xerophytic halophytes deal with salt ions differs markedly. For example, *Ceratoides lanata* behaves in some respects as a nonhalophyte and excludes salt ions at the site of water uptake in the root system while the *Atriplex*

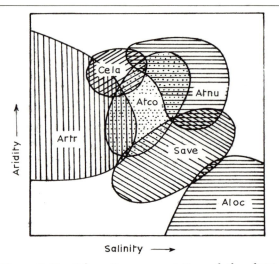

Figure 9.11 Schematic representation of the distribution of cold desert halophytes along gradients of aridity as defined by soil moisture and salinity in Northern Utah. Artr, *Artemisia tridentata*; Cela, *Ceratoides lanata*; Atco, *Atriplex confertifolia*; Atnu, *Atriplex nutallii*; Save, *Sarcobatus vermiculatus*; Aloc, *Allenrolfea occidentalis*. From Breckle (1975).

species allow salt to pass through the plant and then excrete these ions into bladder hairs on the leaves (Breckle, 1975; Moore *et al.*, 1972a). As has been demonstrated in other members of the genus (e.g. Mozafar and Goodin, 1970), the bladder hairs of *Atriplex confertifolia* actively concentrate salt to levels as great as 18% by dry weight (corresponding to -60 MPa osmotic potential, Breckle, 1975). These bladder hairs eventually rupture and salt is then effectively excreted. Breckle calculated that on a community basis, some 0.4 ton salt ha^{-1} annually pass through the small leaf bladder hairs of shadscale. *Atriplex nutallii* (part of the *A. falcata* species complex) also possesses bladder hairs and excretes salt in a similar manner; however, it selectively concentrates potassium to much higher concentrations than sodium (Breckle, 1975).

High concentrations of boron are frequently found in saline soils. The cold desert halophytes described above have been found to tolerate boron concentrations much greater than are normally lethal for most agricultural species (Breckle, 1975; Goodman, 1973). None of the cold desert halophytes appears to exclude boron at the roots, but instead these species seem to tolerate high concentrations in leaf tissues. In sand culture experiments, these three species accumulated boron to levels as high as 1000 ppm by dry weight in the leaves (Breckle, 1975). Growth of *Atriplex* species, especially *A. nutallii*, is less inhibited than growth of *Ceratoides* at high concentrations. The degree to which different levels of soil boron affect growth of these species and alters the competitive balance in different habitats remains to be explored.

9.5 Carbon balance of cold desert shrubs

Moisture deficits, low temperatures, and in some sites, salinity, limit production in the cold desert. Furthermore, periods of carbon gain during the year are restricted primarily to spring and early summer (Fig. 9.8) while energy-consuming processess such as root growth (Fig. 9.9) and respiratory activity occur over a much longer period of the year. The extensive root systems of cold desert species further inflate these carbon costs. The carbon balance of cold desert shrub indicates that a very small dividend of shoot production is realized from annual photosynthetic carbon gain of these species.

Barbour (1973) questioned the long-held assumption that root/shoot ratios of desert species are necessarily large. He summarized data showing root/shoot ratios of many warm desert species to be usually not greater than one. Yet, in cold deserts both of North America and Eurasia, root/shoot ratios in excess of 4 are common (Caldwell *et al.*, 1977b; Pearson, 1965; Rodin and Bazilevich, 1967). The relative importance of the belowground system for cold desert shrubs becomes apparent in a comparison of plant standing carbon pools for an *Atriplex confertifolia* community with those of a mesic hardwood forest ecosystem in Tennessee (Fig. 9.12). The *Atriplex*-dominated community possesses only 2% of the aboveground plant carbon pool of that of the hardwood forest and yet maintains about the same belowground plant carbon pool. Since the estimated annual carbon fixation by this *Atriplex* community is only about 7% of that of the hardwood forest, the commitment to the belowground system is disproportionately large.

The annual carbon fluxes in communities dominated by *Atriplex confertifolia* and *Ceratoides lanata* illustrate the magnitude of carbon invested in root growth of these species (Fig. 9.13). In these

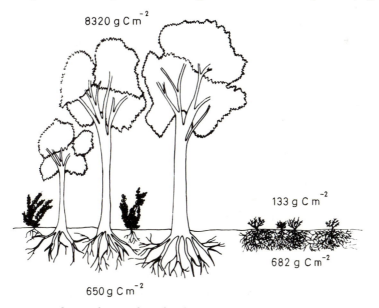

8320 g C m^{-2}

133 g C m^{-2}

650 g C m^{-2}

682 g C m^{-2}

Figure 9.12 Comparative standing carbon pools in the above- and belowground biomass of a mesic hardwood forest ecosystem in Tennessee and a monospecific community of *Atriplex confertifolia*. Adapted from Caldwell and Fernandez (1975).

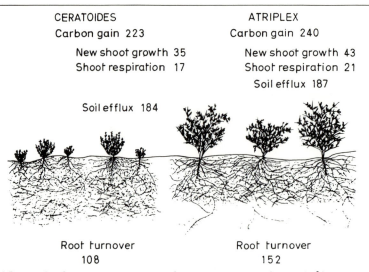

CERATOIDES
Carbon gain 223

New shoot growth 35
Shoot respiration 17

Soil efflux 184

ATRIPLEX
Carbon gain 240

New shoot growth 43
Shoot respiration 21

Soil efflux 187

Root turnover
108

Root turnover
152

Figure 9.13 Annual fluxes of carbon per square meter of community over a 2 yr period incorporating a year of above-average (1973) and below-average (1974) precipitation for monospecific communities of *Atriplex confertifolia* and *Ceratoides lanata*. All carbon fluxes are in units of g carbon m^{-2} community yr^{-1}. Adapted from Caldwell *et al.* (1977b).

established communities, root growth represents root replacement rather than exploration of new regions of the soil profile. This turnover of the root system represents a carbon investment some 3–3.5 times greater than the carbon investment in shoot production. Costs of root system respiration could not be estimated directly in this study, but the magnitude of these costs is also likely very sizeable. Since most of the carbon is utilized by the belowground system, only 17% of annual carbon fixation is invested in shoot production.

Since root maintenance respiration represents a potentially large carbon cost to cold desert species, metabolic adjustments which would result in reduced respiration during part of the year could be of considerable importance to the carbon economy of plants in this environment. Seasonal changes in respiration capacity of individual roots were observed for *Atriplex confertifolia* (Holthausen and Caldwell, 1980). These individual roots were excavated in the field and incubated under standard conditions to determine respiration rates. The roots were not actively growing and thus the values are thought to primarily reflect maintenance respiration capacity. This capacity decreased during the late summer and autumn. During this period the respiration rates determined were as low as any root

respiration rates previously reported and closely coincided with some of the lowest maintenance respiration rates of plant tissues. There is additional evidence that respiratory capacity (as measured at 12 °C) undergoes seasonal shifts at different times for different depths in the soil. Furthermore, the peak respiration capacity at different depths in the profile coincides approximately with periods of maximum root growth (Fig. 9.14). This pulsing of respiratory activity at different depths in the profile along with the general curtailment of respiratory activity to extremely low rates during the late summer and autumn should effect a considerable saving of energy for this large root system. A calculation of this energetic saving is represented in Fig. 9.15. Similar adjustments in root respiration capacity appear to occur in perennial *Agropyron* species of the Great Basin (Richards, unpublished data).

9.6 Nitrogen

As in most desert systems, soil nitrogen concentrations in the Great Basin are low (West and Klemmedson, 1978) and Great Basin species respond to nitrogen fertilization even in years of comparatively low precipitation (James and Jurinak, 1978). Yet, even without fertilization, foliage nitrogen

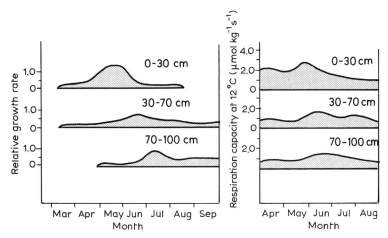

Figure 9.14 Seasonal progression of root growth (adapted from Fig. 9.9) and respiration capacity of individual root elements (adapted from Holthausen and Caldwell, 1980) at three depths in the soil profile.

concentrations can be reasonably high for Great Basin species. For example, *Ceratoides lanata* has nitrogen concentrations on the order of 2–2.5% during the spring period of the year (Cook, 1971). This species has also been found to exhibit very high

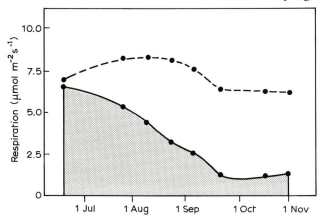

Figure 9.15 Potential root system respiration of *Atriplex confertifolia* expressed on a unit ground area basis. The shaded curve represents values derived by accounting for the seasonal metabolic adjustments of root respiration capacity and temperature dependency of root system respiration at different depths in the profile and the dotted line represents values derived assuming no metabolic adjustments in root respiration capacity but accounting for the normal temperature dependency of root system respiration. Adapted from Holthausen and Caldwell (1980).

nitrate reductase activity (Goodman and Caldwell, 1971). Foliage nitrogen concentrations in excess of 4% have been reported for perennial *Agropyron* grasses of the Great Basin (Caldwell *et al.*, 1981).

Although Great Basin species may be effective in utilization of available soil nitrogen, fixation of atmospheric nitrogen is conducted primarily by free-living organisms, especially soil surface cryptogams such as blue-green algae and lichens with blue-green algal associates (Rychert and Skujins, 1974; Snyder and Wullstein, 1973). Legumes are scarce in the Great Basin and there is little evidence of nodulated nonlegumes. In special circumstances, rhizospheral nitrogen fixation may be important. Wullstein *et al.* (1979) have reported low rates of nitrogen fixation associated with sand grain root sheaths of Great Basin grasses such as *Oryzopsis hymenoides* growing in sand dunes. These root sheaths are thought to form a special low-oxygen, moist microenvironment suitable for bacterial nitrogen fixation. This appears to be primarily a phenomenon of sandy substrates and has not been reported in other soils. The major agent of atmospheric nitrogen fixation for most desert areas is the soil surface cryptogam crust. This crust may fix about 2.5 g nitrogen m^{-2} soil surface annually; however, some 70% of this newly fixed nitrogen may be lost from the system through volatilization and denitrification (West and Skujins, 1977).

9.7 Summary: stress in the cold desert

During the course of the year, cold desert shrubs must endure temperatures ranging between − 40 °C to in excess of 40 °C, physiological drought (as foliage remains on the plant during the winter when soils are frozen), prolonged periods of low soil water potentials often exacerbated by soil salinity, a limited soil nitrogen resource, and in some sites excessive concentrations of boron, chloride and sodium. Timing of physiological activity is crucial in order to compete for resources in the spring of the year and yet these species must tolerate the vicissitudes of weather during this season which periodically subjects physiologically active tissues to stresses such as low temperatures (Fig. 9.16). During this season of activity and growth, herbivory must also be tolerated by some cold desert species. Successful competition in the cold desert apparently requires a massive investment in the heterotrophic belowground system and, thus, aboveground production is quite limited.

The physiological ecology of cold desert plant species has as yet been explored only superficially. Much of the work has centered on North American shrubs. The perennial grass component of the cold desert vegetation has received less attention. The extent to which the physiological ecology of the

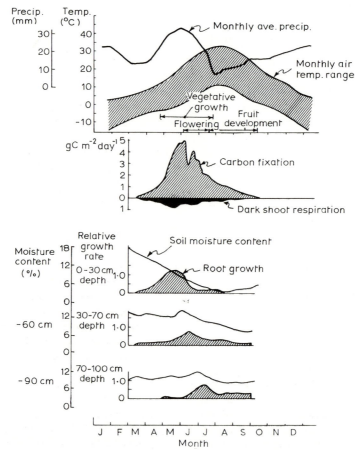

Figure 9.16 Annual progression of photosynthetic carbon gain, shoot respiration and phenological status, and root growth of *Atriplex confertifolia* at different soil depths as these plant processes relate to the average daily temperature minima and maxima, average monthly precipitation, and soil moisture content (from Figs. 9.2, 9.3, 9.8, 9.9, and 9.14).

North American Great Basin is representative of that of the extensive cold deserts of Mongolia and Russia, of South America from Rio Negro to Tierra del Fuego, and of the southern half of the Australian continent remains to be explored.

Acknowledgements

Research conducted with the support of grants from the National Science Foundation (GB 32139 and DEB 7907323) and the Utah Agricultural Experiment Station led to many of the results and concepts developed in this chapter.

References

Barbour, M.G. (1973) Desert dogma reexamined: Root/shoot productivity and plant spacing. *American Midland Naturalist* **89**, 41–57.

Billings, W.D. (1949) The shadscale vegetation zone of Nevada and eastern California in relation to climate and soils. *American Midland Naturalist*, **42**, 87–109.

Björkman, O. (1975) Environmental and biological control of photosynthesis, in: *Environmental and Biological Control of Photosynthesis* (ed. by R. Marcelle), Junk, The Hague, pp. 1–16.

Boutton, T.W., Harrison, A.T. and Smith, B.N. (1980) Distribution of biomass of species differing in photosynthetic pathway along the altitudinal transect in southeastern Wyoming grassland. *Oecologia*, **45**, 287–98.

Breckle, S.W. (1975) *Zur Ökologie und zu den Mineralstoffverhältnissen absalzender und nichtabsalzender Xerohalophyten*, Habilitationsschrift, Universität Bonn, 169 pp.

Caldwell, M.M. and Fernandez, O.A. (1975) Dynamics of Great Basin shrub root systems, in *Environmental Physiology of Desert Organisms* (ed. N.F. Hadley), Dowden, Hutchinson and Ross, Stroudsburg, Pennsylvania, pp. 38–51.

Caldwell, M.M., Osmond, C.B. and Nott, D.L. (1977a) C_4 pathway photosynthesis at low temperature in cold-tolerant *Atriplex* species. *Plant Physiology*, **60**, 157–64.

Caldwell, M.M., White, R.W., Moore, R.T. and Camp, L.B. (1977b) Carbon balance, productivity, and water use of cold-winter desert shrub communities dominated by C_3 and C_4 species. *Oecologia*, **29**, 275–300.

Caldwell, M.M., Richards, J.H., Johnson, D.A., Nowak, R.S. and Dzurec, R.S. (1981) Coping with herbivory: Photosynthetic capacity and resource allocation in two semiarid *Agropyron* bunchgrasses. *Oecologia*, **50**, 14–24.

Campbell, G.S. and Harris, G.A. (1977) Water relations and water use patterns for *Artemisia tridentata* Nutt. in wet and dry years. *Ecology*, **58**, 652–9.

Cook, C. (1971) *Effects of season and intensity of use on desert vegetation*, Utah Agricultural Experiment Station Bulletin **483**, Utah State University, 57 pp.

DePuit, E.J. and Caldwell, M.M. (1973) Seasonal pattern of net photosynthesis of *Artemisia tridentata*. *American Journal of Botany*, **60**, 426–35.

DePuit, E.J. and Caldwell, M.M. (1975) Gas exchange of three cool semi-desert species in relation to temperature and water stress. *Journal of Ecology*, **63**, 835–58.

Detling, J.K. and Klikoff, L.G. (1973) Physiological response to moisture stress as a factor in halophyte distribution. *American Midland Naturalist*, **90**, 307–18.

Fernandez, O.A. and Caldwell, M.M. (1975) Phenology and dynamics of root growth of three cool semi-desert shrubs under field conditions. *Journal of Ecology*, **63**, 703–14.

Goodman, P.J. (1973) Physiological and ecotypic adaptations of plants to salt desert conditions in Utah. *Journal of Ecology*, **61**, 473–94.

Goodman, P.J. and Caldwell, M.M. (1971) Shrub ecotypes in a salt desert. *Nature*, **232**, 571–2.

Hellmuth, O. (1971) Eco-physiological studies on plants in arid and semi-arid regions in western Australia. III. Comparative studies on photosynthesis, respiration and water relations of ten arid zone and two semi-arid zone plants under winter and later summer climatic conditions. *Journal of Ecology*, **59**, 225–60.

Holthausen, R.S. and Caldwell, M.M. (1980) Seasonal dynamics of root system respiration in *Atriplex confertifolia*. *Plant and Soil*, **55**, 307–17.

James, D.W. and Jurinak, J.J. (1978) Nitrogen fertilization of dominant plants in the northeastern Great Basin Desert, in: *Nitrogen in Desert Ecosystems* (eds. N.E. West and J.J. Skujins), Dowden, Hutchinson and Ross, Stroudsburg, Pennsylvania, pp. 219–31.

Johnson, H.B. (1975) Gas-exchange strategies in desert plants, in *Perspectives of Biophysical Ecology*, Ecological Studies 12 (eds. D.M. Gates and R.B. Schmerl), Springer-Verlag, New York, pp. 105–20.

Lange, O.L., Koch, W. and Schulze, E.D. (1969) CO_2-

Gaswechsel und Wasserhaushalt von Pflanzen in der Negev-Wüste am Ende der Trockenzeit. *Berichte Deutschen Botanischen Gesellschaft*, **82**, 39–61.

MacMahon, J.A. (1979) North American deserts: their floral and faunal components, in: *Arid-land Escosystems: Structure, Functioning and Management*, Vol 1 (eds. R.A. Perry and D.W. Goodall), Cambridge University Press, New York, pp. 21–82.

Moore, R.T. and Caldwell, M.M. (1972) Field use of thermocouple psychrometers in desert soils, in *Psychrometry in Water Relations Research* (eds. R.W. Brown and B.P. van Havern), Utah, Agricultural Experiment Station, Logan, Utah, pp. 165–9.

Moore, R.T., Breckle, S.W. and Caldwell, M.M. (1972a) Mineral ion composition and osmotic relations of *Atriplex confertifolia* and *Eurotia lanata*. *Oecologia*, **11**, 67–78.

Moore, R.T., White, R.S. and Caldwell, M.M. (1972b) Transpiration of *Atriplex confertifolia* and *Eurotia lanata* in relation to soil, plant, and atmospheric moisture stresses. *Canadian Journal of Botany*, **50**, 2411–18.

Mozafar, L. and Goodin, J.R. (1970) Vesiculated hairs: A mechanism for salt tolerance in *Atriplex halimus*. *Plant Physiology*, **45**, 62–6.

Pearson, L.C. (1965) Primary production in grazed and ungrazed desert communities of eastern Idaho. *Ecology*, **46**, 278–85.

Rodin, L.E. and Bazilevich, N.I. (1967) *Production and Mineral Cycling in Terrestrial Vegetation*, Oliver and Boyd, London, 288 pp.

Rychert, R.C. and Skujins, J. (1974) Nitrogen fixation by blue-green algae-lichen crusts in the Great Basin Desert. *Soil Science Society of America Proceedings*, **38**, 768–71.

Sammis, T.W. and Gay, L.W. (1979) Evapotranspiration from an arid zone plant community. *Journal of Arid Environments*, **2**, 313–21.

Snyder, J.M. and Wullstein, L.H. (1973) The role of desert cryptogams in nitrogen fixation. *American Midland Naturalist*, **90**, 257–65.

Tieszen, L.L. (1970) Photosynthetic properties of some grasses in eastern South Dakota. *Proceedings South Dakota Academy of Science*, **49**, 78–89.

Troughton, J.H., Wells, P.V. and Mooney, H.A. (1974) Photosynthetic mechanisms and paleoecology from carbon isotope ratios in ancient specimens of C_4 and CAM plants. *Science*, **185**, 610–12.

Walter, H. and Lieth, H. (1960) *Klimadiagramm-Weltatlas*, Gustav Fischer, Jena.

Welkie, G.W. and Caldwell, M.M. (1970) Leaf anatomy of species in some dicotyledon families as related to the C_3 and C_4 pathways of carbon fixation. *Canadian Journal of Botany*, **48**, 2135–46.

West, M. and Mooney, H.A. (1972) Photosynthetic characteristics of three species of sagebrush as related to their distribution patterns in the White Mountains of California. *American Midland Naturalist*, **88**, 479–84.

West, N.E. (1983) Overview of North American temperate deserts and semi-deserts, in *Temperate Deserts and Semi-Deserts* (ed. N.E. West), Elsevier, Amsterdam, pp. 321–30.

West, N.E. and Klemmedson, J.O. (1978) Structural distribution of nitrogen in desert ecosystems, in *Nitrogen in Desert Ecosystems* (eds. N.E. West and J.J. Skujins), Dowden, Hutchinson and Ross, Stroudsburg, Pennsylvania, pp. 1–16.

West, N.E. and Skujins, J. (1977) The nitrogen cycle in North American cold-winter semi-desert ecosystems. *Oecologia Plantarum*, **12**, 45–53.

West, N.E. and Wein, R.W. (1971) A plant phenological index technique. *Bioscience*, **21**, 116–17.

White, R.S. (1976) Seasonal patterns of photosynthesis and respiration in *Atriplex confertifolia* and *Ceratoides lanata*. PhD dissertation Utah State University, Logan, Utah, 124 pp.

Wullstein, L.H., Bruening, M.L. and Bollen, W.B. (1979). Nitrogen fixation associated with sand grain root sheaths (rhizosheaths) of certain xeric grasses. *Physiologia Plantarum*, **46**, 1–4.

10

Chaparral

H. A. MOONEY AND P.C. MILLER

10.1 Extent and general character

The chaparral is but one of a number of plant communities found in the mediterranean-climate region of North America. It is, however, one of the best studied and hence will serve as the focal point for this chapter.

The chaparral has been defined by Cooper (1922) as 'a scrub community, dominated by many species belonging to genera unrelated taxonomically, but of a single constant ecological type, the most important features of which are the root system, extensive in proportion to the size of the plant, the dense rigid branching, and preeminently the leaf, which is small, thick, heavily cutinized, and evergreen' (Fig. 10.1). Herbs are usually lacking in the mature vegetation although they become important in succession following disturbance.

The chaparral is centered in the summer dry region of coastal western North America, extending from Oregon southward down into Baja California (Fig. 10.2). There are elements of the chaparral also found in the summer rainfall regions of Arizona and Mexico. The chaparral in southern California and the chaparral in central Arizona are often regarded as two distinctly different vegetation types primarily because they have been analysed by different workers (Cable, 1975; Carmichael *et al.*, 1977; Major, 1977). The California and Arizona chaparral have a common origin in the Madro-tertiary geoflora (Axelrod, 1958). The separation of the two

types occurred in the late Cenozoic along with topographic changes and the development of the mediterranean-type climate.

The dominant species of the chaparral is *Adenostoma fasciculatum*, or chamise. Also important are species of *Arctostaphylos* and *Ceanothus*. These genera each have over 40 species represented in California.

10.2 Early ecological studies

The first major studies of the Californian chaparral were primarily governmental surveys of the state's various forest reserves (Leiberg, 1899a,b, 1900a,b; Plummer, 1911). Although Hall (1902) discussed some of the adaptive features of chaparral plants in his early study of the San Jacinto mountains, it was Cooper in 1922 who produced the foundation for subsequent work on the physiological ecology of chaparral. Cooper not only described the extent and environmental relations of the type but also outlined successional patterns. Most importantly he made a detailed survey of the anatomical features of the leaves of a large number of sclerophyll species and noted the influence of microenvironment on leaf structure and water loss. Cooper's work was followed by a number of studies which documented the environment, phenology, and water relations of chaparral plants (Bauer, 1936; De Forest and Miller, 1941; Miller, 1947; Shreve, 1927a,b; Whitfield, 1932).

Figure 10.1 Shrub and leaf morphology of two characteristic Californian evergreen sclerophylls. (a) *Heteromeles arbutifolia* a shrub generally found on relatively mesic sites and (b) *Adenostoma fasciculatum* a species found especially on xeric sites (from Thrower and Bradbury, 1977)

(b)

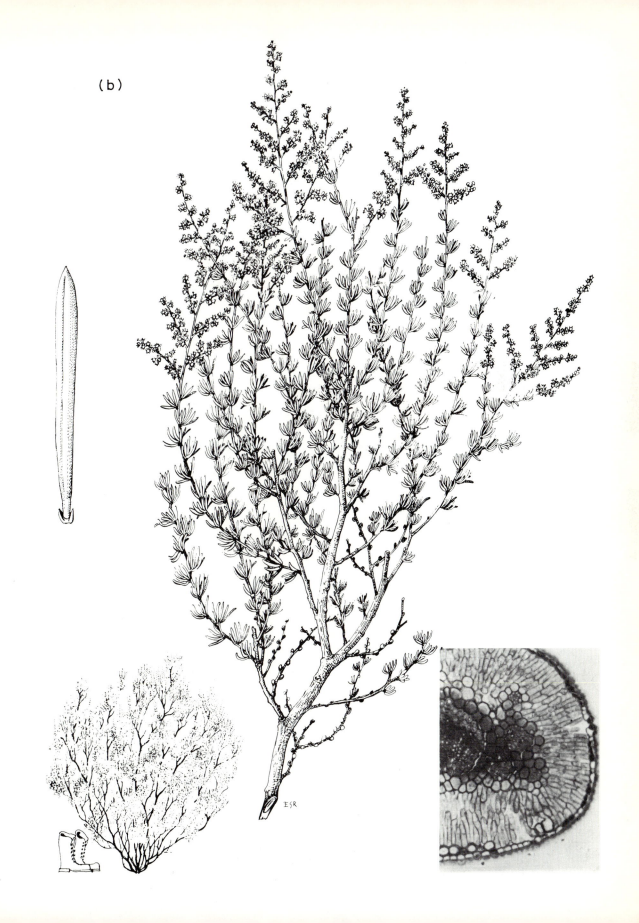

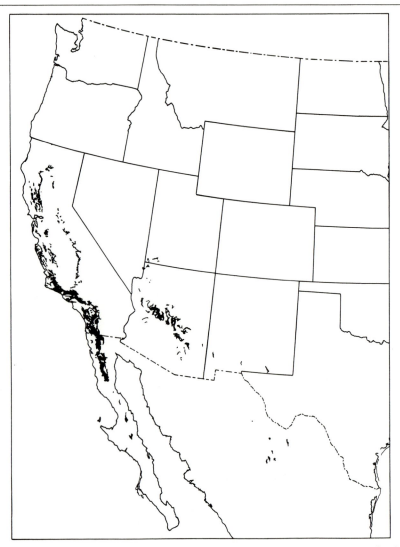

Figure 10.2   Distribution of chaparral (from Brown and Lowe, 1980; Flores *et al.*, 1971; Wieslander and Gleason, 1954; and personal communications with D. Axelrod, J. Rzedowski and J. Vankat). The distribution of this type may be much more extensive in Mexico. Further study is needed to determine the relationships between the northern and Mexican types.

Recent work on the chaparral has been reviewed by Cody and Mooney (1978), Conrad and Oechel (1982), Miller (1981), Mooney (1977), Mooney and Conrad (1977), Thrower and Bradbury (1977).

10.3 Environmental rhythms

The Californian mediterranean-type climate can be divided into three functional seasons: winter, during which cool temperatures and precipitation occur and soil moisture is available; spring, when moderate temperatures prevail and little or no precipitation occurs but soil moisture is still available; and summer, when temperatures are high but no precipitation occurs, and soil moisture is no longer available. The climatic features are related to physiological processes. The dominant features of the interrelations are the seasonal interplay be-

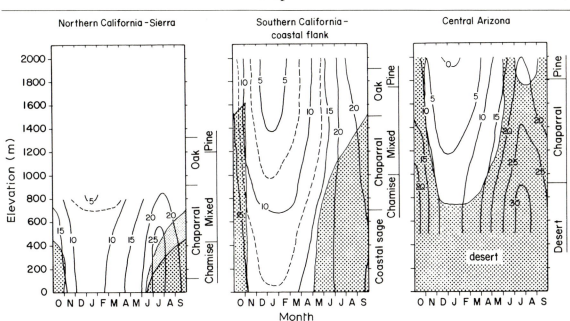

Figure 10.3 Isopleths of monthly mean temperatures and periods of dry soils (shaded) through the year beginning in October and with varying elevation at three locations: (1) the Sierra foothills of Northern California, (2) coastal flank of the Peninsular Ranges in Southern California, and (3) chaparral region of Central Arizona. The elevational range of vegetation types in each area is indicated. The temperature isopleths are not continued past the range of mediterranean-type vegetation. Dashed lines indicate temperatures 2.5 °C from the solid lines. In northern California the lighter shading indicates a shallower soil.

tween temperature–precipitation–soil moisture and the physiological responses of photosynthesis and growth (Miller, 1979). An analysis of climatic data for southern California and central Arizona indicates a similar positioning of chaparral in relation to temperature and precipitation in the two regions (Fig. 10.3). In southern California on the coastal flank of the mountains annual precipitation increases by about $313 \, mm \, yr^{-1} \, km^{-1}$ elevation from $240 \, mm \, yr^{-1}$ at the coast. On the desert flank precipitation increases by about $300 \, mm \, yr^{-1} \, km^{-1}$ elevation from about $50 \, mm \, yr^{-1}$ at sea level. About 2% of the annual precipitation occurs in July, August, and September on the coastal flank of the mountains and about 12% on the desert flank. On the coastal flank, evergreen sclerophylls predominate between 350 and $700 \, mm \, yr^{-1}$ precipitation (400 and 1800 m elevation). On the desert flank the evergreen sclerophylls occur between 900 and 1800 m elevation. On the coastal flank within the evergreen sclerophyll shrub zone, narrow-leaved

shrubs are more common at lower elevations and broad-leaved shrubs are more common at higher elevations. Below the evergreen sclerophyll shrub zone on both flanks, soft-leaved shrubs, grasses, and succulents occur. At higher elevations trees occur.

In central Arizona the annual precipitation also increases at about $300 \, mm \, yr^{-1} \, km^{-1}$ elevation from $180 \, mm \, yr^{-1}$ at sea level although large reductions from this idealized annual precipitation occur because of rain shadow effects. About 35% of the annual precipitation occurs during July, August, and September. The Arizona chaparral occurs where precipitation is between about 400 and $650 \, mm \, yr^{-1}$ (Carmichael *et al.*, 1977) and consists of broad-leaved evergreen shrubs of the same genera as are found in the higher elevation southern Californian chaparral. Certain species such as *Arctostaphylos pungens*, *Ceanothus greggii*, *Rhamnus crocea*, *R. californica*, *Rhus ovata*, and *R. trilobata* occur in both regions. In Arizona the lower elevational limit for the chaparral precipitation regime is about 900 m and the upper elevational limit is about 1800 m (Cable, 1975).

In northern California annual precipitation increases more steeply with elevation than in southern California. Annual precipitation often exceeds annual evapotranspiration. The soil water-holding capacity and infiltration capacity predominates in affecting the occurrence of summer drought. Similar temperatures occur at about 600 m lower elevations, in the interior, and at about 200 m lower elevations along the coast in northern California than in southern California. The marine influence apparently makes the altitudinal–latitudinal shift less abrupt than is predicted by Hopkins bioclimatic law, which states that similar phenological events occur with a 700–910 m elevation change per degree latitude shift (Hopkins, 1920).

Chaparral occurs on highly fractured parent materials which weather into coarse sandy loam or sandy soils. Soil moisture penetrates deeply into the rock cracks where it can be used by deeply rooted plants. On finer textured soils within similar temperature and precipitation regimes, pinyon–juniper occurs in Arizona and grasslands or oak woodlands occur in California. With low annual precipitation (about 400 mm yr^{-1}) a drought occurs during the summer regardless of topographic position and vegetation cover. With high annual precipitation (about 700 mm yr^{-1}) no soil drought occurs. Below 400 mm yr^{-1} the length of the drought is largely independent of annual precipitation and depends more on soil water-holding capacity, aspect differences, and vegetation cover. Species composition and species water relations become overriding factors influencing site moisture conditions (Miller and Poole, 1979, 1980). In Arizona two summer drought periods occur, in May and June and in September and October.

Fire becomes an important feature of the chaparral environment during periods of summer drought. Fires can occur in 3–6 yr intervals if a heavy herbaceous understory is present. As the shrub canopy closes, understory herbs are suppressed, and the occurrence of fire depends on the accumulation of dead woody material and litter. Fires occur in 25–75 yr intervals with shrub cover. The fire frequency is probably shorter in southern California than in Arizona because of the more pronounced summer drought and the occurrence in southern California of the Santa Ana conditions in late summer and early fall. The Santa Ana occurs when a high-pressure air mass system develops over the desert pushing hot, dry air into southern California. The air warms adiabatically as it descends towards the coast creating conditions of extremely low relative humidity and high wind speeds.

Microclimatic conditions alter the influence of the regional climate. Variations in the species composition of mature chaparral are often associated with variations in slope aspect. The influence of slope aspect is probably greatest during the early stages of chaparral recovery after fire when aspect affects shrub establishment and survival. Incident solar irradiance on the soil surface varies with aspect, particularly on unvegetated surfaces. As the shrub canopy grows and becomes closed, the influences of aspect are overridden by the influence of the interception of solar irradiance within the canopy, which is independent of aspect (Miller and Poole, 1980). On an unvegetated chaparral site, annual mean temperatures are about 5 °C higher at the soil surface than in the air (Miller *et al.*, 1981). Daily maximum temperatures are even more extreme. Soil and air temperatures converge as the shrub canopy closes. Since a temperature increase of 5 °C is about equivalent to a 1 km decrease in altitude (Miller *et al.*, 1981), mean temperatures near the soil surface without shrub cover are similar to temperatures at coastal and low elevation sites. These warmer temperatures favor the occurrence of low growing herbaceous species. Soil moisture is also influenced by aspect and vegetation cover. Soil moisture is higher on both north- and south-facing slopes without shrubs than with shrubs. After shrub removal evaporation from south-facing slopes is greater than from north-facing slopes. Surface soils of both slopes become dry during the summer, but soil moisture contents in deeper soil layers are greater on north-facing slopes (Miller and Poole, unpublished). In early stages of chaparral recovery following fire, the topographic effect on soil moisture is apparent as higher moisture contents and less rapid drying of the surface on north-facing slopes. As the shrub vegetation develops, the influence of the shrubs overrides the topographic differences and the soil moisture on north- and south-facing slopes is related to the species compostion on the respective slopes (Miller and Poole, 1979; Ng and Miller, 1980).

10.4 Growth forms and vegetation rhythms

The chaparral is composed of a mixture of growth forms the presence of which is generally separated in time. Mature chaparral consists almost exclusively of evergreen shrubs. Disturbed areas may, however, be covered with a variety of herbs, both annual and perennial, as well as subshrubs which generally have seasonally dimorphic leaves (Westman, 1981) and which reduce their leaf area during the drought period (e.g. *Artemisia californica*, *Diplacus aurantiacus*). A number of the drought deciduous species are green-stemmed (Nilsen and Muller, 1981).

Table 10.1 Mean characteristics of chaparral dominant evergreen sclerophylls growing at Echo Valley, San Diego county, California (from Mooney *et al.*, 1977, and Miller, 1981)

	Species						
Characteristics	*Rhus ovata*	*Ceanothus leucodermis*	*Heteromeles arbutifolia*	*Arctostaphylos glauca*	*Adenostoma fasciculatum*	*Quercus dumosa*	*Ceanothus greggii*
Height (m)*	1.4	1.75	1.68	1.76	1.16	1.52	0.96
Diameter (m)	1.81	1.11	1.56	1.36	0.98	1.37	1.24
Total shoot wt (g)	5462	2078	6271	4839	1312	3007	4093
Total leaf wt (g)	1188	283	1220	1346	221	514	721
Leaf area index ($m^2 m^{-2}$)	1.95	2.11	2.81	3.58	2.1[‡]	2.46	1.61
Root area index[§]	3.1	—	—	4.1	2.1	—	2.4
Leaf size (cm^2)	13.0	1.4	7.2	5.2	0.1	2.1	1.1
Leaf specific wt ($g m^{-2}$)	210	120	230	270	253[‡]	140	370
Leaf nitrogen[†] ($mg g^{-1}$)	7.1	16.1	7.5	5.8	6.5	13.0	12.9
Leaf phosphorus[†] ($mg g^{-1}$)	0.5	0.9	0.4	0.4	0.8	0.5	0.6
Root/shoot[‡] biomass ratio	—	—	0.7	0.9	0.6	—	0.3
Stem growth[‡] duration (d)	31	—	—	48	69	—	68
Maximum leaf[‡] conductance ($mm s^{-1}$)	2.5	—	—	4.5	4.0	3.0	5.0
Minimum xylem water potential (MPa)	-2.5	—	—	< -6.5	-6.0	-5.0	< -6.5
Xylem water potential[‡] at zero conductance (MPa)	-2.0	—	—	-6.0	-5.0	—	-6.0

*Shrub dimension analyses are mean value for the five mature specimens of each species. The area had burned 22 yr prior to harvest.
[†]Mean values for mature leaves.
[‡]From Miller (1981).
[g]From Kummerow *et al.* (1978a)

Many of these same subshrubs are climax species in sites too arid to support evergreen species (Harrison *et al.*, 1971; Kirkpatrick and Hutchinson, 1980).

The evergreen shrubs are between 1 and 3 m in height and have a leaf area index averaging about 2 with a somewhat higher root area index. Leaves are normally small, low in nutrients, and have a high specific weight. Leaves generally constitute about 20% of the total shoot weight. The root biomass is less than that of the shoot (Table 10.1).

The growth cycle of the evergreen shrubs has been studied in considerable detail (Kummerow, 1982; Miller, 1947; Mooney, 1977; Mooney and Kummerow, 1981; Mooney *et al.*, 1974a; Thrower and Bradbury, 1977). Stem growth is restricted principally to spring in all shrubs. Flowering occurs throughout the year with some species flowering prior to and others subsequent to stem growth. Species of *Ceanothus* and *Arctostaphylos* flower in late winter/early spring from the previous years buds whereas other chaparral species generally flower in late spring/summer from current years buds (J. Keeley, 1977a).

The growth cyle of mediterranean-climate plants is tied in large part to soil water potential. Since the soil water potential varies with depth through the seasons and since the various plant growth forms have differing rooting depths, it follows that their growth cycles would be asynchronous. This can be seen in a comparison of a shallow-rooted annual, a deep-rooted evergreen shrub, and a subshrub with a rooting depth intermediate between the annual and evergreen shrub (Fig. 10.4). The annual begins its growth cycle at the beginning of the rainy season whereas the evergreen shrub commences growth toward the end of the rainy period. Conversely the termination of the growth period of these growth form types is similarly displaced. The relationship between rooting depth and growth cycle may not hold for certain trees however (Rundel, 1983).

10.5 Water balance

As noted below there are few differences in the photosynthetic capacities, light and temperature responses of the evergreen sclerophylls of the chap-

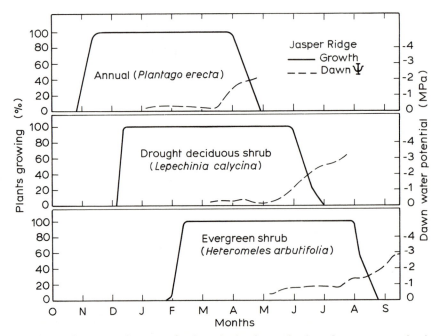

Figure 10.4 Seasonal growth activity of an annual, a drought deciduous shrub, and an evergreen shrub at a northern California study site (Jasper Ridge). Dawn water potential measurements indicate that the available soil moisture for these three types differs, probably reflecting their dissimilar rooting depths (from Mooney, 1983).

arral, although there are differences in the capacities and patterns of CO_2 exchange between sclerophylls and drought deciduous species. On the other hand, considerable differences in the patterns of water use, not only between drought deciduous and evergreen species but also among different evergreen species, have been demonstrated in a number of studies. For example, Miller and Poole (1979) found that the maximum conductance as well as the xylem water potential at which stomatal closure occurred differed by over a factor of 2 among five different sclerophyll shrub species in the chaparral of the San Diego region. As noted earlier the diverse growth forms which occur in the chaparral have dissimilar phenological cycles which relate to their rooting depths and hence use patterns of seasonal water availability. The herbaceous annuals and perennials, which are generally shallow rooted, dry up and become dormant during the seasonal drought. They are generally considered to be drought avoiders. The same is true, to a lesser extent for shallow-rooted drought deciduous subshrubs. The deep-rooted evergreen sclerophylls, on the other hand, are considered to be drought tolerators since they maintain most of their leaf area throughout the seasonal drought period. However, it has been noted recently that the drought avoiders actually withstand the drought and are metabolically active at very low water potentials (Mooney, 1982). For example, *Salvia mellifera*, a shrub which loses all but its terminal leaves during the drought period has positive net photosynthesis at less than -5.0 MPa (Fig. 10.5). Drought-avoiding annuals maintain positive photosynthesis at water potentials of less than -3.0 MPa. Certain deep-rooted shrubs may actually never be subjected to such low water potentials. Burk (1978) has noted how certain evergreen oak species may have dawn water potentials greater than -0.5 MPa even at the end of the drought period whereas co-occuring evergreen shrubs, such as *Ceanothus* may have dawn water potentials less than -5.0 MPa. Roberts (1982)

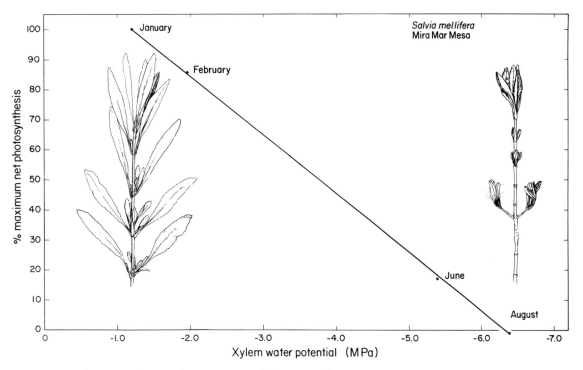

Figure 10.5 Changing photosynthetic capacity of the seasonally dimorphic-leaved *Salvia mellifera*. The drought-enduring leaves have considerably reduced photosynthetic capacity (from Mooney, 1982).

presents evidence that certain evergreen sclerophyll species may show substantial diurnal osmotic adjustment thus maintaining turgor at relatively low midday water potentials. The above findings indicate that neither morphology nor phenology correlate perfectly with water stress tolerance. It is clear, however, that many chaparral species, both herbaceous and woody, can withstand substantial drought stress. The precise mechanisms of tolerance deserve considerably more study.

Competition for water is apparently a primary factor leading to successional mortality in pure stands of *Ceanothus* (Schlesinger and Gill, 1980). Young, presumably shallow-rooted, plants of *C. megacarpus* come under considerably greater drought stress than do older plants (Fig. 10.6).

Species of the woody genus *Ceanothus* show large differences in their water economies which are reflected in their leaf anatomy. The genus *Ceanothus* is composed of two subgenera; Cerastes and Ceanothus. The subgenus Cerastes has stomatal crypts, whereas the species in the subgenus Ceanothus do not. Most members of both subgenera are evergreen. In a study of two species in each subgenus, Barnes (1979) found that those of Cerastes have thicker

leaves with a thicker cuticle and a higher specific weight (Table 10.2). These species have shallower roots and close their stomata at a lower water potential than do species of the Ceanothus subgenus. Species of Cerastes have a larger number of vessels per unit area but the vessels are smaller in diameter than those of the Ceanothus subgenus (Webber, 1936). All of these differences indicate that the more shallow-rooted species of the Cerastes subgenus are adapted to endure greater water stress than the species of the Ceanothus subgenus. Generally species of both sections co-occur in any given chaparral community (Nobs, personal communication) and seldom do members of the same subgenus co-occur.

10.6 Carbon balance

10.6.1 PHOTOSYNTHETIC CAPACITY

Evergreen chaparral shrubs have the capacity to fix carbon year round although they do so at a reduced rate, during the summer drought period (Mooney *et al.*, 1975; Oechel *et al.*, 1981). They all fix carbon through the C_3 pathway (Mooney *et al.*, 1974b).

Carbon fixation rates of evergreen chaparral shrubs are rather low in comparison to herbaceous plants in general, and even in comparison to evergreen plants of such community types as the desert scrub (Mooney and Gulmon, 1979). Mean rates of net photosynthesis of chaparral evergreen shrubs are near 8 μmol CO_2 m^{-2} s^{-1} which is about one-half of that found for drought deciduous shrubs from the same climatic type (Mooney, 1981). Temperatures for optimum photosynthesis extend from about 17–29 °C (Oechel *et al.*, 1981), however there is remarkably little temperature sensitivity in this process; photosynthesis rates at 10 °C are similar to those at 30 °C. This broad temperature optimum for photosynthesis led Mooney *et al.* (1975) to conclude that the principal limitation to winter carbon gain in evergreen sclerophylls is the short photoperiod rather than low temperature. There is little evidence of thermal acclimation in photosynthesis of evergreen sclerophylls (Harrison, 1971). This lack of acclimation might be expected to be due to the broad temperature responses of photosynthesis, as noted

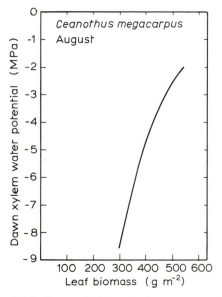

Figure 10.6 Seasonal change in dawn water potential of plants of *Ceanothus megacarpus* growing in pure stands of different ages (from Schlesinger and Gill, 1980).

Table 10.2 Mean comparative leaf characteristics of representatives of two taxonomic sections (Cerastes and Ceanothus) of *Ceanothus* (from Barnes, 1979)

	Cerastes		Ceanothus	
	C. ferrisae	C. ramulosa	C. thyrsiflorus	C. incanus
Assumed rooting depth	shallow	shallow	deep	deep
Leaf duration	evergreen	evergreen	evergreen	evergreen
Stomatal crypts	yes	yes	no	no
Maximum stomatal conductance ($mm\,s^{-1}$)	2.0	2.9	2.5	5.0
Minimum stomatal conductance ($mm\,s^{-1}$)	0.04	0.03	0.36	0.14
Xylem water potential of stomatal closure (MPa)	−5.5	−4.5	−3.0	−2.0
Leaf specific weight ($g\,m^{-2}$)	280	317	163	161
Leaf thickness (μm)	558	605	222	352
Adaxial cuticle thickness (μm)	4.7	8.3	1.6	3.4

above, and to the drought limitations of photosynthesis during the summer high-temperature period.

10.6.2 CARBON ALLOCATION

The patterns of allocation of carbon have been studied in a number of chaparral plants. Mooney and Chu (1974) noted that the flow of carbon to various functional pools changed seasonally in the evergreen shrub *Heteromeles arbutifolia*. Carbon was acquired through photosynthesis year round. In the spring primary allocation was to canopy development, and in the winter to storage, reproduction, and herbivore defense. Oechel and Lawrence (1981) give detailed carbon budgets for a number of chaparral shrubs noting that in some species root respiration accounts for nearly a quarter of the annual carbon budget (Table 10.3).

The leaves of evergreen sclerophylls are generally costly to produce in carbon units because of their high content of lignin and secondary products with lipid-like structures. Some of the drought deciduous species also have costly leaves due to their high contents of resins and terpenes (Merino *et al.*, 1982). Any cost/benefit analysis of leaves must consider leaf duration. The leaves of the evergreen chaparral species generally last two seasons (Jow *et al.*, 1980) although under favorable conditions they may last as long as 6 yr (unpublished).

10.6.3 COMMUNITY PRODUCTIVITY AND PRODUCTIVE STRUCTURE

Ehleringer and Mooney (1982) give composite data for a less than 20 yr old chaparral stand which indicated total biomass of approximately

Table 10.3   Annual carbon budget of four evergreen chaparral shrubs. Data for shrubs growing at Echo Valley, San Diego County (from Oechel and Lawrence, 1981)

	Standing crop	New biomass	Maintenance respiration	Growth respiration	% Allocation
Arctostaphylos glauca					
$1439g$ dwt m^{-2} yr^{-1}					
Leaves	568	235	40	164	31
Stems	2300	74	396	22	34
Main roots	1020	22	139	7	12
Absorbing roots	206	194	88	58	24
Growth =		525			
Respiration =				914	
Adenostoma fasciculatum					
$1213g$ dwt m^{-2} yr^{-1}					
Leaves	222	72	384	50	42
Stems	1370	190	141	57	32
Main roots	760	94	80	28	17
Absorbing roots	74	67	30	20	10
Growth =		423			
Respiration =				790	
Ceanothus greggii					
$1058g$ dwt m^{-2} yr^{-1}					
Leaves	400	181	24	127	31
Stems	2810	58	248	17	31
Main roots	1740	25	159	8	18
Absorbing roots	135	108	71	32	20
Growth =		372			
Respiration =				686	
Rhus ovata					
665 *g* dwt m^{-2} yr^{-1}					
Leaves	343	109	26	76	32
Stems	1640	64	101	19	28
Main roots	986	31	63	9	15
Absorbing roots	93	57	93	17	25
Growth =		261			
Respiration =				404	

$4600\,g\,m^{-2}$; half of which was shoot weight (Table 10.4). This stand, located in San Diego County, was dominated by *Adenostoma fasciculatum* and *Ceanothus greggii*. Annual shoot production was approximately $400\,g\,m^{-2}$ with about $130\,g\,m^{-2}$ accumulating as standing biomass per year. Schlesinger and Gill (1980) indicate that pure stands of *Ceanothus megacarpus* may have net annual aboveground productivity values as high as $850\,g\,m^{-2}\,yr^{-1}$.

10.7 Nutrient balance

Chaparral soils are generally nutrient deficient (Hellmers *et al.*, 1955). Not surprisingly, chaparral shrubs show a number of characteristics which presumably adapt them to these low nutrient conditions. A number of these shrubs, including *Ceanothus* spp. and *Cercocarpus betuloides*, are nonleguminous nitrogen fixers (Delwiche *et al.*, 1965; Hellmers and Kelleher, 1959; Kummerow *et al.*,

Table 10.4 Productive structure of a chaparral scrub communty (composite data compiled from various sources, see Ehleringer and Mooney, 1982).

Stand age (year)	17–18
Height (m)	1.5
Leaf area index (m²m⁻²).	2.5
Biomass (g m⁻²)	
Shoot	2039
Wood	1672
Leaves	367
Roots	1223
Litter	1359
Total biomass	4621
Allocation (%)	
Shoot	
Wood	82.0
Leaves	18.0
Shoot	62.5
Root	37.5
Root/Shoot ratio	0.6
Production (g m⁻² yr⁻¹)	
Aboveground biomass accumulation	130
Litter fall	282
Shoot, net	412
Decomposition (g m⁻² yr⁻¹)	
Litter	264

1978b; Vlamis *et al.*, 1958). The early successional leguminous subshrub, *Lotus scoparius*, may be a nitrogen fixer (Nilsen and Schlesinger, 1981). However, Schlesinger and Hasey (1980) calculate that, on an area basis, atomospheric desposition of nitrogen can far exceed fixation by chaparral shrubs in dry regions. In a summary of the nitrogen balance of chaparral ecosystems, Schlesinger *et al.* (1982b) indicate 1.5 kg N ha⁻¹ input through precipitation and only 0.1 kg ha⁻¹ by symbiotic nitrogen fixation. However, 1.0 kg ha⁻¹ was contributed by asymbiotic nitrogen fixation. The amounts of these inputs become small compared to the potential loss through volatilization during fire as discussed below.

Nitrogen concentrations in the leaves of evergreen shrubs, which is generally not much above 1% by weight, varies somewhat with both age and season (Mooney *et al.*, 1977; Shaver, 1981). There is some evidence of uptake and storage of nutrients by evergreen leaves of chaparral shrubs

during the non-growing period (Mooney and Rundel, 1979; Shaver, 1981). It has also been noted that the leaves of post-fire resprout plants have higher concentrations of nitrogen, phosphorus, and potassium for the first few years following a fire (Rundel and Parsons, 1980). These various patterns indicate uptake and storage patterns for certain nutrients which are presumably adaptive in these nutrient-poor habitats. The generally low nutrient content of the leaves of chaparral plants along with their high content of secondary chemicals has been proposed to be responsible for their relatively low herbivory rate (Mooney and Dunn, 1970). Pure stands of *Ceanothus*, however, have been reported to be nutrient unlimited. This may be due to nitrogen fixation, efficient nutrient recycling or the maritime climate in which the study was conducted (Schlesinger *et al.*, 1982a).

10.8 Fire ecology

Since the early work of Brandegee (1891) a vast literature has accumulated on the fire ecology of the chaparral, much of which has recently been reviewed (Mooney and Conrad, 1977). The post-fire successional sequences in the chaparral are well known. A large fraction of the dominant shrubs resprout after burning. During the interval before these shrubs regain dominance, an herbaceous flora may flourish including many 'fire annuals' which appear only after fire. Although these general trends have been thoroughly documented (Christensen and Muller, 1975a; Horton and Kraebel, 1955; S. Keeley, 1977; Keeley *et al.*, 1981; Sampson, 1944; Sweeney, 1956; Vogl and Schorr, 1972), the exact mechanisms responsible for the observed patterns still need clarification.

Sweeney (1956), in his classic study of the fire responses of chaparral herbs, concluded that the seeds of many herb species are stimulated to germinate by the high temperatures which accompany fire, although he was unable to demonstrate this for many species. He found that these seeds are present in the soil at the time of the fire and that they maintain their viability on site for decades between fire episodes.

McPherson and Muller (1969) and Christensen

and Muller (1975a) found that in addition to endogenous inhibitors, which restrict germination of herb seeds until heat treatment, there are apparently exogenous inhibitors which are produced by the leaves of shrubs which restrict herb germination until the overstory shrubs are removed, mechanically, or by fire. Kaminsky (1981), however, suggested that microbial associates of *Adenostoma* are responsible for the inhibition of herb germination. Seeds of *Emmenanthe penduliflora*, a characteristic fire annual, are stimulated to germinate by contact with burned, but not ashed, stems of the shrub *Adenostoma fasciculatum* (Jones and Schlesinger, 1980; Wicklow, 1977). This phenomenon is known from at least a dozen other chaparral herbs (Keeley, unpublished). The mechanism of this response is not known.

Christensen and Muller (1975b) concluded that even if seeds of herbs did germinate under a mature chaparral canopy, their growth would be limited by nutrient deficiencies. Such limitations are removed by the release of minerals from the shrubs that burn during a fire. They further concluded that even if the herb seeds germinated and made good growth under the mature chaparral canopy, there would be little chance of survival to seed set because of high herbivore activity. Such activity would be an additional factor selecting for efficient dormancy mechanisms in the seeds of herb species.

These studies and others (Keeley *et al.*, 1981) make it clear that the factors controlling the dramatic fluctuating herb population cycles through time are complex and need further study. It is of interest that the fire herb cycle found in the California chaparral is not present in the shrub zone of other mediterranean-climate areas (Keeley and Johnson, 1977).

Chaparral shrubs respond differentially to fire. Some resprout after fire from 'burls' whereas others do not. Within the genera *Arctostaphylos* and *Ceanothus* there are both sprouters and non-sprouters. Apparently non-sprouters have evolved from the sprouting species (Wells, 1969). J. Keeley (1977b) argues that conditions that produce intense fires, such as long fire-free periods, have selected for the obligate-seedling, non-sprouting species.

The range of fire responses among chaparral

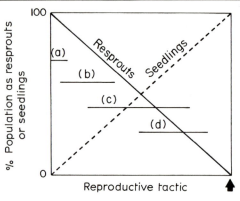

Figure 10.7 Reproductive characteristics of different chaparral shrubs and trees. Some species such as *Quercus* spp. do not have fire-stimulated germination although they resprout (a). Others such as *Adenostoma* have both resprouts and seedlings after a fire (c). Most chaparral species fall into category (b) and certain *Ceanothus* species into (d). Finally shrubs or certain species of *Arctostaphylos* and *Ceanothus* are killed by fire but they produce abundant seedlings (indicated by the arrow) (from Keeley, 1977b).

shrubs extends from species which resprout after fire and produce very few seedlings to species which only produce seedlings (J. Keeley, 1977b; Fig. 10.7). The populational implications of these various reproductive strategies have been studied in some detail (J. Keeley, 1977b; Keeley and Keeley, 1977; Keeley and Zedler, 1978). Less information is available on the adaptive physiology of fire sprouting and fire-stimulated germination of shrubs. Heat stimulation of seed germination in shrubs, particularly in *Ceanothus* and *Rhus* species, has been well documented (Hadley, 1961; Quick and Quick, 1961; Stone and Juhren, 1951). Charred wood stimulates germination of some shrub species (Keeley, unpublished). Apparently there is some variability in response to heat treatment even within seeds produced by a single species. *Adenostoma fasciculatum* produces some seed which only germinates after heat treatment, some that does not require heat but which germinates if suitable moisture and temperature conditions prevail (Stone and Juhren, 1953).

The resprouting potential of *Adenostoma* is directly related to the status of carbohydrate storage in the roots which varies seasonally. During periods when the roots contain large quantities of starch, such as in winter, post-fire resprouting potential is high (Jones and Laude, 1960). New resprouts generally are under less water stress and have leaves with a higher photosynthetic capacity than tissue of mature plants (Oechel and Hastings, 1982; Radosevich *et al.*, 1977).

The role of post-fire herbs and subshrubs in nutrient conservation in chaparral ecosystems is now receiving some attention (Nilsen and Schlesinger, 1981). Large amounts of nutrients are ashed in chaparral fires (Christensen and Muller, 1975a) and can be lost in run off. DeBano and Conrad (1978) estimate N losses of 146 kg ha^{-1} by volatilization during fire and erosion during the first year after fire.

10.9 Summary

The chaparral, or evergreen scrub, is only one of a number of distinctive plant communities associated with summer drought climates; however, it has received the most research focus. Its distribution is centered in California with extension into Mexico and Arizona.

The evergreen shrub is the dominant growth form of mature chaparral. A variety of growth forms, including annuals, herbaceous perennials, and drought deciduous subshrubs, also occur within the chaparral region but generally as successional species.

Water is the principal limiting factor to the growth cycle of chaparral plants. Rooting depth varies among and within growth forms and in part determines the seasonality of growth activity.

Although most chaparral plants of all growth forms can tolerate drought, some species are particularly resistant. Growth form or leaf type itself is not a universal predictor of the drought stress a species may experience. Species which are primarily drought deciduous may maintain leaves through the drought which are photosynthetically active at < -5.0 MPa. On the other hand, some sclerophylls may actually experience little drought stress because of deep roots. However, in certain groups, particularly among species of *Ceanothus*, leaf type and growth form relate to drought tolerance, which again relates to rooting type. It would appear that a

principal key to understanding species patterning and seasonality within the chaparral is comparative water relations. However, a comprehensive and synthetic treatment of chaparral water relations is still lacking.

The photosynthetic capacity of chaparral plants differs among growth forms with the evergreen shrubs having lower rates than drought deciduous shrub or herbaceous species. The evergreen shrubs differ little among species in rate capacity, all having relatively low rates on a leaf area basis. These shrubs, which maintain photosynthetic capacity year round, show broad temperature responses and little thermal acclimation.

The leaves of sclerophylls are generally costly to produce in carbon terms because of lignins and various secondary chemicals. At the same time they experience relatively low herbivory levels. Allocation of carbon to various plant parts and processes differs among species and among seasons for any given species. A considerable carbon cost is devoted to growth and maintenance respiration.

Nutrients appear limiting in most chaparral communities and may in part explain the relatively low photosynthetic capacity of the resident plants. In addition to water limitations, low nutrient availability may be responsible for the generally low productivity of most chaparral woody communities.

Fire occurs in chaparral communities with regular frequency. The nutrient, water, and energy balance of burned sites is drastically changed from that of mature chaparral stands. These changes are responsible, in part, for the shifts in growth forms which occur through post-fire succession. There has been considerable work on the ecology of germination of chaparral herbs and shrubs after fire. Many plants require fire for germination through heat stimulation and also apparently through stimulation by charcoal directly. Less work has been directed toward the physiological ecology of resprouting shrubs although what work is available indicates different physiological responses between resprout and mature branches. This relates to different resource availability patterns.

The key to the management of chaparral is an understanding of the control of the flows of water and nutrients by plants through time as well as knowledge of the ecophysiology and population biology of the component species. A considerable amount of information on the ecophysiology of the chaparral has accumulated this past decade. This information needs to be interfaced with knowledge from population biology and ecosystem science in order to understand the structure and functioning of this unique ecosystem type.

Acknowledgements

We thank D. Axelrod, J. Rzedowski, and J. Vankat for comments on chaparral distribution. J. and S. Keeley and P. Rundel kindly reviewed the manuscript. Many of the results summarized here were supported by grants from the National Science Foundation to both of us.

References

Axelrod, D.I. (1958) Evolution of Madrotertiary geoflora. *Botanical Review*, **24**, 433–509.

Barnes, F.J. (1979) Water relations of four species of *Ceanothus*, MA thesis, San Jose State University, 44 pp.

Bauer, H.L. (1936) Moisture relations in the chaparral of the Santa Monica Mountains, California. *Ecological Monographs*, **6**, 409–54.

Brandegee, T.S. (1891) The vegetation of 'burns'. *Zoe*, **2**, 118–22.

Brown, D.F. and Lowe, C.H. (1980) *Biotic communities of the southwest*. Forest Service General Technical Report. USDA For. Ser Gen. Tech. Rept. RM–78.

Burk, J. (1978) Seasonal and diurnal water potentials in selected chaparral shrubs. *American Midland Naturalist*, **99**, 244–8.

Cable, D.R. (1975) Range management in the chaparral type and its ecological basis: The status of our knowledge. US Forest Service Research Paper RM-202, 30 pp.

Carmichael, R.S., Knipe, D.D., Pase, C.P. and Brady, W.W. (1977) Arizona chaparral: plant associations and ecology, Research Paper RM-202, Rocky Mountain Forest and Range Experiment Station, Forest Service USDA.

Cody, M.L. and Mooney, H.A. (1978) Convergence versus nonconvergence in mediterranean-climate ecosystems. *Annual Review Ecology Systems*, **9**, 265–321.

Conrad, C.E. and Oechel, W.C. (Tech. Coordinators) (1982) Dynamics and Management of Mediterranean-Type Ecosystems, USDA, Pacific Southwest Forest and Range Experiment Station General Technical Report, PSW-58.

Cooper, W.S. (1922) *The broad-sclerophyll vegetation of California. An ecological study of the chaparral and its related communities*, Carnegie Insititution of Washington, Publ. 319, 124 pp.

Christensen, N.L. and Muller, C.H. (1975a) Effects of fire on factors controlling plant growth in *Adenostoma* chaparral. *Ecological Monographs*, **45**, 29–55.

Christensen, N.L. and Muller, C.H. (1975b) Relative importance of factors controlling germination and seedling survival in *Adenostoma* chaparral. *American Midland Naturalist*, **93**, 71–8.

De Forest, J. and Miller, E. Jr (1941) *Some environmental conditions of a southern California chaparral*, General Biology Series 1, University of Southern California Press, Los Angeles.

DeBano, L.F. and Conrad, C.E. (1978) The effect of fire on nutrients in a chaparral ecosystem. *Ecology*, **59**, 489–97.

Delwiche, C.C., Zinke, P.Z. and Johnson, C.M. (1965) Nitrogen fixation by *Ceanothus*. *Plant Physiology*, **40**, 1045–7.

Ehleringer, J. and Mooney, H.A. (1982) Photosynthesis and productivity of desert and mediterranean-climate plants, in *Physiological Plant Ecology*, IV (eds O.L. Lange, P.S. Nobel, C.B. Osmond and H. Ziegler), Springer Verlag. Berlin, Heidelberg, NY.

Flores, M.G., Jimenez, L.J., Madrigal, S.X., Moncayo, R.F. and Takaki, T.F. (1971) Mapa y descripcion de los tipos de vegetacion de la Republica Mexicana, Subsecretaria de Planeacion direcction General de Estudios, Direcction de Agrologia, 4 RH, 59 pp.

Hadley, E.B. (1961) Influence of temperature and other factors on *Ceanothus megacarpus* seed germination. *Madroño*, **16**, 132–8.

Hall, H.M. (1902) A botanical survey of San Jacinto Mountain. *University of California Publications in Botany*, **1**, 1–144.

Harrison, A.T. (1971) Temperature-related effects on photosynthesis in *Heteromeles arbutifolia* M. Roem, PhD thesis, Stanford University.

Harrison, A.T., Small, E. and Mooney, H.A. (1971)

Drought relationships and distribution of two mediterranean climate California plant communities. *Ecology*, 52, 869–75.

Hellmers, H. and Kellehar, J.M. (1959) *Ceanothus leucodermis* and soil nitrogen in southern California mountains. *Forestry Science*, 5, 275–8.

Hellmers, H., Bonner, J.F. and Kelleher, J.M. (1955) Soil fertility: a watershed management problem in the San Gabriel mountains of southern California. *Soil Science*, 80, 185–97.

Hopkins, A.D. (1920) The Bioclimatic Law. *Journal of the Washington Academy of Science*, 10, 34–40.

Horton, J.S. and Kraebel, C.J. (1955) Development of vegetation after fire in the chamise chaparral of southern California. *Ecology*, 36, 244–62.

Jones, C.S. and Schlesinger, W.H. (1980) *Emmenanthe penduliflora* (Hydrophyllaceae): further consideration of germination response. *Madroño*, 27, 122–5.

Jones, M.B. and Laude, H.M. (1960) Relationship between sprouting in chamise and the physiological condition of the plant, *J. Range Management*, 13, 210–14.

Jow, W.M., Bullock, S.H. and Kummerow, J. (1980) Leaf turnover rates of *Adenostoma fasciculatum* (rosaceae). *American Journal of Botany*, 67, 256–61.

Kaminsky, R. (1981) The microbial origin of the allelopathic potential of *Adenostoma fasciculatum* H & A. *Ecological Monographs*, 51, 365–82.

Keeley, J. (1977a) Fire-dependent reproductive strategy in *Arctostaphylos* and *Ceanothus*, in *Proceedings of the Symposium on the Environmental Consequences of Fire and Fuel Management in Mediterranean Ecosystems*. (H.A. Mooney and C.E. Conard, Technical Coordinators) USDA, Forest Service, General Technical Report WO-3, pp. 391–6.

Keeley, J. (1977b) Seed production, seed populations in soil, and seedling production after fire for two congeneric pairs of sprouting and nonsprouting chaparral shrubs. *Ecology*, 58, 820–9.

Keeley, J. and Keeley, S. (1977) Energy allocation patterns of a sprouting and a nonsprouting species of *Arctostaphylos* in the California chaparral. *American Midland Naturalist*, 98, 1–10.

Keeley, J. and Zedler, P.H. (1978) Reproduction of chaparral shrubs after fire: a comparison of sprouting and seed strategies. *American Midland Naturalist*, 99, 142–61.

Keeley, S.C. (1977) The relationship of precipitation to post-fire succession on southern California chaparral, in *Proceedings of the Symposium on the Environmental Consequences of Fire and Fuel Management in Mediterranean Ecosystems* (H.A. Mooney and C.E. Conard,

Technical Coordinators), USDA, Forest Service, General Technical Report WO-3.

Keeley, S.C. and Johnson, A.W. (1977) A comparison of the pattern of herb and shrub growth in comparable sites in Chile and California. *American Midland Naturalist*, 97, 20–132.

Keeley, S.C., Keeley, J.E., Hutchinson, S.M. and Johnson, A.W. (1981) Postfire succession of the herbaceous flora in southern California. *Ecology*, 62, 1608–21.

Kirkpatrick, J.B. and Hutchinson, C.F. (1980) The environmental relationships of Californian coastal sage scrub and some of its component communities and species. *Journal Biogeography*, 7, 23–38.

Kummerow, J. (1982) Comparative phenology of mediterranean-type plant communities, in *Mediterranean-Type Ecosystems. The Role of Nutrients* (eds F.J. Kruger, D.T. Mitchell, and J.U.M. Jarvis), Springer Verlag, NY, Heidelberg, Berlin.

Kummerow, J., Krause, D. and Jow, W. (1978a) Seasonal changes in fine root density in the southern Californian chaparral. *Oecologia*, 37, 201–12.

Kummerow, J., Alexander, J.V., Neel, J.W. and Fishbeck, K. (1978b) Symbiotic nitrogen fixation in *Ceanothus* roots. *American Journal of Botany*, 65, 63–9.

Leiberg, J.B. (1899a) *San Bernardino Forest Reserve*, US Geological Survey, 19th Annual Report Pt. 5. pp 359–65.

Leiberg, J.B. (1899b) *San Jacinto Forest Reserve*, US Geological Survey, 19th Annual Report, Pt. 5, pp. 351–7.

Leiberg, J.B. (1899c) *San Gabriel Forest Reserve*, US Geological Survey, 19th Annual Report, Pt. 5, pp. 367–71.

Leiberg, J.B. (1900a) *San Bernardino Forest Reserve*, US Geological Survey, 20th Annual Report, Pt. 5, pp. 429–54.

Leiberg, J.B. (1900b) *San Jacinto Forest Reserve*, US Geological Survey, 20th Annual Report, Pt. 5, pp. 455–78.

Leiberg, J.B. (1900c) *San Gabriel Forest Reserve*, US Geological Survey, 20th Annual Report, Pt. 5, pp. 409–28.

Major, J. (1977) California climate in relation to vegetation, in *Terrestrial Vegetation of California* (eds M.C. Barbour and J. Major), John Wiley and Sons, New York, pp. 11–74.

McPherson, J.K. and Muller, C.H. (1969) Allelopathic effect of *Adenostoma fasciculatum*, 'chamise', in the California chaparral. *Ecological Monographs*, 39, 177–98.

Merino, J., Field, C. and Mooney, H.A. (1982) Construction and maintenance costs of mediterranean-climate evergreen and deciduous leaves. I. Growth and CO_2 exchange analysis. *Oecologia*, **53**, 208–13.

Miller, E.H., Jr (1947) Growth and environmental conditions in southern California chaparral. *American Midland Naturalist*, **37**, 379–420.

Miller, P.C. (1979) Quantitative plant ecology, in *Analysis of Ecosystems* (eds D. Horn, G.R. Stairs and R.D. Mitchell), Ohio State University Press, Columbus, pp. 179–232.

Miller, P.C. (ed.) (1981) *Resource Use by Chaparral and Matorral*, Springer Verlag. New York, Heidelberg, Berlin, 455 pp.

Miller, P.C. and Poole, D.K. (1979) Patterns of water use by shrub in southern California. *Forest Science*, **25**, 84–97.

Miller, P.C. and Poole, D.K. (1980) Partitioning of solar and net irradiance in mixed and chaparral in southern California. *Oecologia*, **47**, 328–32.

Miller, P.C., Bradbury, D.E., Hajek, E., La Marche, V. and Thrower, N.J.W. (1977) Past and present environment, in *Convergent evolution in Chile and California mediterranean climate ecosystems* (ed. H.A. Mooney), Dowden, Hutchinson and Ross, Stroudsburg, PA, pp. 27–72.

Miller, P.C., Hajek, E., Poole, D.K. and Roberts, S.W. (1981) Microclimate and energy exchange, in *Resource use by chaparral and matorral: A comparison of vegetation function in two mediterranean type ecosystems*, (ed. P.C. Miller), Springer-Verlag, New York, Heidelberg, Berlin.

Mooney, H.A. (ed.) (1977) *Convergent Evolution in Chile and California. Mediterranean Climate Ecosystems*, Dowden, Hutchinson and Ross, Stroudsburg, PA, 224 pp.

Mooney, H.A. (1981) Primary production in mediterranean-climate regions, in *Ecosystems of the World*, Vol. XI, *Mediterranean Type Shrublands*, (eds F. di Castri, D.W. Goodall and R.L. Specht), Elsevier Scientific, Amsterdam.

Mooney, H.A. (1982) Habitat, plant form, and plant water relations in mediterranean-climate regions, in Définition et localisation des Ecosystémes méditerraneéns terrestres, *Ecologia Mediterranea VIII*, (ed. P. Quezel), Marseille.

Mooney, H.A. (1983) Carbon-gaining capacity and allocation patterns of mediterranean-climate plants, in *Mediterranean Type Ecosystems. The Role of Nutrients*, (ed. F. Kruger, D.T. Mitohell and J.U.M. Jarvis), Springer Verlag, NY, Heidelberg, Berlin, pp. 103–19.

Mooney, H.A. and Chu, C. (1974) Seasonal allocation in *Heteromeles arbutifolia*, a California evergreen shrub. *Oecologia*, **14**, 295–306.

Mooney, H.A. and Conrad C.E., (Technical Coordinators) (1977) *Proceedings of the Symposium on the Environmental Consequences of Fire and Fuel Management in Mediterranean Ecosystems*, USDA Forest Service. General Technical Report WO-3, 498 pp.

Mooney, H.A. and Dunn, E.L. (1970) Convergent evolution of mediterranean climate evergreen sclerophyll shrubs. *Evolution*, **24**, 292–303.

Mooney, H.A. and Gulmon, S.L. (1979) Environmental and evolutionary constraints in the photosynthetic characteristics of higher plants, in *Topics in Plant Population Biology* (eds O.T. Solbrig, S. Jain, G.B. Johnson and P.H. Raven), Columbia University Press, NY, pp. 316–37.

Mooney, H.A. and Kummerow, J. (1981) Phenological development of plants in mediterranean climate plants, in *Ecosystems of the World*, Vol XI, *Mediterranean Type Shrublands* (eds F. di Castri, D.W. Goodall and R.L. Specht), Elsevier Scientific, Amsterdam.

Mooney, H.A. and Rundel, P.W. (1979) Nutrient relations of the evergreen shrub, *Adenostoma fasciculatum*, in the California chaparral, *Botanical Gazette*, **140**, 109–13.

Mooney, H.A., Troughton, J. and Berry, J. (1974b) Arid climates and photosynthetic systems. *Carnegie Institute Year Book*, **73**, 757–67.

Mooney, H.A., Harrison, A.T. and Morrow, P.A. (1975) Environmental limitations of photosynthesis on a California evergreen shrub. *Oecologia*, **19**, 293–301.

Mooney, H.A., Kummerow, J., Johnson, A.W., Parsons, D.J., Hoffmann, A., Hays, R.I., Giliberto, J. and Chu, C. (1977) *The producers – their resources and adaptive responses*, Dowden, Hutchinson and Ross, Stroudsburg, PA.

Mooney, H.A., Parsons, D.J. and Kummerow, J. (1974a) Plant development in mediterranean climates, in *Phenology and Seasonality Modeling* (ed. H. Lieth), Springer Verlag, NY, Heidelberg, Berlin, pp. 255–67.

Ng, E. and Miller, P.C. (1980) Soil moisture relations in the southern California chaparral. *Ecology*, **61**, 98–107.

Nilsen, E. and Muller, W.H. (1981) Phenology of *Lotus scoparius*. *Ecological Monographs*, **51**, 323–41.

Nilsen, E. and Schlesinger, W.H. (1981) Phenology, productivity and nutrient accumulation in the post-fire shrub *Lotus scoparius*. *Oecologia*, **50**, 217–24.

Oechel, W. and Hastings, S. (1982) The effects of fire on photosynthesis in chaparral resprouts, in *Mediterranean-Type Ecosystems. The Role of Nutrients* (eds F.J. Kruger, D.T. Mitchell, and J.U.M. Jarvis), Springer Verlag, NY, Heidelberg, Berlin.

Oechel, W. and Lawrence, W. (1981) Carbon allocation

and utilization, in *Resource Use by Chaparral and Matorral* (ed. P.C. Miller), Springer Verlag, Heidelberg, Berlin.

Oechel, W., Lawrence, W., Mustafa, J. and Martinez, J. (1981) Energy and carbon acquisition, in *Resource Use by Chaparral and Matorral* (ed. P.C. Miller), Springer Verlag, Heidelberg, Berlin.

Plummer, F.G. (1911) *Chaparral*, US Department of Agriculture, Forest Service Bulletin 85.

Quick, C.R. and Quick, A.S. (1961) Germination of *Ceanothus* seeds. *Madrono*, **16**, 23–30.

Radosevich, S.R., Conard, S.G. and Adams, D.R. (1977) Regrowth responses of chamise following fire, in *Proceedings of the Symposium on the Environmental Consequences of Fire and Fuel Management in Mediterranean Ecosystems* (H.A. Mooney, and C.E. Conrad, Technical Coordinators), USDA, Forest Service, General Technical Report WO–3, pp. 378–82.

Roberts, S.W. (1982) Some recent aspects and problems of chaparral plant water relations, in *Dynamics and Management of Mediterranean-Type Ecosystems* (C.E. Conrad and W.C. Oechel, Technical Coordinators), USDA, Pacific Southwest Forest and Range Experiment Station, General Technical Report PSW-58.

Rundel, P.W. (1983) Ecological relationships of foothill woodland and chaparral communities in the southern Sierra Nevada, California, in *Definition et localisation des Ecosystemes mediterraneens terrestres, Ecologia Mediterranea VIII*, (ed. P. Quezel), Marselle.

Rundel, P.W. and Parsons, D.J. (1980) Nutrient changes in two chaparral shrubs along a fire-induced age-gradient. *American Journal Botany*, **67**, 51–8.

Sampson, A.W. (1944) *Plant Succession of burned chaparral lands in northern California*, California Agricultural Experimental Station Bulletin 635, 145 pp.

Schlesinger, W.H. and Gill, D.S. (1980) Biomass, production, and changes in the availability of light, water and nutrients during the development of pure stands of the chaparral shrub, *Ceanothus megacarpus*, after fire. *Ecology*, **61**, 781–9.

Schlesinger, W.H. and Hasey, M.M. (1980) The nutrient content of precipitation, dry fallout and intercepted aerosols in the chaparral of southern California. *American Midland Naturalist*, **103**, 114–22.

Schlesinger, W.H., Gray, J.T., Gill, D.S. and Mahall, B.E. (1982a) *Ceanothus megacarpus* chaparral: A synthesis of ecosystem processes during development and growth. *Botanical Review*, **48**, 71–117.

Schlesinger, W.H., Gray, J.T. and Gilliam, F.S. (1982b) Atmospheric deposition processes and their importance as sources of nutrients in a chaparral ecosystem and southern California. *Water Resources Research*, **18**, 623–9.

Shaver, G. (1981) Mineral nutrient and nonstructural carbon utilization, in *Resource Use by Chaparral and Matorral*, (ed. P.C. Miller), Springer Verlag, NY, Heidelberg, Berlin.

Shaver, G. (1982) Mineral nutrient and nonstructural carbon pools in shrubs from mediterranean-type ecosystems of California and Chile, mss.

Shreve, F. (1927a) The vegetation of a coastal mountain range. *Ecology*, **8**, 27–45.

Shreve, F. (1927b) The physical conditions of a coastal mountain range. *Ecology*, **8**, 398–414.

Stone, E.C. and Juhren, G. (1951) The effect of fire on the germination of the seed of *Rhus ovata*. *American Journal of Botany*, **38**, 368–72.

Stone, E.C. and Juhren, G. (1953) Fire stimulated germination. *California Agriculture*, **7**, 13–14.

Sweeney, J.R. (1956) Responses of vegetation to fire. *University of California Publications in Botany*, **28**, 143–250.

Thrower, N.J.W. and Bradbury, D.E. (eds) (1977) *Chile-California Mediterranean Scrub Atlas. A Comparative Analysis*, Dowden, Hutchinson and Ross, Stroudsburg, PA, 237 pp.

Vlamis, J.M., Schultz, A.M. and Biswell, H. (1958) Nitrogen-fixation by deerbrush. *California Agriculture*, **12**, 11, 15.

Vogl, R.J. and Schorr, P.K. (1972) Fire and manzanita chaparral in the San Jacinto mountains, California. *Ecology*, **53**, 1179–88.

Webber, I.E. (1936) The woods of sclerophyllous and desert shrubs of California. *American Journal of Botany*, **23**, 181–8.

Wells, P.V. (1969) The relation between mode of reproduction and extent of speciation in woody genera of the California chaparral. *Evolution*, **23**, 264–7.

Westman, W.F. (1981) Seasonal dimorphism of foliage in Californian coastal sage scrub. *Oecologia*, **51**, 385–8.

Wieslander, A.E. and Gleason, C.H. (1954) *Major brushland areas of the Coast Ranges and Sierra-Cascade foothills in California*, Californian Forest and Range Experimental Station Miscellaneous Paper 15.

Whitfield, C.J. (1932) Osmotic concentration of chaparral coastal sagebrush and dune species of southern California. *Ecology*, **13**, 279–85.

Wicklow, D.T. (1977) Germination response in *Emmenanthe penduliflora* (Hydrophyllaceae). *Ecology*, **58**, 201–5.

11

Grasslands

PAUL G. RISSER

11.1 Introduction

Grasslands are biological communities containing few trees and characterized by mixed herbaceous vegetation dominated by grasses. Although grassland types range from the dense bamboo of the tropics to the northern continental steppes, from dry plains to arctic grasslands, and though they occur on every continent and large island throughout the world, certain abiotic and biotic features are common to most grasslands.

Regional climate directly controls the biotic components and ultimately the associated soils and their development. In general, natural grasslands occur in localities which are too arid to support fully developed forest but are not so adverse as to preclude the development of a closed perennial herbaceous layer (Coupland, 1979). Precipitation patterns are markedly seasonal so that even though annual rainfall ranges from 200 to as much as 1000 mm, droughts of several weeks to several months are typical. Many grassland climates also are characterized by high temperatures and wind, thus exacerbating the dry conditions (Risser *et al.*, 1981). In temperate grassland soils, leaching is restricted because of the scarcity of rainfall, impermeable layers may be formed at the depth of normal percolation, and organic matter may accumulate because of the relatively slow rate of decomposition under dry conditions.

Grasslands in North America contain a total of approximately 600 genera and 7500 species of grasses (Gramineae); additionally, numerous grasslike plants (sedges and rushes), forbs, and woody plants are characteristic of these communities. Composites (Compositae), sedges (Cyperaceae), and legumes (Leguminosae) are important families in the grassland. In fact, usually far more species of these latter taxa are present, but grasses constitute most of the biomass.

By the late Cretaceous period, when grasses underwent major diversification, mammals had already differentiated from their reptilian ancestors, but it was not until the Paleocene that several large groups of herbivorous mammals emerged (Van Valen and Sloan, 1966). The important point, however, is that grasses have been subjected to grazing pressure for millions of years (Singh *et al.*, 1983). Grasslands have also been subjected to fires and burning caused by lightning and humans (Daubenmire, 1968).

North American grasslands have been subdivided in various ways, but it is useful to distinguish at least six types. In the Great Plains and Central Lowlands, the tallgrass prairie occupies the eastern portion, which receives more rainfall than the mixed prairie and the shortgrass steppe further west. The tallgrass prairie is dominated by big bluestem (*Andropogon gerardii*), little bluestem (*Schizachyrium scoparium*), Indiangrass (*Sorghastrum nutans*), and switchgrass (*Panicum virgatum*). The shortgrass steppe occupies the driest portion of the Great Plains and is dominated by blue grama (*Bouteloua gracilis*) and buffalo grass (*Buchloe dactyloides*). In between, the mixed-grass prairie contains species from the tallgrass prairie and shortgrass steppe but also includes dominant species of the genera *Stipa* and *Agropyron*. Desert grassland dominants are frequently grama grasses but include woody species, such as mesquite (*Prosopis* sp.) and creosote bush (*Larrea tridentata*). The annual grassland occurs in the Central Valley of California and is largely dominated by annual species of *Bromus* and *Avena*. In the arid north-western United States, the bunchgrass steppe is characterized by bluebunch wheatgrass (*Agropyron spicatum*) and sagebrush (*Artemisia* sp.). Other sub-divisions, e.g. mountain grassland, alpine meadow, and coastal prairie, have been distinguished in various classifications (Risser *et al.*, 1981; Singh *et al.*, 1983).

The environmental 'demands' on grassland species have required adaptations allowing plants to persist in a complex ecosystem subjected to grazing, burning, frequently large temperature extremes, and always the climatic and soil conditions associated with drought. This chapter discusses the physiological adaptive strategies that have evolved in plants of these grassland habitats, with an emphasis on north temperate grasslands.

11.2 Plant response to environmental conditions

There are numerous aspects of the environment with which grassland plants must contend. For convenience, these aspects are typically categorized as variables that we can measure and for which a plant response can be shown. This simplistic logic will be used in the following pages although it must be understood that these factors are defined by man, that they interact in complex ways which change during the ontogeny of the individual, and finally, that the factors are influenced by both the biotic and abiotic components of the ecosystem. The sub-sequent sections of this chapter deal with physiological mechanisms which have evolved to permit grassland plants to survive under specific conditions of water, solar radiation and temperature, soil nutrients, grazing, and burning.

11.2.1 WATER

The climates and habitats of grasslands are usually characterized either by recurring drought periods or by soils of limited water storage capacity. The exceptions are the wet meadows and marshlands, which are composed of vegetation capable of withstanding wet conditions for prolonged periods. Annual precipitation ranges from about 200–1000 mm yr^{-1}, and potential evapotranspiration may range up to 1500 mm yr^{-1}, or five times greater than precipitation (Webb *et al.*, 1983). For the purpose of this review, emphasis will be placed on the morphological and physiological mechanisms which permit the persistence of grassland plants. There are a number of excellent general reviews of the relations between water deficits and various physiological processes (Bewley, 1979; R.W. Brown, 1977; Hsiao, 1973; Kozlowski, 1972; Levitt, 1972).

Available soil moisture affects the success of seeds and seedlings even more than it affects mature plants (Knipe, 1968; LaGory *et al.*, 1982). The magnitude of reproduction by seeds is variable in grasslands (Quinton *et al.*, 1982), even those dominated by annuals (Jain, 1979), although most species produce many seeds (Canode and Law, 1975; Weaver, 1954). Various strategies for seed germination have been discussed (Angevine and Chabot, 1979; Grime, 1979; Jain, 1979), but seed dormancy and delayed germination are characteristics of most grassland species. In greenhouse experiments, Robocker *et al.* (1953) found that seeds of some species germinated best in the first year after harvest, e.g. little bluestem and sideoats grama (*Bouteloua curtipendula*); some in the second year, e.g. Canada wildrye (*Elymus canadensis*), Virginia wildrye (*E. virginicus*); and some in the third, e.g. green needlegrass (*Stipa viridula*), sand dropseed (*Sporobolus cryptandrus*); Indiangrass germinated equally well when planted in the first, second, or third year after harvest. For all 10 species used in the experiment, seedlings emerged from less than 50% of the apparently sound seed planted.

Although the ultimate success of a seedling relates to a number of considerations, ranging from vegetation structure to soil conditions (Blake, 1935;

Havercamp and Whitney, 1983; Rabinowitz, 1978; Rabinowitz *et al.*, 1979), the initial germination process requires hydration of the seed and sufficient soil moisture for seedling growth (McDonough, 1977; Oomes and Elberse, 1976). In addition to seed coat properties, both soil osmotic and matric water potentials can be important in determining the availability of water to seeds. Halophytes may tolerate high osmotic stress, but in most rangeland plants germination is reduced at high solute concentration though not all species are equally susceptible (Knipe, 1973; Young *et al.*, 1970). Alternate wetting and drying cycles may increase the germination of a number of range species (Griswald, 1936; Maynard and Gates, 1963). Decreasing soil matric water potential may also decrease germination (Harper and Benton, 1966), but this effect is influenced by seed size and shapes as well as the composition of the seed coat (McDonough, 1977).

On the high plains of North America, considerable difficulty has been experienced in establishing stands of blue grama (Hyder *et al.*, 1971; Reichhardt, 1982; Wilson and Sarles, 1978) for several reasons. Blue grama originates adventitious roots near the soil surface where moisture conditions seldom favor root development; its thread-like subcoleoptile internode and the seminal primary root have a limited capacity for water uptake; seminal lateral roots are absent; and the seedling leaf area overexpands during cool, humid weather. Further, when transpiration exceeds the rate of water supply, desiccation and death of seedlings results. Although blue grama seedlings normally are not very successful, established seedlings can survive in soils with a water potential of from -2.0 to -4.0 MPa (Wilson and Sarles, 1978). Blue grama seedlings, though they may extend roots to a depth of 52 cm in 50 days (Wilson and Briske, 1978), do not withstand drought unless adventitious roots are developed. However, the development of adventitious roots may require several days with a moist soil surface at temperatures of about 15 °C (Briske and Wilson, 1977). In contrast to the shallow origin of blue grama adventitious roots, those of crested wheatgrass (*Agropyron cristatum*) arise deeper on the seedling axis where soil moisture is more

dependable (Wilson and Briske, 1978).

Although drought stress frequently favors root growth relative to shoot growth (Davidson, 1969; Sosebee and Wiebe, 1971), and it has been suggested that deep or extensive root systems enhance a plant's competitive ability (Bookman and Mack, 1982) by increasing its water and nutrient uptake (Aspinall, 1960; Berendse, 1979, 1981), grassland species behave differently with respect to growth in dry soils (Koshi *et al.*, 1982; Weaver, 1982; Weaver and Darland, 1949). At or below the permanent wilting point, the roots of sideoats grama and weeping lovegrass (*Eragrostis curvula*) penetrate the soil extensively. Using field observation chambers, Ares (1976) found that blue grama roots began to grow and differentiate a short time before shoot growth was apparent. The drying of the soil during the summer resulted in the death of 30–60% of the roots which were formed in that growing season. Massive root growth occurred again when soil water was abundant near the end of the growing season. He also found that the young, nonsuberized roots, which are particularly active in water uptake, were concentrated in the regions of the soil profile where soil water potential was highest throughout the season. Diurnal patterns also occur in the absorption of water from various layers in the soil profile. Sala *et al.* (1981) concluded that when the upper soil layers are depleted of water, the amount of root extension in the deeper soil layers, not axil root resistance, limited water absorption.

In earlier descriptive studies, Weaver and Albertson (1956) found that western wheatgrass had a coarse, profusely branched root system extending to a depth of 1.8 m, whereas blue grama had a fine root system with a depth of about 1.0 m. The authors suggested that these differences in rooting pattern resulted in an underground niche separation, western wheatgrass roots using deeper soil moisture, perhaps stored from winter precipitation and available in the early spring, and roots of blue grama utilizing surface moisture distributed later in the growing season after convectional storms. Other studies have shown a stratification of rooting depths of associated species and differences in root phenology (Berendse, 1979, 1981; Bookman and Mack, 1982; Dalrymple and Dwyer, 1967; Harris,

1977; Parrish and Bazzaz, 1976; Weaver and Darland, 1949).

Short grasses might be expected to persist on clay soils because their fine roots would extract water which is held tightly on soil particles. However, White and Lewis (1969) found that the short grasses, blue grama and buffalo grass, did not persist on dense clays, whereas a species of the mixed-grass prairie, green needlegrass, was successful. In general, dominant grasses of the shortgrass steppe have a fine, spreading root system near the soil surface, and species of the mixed-grass prairie are likely to have longer, more deeply placed roots which are more vertically oriented. Thus, as discussed above, the mixed grasses are better able to use the subsoil moisture. Also, the fine roots of the short grasses are apparently less likely to withstand the shrinking and swelling forces of the dense clays as the soil moisture changes throughout the year.

Drought may increase carbohydrate reserves in some grasses (Brown and Blaser, 1965) but decrease reserves in other species (Bukey and Weaver, 1939; Dina and Klikoff, 1973). Trlica and Cook (1972) found that irrigation decreased carbohydrate reserves in the crowns of crested wheatgrass and Russian wild rye (*Elymus junceus*). By contrast, blue grama seedlings grown in growth chambers (Bokhari *et al.*, 1974) showed a many-fold net increase in both labile (nonstructural) and nonlabile (structural) carbohydrates after the addition of water. There was a decline in labile components of the shoot, with the increases occurring in the root system due primarily to the translocation of current photosynthate belowground. These carbohydrate responses indicate reserves susceptible to use under stress conditions, and also demonstrate regrowth capabilities once the water shortage is alleviated.

High positive correlations have consistently been found between available soil moisture and forage production (Cable, 1971; Mohammad *et al.*, 1982; Shiflet and Dietz, 1974), which relate to the decrease in net photosynthesis with decreasing leaf water potentials (Davies and Kozlowski, 1977; Doley and Trirett, 1974; Kemp and Williams, 1980; Ludlow and Ng, 1976). The relationship between water stress and photosynthesis is complicated by phe-nology, since many species (e.g. squirreltail, *Sitanion hystrix*) are adapted for growth during the most favorable time of year, thus reducing the likelihood of encountering severe moisture stress (Clary, 1975).

Perhaps the dominant theme in the last decade in research concerning physiological adaptations of range plants has been the discovery and subsequent investigations of the C_3 and C_4 carbon-fixation pathways. Comparisons of these pathways have been adequately discussed in numerous papers and will not be exhaustively treated here (Björkman, 1976; Boutton *et al.*, 1980. Ehleringer, 1978; Ehleringer and Björkman, 1977; Waller and Lewis, 1979). Surprisingly few unambiguous generalizations are possible about the ecological significance of the two pathways. Although this situation may be in part because of the variations in enzyme activities and leaf anatomy (W.V. Brown, 1977; Hatch *et al.*, 1975; Kemp *et al.*, 1983), it is important to recognize that the photosynthetic pathway is only one of a myriad of adaptive strategies employed by rangeland plants. The C_4 pathway serves as a mechanism to concentrate CO_2 within the chloroplast environment, where the carboxylation of ribulose biphosphate (RuBP) by RuBP carboxylase – oxygenase occurs (Björkman, 1971; Ehleringer, 1978; Hatch, 1971). It would be expected that the advantage of the C_4 pathway over the C_3 pathway would be greatest under conditions where photosynthesis is limited by intercellular CO_2 concentration: high light intensities, high leaf temperatures, and reduced stomatal conductances.

Big galleta (*Hilaria rigida*) is a hardy grass of the western plains that can be used as an example. This species displays numerous C_4 characteristics, such as Kranz anatomy, a low CO_2 compensation point, lack of light saturation at full sunlight, and high photosynthetic ($67 \, mol \, CO_2 \, m^{-2} \, s^{-1}$) rates under optimal field conditions (Nobel, 1980). Mesophyll cells of C_4 plants contain phosphoenolpyruvate (PEP) carboxylase, which has a high affinity for CO_2, and as a result, decarboxylation conductance may be relatively unimportant in cell conductance. Since (PEP) carboxylase is located in the cytoplasm of C_4 mesophyll cells, the diffusion barrier of the chloroplast membrane does not have to be overcome, and thus, maximal cell conductance rates would be higher than those expected for C$_3$ plants. Nobel (1980) reports that big galleta, a C$_4$ plant, has the highest cell conductance so far reported in the literature.

Greater ratios of assimilation rate to transpiration rate, generally referred to as water use efficiencies, have been found in many C$_4$ plants as compared with those of C$_3$ plants (Black, 1971, 1973; Downes, 1971; El-Sharkawy and Hesketh, 1965; Imbamba and Tieszen, 1977; Shantz and Pimeisel, 1927; Slatyer, 1976; Weaver, 1949). This finding suggests that C$_4$ species are better adapted than are C$_3$ species for rangeland regions with extended periods of drought. It should be noted, however, that even C$_3$ grass species can exhibit high leaf diffusion resistances yet maintain relatively high rates of photosynthesis (Frank and Barker, 1976) to produce high water use efficiency.

Some authors have noted that the C$_3$ and C$_4$ pathways may be adaptations to deficient moisture conditions, as indicated by their different geographical and habitat distributions (Downton *et al.*, 1969; Mooney *et al.*, 1974; Teeri and Stowe, 1976; Tieszen *et al.*, 1979). These authors and others have recognized that the potential adaptive advantage of C$_4$ plants in dry habitats has exceptions, and different taxa may be afforded different advantages by the two pathways (Baskin and Baskin, 1978; Doliner and Jolliffe, 1979; Stowe and Teeri, 1978). In fact, several authors have concluded that the C$_4$ pathway does not offer any particular ability to withstand low leaf water potentials (Boyer, 1970; Mederski *et al.*, 1975).

The effect of water stress on net photosynthesis in blue grama (C$_4$) and two C$_3$ species, green needlegrass and sand dropseed, illustrates the difficulty of drawing ecological generalizations based on photosynthetic pathways. In blue grama and green needlegrass, net photosynthesis declined to 25% of the maximum rate as moisture stress increased to -3.0 MPa soil water potential (Redmann, 1971). At the same soil water potential, net photosynthesis in sand dropseed was zero. Kemp and Williams (1980) found that the photosynthetic mechanism of blue grama (C$_4$) was not more resistant to low leaf water potential than was that of western wheatgrass (C$_3$). If the plants were not water stressed, net photosyn-

thesis was limited primarily by stomatal conductance in blue grama and by biochemical factors in western wheatgrass. Photosynthesis decreased exponentially with decreasing leaf water potential in both species, and the photosynthetic decrease resulted from decreases in both internal leaf and stomatal conductances. Comparisons of other C$_3$ and C$_4$ species have shown that the control of gas exchange is dominated by internal conductance in C$_3$ species and by stomatal conductance in C$_4$ species (Gifford, 1974; Rawson *et al.*, 1977).

Under cycles of changing soil water potential, the photosynthetic rate of western wheatgrass (C$_3$) was more sensitive to water stress than was that of blue grama, a C$_4$ species (Brown and Trlica, 1977). In these studies, photorespiration (CO$_2$ efflux into CO$_2$-free air) rates decreased as leaf water potential decreased. Kemp and Williams (1980) also found a decline of photorespiration as the leaf water potential decreased in western wheatgrass. A number of other C$_3$ species show decreases in photorespiration with decreasing leaf water potential (Bunce and Miller, 1976; Lawlor, 1976), but some species native in arid habitats show increases in photorespiration with declining leaf water potential. Bunce and Miller (1976) suggest that this increased photorespiration may be related to maintaining high chloroplast activity in plants adapted to arid enviornments. Since photorespiration is not conveniently detectable in C$_4$ plants by gas exchange methods, there have been no studies on the effects of water stress on photorespiration in C$_4$ species. Brown and Trlica (1977) and Kemp and Williams (1980) showed that in blue grama and western wheatgrass dark respiration decreased with decreasing leaf water potential. Although this may be an adaptive response to low water availability, as both the C$_3$ and C$_4$ species responded similarly, this characteristic was not judged to be important in the niche separation of these two species (Kemp and Williams, 1980).

The capacity for CO$_2$ fixation by roots has been demonstrated in many C$_3$ and C$_4$ species, and the relatively small amount of CO$_2$ uptake may be important in root growth and metabolism (Splittstoesser, 1966). The net CO$_2$ uptake in the roots of some blue grama plants under water stress could be a direct result of increased CO_2 fixation supplying carbon for roots as the CO_2 from the leaves decreases with increasing plant water stress. However, the relative increase in CO_2 uptake might result indirectly from the large reduction in mitochondrial respiration that occurs in water-stressed tissue (Gerwick and Williams, 1978; Greenway and West, 1973; Kemp and Williams, 1980).

Rangeland plants demonstrate a wide array of physiological and morphological adaptions to drought stress. Leaves may have heavy pubescence and few stomata and may curl or roll inwardly under dry conditions; roots may penetrate deeply in the soil or exist as a fibrous mat efficiently using even small amounts of precipitation. The physiological processes, especially photosynthesis and respiration, may function even though soil water potential is as low as -3.0 MPa. While many of these adaptations have been found in a number of plant species, no one suite of morphological or physiological adaptations is found in all grassland plants.

11.2.2 SOLAR RADIATION AND TEMPERATURE

Although cultivated crops may have a solar energy capture efficiency of more than 5%, most native rangelands have an efficiency of no more than 1–2% (Sims and Singh, 1978). Incident growing-season solar radiation ($J\,m^{-2}$) for grasslands throughout the western United States ranges from about $33-75 \times 10^8$. Comparisons of energy capture efficiencies are sometimes confused, because energy input and capture are based on different measurements; that is, energy input has been variously calculated as total year-long solar radiation, growing-season radiation, photosynthetically active radiation, etc. Nevertheless, in comparing growing-season photosynthetically active radiation, Sims and Singh (1978) found that the efficiency of energy capture in total net production ranged from 0.12 to 1.40% for both grazed and ungrazed grasslands across the western United States. Plant communities dominated by cool-season plants were comparable to or more efficient in energy capture than were communities dominated by warm-season plants. Grasslands having higher water use efficiencies for total net production also had greater efficiency of energy capture. Comparison of

grasslands indicated an apparent linear increase in annual net production with increasing precipitation up to approximately $500\,mm\,yr^{-1}$. Likewise, annual net production increased linearly with increases in growing season and annual actual evapotranspiration.

Solar radiation capture by individual plants is a function of several characteristics, including leaf size, angle of display, pubescence, age, and physiological conditions. In general, temperate C_3 plants usually show light saturation at lower light intensities, whereas C_4 species may continue to show an increase in photosynthesis as light intensity increases to high values (Cooper and Tainton, 1968). Early evaluations of the C_3–C_4 pathways suggested that, in general, C_4 species have maximum photosynthetic rates exceeding those of C_3 species. However, C_3 species do exist which possess maximum photosynthetic rates on a leaf-area basis as high as those known for C_4 species (Ehleringer *et al.*, 1976; McNaughton and Fullem, 1970; Mooney *et al.*, 1976).

Several observations indicate that temperature may determine the relative abundances of C_3 and C_4 species in different regions. At low leaf temperatures, quantum yields of C_3 plants are greater than those of C_4 plants, reflecting the lower intrinsic energy cost of the C_3 pathway (Ehleringer and Björkman, 1977). Unlike single leaves, primary production of an entire canopy is often light limited. Thus, lower quantum yields at low temperatures may be quite disadvantageous. The lower quantum yields in C_4 species at low and moderate temperatures are probably an important factor in limiting the geographical distribution of C_4 grasses (Ehleringer, 1978). Teeri and Stowe (1976) found that high average July minimum temperature had the strongest correlation with the relative abundance of C_4 grasses in a regional flora. The fact that locations with a normal July minimum temperature below about 8°C appear to have few or no C_4 grass species further supports the hypothesis that intolerance of low temperature during growth may be a general property of C_4 taxa (Slack *et al.*, 1974). Temperature was a better predictor of C_4 species distributions in Europe and California than were a number of other factors, such as light, soil nitrogen, soil salinity, and continentality of climate (Doliner and Jolliffe, 1979). Hartley (1973) demonstrated a correlation between decreasing mean July temperature and increasing percentages of Festucoid (C_3) grasses in many regions of the United States, but there are some exceptions to these general trends (Allen, 1982a, Baskin and Baskin, 1978).

Surveys of photosynthetic response to temperature (Cooper and Tainton, 1968) have shown that in most temperate Festucoid grasses (e.g. orchard grass, *Dactylis glomerata;* colonial bengrass, *Agrostis tenuis,* and timothy, *Phleum pratense*) the optimum temperature for growth is between 20 and 25 °C and that growth rate diminishes rapidly below about 10 °C. Even with ample water supply, growth rate is reduced at temperatures above 25 °C and frequently ceases above 35 °C. Subtropical and tropical grasses, which are mainly non-Festucoid (e.g. Bermuda grass, *Cynodon dactylon,* axonopus, *Axonopus affinis;* Dallis grass, *Paspalum dilatatum;* and weeping lovegrass, *Eragrostis curvala*) have a much higher optimum temperature for growth of 30–35 °C. In contrast, their growth at 10–15 °C is quite slow. Under laboratory conditions, optimum temperatures for some species are subject to acclimation. For example, the optimum temperature for photosynthesis in big galleta was raised from 29 to 43 °C (Nobel, 1980). It has been shown that orchard grass not only shifts the optimum temperature for production, but at higher temperatures reduces the proportion of reducing carbohydrates (Baker and Jung, 1972). It is not clear whether temperature influenced the type of carbohydrate synthesized or the conversion of reducing to non-reducing carbohydrate (frutose to fructosan) or both. Some species, such as western wheatgrass and blue grama, do not undergo physiological acclimation (Kemp and Williams, 1980).

Temperature also affects the rate of photosynthate transport. Plants using the C_4 photosynthetic pathway export photosynthates more rapidly from their leaves than do C_3 plants (Lush and Evans, 1974; Stephenson *et al.,* 1976) although the C_4 plants store a proportion of the daily photosynthate in the bundle-sheath cells as starch, which is mobilized during the following night. With low night temperatures many C_4 plants have reduced growth rates (e.g. in a tropical crabgrass, *Digitaria decumbens,* at night temperatures of 10 °C) associated with failure to mobilize the starch in the bundle-sheath chloroplasts. This failure to mobilize the starch is followed by a reduction in the net rate of photosynthesis even at day temperatures of 30 °C (Hilliard and West, 1970). Species not well adapted to cool growing seasons seem to retain a large fraction (30–40%) of current photosynthate in their leaves to be translocated during the night period (Lush and Evans, 1974).

Kemp and Williams (1980) grew western wheatgrass and blue grama in warm (35–15 °C) and cool (20–15 °C) temperature regimes, and after 8 weeks, western wheatgrass had accumulated twice as much biomass at the cool temperature as it had at the warmer temperature. Blue grama had accumulated 2.7 times more biomass at the warmer temperature, and the difference between growth at the two temperatures was due largely to the shoot biomass. In this experiment, the proportion of root: shoot biomass was not affected by the prevailing temperatures. Others have argued that on a geographical basis, root growth is increased relatively at lower temperatures and shoot growth at higher temperatures, the shoot:root ratio steadily increasing with temperatures (Cooper and Tainton, 1968).

Rates of root respiration in western wheatgrass and blue grama (Kemp and Williams, 1980) corresponded to photosynthetic activities, that is, western wheatgrass root respiration was comparatively greater at cool temperatures, while blue grama root respiration was greater at the warmer temperatures. Changes in the aboveground environment can affect root respiration (Osman, 1971) and may be related to photosynthesis and the subsequent flow of assimilate to the roots (Pearson and Hunt, 1972). Therefore, greater rates of root respiration of blue grama and western wheatgrass at optimum photosynthesis temperatures may be a function of greater assimilate transport to the roots (Kemp and Williams, 1980). This assimilate transport to the roots would permit maximum root growth from respiratory energy generated in the roots, which is necessary for nutrient uptake and the turnover of cellular compounds (Penning de Vries, 1975).

During the early seasonal growth of blue grama,

most of the current photosynthate is used for the structural development of the plant, as indicated by high nonlabile: labile carbohydrate ratios (Bokhari *et al.*, 1974). After the maturing plants attain their optimum photosynthetic capacity, there is some storage of surplus carbohydrates, presumably to be used for maintenance as photosynthesis declines later in the season. Most of the carbohydrates are stored in a labile form, immediately available for regrowth. There is an accumulation of labile energy in all blue grama plant parts, as exhibited by decreasing nonlabile: labile ratios at higher temperatures. The addition of water and nitrogen also increases labile carbohydrates at all the tested temperatures, thus increasing both the amount and quality of the forage regardless of the temperature regime.

Clearly, physiological responses to soil moisture, solar radiation, and temperature are interrelated. In some instances, one environmental factor appears to be clearly dominant. In their studies of western wheatgrass and blue grama, Kemp and Williams (1980) concluded that different responses to temperature were particularly important in establishing habitat preference within the grassland biome, since both species responded in a similar manner to water stress. The importance of temperature in affecting the latitudinal abundance of C_3 and C_4 species was noted earlier. However, these interactions occur within the context of the general drought adaptedness of most grassland species.

11.2.3 NUTRIENTS

Over a long period of time numerous studies have been conducted on the growth response of grassland species to the application of various nutrients (e.g. Charley, 1977; Daniel, 1932; Harper *et al.*, 1933; Rauzi and Fairbourn, 1983; White *et al.*, 1983), but relatively few studies have dealt with the physiological behavior of individual species (Eckert and Spencer, 1982; Everitt *et al.*, 1982; Hinnant and Kothmann, 1982; Pemadasa, 1981; Rice, 1950; Weiner, 1980). Even fewer investigations have addressed comparative physiological adaptations of grassland species within a given community (e.g. Allen, 1982b; Christie and Detling, 1982).

Although water is frequently considered to be the principal controlling variable in grasslands, nitrogen has been reported as the critically limiting factor in bluestem range production in the Flint Hills in eastern Kansas (Owensby *et al.*, 1970a). Therefore, it is useful to consider the physiological strategies characteristic of rangeland plants which permit individuals to exist in various nutrient regimes.

Nutrient uptake by plants does not occur uniformly throughout the year (Hickman, 1975). Big bluestem and Indiangrass accumulate 85% and 95%, respectively, of their annual herbage nitrogen requirements between 15 May and 15 June (McKendrick *et al.*, 1975). As might be expected from this differential uptake pattern, there are considerable intraseasonal differences in the nutrient contents of plant parts (Charley, 1977; Rauzi, 1975). Nutrient uptake rates and contents are a function not only of phenology, but of weather conditions, microbial activity (Clark, 1977), and associated vegetation (Charley and West, 1975). In a field irrigation and fertilization study, Power (1971) found that smooth brome (*Bromus inermis*) growth responded to added nitrogen more than to irrigation in a soil poor in nitrogen, but responded more to irrigation than to nitrogen fertilization in a soil high in nitrogen. Lauenroth *et al.* (1978) found that both western wheatgrass and blue grama increased in biomass after irrigation in a shortgrass prairie, but responded only slightly to nitrogen fertilization unless water was also added, suggesting a relationship between water and nutrients in the shortgrass prairie.

In the tallgrass prairie, Owensby *et al.* (1970a) found that added moisture alone generally failed to increase total herbage yields. However, supplemental nitrogen increased yields and also the efficiency of water use. In fact, other investigators (Viets, 1962; Wight and Black, 1972) have reported that fertilization frequently increases the water use efficiency of native plants, especially when water is nonlimiting, due to increased growth.

Brown (1978) reported that C_4 plants have a greater nitrogen use efficiency as compared with that of C_3 plants, which may give C_4 plants an adaptive advantage, especially on sites relatively low in nitrogen. There is some evidence of a greater abundance of C_4 plants on low fertility range soils (Waller and Lewis, 1979; White, 1961).

Substantial quantities of nitrogen can enter plant communities through symbiotic fixation of nitrogen by legumes (DuBois and Kapustka, 1983). In agronomic plant communities, this nitrogen fixation rate may range between 64 and 648 kg N ha^{-1} yr^{-1} (Phillips, 1980; Williams *et al.*, 1964). Using lysimeter studies, Jones *et al.* (1974) found that a subterranean clover grass mixture yielded 180 kg N ha^{-1} yr^{-1} averaged over 4 yr. The grass growing alone yielded only 50 kg N ha^{-1}; thus, the legumes apparently fixed 130 kg N ha^{-1} and were an important contributor of available N to the system. In another experiment with plants at a density of 10 plants m^{-2}, clover obtained 50% of its nitrogen from N$_2$ when grass was absent. When an identical density of plants was composed of a 50% grass and 50% clover mixture, the clover obtained 84% of its nitrogen from N$_2$, but the total N$_2$ fixation decreased from 58 to 21 kg N ha^{-1}. So, while symbiotic fixation of nitrogen is an important process in community production (Dobson and Beaty, 1977), it is clear that over the long term, grasses outcompete clover for soil nitrogen. The excretory products of grazing animals are also important in transferring nitrogen from legumes to nonlegumes. Although there is some evidence of the transfer of fixed N$_2$–N through the roots of legumes to grass, most of the nitrogen is probably released by dying or dead plant parts. In general, living legume plant parts release little nitrogen into the rhizosphere (Simpson, 1965). NH$_3$ as a decay product may constitute a significant portion of the grassland's total notrogen intake (Denmead *et al.*, 1976). In highly productive Australian grasslands with large inputs of nitrogen from subterranean clover, significant amounts of NH$_3$ were lost from the soils, but most of this gas was absorbed by the foliage as it diffused through the plant canopy. As noted previously, however, nitrogen uptake and response to nitrogen are frequently influenced by soil moisture and fertility. Jones (1967) found that the uptake of nitrogen was 230 kg ha^{-1} when subterranean clover was fertilized with phosphorus and sulfur in a wet year, but only 85 kg ha^{-1} in a dry year.

The addition of nitrogen has been shown to increase, decrease, or have no effect on carbohydrate reserves (Trlica, 1977). For example, many reports (Colby *et al.*, 1965; Drake *et al.*, 1963) indicate that fertilization reduces carbohydrate concentration, a fact that has been attributed to increased use of the carbon skeleton for amino acid and protein synthesis. By contrast, others (Adegbola and McKell, 1966; Bokhari and Dyer, 1974; Murata, 1969) have reported that nitrogen fertilization increased carbohydrate reserves in many grasses. The nonstructural carbohydrates are the major energy source for growth, maintenance, and grazing-induced regrowth in perennial grasses.

In the shortgrass prairie, supplemental nitrogen caused greater growth in the aboveground components, but the root system always had a higher concentration of total nonstructural carbohydrates than the aboveground components. The greater absolute storage of total nonstructural carbohydrates in the aboveground biomass brought about by increased nitrogen was the result of higher biomass, not a greater concentration of total nonstructural carbohydrates (Bokhari, 1978). With increasing nitrogen fertilizer, the concentration of sucrose and fructosan in stems, stolons, roots, and rhizomes of Bermudagrass decreased (Adegbola and McKell, 1966). These authors further noted that the regrowth potential was positively associated with the level of carbohydrate reserves available at the beginning of the growing season.

Decreased total available carbohydrate reserves in buffalo grass during vegetative and reproductive growth were found after application of nitrogen fertilizer (Pettit and Fagan, 1974). However, available carbohydrates accumulated more rapidly during later phenological stages in plants that had received heavy nitrogen applications. With nitrogen additions, total nonstructural carbohydrates in stem bases of green needlegrass decreased, but this decrease occurred only from the time of growth initiation until the second leaf was formed (White *et al.*, 1972). So it appears that nitrogen fertilization results in decreased carbohydrate reserves during early season growth, but carbohydrate reserves increase later in the season after photosynthetic tissue has been added (Trlica, 1977).

The nitrogen that enters green herbage in the

years following fertilization is either mineralized from organic compounds or has overwintered in the roots and crowns and then is translocated to the tops of the plants. The organic material mineralization pathway may be a very efficient mechanism for shunting nitrogen back to the plant although internal translocation is even more efficient. The rhizomes of big bluestem more than doubled in nitrogen content between July (0.9%) and the following January (2.0%), but then decreased to 0.5% by the spring, thus indicating the dynamic internal cycling of nitrogen accumulation during the growing season and then the depletion during early spring growth of the nitrogen stored in the rhizomes (McKendrick *et al.*, 1975). In blue grama, under a much drier climate, the quantity of nitrogen in the aboveground parts, including live and dead components, decreased by an average of 33% from June to October over a 5 yr period. This change was ascribed to translocation because the leachates of graminoid leaves show negligible amounts of nitrogen in any form and only small amounts of leachable organic compounds (Clark, 1977).

As the pervious discussion indicates, most of the studies of nutrients in grasslands have been aimed at nitrogen, with some attention being paid to phosphorus, especially as the latter contributes to nitrogen fixation in legumes. Both phosphorus and nitrogen may produce a significant yield response, though this response is usually mediated by adequate moisture and interacts with seasonal growth patterns. Because of the general nitrogen limitation, grasses are highly efficient at acquiring nitrogen from the soil and retaining it through internal translocation from senescing to growing parts.

11.2.4 GRAZING

Ecologists now recognize that plants possess a wide array of antiherbivore devices against livestock and other herbivores (Caswell *et al.*, 1973; Denno *et al.*, 1980; Ehrlich and Raven, 1964; Fraenkel, 1959; Freeland and Janzen, 1974; Mooney and Gulmon, 1982). In addition to chemical barriers found in many dicots, grasses frequently contain high amounts of silica which is abrasive to the grinding teeth of grazers; lignin, which limits digestion; and

phenolics, which may be toxic to the grazers themselves or to their internal microbial activities (Singh *et al.*, 1983). There are examples of chemical stimulation of plants by the grazers (Dyer, 1980; Dyer and Bokhari, 1976; Reardon *et al.*, 1972. 1974), but to date the evidence for this process is limited. While it is true that plants and herbivores have evolved a number of apparently mutually advantageous characteristics, that this system represents a highly co-evolved, mutualistic relationship (Owen, 1980; Owen and Wiegert, 1981, 1982) is open to discussion (Herrera, 1982; Silverton, 1982). This chapter addresses only morphological and physiological plant adaptations for livestock grazing.

The apical meristems in many range plants are inaccessibile to grazing (Dahl and Hyder, 1977). Furthermore, in most grasses the leaf intercalary meristems remain in a basal position, permitting regrowth after grazing. Grass species such as Bermudagrass, Kentucky bluegrass (*Poa pratensis*), squirreltail grass, prairie sandreed (*Calamovilia longifolia*), *Stipa* species, and most *Bouteloua* species are characterized by short growing shoots, and when being grazed on, these species can maintain a much higher proportion of photosynthetic tissue than can species with long shoots. In plants with long shoots, such as Indiangrass, western wheatgrass, switchgrass, and Johnsongrass (*Sorghum halepense*), the entire leaf complement is elevated and vulnerable to grazing.

Another morphological characteristic which determines grazing impact is the ratio of vegetative to reproductive shoots. Because of their importance to regrowth, those species producing a high proportion of reproductive shoots have been considered to be less resistant to continuous heavy grazing than those species where a high proportion of the shoots remain vegetative (Dahl and Hyder, 1977). While most grassland plant populations are maintained vegetatively, reproduction in some grasslands is by seeds, and Hyder (1972) has emphasized that reproductive shoots are adapted for seed production rather than for tolerance to defoliation.

Grazing stress on particular species can be partially determined by relative palatabilities of species in a community. Both sideoats grama and June grass

(*Koleria cristata*) are quite palatable, as are big and little bluestem, but at maturity, the coarse seedstalks of the latter two species reduce the grazing pressure on the bluestems. Rogler (1944) reported that steers generally preferred warm-season grasses over cool-season grasses. In Kansas, cattle demonstrated a positive grazing selection for big and little bluestem whenever these species were present, but preference for western wheatgrass varied with the site (Tomanek *et al.*, 1958). Also, susceptibility to grazing may be influenced by spatial distribution; for example, palatable species growing in close association with unpalatable species may receive less grazing pressure than if the palatable species were growing alone (McNaughton, 1978).

Caswell *et al.* (1973) suggested that C_4 plants are generally inferior food sources for herbivores, primarily insects, as compared with C_3 plants. In further tests (Caswell and Reed, 1975), *Melanoplus confusus* and other species of grasshoppers (Caswell and Reed, 1976) were capable of digesting C_3 grass material, but were unable to digest totally the thick-walled bundle sheath cells of C_4 grasses. Crampton and Harris (1969) noted that C_4 plants were generally lower in digestible nutrients for cattle and sheep than were C_3 plants at the same stage of maturity. Akin and Burdick (1977) postulated a 'sheath barrier' to the utilization of starch and other nutrients in the parenchyma bundle sheath cells, which are decomposed slowly by rumen bacteria.

Because of the difficulties of using herbivores for grazing experiments (e.g. maintaining and managing livestock, uneven grazing, secondary effects, such as trampling), most studies of grazing impacts have employed clipping to simulate grazing. These clipping studies have resulted in a large body of information about the responses of range species to defoliation. Although there are exceptions, defoliation results in a number of predictable responses. Many grasses respond to defoliation by increasing the assimilates allocated to young leaves or regrowing tillers (Briske and Stuth, 1982; Burleson and Hewitt, 1982; Cook and Stoddart, 1953; Detling *et al.*, 1979; Gifford and Marshall, 1973; Hart and Balla, 1982; Hutchison, 1971; Jameson, 1963; Jameson and Huss, 1959; McNaughton, 1979; Parker and Sampson, 1931; Ryle and Powell,

1975). When shoots are clipped frequently or severely or grazing intensity is great, root mortality usually increases along with a decrease in root extension and branching (Biswell and Weaver, 1933; Crider, 1955; Davidson and Milthorpe, 1966a,b; Detling *et al.*, 1979; Dyer and Bokhari, 1976; Hodgkinson and Baas-Becking, 1977; Oswalt *et al.*, 1959). The detrimental effects on plant growth increase when the frequency and/or degree of defoliation increases (Buwai and Trlica, 1977; Dwyer *et al.*, 1963; Jameson, 1964; Westoby, 1980). However, in the shortgrass of Colorado, Bartos and Sims (1974) did not find significant differences in the root mass among four grazing intensity treatments. Not only may defoliation reduce root growth, but frequently there is a pronounced diversion of carbohydrates and nitrogen from the roots (Bahrani *et al.*, 1983; Bukey and Weaver, 1939; Gifford and Marshall, 1973; Kinsinger and Shaulis, 1961; McNaughton, 1979; Owensby *et al.*, 1974; Pierre and Bertram, 1929; Ryle and Powell, 1975; White, 1973).

Studies of the impact of timing of defoliation have given ambiguous results. Owensby *et al.* (1977) compared grazing systems in the tallgrass prairie, one with continuous season-long grazing and one where the rangeland was intensively stocked in the early portion of the season. Although during mid-summer the carbohydrate reserves were lower on the intensively early stocked pasture (Smith and Leinweber, 1971), there were no differences in the nitrogen reserve cycle or the vigor or regrowth potential of big bluestem. Kinsinger and Shaulis (1961) reported that big bluestem carbohydrate reserve storage was not adversely affected by a single year's intensive clipping but that two years of intensive clipping reduced reserve storage. When black grama (*Bouteloua eriopoda*) was subjected to various combinations of clipping frequencies, total nonstructural carbohydrates and crown diameters showed little response to season of clipping, but those plants clipped during or after flowering or continuously through the growing season produced less herbage in the following year than those plants clipped only during the vegetative state. The removal of 65% of the current year's growth any time during the season significantly reduced the number of stolons produced in the following year (Miller and Donart, 1979). Wardlaw (1969) found 30% more than normal assimilate deposition in crowns of partially defoliated plants of ryegrass, *Lolium temulentum*, and 60% more deposition in roots of partially defoliated plants under mild water stress.

Numerous field studies show that moderate clipping stimulates aboveground production (Gay and Dwyer, 1965; Hyder, 1972). Several studies have shown that carbohydrates stored in the roots are used for regrowth following clipping (e.g. Bokhari, 1977). Carbohydrate reserves decrease following herbage removal (Chung and Trlica, 1980) and continue to decline until the shoots develop adequate photosynthetically active leaf areas (May, 1960). The reported exceptions to the use of belowground reserves (Jameson, 1963; Weinmann, 1948) may be due to experimental procedures (White, 1973). Ward and Blaser (1961) reported that carbohydrate reserves stimulated dry matter production in orchardgrass for the first 25 d after partial or complete aboveground herbage removal, but Davidson and Milthorpe (1966a,b) found that orchardgrass shoot regrowth following clipping depended on carbohydrate reserves only for the first 2–4 d.

Using western wheatgrass in a hydroponic system, Bokhari and Singh (1974) found that, in general, clipping induced shoot growth but decreased root growth. However, there was a greater relative accumulation of assimilates in the belowground organs of moderately clipped plants. In addition, greater percentage increase in dry weight of crowns at warmer temperatures (26–18 °C) and in roots at cooler temperatures (13–7 °C) were observed. Total dry matter production and the proportion of younger, more photosynthetically active tissue were greater in moderately clipped plants but were reduced in severely clipped plants.

A number of studies have investigated the effects of partial defoliation on photosynthesis and respiration rates. Simulated or actual grazing by insects on a single leaf often reduces photosynthetic rates in the remaining tissue of the grazed leaf if tissue damage exceeds a threshold level (Poston *et al.*, 1976). On the other hand, moderate defoliation may temporarily enhance photosynthesis in remaining leaves or at least retard photosynthetic

decline with age (Hodgkinson, 1974; Hodgkinson *et al.*, 1972; Sanders *et al.*, 1977). Davidson and Milthorpe (1966b) found that respiration declined markedly within several hours of defoliation and that the rate of decline increased and the rate of recovery decreased as the severity of defoliation increased. Others (Hansen, 1977, 1978; Osman, 1971) have also shown the same response of respiration and subsequent photosynthate translocation to roots and crowns.

Detling *et al.* (1979) grew blue grama seedlings hydroponically for 40 d and monitored CO_2 exchange rates over the next 10 d. The plants were then clipped at 4 cm (removing 62% of the remaining leaf blade and 45% of the total shoot biomass). Immediately after clipping, photosynthesis in the leaves decreased to only 40% of that of the control plants, but three days after clipping, photosynthesis had increased to 21% greater (per unit leaf area) than that of the controls. No statistically different changes occurred in respiration. The biomass of unclipped plants nearly doubled during the 10 d study period, while that of defoliated plants increased 67%. Over half of the new growth of defoliated plants was allocated to new leaf blades and 18% to new roots; in the control plants 33% of the new growth was in the leaf blades, but 29% went to new roots. As a consequence of increased photosynthetic rates and increased carbon allocation to the synthesis of additional photosynthetic tissue following defoliation, net CO_2 uptake per treated plant increased from 9% to 80% of that of the control plants from day 1 to day 10.

Results from hydroponic experiments require some caution in their interpretation and extrapolation to field situations (Detling *et al.*, 1980). In the field, plant growth may be stimulated by increased nutrient cycling brought about by grazing (Lotero *et al.*, 1966; Parton and Risser, 1979). Hodgkinson and Baas-Becking (1977) have suggested that defoliation during periods of low soil water availability could accelerate plant death because of the inability of the roots to elongate as fast as the movement of the receding drying front in the soil profile. On the other hand, soil water will be conserved after grazing or clipping because of the reduced water loss from transpiration. Sosebee and Wiebe (1971) found that defoliation and soil mosture interact to determine patterns of assimilate translocation. Reduced water supply increased translocation to roots and crowns, while partial defoliation increased translocation to younger leaves. Therefore, the balance between root and crown storage and the use of leaf photosynthate is influenced by the severity of defoliation and soil water potential.

Grassland species have evolved under the influence of grazing (Roe, 1951), and it should not be surprising that plants are adapted to grazing by virtue of morphology, phenology, and physiological processes which reduce or compensate for grazing impacts. McNaughton (1979) summarized the compensation mechanisms demonstrated by plants in response to herbivore activities:

1. increased photosynthetic rates of the remaining green tissues;

2. reallocation of assimilates and nutrients from elsewhere in the plants;

3. Removal of older tissue not operating at maximum photosynthetic level and the consequent increase in light intensities for the remaining more active tissues;

4. hormonal control of meristems, promoting leaf growth, tillering, and reduction of leaf senescence;

5. conservation of soil moisture by reduction of the transpiration surface and reduction of mesophyll resistance relative to stomatal resistance;

6. increased nutrient recycling from dung and urine;

7. promotion of regrowth from substrates in ruminant saliva.

Indeed, partial defoliation may stimulate growth (Cable, 1982; Heitschmidt *et al.*, 1982), and the response may be interrelated to such conditions as nutrient status and the presence of mycorrhizal fungi (Wallace, 1981).

11.2.5 BURNING

Most grasslands have been periodically subjected to fires, either of natural or human origin. In general, there is an increase in flowering and herbage production in the first year or so after burning (Daubenmire, 1968; Gay and Dwyer, 1965; Hadley,

1970; Hulbert, 1967; Old, 1969; Owensby and Smith, 1973; Parton and Risser, 1979; Rice and Parenti, 1978) although this response varies with precipitation and season of burning (Adams *et al.*, 1982; Kelting, 1957; Owensby and Smith, 1973; Steuter and Wright, 1983; White and Currie, 1983). Though not so common in the shortgrass steppe, this increased productivity occurs in most grassland types, ranging from the tallgrass prairie (Adams and Anderson, 1978; Launchbaugh and Owensby, 1978) to mountain grasslands (Merrill *et al.*, 1980). Various measures of forage quality (e.g. total nonstructural carbohydrate, crude fiber, protein) characteristically increase in the regrowth after fire (McGinty *et al.*, 1983; Rains *et al.*, 1975; Smith and Young, 1959; Smith *et al.*, 1960). Although Merrill *et al.* (1980) did not find a change in the mineral content of mountain grassland herbage after fire, in the tallgrass region of eastern Kansas Owensby *et al.* (1970b) found that total nitrogen and nitrogen concentration in the tissue were higher on burned plots. Peet *et al.* (1975) used gas exchange methods to measure photosynthesis and respiration in clumps of big bluestem removed from unburned and burned Wisconsin prairie. There were no differences in the physiological performance of plants in the two sites, but total carbon gain was greater on the burned site, primarily because more photosynthetic leaf and shoot area were produced on the burned site earlier in the growing season. This more rapid growth rate was a consequence of the altered microclimate of the burned area.

11.2.6 PHYSIOLOGICAL STRATEGIES IN THE CONTEXT OF OTHER ECOLOGICAL PROCESSES

The preceding sections have discussed a myriad of physiological responses to various environmental conditions, particularly water, solar radiation and temperature, nutrients, grazing, and burning. Alternatively, these physiological strategies could also have been discussed in the context of various ecological processes (McNaughton *et al.*, 1983). Although few comments about these processes will be made below, most of the adaptive strategies occur in response to environmental conditions operating within more complex ecological processes.

A number of competitive relationships among grassland species have been described (Black *et al.*, 1969; Risser, 1969), and indeed, much of the previous discussion of C$_3$ and C$_4$ plants has involved competitive relationships between groups of plants (e.g. Baskin and Baskin, 1978). The work of Kemp and Williams (1980) exemplifies the relationship between plant competition and the response of rangeland plants to environmental conditions. After examining a number of physiological characteristics, they concluded that the phenological displacement of growth in western wheatgrass and blue grama is a function of niche separation with respect to seasonal temperature. All aspects of carbon metabolism were greatly reduced in both species by plant water stress, but the optimum photosynthesis of each species showed a different response to temperature. Therefore, there appears to be a niche separation based on divergent growth responses to temperature as a mechanism for reducing competition for limited water resources in the shortgrass prairie. Similarly, considered as a community, the grassland demonstrates numerous displacements in phenology of growth (Ahshapanek, 1962; Coughenour *et al.*, 1979; Dickinson and Dodd, 1976; Rice, 1950; Tieszen, 1970) and space, both in terms of habitat preference (Weaver and Albertson, 1956) and stratification of canopy and roots (Berendse, 1979, 1981; Weaver, 1958).

A number of physiological strategies have been related to succession although some commonly assumed adaptations may be only plastic responses to environmental conditions (Clary, 1979; Roos and Quinn, 1977). Bazzaz (1979) has discussed the physiological ecology of plant succession though the major portion of his discussion relates to forest communities. A number of ideas pertaining to the life history strategies are presented, especially with respect to seed germination, seedling growth, and annual versus perennial growth strategies. Because Bazzaz effectively summarizes physiological characteristics of early and late successional plants, no attempt will be made to extract those characteristics which might be particularly typical of grassland plants nor to repeat the discussion of old-field successional grasslands.

11.3 Summary

It is apparent that a considerable amount of research has been conducted on the morphological and physiological adaptations found in plants characteristic of the North American grasslands. From these studies, it is possible to distill several adaptive strategies which can be found in some generality (Table 11.1). The following pages discuss the specific studies that are used as the bases for this enumeration. A review of this list, however, provides little more than a menu for selecting potential adaptations. With a few exceptions, it is impossible to erect a generalized model of morphological–physiological characteristics for either a single plant species or the grassland community.

Early studies of the physiological ecology of grasslands were done primarily by John Weaver and his students in the mixed-grass and tallgrass prairies. While some of these studies were physiological in nature, e.g. osmotic pressure of stems and roots, most of these investigations dealt with morphology, growth habit, and apparent responses to drought. These morphological-adaptation studies were followed in the 1940s and 1950s with a thorough description of range plants, particularly the response of these plants to grazing pressure. From these descriptive studies come our basic understanding of growth and morphological characteristics, which permits the prediction of the probability of a plant's persistence under a grazing regime of ungulates. During this period, there were also studies which consider range productivity and methods for measuring range condition (Risser, 1980).

In the last twenty years, relatively few studies have been aimed at refining our understanding of morphological adaptations of grassland species. Rather, most of the research has addressed gas exchange rates, energy storage dynamics, and the water status of plants, primarily of the shortgrass or dry prairies. The impetus for these investigations has arisen from several quarters: the discovery and comparative examination of the C$_3$–C$_4$ carbon fixation pathways, the development of reasonably reliable and inexpensive equipment to measure gas exchanges and plant water status, and the questions

Table 11.1 Physiological–morphological adaptive strategies of grassland plants.

Drought

Mechanisms, such as closing of stomata, leaf curling, pubescence, and paraheliotropism reduce water loss.

More belowground labile carbohydrates are stored under adequate moisture, thus reducing vulnerability of energy storage compounds.

Dark respiration decreases under drought stress conditions, thereby preserving substrate.

Gas exchange processes are maintained under water potentials of -2.0 to -4.0 MPa.

Water use efficiency is increased with C_4 pathway, generally favored in drier climates.

Earlier seasonal growth, when moisture is adequate, is characterized by major investment in structural carbohydrates.

Dormancy avoids adverse season.

Germination is delayed over more than 1 yr.

Has ability to germinate under relatively dry soil conditions.

Primary root grows rapidly and adventitious roots develop rapidly.

Root strength withstands shrink–swell characteristics of clay soils.

Temperature

Photosynthetic optima are related to carbon fixation pathway and coincident with prevailing temperature regime.

Acclimatizes to photosynthetic temperature optimum.

Root respiration temperature optima are correlated with optimum temperatures for photosynthesis.

Nutrients

Rapid nutrient uptake occurs during season when nutrients available and soil moisture adequate.

Internally stores and recycles nutrients.

Nutrient uptake and consequent increase in forage quality

occurs during season when plant most tolerant to grazing.

Legumes harbor nitrogen fixation mechanism, and while associated vegetation acquires some nitrogen from legumes, is more efficient than legumes at scavenging soil nitrogen.

Ammonia lost from soil surface absorbed by plant canopy.

Minimal loss of nutrients occurs via leaching from grass leaves.

Grazing

Plants offer various antiherbivore devices, such as toxic compounds, low palatability, coarse seed stalks, and short growth cycle.

Intercalary meristems permit continuous regrowth, and buds may be near soil surface and thereby protected from grazing.

Increase in total herbage production and rate of photosynthesis occurs with light to moderate grazing, and reduction of inefficient photosynthetic tissue coincides with grazing pressure.

Increases allocation of assimilates to young leaves and regrowing tillers, which increases photosynthetic capacity but may account for reduced root growth.

Reduced water supply may increase energy translocation to roots and crowns, but grazing increases energy translocation to young leaves.

Burning

Meristimatic tissue is protected from burning.

Has rapid regrowth potential, especially in response to elevated soil surface temperatures.

Seed production increases following burning.

---

which were asked when formulating mathematical models used to simulate and predict grassland production.

As noted previously, some of the evidence concerning a number of the adaptive mechanisms is still contradictory. Part of the uncertainty arises from the difficulty of measuring the relative importance of several presumed adaptive advantages, especially as a plant confronts a seasonally changing environment and an array of competing organisms. In addition, so few plant species have been investigated thoroughly that any comprehensive discussion of

the adaptive significance of generalized phenomena quickly focuses on relatively few species.

As an aim, we should strive for the acquisition and organization of information that will enable us confidently to describe the morphological–physiological adaptation of plant species. Such a treatment exists for only a few grassland species, such as blue grama (Detling, 1979). Secondly, it will ultimately be necessary to describe the suite of morphological–physiological adaptations of species in various grassland communities. To the present time, attempts at such descriptions have dealt largely with growth and morphological adaptations (Grime, 1979) and have not been coupled with physiological processes in the ecosystem context. Only on the basis of organized information can we begin to develop theories about grassland adaptive strategies which are more than descriptive generalities.

Acknowledgements

The author wishes to express his appreciation to Beverly Richie, Pat Duzan, and Alice Adams for typing the manuscript; to Robert Zewadski for editorial assistance; and to Dwight E. Adams, Scott L. Collins, Karen L. Dooley, Louis Iverson, Larry L. Tieszen, and George J. Williams, III, who provided helpful reviews.

References

Adams, D.E. and Anderson, R.C. (1978) The response of central Oklahoma grassland to burning. *Southwestern Naturalist*, **23**, 623–32.

Adams, D.E., Anderson, R.C. and Collins, S.L. (1982) Differential response of woody and herbaceous species to summer and winter burning in an Oklahoma grassland. *Southwestern Naturalist*, **27**, 55–61.

Adegbola, A.A. and McKell, C.M. (1966) Regrowth potential of coastal Bermuda grass as related to previous nitrogen fertilization. *Agronomy Journal*, **58**, 145–6.

Ahshapanek, D. (1962) Phenology of a native tall-grass prairie in central Oklahoma. *Ecology*, **43**, 135–8.

Akin, D.E. and Burdick, D. (1977) Rumen microbial degradation of starch-containing bundle sheath cells in warm-season grasses. *Crop Science*, **17**, 529–33.

Allen, E.B. (1982a) Germination and competition of *Salsola kali* with native C_3 and C_4 species under three temperature regimes. *Bulletin of the Torrey Botanical Club*, **109**, 39–46.

Allen, E.B. (1982b) Water and nutrient competition between *Salsola kali* and two native grass species (*Agropyron smithii* and *Bouteloua gracilis*). *Ecology*, **63**, 732–41.

Angevine, M.W. and Chabot, B.F. (1979) Seed germination syndromes in higher plants, in *Topics in Plant Population Biology* (eds O.T. Solbrig, S. Jain, G.J. Johnson and P.H. Raven), Columbia University Press, New York, pp. 188–206.

Ares, J. (1976) Dynamics of the root system of blue grama. *Journal of Range Management*, **29**, 208–13.

Aspinall, D. (1960) An analysis of competition between barley and white persicaria. II. Factors determining the course of competition. *Annals of Applied Biology*, **48**, 637–54.

Bahrani, J., Beaty, E.R. and Tan, K.H. (1983) Relationship between carbohydrate, nitrogen contents, and regrowth of tall fescue tillers. *Journal of Range Management*, **36**, 234–5.

Baker, B.S. and Jung, G.A. (1972) Growth and metabolic changes occurring in orchard grass during temperature acclimation. *Botanical Gazette*, **133**, 120–6.

Bartos, D.L. and Sims, P.L. (1974) Root dynamics of a shortgrass ecosystem. *Journal of Range Management*, **27**, 33–6.

Baskin, J.M. and Baskin, C.C. (1978) A discussion of the growth and competitive ability of C_3 and C_4 plants. *Castanea*, **43**, 71–6.

Bazzaz, F.A. (1979) The physiological ecology of plant succession. *Annual Review of Ecology and Systematics*, **10**, 351–71.

Berendse, F. (1979) Competition between plant populations with different rooting depths. *Oecologia*, **43**, 19–26.

Berendse, F. (1981) Competition between plant populations with different rooting depths. II. Pot experiments. *Oecologia*, **48**, 334–41.

Bewley, J.D. (1979) Physiological aspects of desiccation tolerance. *Annual Review of Plant Physiology*, **30**, 195–238.

Biswell, H.H. and Weaver, J.E. (1933) Effects of frequent clipping on the development of roots and tops of grasses in prairie sod. *Ecology*, **14**, 368–90.

Björkman, O. (1971) Comparative photosynthetic CO_2 exchange in higher plants, in *Photosynthesis and Photorespiration* (eds M.D. Hatch, C.B. Osmond and R.O. Slatyer), Wiley Interscience, New York, pp. 18–32.

Björkman, O. (1976) Adaptive and genetic aspects of C_4 photosynthesis, in CO_2 *Metabolism and Plant Productivity* (eds R.H. Burris and C.C. Black), University Park Press, Baltimore, Maryland, pp. 287–309.

Black, C.C. (1971) Ecological implications of dividing plants into groups with distinct photosynthetic production capacities. *Advances in Ecological Research*, **7**, 87–114.

Black, C.C. (1973) Photosynthetic carbon fixation in relation to net CO_2 uptake. *Annual Review of Plant Physiology*, **24**, 253–86.

Black, C.C., Chen, T.M. and Brown, R.H. (1969) Biochemical basis for plant competition. *Weed Science*, **17**, 338–44.

Blake, A. (1935) Viability of germination of seeds and early life history of prairie plants. *Ecological Monographs*, **5**, 405–60.

Bokhari, U.G. (1977) Regrowth of western wheatgrass utilizing ^{14}C-labeled assimilates stored in belowground parts. *Plant and Soil*, **48**, 115–27.

Bokhari, U.G. (1978) Total nonstructural carbohydrates in the vegetation components of a shortgrass prairie ecosystem under stress conditions. *Journal of Range Management*, **31**, 224–30.

Bokhari, U.G. and Dyer, M.I. (1974) *Carbohydrate reserves of the blue grama (Bouteloua gracilis) at various environmental factors under growth chamber experiments*, US International Biological Program Grassland Biome Technical Report Number 255, Colorado State University, Fort Collins, Colorado.

Bokhari, U.G. and Singh, J.S. (1974) Effects of temperature and clipping on growth, carbohydrate reserves and root exudation of western wheatgrass in hydroponic culture. *Crop Science*, **14**, 790–4.

Bokhari, U.G., Dyer, M.I. and Singh, J.S. (1974) Labile and nonlabile energy in blue grama (*Bouteloua gracilis*) as influenced by temperature, water stress, and fertilizer treatments. *Canadian Journal of Botany*, **52**, 2289–98.

Bookman, P.A. and Mack, R.N. (1982) Root interaction between *Bromus tectorum* and *Poa pratensis*: a three-dimensional analysis. *Ecology*, **63**, 640–6.

Boutton, T.W., Harrison, A.T. and Smith, B.N. (1980) Distribution of biomass of species differing in photosynthetic pathway along the altitudinal transect in southeastern Wyoming grassland. *Oecologia*, **45**, 287–98.

Boyer, J.S. (1970) Differing sensitivity of photosynthesis to low leaf water potentials in corn and soybean. *Plant Physiology*, **46**, 236–9.

Briske, D.D. and Stuth, J.W. (1982) Tiller defoliation in moderate and heavy grazing regime. *Journal of Range Management*, **35**, 511–14.

Briske, D.D. and Wilson, A.M. (1977) Temperature effects on adventitious root development in blue grama seedlings. *Journal of Range Management*, **30**, 276–80.

Brown, L.F. and Trlica, M.J. (1977) Interacting effects of soil water, temperature, and irradiance on CO_2 ex-

change rates of two dominant grasses of the shortgrass prairie. *Journal of Applied Ecology*, **14**, 197–204.

Brown, R.H. (1978) A difference in N use efficiency in C_3 and C_4 plants and its implications in adaptation and evolution. *Crop Science*, **18**, 93–8.

Brown, R.H. and Blaser, R.E. (1965) Relationships between reserve carbohydrate accumulation and growth rate in orchardgrass and tall fescue. *Crop Science*, **5**, 577–82.

Brown, R.W. (1977) Water relations in range plants, in *Rangeland Plant Physiology* (ed. R.E. Sosebee), Range Science Series Number 4, Society for Range Management, Denver, Colorado, pp. 97–140.

Brown, W.V. (1977) The Kranz syndrome and its subtypes in grass systematics. *Bulletin of the Torrey Botanical Club*, **23**, 1–97.

Bukey, F.S. and Weaver, J.E. (1939) Effects of frequent clipping on the underground food reserves of certain prairie grasses. *Ecology*, **20**, 246–52.

Bunce, J.A. and Miller, L.N. (1976) Differential effects of water stress on respiration in the light in woody plants from wet and dry habitats. *Canadian Journal of Botany*, **54**, 2457–64.

Burleson, W.H. and Hewitt, G.B. (1982) Response of needle-and-thread and western wheatgrass to defoliation by grasshoppers. *Journal of Range Management*, **35**, 223–6.

Buwai, M. and Trlica, M.J. (1977) Multiple defoliation effects on herbage yield, vigor and total nonstructural carbohydrates of five range species. *Journal of Range Management*, **30**, 164–71.

Cable, D.R. (1971) Growth and development of Arizona cottontop (*Trichachne californica* [Benth.] Chase). *Botanical Gazette*, **132**, 119–45.

Cable, D.R. (1982) Partial defoliation stimulates growth of Arizona cottontop. *Journal of Range Management*, **35**, 591–3.

Canode, C.L. and Law, A.G. (1975) Seed production of Kentucky bluegrass associated with age of stand. *Agronomy Journal*, **67**, 790–4.

Caswell, H. and Reed, F.C. (1975) Indigestibility of C_4 bundle sheath cells by the grasshopper, *Melanoplus confusus*. *Annals of the Entomological Society of America*, **68**, 686–8.

Caswell, H. and Reed, F.C. (1976) Plant-herbivore interactions: the indigestibility of C_4 bundle sheath cells by grasshoppers. *Oecologia*, **26**, 151–6.

Caswell, H., Reed, F., Stephenson, S.N. and Werner, P.A. (1973) Photosynthetic pathways and selective herbivory: a hypothesis. *American Naturalist*, **107**, 465–80.

Charley, J.L. (1977) Mineral cycling in rangeland ecosystems, in *Rangeland Plant Physiology* (ed. R.E. Sosebee), Range Science Series Number 4, Society for Range Management, Denver, Colorado, pp. 215–56.

Charley, J.L. and West, N.E. (1975) Plant-induced soil chemical patterns in some desert shrub-dominated ecosystems of Utah. *Journal of Ecology*, 63, 945–64.

Christie, E.K. and Detling, J.K. (1982) Analysis of interference between C_3 and C_4 grasses in relation to temperature and soil nitrogen supply. *Ecology*, 63, 1277–84.

Chung, H.H. and Trlica, M.J. (1980) ^{14}C distribution and utilization of blue grama as affected by temperature, water potential and defoliation regimes. *Oecologia*, 47, 190–5.

Clark, F.E. (1977) Internal cycling of [15]nitrogen in shortgrass prairie. *Ecology*, 58, 1322–33.

Clary, W.P. (1975) Ecotypic adaptation in *Sitanion hystrix*. *Ecology*, 56, 1407–15.

Clary, W.P. (1979) Variation in leaf anatomy and CO_2 assimilation in *Sitanion hystrix* ecotypes. *Great Basin Naturalist*, 39, 427–32.

Colby, W.G., Drake, M., Field, D.L. and Kreowski, G. (1965) Seasonal pattern of fructosan in orchardgrass stubble as influenced by nitrogen and harvest management. *Agronomy Journal*, 57, 169–73.

Cook, C.W. and Stoddart, L.A. (1953) Some growth responses of crested wheatgrass following herbage removal. *Journal of Range Management*, 6, 267–70.

Cooper, J.P. and Tainton, N.M. (1968) Light and temperature requirements for the growth of tropical and temperate grasses. *Herbage Abstracts*, 38, 167–76.

Coughenour, M.B., Dodd, J.L., Coleman, D.C. and Lauenroth, W.K. (1979) Partitioning of carbon and SO_2 – sulfur in a native grassland. *Oecologia*, 42, 229–40.

Coupland, R.T. (ed.) (1979) *Grassland ecosystems of the world: analysis of grasslands and their uses*, International Biological Programme 18, Cambridge University Press, Cambridge, UK.

Crampton, E.W. and Harris, L.E. (1969) *Applied Animal Nutrition*, W.H. Freeman and Company, San Francisco, California.

Crider, F.J. (1955) *Root-growth stoppage resulting from defoliation of grass*, US Department of Agriculture Technical Bulletin 1102.

Dahl, B.E. and Hyder, D.N. (1977) Developmental morphology and management implications, in *Rangeland Plant Physiology* (ed. R.E. Sosebee), Society for Range Management, Denver, Colorado, pp. 257–90.

Dalrymple, R.L. and Dwyer, D.D. (1967) Root and shoot growth of range grasses. *Journal of Range Management*, 20, 141–5.

Daniel, H. (1932) A study of certain factors which affect the calcium, phosphorus and nitrogen content of prairie grass. *Proceedings of the Oklahoma Academy of Science*, 12, 42–5.

Daubenmire, R. (1968) The ecology of fire in grasslands. *Advances in Ecological Research*, 5, 209–66.

Davidson, R.L. (1969) Effects of soil nutrients and moisture on root/shoot ratios in *Lolium perenne* L. and *Trifolium repens* L. *Annals of Botany*, 33, 571–7.

Davidson, J.L. and Milthorpe, F.L. (1966a) Leaf growth in *Dactylis glomerata* following defoliation. *Annals of Botany*, 30, 173–84.

Davidson, J.L. and Milthorpe, F.L. (1966b) The effect of defoliation on the carbon balance in *Dactylis glomerata*. *Annals of Botany*, 30, 185–98.

Davies, W.J. and Kozlowski, T.T. (1977) Variations among woody plants in stomatal conductance and photosynthesis during and after drought. *Plant and Soil*, 46, 435–44.

Denmead, O.T., Freny, J.R. and Simpson, J.R. (1976) A closed ammonium cycle within a plant community. *Soil Biology and Biochemistry*, 8, 161–4.

Denno, R.F., Raupp, M.J., Tallamy, D.W. and Reichelderfer, C.F. (1980) Migration in heterogeneous environments: differences in habitat selection between the wing forms of the dimorphic planthopper *Prokelisia marginata* (Homoptera: Delphacidae). *Ecology*, 61, 859–67.

Detling, J.K. (1979) Processes controlling blue grama production on the shortgrass prairie, in *Perspectives in Grassland Ecology* (ed. N.R. French), Springer-Verlag, New York, pp. 25–42.

Detling, J.K., Dyer, M.I. and Winn, D.T. (1979) Net photosynthesis, root respiration, and regrowth of *Bouteloua gracilis* following simulated grazing. *Oecologia*, 41, 127–34.

Detling, J.K., Winn, D.T., Proctor-Gregg, C. and Painter, E.L. (1980) Effects of simulated grazing by belowground herbivores on growth, CO_2 exchange, and carbon allocation patterns of *Bouteloua gracilis*. *Journal of Applied Ecology*, 17, 771–8.

Dickinson, C.E. and Dodd, J.L. (1976) Phenological pattern in the shortgrass prairie. *American Midland Naturalist*, 96, 367–78.

Dina, S.J. and Klikoff, L.G. (1973) Effect of plant moisture stress on carbohydrate and nitrogen content of big sagebrush. *Journal of Range Management*, 26, 207–9.

Dobson, J.W. and Beaty, E.R. (1977) Forage yields of five perennial grasses with and without white clover at four nitrogen rates. *Journal of Range Management*, **30**, 461–5.

Doley, D. and Trirett, N.B.A. (1974) Effects of low water potentials on transpiration and photosynthesis in Mitchell grass (*Astrebla lappacea*). *Australian Journal of Plant Physiology*, **1**, 539–50.

Doliner, L.H. and Jolliffe, P.A. (1979) Ecological evidence concerning the adaptive significance of the C$_4$ dicarboxylic acid pathway of photosynthesis. *Oecologia*, **38**, 23–34.

Downes, R.W. (1971) Adaptation of sorghum plants to light intensity: its effect on gas exchange in response to changes in light, temperature and CO$_2$. in *Photosynthesis and Photorespiration* (eds M.D. Hatch, C.B. Osmond and R.O. Slatyer), Wiley Interscience, New York, pp. 57–62.

Downton, W.J.S., Berry, J. and Tregunna, E.B. (1969) Photosynthesis: temperate and tropical characteristics within a single grass genus. *Science*, **163**, 78–9.

Drake, M., Colby, W.C. and Bredakis, E. (1963) Yield of orchardgrass as influenced by rates of nitrogen and harvest management. *Agronomy Journal*, **55**, 361–2.

DuBois, J.D. and Kapustka, L.A. (1983) Biological nitrogen influx in an Ohio relict prairie. *American Journal of Botany*, **70**, 8–16.

Dwyer, D.D., Elder, W.C. and Single, G. (1963) *Effects of height and frequency of clipping on pure stands of range grasses in north central Oklahoma*, Oklahoma Agriculture Experiment Bulletin B-614, Stillwater, Oklahoma.

Dyer, M.I. (1980) Mammalian epidermal growth factor promotes plant growth. *Proceedings of the National Academy of Science, USA*, **77**, 4836–7.

Dyer, M.I. and Bokhari, U.G. (1976) Plant–animal interactions: studies of the effects of grasshopper grazing on blue grama grass. *Ecology*, **57**, 762–72.

Eckert, R.E., Jr and Spencer, J.S. (1982) Basal-area growth and reproductive responses of thurber needlegrass and squirreltail to weed control and nitrogen fertilization. *Journal of Range Management*, **35**, 610–14.

Ehleringer, J.R. (1978) Implications of quantum yield differences on the distributions of C$_3$ and C$_4$ grasses. *Oecologia*, **31**, 255–67.

Ehleringer, J.R. and Björkman, O. (1977) Quantum yields for CO$_2$ uptake in C$_3$ and C$_4$ plants: dependence on temperature, CO$_2$ and O$_2$ concentration. *Plant Physiology*, **59**, 86–90.

Ehleringer, J.R., Björkman, O. and Mooney, H.A. (1976) Leaf pubescence: effects on absorptance and photosynthesis in a desert shrub. *Science*, **192**, 376–7.

Ehrlich, P.R. and Raven, P.H. (1964) Butterflies and plants: a study in coevolution. *Evolution*, **18**, 586–608.

El-Sharkawy, M.A. and Hesketh, J.K. (1965) Photosynthesis among species in relation to characteristics of leaf and CO$_2$ diffusion resistance. *Crop Science*, **5**, 517–21.

Everitt, J.H., Alaniz, M.A. and Gerbermann, A.H. (1982) Chemical composition of native range grasses growing on saline soils of the South Texas Plains. *Journal of Range Management*, **35**, 43–5.

Frank, A.B. and Barker, R.E. (1976) Rates of photosynthesis and transpiration and diffusive resistance of six grasses grown under controlled conditions. *Agronomy Journal*, **68**, 487–90.

Fraenkel, G.S. (1959) The raison d'etre of secondary plant substances. *Science*, **129**, 1466–70.

Freeland, W.J. and Janzen, D.H. (1974) Strategies in herbivory by mammals: the role of plant secondary compounds. *American Naturalist*, **108**, 269–89.

Gay, C.W. and Dwyer, D.D. (1965) Effect of one year's nitrogen fertilization on native vegetation under clipping and burning. *Journal of Range Management*, **18**, 273–7.

Gerwick, B.C. and Williams, B.C., III (1978) Temperature and water regulation of gas exchange of *Opuntia polyacantha*. *Oecologia*, **35**, 149–59.

Gifford, R.M. (1974) A comparison of potential photosynthesis, productivity and yield of plant species with differing photosynthetic metabolism. *Australian Journal of Plant Physiology*, **1**, 107–17.

Gifford, R.M. and Marshall, C. (1973) Photosynthesis and assimilate distribution in *Lolium multiflorum* Lam following differential tiller defoliation. *Australian Journal of Biological Science*, **26**, 517–26.

Greenway, H. and West, R.K. (1973) Respiration and mitochondrial activity in *Zea mays* roots as affected by osmotic stress. *Annals of Botany*, **37**, 21–35.

Grime, J.P. (1979) *Plant Strategies and Vegetation Processes*, John Wiley and Sons, New York.

Griswald, S.M. (1936) Effects of alternate moistening and drying on germination of seeds of western range grasses. *Botanical Gazette*, **98**, 243–69.

Hadley, E.B. (1970) Net productivity and burning responses of native eastern North Dakota prairie communities. *American Midland Naturalist*, **84**, 121–35.

Hansen, G.K. (1977) Adaptation to photosynthesis and diurnal oscillation of root respiration rates for *Lolium multiflorum*. *Physiologia Plantarum*, **39**, 275–9.

Hansen, G.K. (1978) Utilization of photosynthates for growth, respiration, and storage in tops and roots of *Lolium multiflorum*. *Physiologia Plantarum*, **42**, 5–13.

Harper, J.L. and Benton, R.A. (1966) The behavior of seeds in soil. II. The germination of seeds on the surface of a water supplying substrate. *Journal of Ecology*, **54**, 151–66.

Harper, H., Daniel, H. and Murphy, H. (1933) The total nitrogen, phosphorous and calcium of common weeds and native grasses in Oklahoma. *Proceedings of the Oklahoma Academy of Science*, **4**, 36–44.

Harris, G.A. (1977) Root phenology as a factor of competition among grass seedlings. *Journal of Range Management*, **30**, 172–7.

Hart, R.H. and Balla, E.F. (1982) Forage production and removal from western and crested wheatgrass under grazing. *Journal of Range Management*, **82**, 362–6.

Hartley, W. (1973) Studies on the origin, evolution and distribution of the Gramineae. V. The subfamily Festucoideae. *Australian Journal of Botany*, **21**, 201–34.

Hatch, M.D. (1971) Mechanism and function of the C_4-pathway of photosynthesis, in *Photosynthesis and Photorespiration* (eds M.D. Hatch, C.B. Osmond and R.O. Slatyer), Wiley Interscience, New York, pp. 132–52.

Hatch, M.D., Kagawa, T. and Craig, S. (1975) Subdivisions of C_4-pathway species based on differing C_4 acid carboxylating systems and ultrastructural features. *Australian Journal of Plant Physiology*, **2**, 111–28.

Havercamp, J. and Whitney, G.G. (1983) The life history characteristics of three ecologically distinct groups of forbs associated with the tallgrass prairie. *American Midland Naturalist*, **109**, 105–19.

Heitschmidt, R.K. Price, D.L., Gordon, R.A. and Frasure, J.R. (1982) Short duration grazing at the Texas Experimental Ranch: effects on aboveground net primary production and seasonal growth dynamics. *Journal of Range Management*, **35**, 367–71.

Herrera, C.M. (1982) Grasses, grazers, mutualism and coevolution: a comment. *Oikos*, **38**, 254–8.

Hickman, O.E. (1975) *Seasonal trends in the nutritive content of important range forage species near Silver Lake, Oregon*, US Department of Agriculture, Forest Service Research Paper, PNW-187.

Hilliard, J.H. and West, S.H. (1970) Starch accumulation associated with growth reduction at low temperatures in a tropical plant. *Science*, **168**, 494–6.

Hinnant, R.T. and Kothmann, M.M. (1982) Potassium content of three grass species during winter. *Journal of Range Management*, **35**, 211–13.

Hodgkinson, K.C. (1974) Influence of partial defoliation on photosynthesis, photorespiration, and transpiration by lucerne leaves of different ages. *Australian Journal of Plant Physiology*, **1**, 561–78.

Hodgkinson, K.C. and Baas-Becking, H.G. (1977) Effect of defoliation on root growth of some arid zone perennial plants. *Australian Journal of Agricultural Research*, **29**, 31–42.

Hodgkinson, K.C., Smith, N.G. and Miles, G.E. (1972) The photosynthetic capacity of stubble leaves and their contribution to growth of the lucerne plant after high level of cutting. *Australian Journal of Agricultural Research*, **23**, 225–38.

Hsiao, T.C. (1973) Plant responses to water stress. *Annual Review of Plant Physiology*, **24**, 519–70.

Hulbert, L.C. (1967) Fire and litter effects on undisturbed bluestem prairie. *Ecology*, **50**, 874–7.

Hutchison, K.J. (1971) Productivity and energy flow in grazing/fodder conversion systems. *Herbage Abstracts*, **41**, 1–10.

Hyder, D.N. (1972) Defoliation in relation to vegetative growth, in *The Biology and Utilization of Grasses* (eds V.B. Younger and C.M. McKell), Academic Press, New York, pp. 304–17.

Hyder, D.N., Everson, A.C. and Bement, R.E. (1971) Seedlings morphology and seeding failures with blue grama. *Journal of Range Management*, **24**, 287–92.

Imbamba, S.K. and Tieszen, L.L. (1977) Influence of light and temperature on photosynthesis and transpiration in some C_3 and C_4 vegetable plants from Kenya. *Physiologia Plantarum*, **39**, 311–16.

Jain, S. (1979) Adaptive strategies: polymorphism, plasticity, and homeostasis, in *Topics in Plant Population Biology*, (eds O.T. Solbrig, S. Jain, G.J. Johnson and P.H. Raven), Columbia University Press, New York, pp. 160–87.

Jameson, D.A. (1963) Responses of individual plants to harvesting. *Botanical Review*, **29**, 532–94.

Jameson, D.A. (1964) Effects of defoliation on forage plant physiology, in *Forage Plant Physiology and Soil Relationships* (eds W. Keller and T.A. Ronningen), American Society of Agronomy Special Publication Number 5, pp. 67–80.

Jameson, D.A. and Huss, D.L. (1959) The effect of clipping leaves and stems on number of tillers, herbage weights, root weights and food reserves of little bluestem. *Journal of Range Management*, **12**, 122–6.

Jones, M.G. (1967) Forage and nitrogen production on nitrogen-fertilized California grasslands compared with a subclover-grass association. *Agronomy Journal*, **59**, 209–14.

Jones, M.G., Street, J.E. and Williams, W.A. (1974) Leaching and uptake of nitrogen applied to annual grass and clover-grass mixtures in lysimeters. *Agronomy Journal*, **66**, 256–8.

Kelting, R.W. (1957) Winter burning in central Oklahoma grasslands. *Ecology*, **38**, 520–2.

Kemp, P.R. and Williams, G.J. III (1980) A physiological basis for niche separation between *Agropyron smithii* (C₃) and *Bouteloua gracilis* (C₄). *Ecology*, **61**, 846–58.

Kemp, P.R., Cunningham, G.L. and Adams, H.P. (1983) Specialization of mesophyll morphology in relation to C₄ photosynthesis in the Poaceae. *American Journal of Botany*, **70**, 349–54.

Kinsinger, F.E. and Shaulis, N. (1961) Carbohydrate content of underground parts of grasses as affected by clipping. *Journal of Range Management*, **14**, 9–12.

Knipe, O.D. (1968) Effects of moisture stress on germination of alkali sacaton, galleta, and blue grama. *Journal of Range Management*, **21**, 3–4.

Knipe, O.D. (1973) Western wheatgrass germination as related to temperature, light, and moisture stress. *Journal of Range Management*, **26**, 68–9.

Koshi, P.T., Stubbendieck, J., Eck, H.V. and McCully, W.G. (1982) Switchgrass: forage yield, forage quality, and water-use efficiency. *Journal of Range Management*, **35**, 623–7.

Kozlowski, T.T. (ed.) (1972) Water deficits and plant growth, Vol. III, *Plant Responses and Control of Water Balance*, Academic Press, New York.

LaGory, K.E., LaGory, M.K. and Perino, J.V. (1982) Response of big and little bluestem (*Andropogon*) seedlings to soil and moisture conditions. *Ohio Journal of Science*, **82**, 19–23.

Lauenroth, W.K., Dodd, J.L. and Sims, P.L. (1978) The effects of water- and nitrogen-induced stresses on plant community structure in a semi-arid grassland. *Oecologia*, **36**, 211–22.

Launchbaugh, J.L. and Owensby, C.E. (1978) *Kansas rangelands: their management based on a half century of research*, Agriculture Experiment Station Bulletin 622. Kansas State University, Manhattan, Kansas.

Lawlor, D.W. (1976) Water stress induced changes in photosynthesis, photorespiration, respiration and CO₂ compensation concentration of wheat. *Photosynthetica*, **10**, 378–87.

Levitt, J. (1972) *Response of Plants to Environmental Stresses*, Academic Press, New York.

Lotero, J., Woodhouse, W.W. and Petersen, R.G. (1966) Local effect on fertility of urine voided by grazing cattle. *Agronomy Journal*, **58**, 262–5.

Ludlow, M.M. and Ng, T.T. (1976) Effect of water deficit on carbon dioxide exchange and leaf elongation rate of *Panicum maximum* var. *trichoglume*. *Australian Journal of Plant Physiology*, **3**, 401–13.

Lush, W.M. and Evans, L.T. (1974) Translocation of photosynthetic assimilate from grass leaves, as influenced by environment and species. *Australian Journal of*

Plant Physiology, **1**, 417–31.

May, L.H. (1960) The utilization of carbohydrate reserves in pasture plants after defoliation. *Herbage Abstracts*, **30** 239–45.

Maynard, M.L. and Gates, D.H. (1963) Effects of wetting and drying on germination of crested wheatgrass seed. *Journal of Range Management*, **16**, 119–21.

McDonough, W.T. (1977) Seed Physiology, in *Rangeland Plant Physiology* (ed. R.E. Sosebee), Range Science Series Number 4, Society for Range Management, Denver, Colorado, pp. 155–84.

McGinty, A., Smeins, F.E. and Merrill, L.B. (1983) Influence of spring burning on cattle diets and performance on the Edwards Plateau. *Journal of Range Management*, **36**, 175–8.

McKendrick, J.D., Owensby, C.E. and Hyde, R.M. (1975) Big bluestem and Indiangrass vegetative reproduction and annual reserve carbohydrate and nitrogen cycles. *Agro-Ecosystems*, **2**, 75–93.

McNaughton, S.J. (1978) Serengeti ungulates: feeding selectivity influences the effectiveness of plant defence guilds. *Science*, **99**, 806–7.

McNaughton, S.J. (1979) Grazing as an optimization process: grass-ungulate relationships in the Serengeti. *American Naturalist*, **113**, 691–703.

McNaughton, S.J. and Fullem, L.W. (1970) Photosynthesis and photorespiration in *Typha latifolia*. *Plant Physiology*, **45**, 703–7.

McNaughton, S.J., Wallace, L.L. and Coughenour, M.G. (1983) Plant adaptation in an ecosystem context: effects of defoliation, nitrogen, and water on growth of an African C₄ sedge. *Ecology*, **64**, 307–18.

Mederski, H.J., Chen, L.H. and Curry, R.B. (1975) Effect of leaf water deficit on stomatal and nonstomatal regulation of net carbon dioxide assimilation. *Plant Physiology*, **55**, 589–93.

Merrill, E.H., Mayland, H.F. and Peek, J.M. (1980) Effects of a fall wildfire on herbaceous vegetation on xeric sites in the Selway – Bitterroot Wilderness, Idaho. *Journal of Range Management*, **33**, 363–7.

Miller, R.F. and Donart, G.B. (1979) Response of *Bouteloua eriopoda* (Torr.) Tor. and *Sporobolus flexuosus* (Thumb.) Rydb. to season of defoliation. *Journal of Range Management*, **32**, 63–7.

Mohammad, N., Dwyer, D.D. and Busby, F.E. (1982) Responses of crested wheatgrass and Russian wildrye to water stress and defoliation. *Journal of Range Management*, **35**, 227–30.

Mooney, H.A. and Gulmon, S.L. (1982) Constraints on leaf structure and function in reference to herbivory. *BioScience*, **32**, 198–206.

Mooney, H.A., Ehleringer, J.R. and Berry, J.A. (1976) High photosynthetic capacity of a winter annual in Death Valley. *Science*, **194**, 322–4.

Mooney, H.A., Troughton, J.H. and Berry, J.A. (1974) Arid climates and photosynthetic systems. *Carnegie Institution of Washington Year Book*, **73**, 793–805.

Murata, Y. (1969) Physiological responses to nitrogen in plants, in *Physiological Aspects of Crop Yield* (eds. J.D. Estin, F.A. Haskins, C.Y. Sullivan and C.H.M. Van Bavel), American Society of Agronomy, Madison, Wisconsin.

Nobel, P.S. (1980) Water vapor conductance and CO_2 uptake for leaves of a C_4 desert grass. *Ecology*, **61**, 252–8.

Old, S.M. (1969) Microclimates, fire and plant production in an Illinois prairie. *Ecological Monographs*, **39**, 355–84.

Oomes, M.J.M. and Elberse, W.T. (1976) Germination of six grassland herbs in microsites with deficient water contents. *Journal of Ecology*, **64**, 745–55.

Osman, A.M. (1971) Root respiration of wheat plants as influenced by age, temperature, and irradiance of shoots. *Photosynthetica*, **10**, 378–87.

Oswalt, D.L., Bertrand, A.R. and Teel, M.R. (1959) Influence of nitrogen fertilization and clipping on grass roots. *Proceedings of the Soil Science Society of America*, **23**, 228–30.

Owen, D.F. (1980) How plants may benefit from the animals that eat them. *Oikos*, **35**, 230–5.

Owen, D.F. and Wiegert, R.G. (1981) Mutualism between grasses and grazers: an evolutionary hypothesis. *Oikos*, **36**, 376–8.

Owen, D.F. and Wiegert, R.G. (1982) Grasses and grazers: is there a mutualism? *Oikos*, **38**, 258–9.

Owensby, C.E. and Smith, E.F. (1973) Burning true prairie, *Proceedings of the Third Midwest Prairie Conference* (ed. L.C. Hulbert), Division of Biology, Kansas State University, Manhattan, Kansas.

Owensby, C.E., Hyde, R.M. and Anderson, K. (1970a) Efffects of clipping and supplemental nitrogen and water on loamy upland bluestem range. *Journal of Range Management*, **23**, 341–6.

Owensby, C.E., Paulson, G.M. and McKendrick, J.D. (1970b) Effect of burning and clipping on big bluestem reserve carbohydrates. *Journal of Range Management*, **23**, 358–62.

Owensby, C.E., Rains, J.R. and McKendrick, J.D. (1974) Effects of one year of intensive clipping on big bluestem. *Journal of Range Management*, **27**, 341–3.

Owensby, C.E., Smith, E.F. and Rains, J.R. (1977) Carbohydrate and nitrogen reserve cycles for continuous, season-long and intensive-early stocked Flint Hills bluestem range. *Journal of Range Management*, **30**, 258–60.

Parker, K.W. and Sampson, A.W. (1931) Growth and yield of certan Gramineae by reduction of photosynthetic tissue. *Hilgardia*, **5**, 361–81.

Parrish, J.A.D. and Bazzaz, F.A. (1976) Underground niche separation in successional plants. *Ecology*, **57**, 1281–8.

Parton, W.J. and Risser, P.G. (1979) Simulated impact of management practices upon the tallgrass prairie, in *Perspectives in Grassland Ecology* (ed. N.R. French), Springer-Verlag, New York.

Pearson, D.J. and Hunt, L.A. (1972) Studies on the daily course of carbon exchange in alfalfa plants. *Canadian Journal of Botany*, **50**, 1377–84.

Peet, M.R., Anderson, R.C. and Adams, M.S. (1975) Effect of fire on big bluestem production. *American Midland Naturalist*, **94**, 15–26.

Pemadasa, M.A. (1981) The mineral nutrition of the vegetation of a montane grassland in Sri Lanka. *Journal of Ecology*, **69**, 125–34.

Pettit, R.D. and Fagan, R.E. (1974) Influence of nitrogen and irrigation on carbohydrate reserves of buffalograss. *Journal of Range Management*, **27**, 279–82.

Phillips, D.A. (1980) Efficiency of symbiotic nitrogen fixation in legumes. *Annual Review of Plant Physiology*, **31**, 29–49.

Pierre, W.H. and Bertram, F.D. (1929) Kudzu production with special response to influence of frequency of cutting on yields and formation of root reserves. *Journal of American Society of Agronomy*, **21**, 1079–101.

Poston, F.L., Pedigo, L.P., Pearce, L.P. and Hammond, R.B. (1976) Effects of artificial and insect defoliation on soybean net photosynthesis. *Journal of Economic Entomology*, **69**, 109–12.

Power, J.F. (1971) Evaluation of water and nitrogen stress on bromegrass growth. *Agronomy Journal*, **63**, 726–8.

Quinton, D.A., McLean, A. and Stout, D.G. (1982) Vegetation and reproductive growth of bluebunch wheatgrass in interior British Columbia. *Journal of Range Management*, **35**, 46–51.

Rabinowitz, D. (1978) Abundance and diaspore weight in rare and common prairie species. *Oecologia*, **37**, 213–20.

Rabinowitz, D., Bassett, B.K. and Renfro, G.E. (1979) Abundance and neighborhood structure for sparce and common grasses in a Missouri prairie. *American Journal of Botany*, **66**, 867–9.

Rains, J.R., Owensby, C.E. and Kemp, K.E. (1975) Effects of nitrogen fertilization, burning, and grazing on reserve constituents of big bluestem. *Journal of Range Management*, **28**, 358–62.

Rauzi, F. (1975) Seasonal yield and chemical composition of crested wheatgrass in southeastern Wyoming. *Journal of Range Management*, **28**, 219–21.

Rauzi, F. and Fairbourn, M.L. (1983) Effects of annual applications of low N fertilizer rates on a mixed grass prairie. *Journal of Range Management*, **36**, 359–62.

Rawson, H.M., Gegg, J.E. and Woodward, R.G. (1977) The effect of atmospheric humidity on photosynthesis, transpiration and water use efficiency of leaves of several plant species. *Planta*, **134**, 5–10.

Reardon, P.O., Leinweber, C.L. and Merrill, L.B. (1972) The effect of bovine saliva on grasses. *Journal of Animal Science*, **34**, 897–8.

Reardon, P.O., Leinweber, C.L. and Merrill, L.B. (1974) Response of sideoats grama to animal saliva and thiamine. *Journal of Range Management*, **27**, 400–1.

Redmann, R.E. (1971) Carbon dioxide exchange by native Great Plains grasses. *Canadian Journal of Botany*, **49**, 1341–4.

Reichhardt, K.L. (1982) Succession of abandoned fields on the shortgrass prairie, northeastern Colorado. *Southwestern Naturalist*, **27**, 299–304.

Rice, E.L. (1950) Growth and floral development of five species of range grasses in central Oklahoma. *Botanical Gazette*, **111**, 361–77.

Rice, E.L. and Parenti, R.L. (1978) Causes of decreases in productivity in undisturbed tall-grass prairie. *American Journal of Botany*, **65**, 1091–7.

Risser, P.G. (1969) Competitive relationships among herbaceous grassland plants. *Botanical Review*, **35**, 251–84.

Risser, P.G. (1980) *Methods for inventory and monitoring of vegetation, litter and soil surface condition*, National Academy of Science, Washington, DC.

Risser, P.G., Birney, E.C., Blocker, H.D., May, S.W., Parton, W.J. and Wiens, J.A. (1981) *The True Prairie Ecosystem*, Dowden, Hutchinson and Ross, Stroudsburg, Pennsylvania.

Robocker, W.C., Curtis, J.T. and Ahlgren, H.L. (1953) Some factors affecting emergence and establishment of native grass seedlings in Wisconsin. *Ecology*, **34**, 194–9.

Roe, F.E. (1951) *The North American Buffalo: A Critical Study of the Species in its Wild State*, University of Toronto Press, Toronto, Canada.

Rogler, G.A. (1944) Relative palatabilities of grasses under cultivation on the Northern Great Plains. *Journal of the American Society of Agronomy*, **36**, 487–96.

Roos, F.H. and Quinn, J.A. (1977) Phenology and reproductive allocation in *Andropogon scoparius* (Gramineae)

populations in communities of different successional stages. *American Journal of Botany*, **64**, 535–46.

Ryle, G.J.A. and Powell, C.E. (1975) Defoliation and regrowth in the Graminacous plant: the role of current assimilate. *Annals of Botany (London)*, **39**, 297–310.

Sala, O.E., Lauenroth, W.K., Parton, W.J. and Trlica, M.J. (1981) Water status of soil and vegetation in a shortgrass steppe, *Oecologia*, **48**, 317–31.

Sanders, T.H., Ashley, D.A. and Brown, R.H. (1977) Effects of partial defoliation on petiole phloem area, photosynthesis, and ^{14}C translocation in developing soybean leaves. *Crop Science*, **17**, 548–50.

Shantz, H.L. and Pimeisel, L.N. (1927) The water requirements of plants of Akron, Colorado. *Journal of Agricultural Research*, **34**, 1093–189.

Shiflet, N.N and Dietz, H.E. (1974) Relationship between precipitation and annual rangeland herbage production in southeastern Kansas. *Journal of Range Management*, **27**, 272–6.

Silverton, J.W. (1982) No evolved mutualism between grasses and grazers. *Oikos*, **38**, 253–4.

Simpson, J.R. (1965) The transference of nitrogen from pasture legumes to an associated grass under several systems of management in pot culture. *Australian Journal of Agriculture Research*, **16**, 915–26.

Sims, P.L. and Singh, J.S. (1978) The structure and function of ten western North American grasslands, III. Net primary production, turnover and efficiencies of energy capture and water use. *Journal of Ecology*, **66**, 673–97.

Singh, J.S., Lauenroth, W.K. and Milchunas, D.G. (1983) Geography of grassland ecosystems, in *Progress in Physical Geography*, Vol. 7, pp. 46–80.

Slack, C.R., Roughan, P.G. and Bassett, H.C.M. (1974) Selective inhibition of mesophyll chloroplast development in some C_4 pathway species by low night temperature. *Planta*, **118**, 57–73.

Slatyer, U.G. (1976) Comparative photosynthesis, growth and transpiration of two species of *Atriplex*. *Planta*, **93**, 175–89.

Smith, A.E. and Leinweber, D.L. (1971) Relationship of carbohydrate trend and morphological development of little bluestem tillers. *Ecology*, **52**, 1052–7.

Smith, E.F. and Young, V.A. (1959) The effect of burning on the chemical composition of little bluestem. *Journal of Range Management*, **12**, 139–40.

Smith, E.F., Young, V.A. Anderson, K.L., Ruliffson, W.S. and Rogers, S.N. (1960) The digestibility of forage with burned and nonburned bluestem pasture as determined with grazing animals. *Journal of Animal Science*, **19**, 388–91.

Sosebee, R.E. and Wiebe, H.H. (1971) Effect of water stress

and clipping on photosynthate translocation in two grasses. *Agronomy Journal*, **63**, 14–19.

Splittstoesser, W.E. (1966) Dark CO_2 fixation and its role in the growth of plant tissue. *Plant Physiology*, **41**, 755–9.

Stephenson, R.A., Brown, R.H. and Ashley, D.A. (1976) Translocation of ^{14}C-labeled assimilate and photosynthesis in C_3 and C_4 species. *Crop Science*, **16**, 285–8.

Steuter, A.A. and Wright, H.A. (1983) Spring burning effects on redberry juniper-mixed grass habitats. *Journal of Range Management*, **36**, 161–64.

Stowe, L.G. and Teeri, J.A. (1978) The geographic distribution of C_4 species of the Dicotyledonae in relation to climate. *American Naturalist*, **112**, 609–23.

Teeri, J.A. and Stowe, L.G. (1976) Climatic patterns and the distribution of C_4 grasses in North America. *Oecologia*, **23**, 1–2.

Tieszen, L.L. (1970) Photosynthetic properties of some grasses in eastern South Dakota. *Proceedings of the South Dakota Academy of Science*, **49**, 78–89.

Tieszen, L.L., Senyimba, M.M., Imbamba, S.K. and Troughton, J.H. (1979) The distribution of C_3 and C_4 grasses and carbon isotope discrimination along an altitudinal and moisture gradient in Kenya. *Oecologia*, **37**, 337–50.

Tomanek, G.W., Martin, E.P. and Albertson, F.W. (1958) Grazing preference comparisons of six native grasses in the mixed prairie. *Journal of Range Management*, **11**, 191–3.

Trlica, M.J. (1977) Distribution and utilization of carbohydrate reserves in range plants, in *Rangeland Plant Physiology* (ed. R.E. Sosebee), Range Science Series Number 4, Society for Range Management, Denver, Colorado, pp. 73–96.

Trlica, M.J. and Cook, C.W. (1972) Carbohydrate reserves of crested wheatgrass and Russian wildrye as influenced by development and defoliation. *Journal of Range Management*, **25**, 430–5.

Van Valen, L. and Sloan, R.E. (1966) The extinction of the multi-tuberculates. *Systematic Zoology*, **15**, 261–78.

Viets, F.G. (1962) Fertilizers and the efficient use of water. *Advances in Agronomy*, **14**, 223–64.

Wallace, L.L. (1981) Growth, morphology and gas exchange of mycorrhizal and nonmycorrhizal *Panicum coloratum* L., a C_4 grass species, under different clipping and fertilization regimes. *Oecologia*, **49**, 272–8.

Waller, S.S. and Lewis, J.K. (1979) Occurrence of C_3 and C_4 photosynthetic pathways in North American grasses. *Journal of Range Management*, **32**, 12–28.

Ward, C.Y. and Blaser, R.E. (1961) Carbohydrate food reserves and leaf area in regrowth of orchardgrass. *Crop Science*, **1**, 366–70.

Wardlaw, I.F., (1969) The effect of water stress on translocation in relation to photosynthesis and growth. II. Effect during leaf development in *Lolium temulentum* L. *Australian Journal of Biological Sciences*, **22**, 1–6.

Weaver, J.E. (1954) *North American Prairie*, Johnsen Publishing Company, Lincoln, Nebraska.

Weaver, J.E. (1958) Summary and interpretation of underground development in natural grassland communities. *Ecological Monographs*, **28**, 55–78.

Weaver, J.E. and Albertson, F.W. (1956) *Grasslands of the Great Plains*. Johnsen Publishing Company, Lincoln, Nebraska.

Weaver, J.E. and Darland, R.W. (1949) Soil-root relationships of certain native grasses in various soil types. *Ecological Monographs*, **19**, 303–38.

Weaver, R.J. (1949) Water usage of certain native grasses in prairie and pasture, *Ecology*, **22**, 175–91.

Weaver, T. (1982) Distribution of root biomass in well-drained surface soils. *American Midland Naturalist*, **107**, 393–5.

Webb, W.L., Lauenroth, W.K., Szarek, S.R. and Kinerson, R.S. (1983) Primary production and abiotic controls in forests, grasslands, and desert ecosystems in the United States. *Ecology*, **64**, 134–51.

Weiner, J. (1980) The effects of plant density, species proportion and potassium-phosphorous fertilization on interference between *Trifolium incarnatum* and *Lolium multiflorum* with limited nitrogen supply. *Journal of Ecology*, **68**, 969–80.

Weinmann, H. (1948) Underground development and reserves of grasses: a review. *Journal of the British Grassland Society*, **3**, 115–40.

Westoby, M. (1980) Relations between genet and tiller population dynamics: survival of *Phalaris tuberosa* tillers after clipping. *Journal of Ecology*, **68**, 863–70.

White, E.M. (1961) A possible relationship of little bluestem distribution to soils. *Journal of Range Management*, **14**, 243–7.

White, E.M. and Lewis, J.K. (1969) Ecological effect of a clay soil's structure on some native grass roots. *Journal of Range Management*, **22**, 401–7.

White, E.M., Gartner, F.R. and Butterfield, R. (1983) Blue grama (*Bouteloua gracilis*) response to fertilization of a claypan soil in the greenhouse. *Journal of Range Management*, **36**, 232–3.

White, L.M. (1973) Carbohydrate reserves of grasses: a review. *Journal of Range Management*, **26**, 13–18.

White, L.M., Brown, J.H. and Cooper, C.S. (1972) Nitrogen fertilization and clipping effects on green needlegrass (*Stipa viridula* Trin.): III. Carbohydrate reserves. *Agronomy Journal*, **64**, 824–8.

White, R.S. and Currie, P.O. (1983) Prescribed burning in

the northern Great Plains: yield and cover responses of 3 forage species in the mixed grass prairie. *Journal of Range Management*, **36**, 179–83.

Wight, J.R. and Black, A.L. (1972) Energy fixation and precipitation use efficiency in a fertilized rangeland ecosystem of the northern Great Plains. *Journal of Range Management*, **25**, 376–80.

Williams, W.A., McKell, C.M. and Reppert, J.M. (1964) Sulfur fertilization of an annual-range soil during years below normal rainfall. *Journal of Range Management*, **17**, 1–5.

Wilson, A.M. and Briske, D.D. (1978) Drought and temperature effects on the establishment of blue grama seedlings. *Proceedings of the First International Rangeland Congress*, (ed. D.N. Hyder), Society for Range Management, Denver, Colorado, pp. 359–61.

Wilson, A.M. and Sarles, J.A. (1978) Quantification of growth drought tolerance and avoidance of blue grama seedlings. *Agronomy Journal*, **70**, 231–7.

Young, J.A., Evans, R.A. and Kay, B.L. (1970) Germination characteristics of range legumes. *Journal of Range Management*, **23**, 98–103.

12

Deciduous forest

DAVID J. HICKS AND BRIAN F. CHABOT

12.1 Introduction

Coves of the southern Appalachians contain what may be regarded as prototypes of the deciduous forest association in North America. A diverse community containing more than ten dominant tree species forms a continuous canopy overhead. The largest trees have a basal diameter in excess of 1 m. Seedlings of the canopy tree species are mixed in with a rich herb and shrub layer, suggesting to the practiced eye that the forest is perpetuating itself. In such a scene ecologists can well focus on the many elements which have led to the evolution of one of the richest associations of plant species on the North American continent. Competition for light, nutrients, and space, the impacts of herbivores and pathogens, and opportunities for mutualistic relationships with pollinators and dispersers have all led to a diverse set of adaptations. There are still places, little touched by man, which suggest a certain timelessness to the forest. But it is change which is the most pervasive force.

During the Tertiary period, deciduous forests occupied more extensive continental areas in temperate portions of the northern hemisphere. Subsequently, climatic changes, due in part to the uplift of western mountain ranges in both North America and Asia, greatly reduced the extent of deciduous forests. This formation now occurs in fairly restricted areas of North America, eastern Asia, and Europe. Because of the Tertiary interchange, there are strong floristic and ecological similarities between the forests of these regions (Graham, 1972). Relict populations of deciduous forest species, isolated during the Tertiary and Quaternary, also occur in the western United States, the central grasslands, and the highlands of Mexico.

In the northern part of the deciduous forest region, climatic disruption during the Pleistocene and glacial modification of the soils have left a lasting impact on the forests. At full glacial, many of the species present in northern regions today occurred much further south (Delcourt and Delcourt, 1979). Evidence is accumulating that species which commonly occur together in the forest today were in widely separated refugia during full glacial time (Davis, 1969). Modern communities have been assembled by migration of species from these refugia only in the past few thousand years, or even more recently in the north. Consequently, most of the species have been ecologically associated for only a few generations.

Changes produced by biotic forces have been no less profound. The greatest of these has been land clearing for agriculture. More than 80% of the original forest was removed over vast areas. It is difficult to find even a small tract of mature forest where human impact is not evident. Insects and pathogens – such as the chestnut blight, Dutch elm disease, spruce budworm – have taken their toll.

Beyond those changes which are barely perceptible on the scale of human lifespans, there are others, more dramatic to our eyes, which occur with the passing of seasons and with variations in the landscape. All of these changing forces have a profound influence in constantly reshaping the structure of the forest. It is not possible to understand the adaptive ecology of forest plants without considering the dynamic aspects of the environment in which they operate. Beyond reviewing the status of research on physiological ecology of deciduous forest species, we also hope to integrate what we know about cycles of activity in different plant species and why they occur.

12.2 Geography and vegetation

Much environmental and vegetational differentiation occurs in the deciduous forest region of North America. Several geographic features are important (Hunt, 1974). Unlike western North America, the Atlantic coast is bordered by a broad, flat coastal plain. The vegetation here is classified as Southeastern Coniferous Forest. Inland from the coastal plain are the Piedmont Plateau and the Appalachian Mountains. Geologically, these areas are rather complex with the underlying metamorphic and sedimentary rocks having given rise to a fairly wide range of soil types. The Piedmont has rolling to hilly topography and the Appalachians are rugged in some sections. The highest peaks are Mount Mitchell in the South at 2038 m and Mount Washington in the north at 1916 m. With these elevations, the Appalachians are not as significant a factor in altering air mass movement and climate as are either the Rocky Mountains or the Sierra Nevada-Cascade Mountains. West of the Appalachians, the deciduous forest occupies areas that are geologically classified as plateaus, although erosion has produced a rolling or even mountainous topography. The coastal plain extends up the Mississippi Valley and divides the plateau into two lobes at its southern margin. The rocks of the plateau are sedimentary and include much limestone and dolomite, a substrate relatively rare in other geological provinces.

Climatic gradients are strong in both north and south and east to west directions because of changes in the degree of influence of oceanic and continental

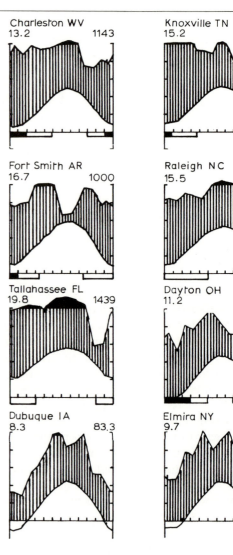

Figure 12.1 Climatic patterns for stations within the Eastern Deciduous Forest. Mean annual temperature (°C) and annual precipitation (mm) are in upper left and right of each monthly temperature; upper curve is mean monthly precipitation. Charleston is in the mixed mesophytic association; Knoxville is in the oak–chestnut association; Fort Smith is in the oak–hickory association; Raleigh is in the oak–pine association; Tallahassee is in the Southern mixed hardwoods association; Dayton is in the beech–maple association; Dubuque is in the maple–basswood association; Elmira is in the Northern hardwoods association (after Vankat, 1979; diagrams from Walter and Lieth, 1967).

air masses and in latitudinal changes in temperature (Fig. 12.1). From north to south, temperatures, growing season, and precipitation increase. From east to west at a given latitude the most significant environmental change is precipitation. Rainfall increases to a maximum at higher elevations in the southern Appalachian Mountains where over 250 cm per year have been recorded, then decreases toward the interior of the continent.

The boundaries of the deciduous forest region are determined by climatic and edaphic factors. At northern latitudes and high elevations, the decreasing temperature and growing season favor evergreen over deciduous trees. The deciduous forest grades into boreal forest dominated by species of spruce (*Picea*), fir (*Abies*), and pine (*Pinus*). The long growing season, mild climates, and impoverished soils of the coastal plain appear to be factors favoring the evergreen needle-leaf and broadleaf species of the Southeastern Coniferous Forest. In the Midwest the forest is invaded by a broad wedge of

steppe, the Prairie Peninsula. The existence of this wedge of prairie, as well as the position of the forest-steppe boundary, is partially controlled by climate. Decreasing precipitation and increasing summer temperatures favor grasses and herbs over trees. However, at any particular point on the boundary, soils, fire, and grazing by large animals also are, or were, important factors influencing the extent of the deciduous forest. The deciduous forest extends across the steppe as gallery forests along rivers. Such forests are composed mostly of flood-tolerant species, not those of the mesic uplands.

Several vegetation types have been recognized within the deciduous forest region (Fig. 12.2). Detailed descriptions are provided by Braun (1950), Oosting (1956), Daubenmire (1978), and Vankat (1979). The mixed mesophytic forest association occupies a central position dominating the Appalachain and Cumberland Plateaus. Occurring primarily on well-developed soils on mesic sites, one or more of ten major species can dominate on any

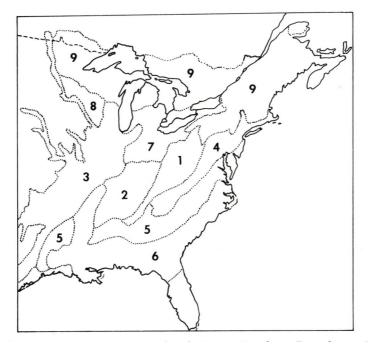

Figure 12.2 Major plant community association within the Eastern Deciduous Forest biome: (1) mixed mesophytic forest association; (2) Western mesophytic forest association; (3) oak–hickory association; (4) oak–chestnut association; (5) oak–pine association; (6) Southern mixed hardwood association; (7) beech–maple association; (8) maple–basswood association, (9) Northern hardwoods association. (Braun, 1950; Vankat, 1979)

particular site. Many of these species decrease in importance toward the north, leaving beech (*Fagus grandifolia*) and sugar maple (*Acer saccharum*) as the dominant species. In the northwest, basswood (*Tilia americana*) replaces beech in association with maple. The northern hardwood association represents a transition between the deciduous and boreal forests. Hemlock (*Tsuga canadensis*), yellow birch (*Betula alleghaniensis*), white ash (*Fraxinus americana*), and white pine (*Pinus alba*) become important members along with beech and sugar maple. Eastward from the central plateaus, the oak–chestnut association replaces the mixed mesophytic forest. With the elimination of chestnut in this century, the forest composition is undergoing a readjustment. Within the deciduous forest region, xeric sites are usually dominated by the oak–hickory association. This involves several different species of oak and hickory.

12.3 Forest structure

Despite the different associations of species within the deciduous forest region, it is useful to describe the structure of a 'typical' deciduous forest. Our description applies to climax forests on mesic, fairly rich soils. The trees are 25–30 m in maximum height. A completely closed canopy may be broken by gaps where trees have recently died. Trunk diameters include all sizes and reach a maximum of 1–1.5 m. The deciduous species which dominate the canopy are in leaf from late May to October. There may be a few evergreen conifers among the canopy species. These are long-lived successional trees, such as white pine, or shade-tolerant climax species, such as the eastern hemlock. The shrub and small tree strata are composed both of saplings of the canopy species and of understory specialists which are not capable of reaching the canopy. Herb and moss strata are extremely variable in structure, but would have less than 100% cover in most mesic forests. The herbaceous community undergoes pronounced seasonal changes. Some species are active only in the early spring before the forest canopy has leafed out. Other species have maximum growth and flowering in early or late summer after canopy closure.

In environments that are edaphically or climati-

cally xeric, the canopy is shorter and more open, and drought-tolerant conifers are more important. Because of the greater transmission of light through the tree canopy in such situations, shrub and herb layers have higher cover and more diversity. Deciduous and evergreen heath shrubs (Ericaceae) often dominate the shrub strata on dry, nutrient-poor sites. Forests on flooded soils also usually have a more open canopy in comparison with mesic forests. Individual trees in floodplain forests can be larger in diameter and height with herb and shrub layers less well developed depending on the depth and duration of flooding.

12.4 Plant response to seasonal environments

12.4.1 GENERAL CONCEPTS

Several concepts have proven to be useful in understanding the adaptive ecology of plants. One of these is the concept of strategy. The importance of individual traits is often best understood when they are considered as part of more general, integrated, syndromes. The most general strategic alternatives are those of avoidance and tolerance. Deciduous forest plants must be able to either avoid or tolerate the stresses that exist under a closed canopy. Avoidance may be in seasonal time, such as the spring ephemerals which take advantage of the period when canopy trees are leafless and largely inactive, or avoidance in successional time by taking advantage of relatively short-lived canopy gaps that result from tree death. Forest herbs cannot grow away from the zone of maximum shade stress at the forest floor, so they must either be tolerators or seasonal avoiders. Tolerance and avoidance both involve distinct sets of traits relating to physiology, development rates, resource allocation patterns, use of environmental cues, pollinator and disperser syndromes, all of which together permit individual plants to integrate their activity with that of the community as a whole.

The calculation of whole plant carbon budgets appears to be a useful mechanism for integrating and evaluating the effects of specific environmental factors in the functioning of individual plants and in evaluating the importance of specific adaptive traits in moderating environmental stress. Though the conceptual approach was outlined by Mooney in 1972, it has as yet had only scattered application. We will be using this approach wherever possible in summarizing information on physiological ecology of deciduous forest species.

Seasonal change in the forest creates adaptive opportunities which might not otherwise exist. Individual species are tied closely to environmental change. Indeed community structure is based upon either complementary or out of phase interactions between a larger number of species. There are also subtle microtopographic features produced by treefalls which influence seedling establishment patterns (Stone, 1975; Thompson, 1980). Because of the importance of seasonal change, we will be placing our review in the context of the sequential environmental challenges which different groups of species face during the course of a year.

12.4.2 INITIATION OF GROWTH

Herbaceous species are usually the first to begin active growth in the spring. Herbaceous plants respond phenologically and physiologically to the microclimate at the forest floor which, in turn, is dependent upon the activities of the canopy tree species. The herbaceous flora of the deciduous forest is dominated by species that overwinter as rosettes or underground perennating organs. A floristic analysis of eight forests (Cain, 1950, cf. also Struik and Curtis, 1962) shows that the herb stratum is dominated by hemicryptophytes (58% of the species) and cryptophytes (27%). Chamaephytes, with their buds at the soil surface, comprise only 8% of the species and less than 6% are annuals reproducing from seeds.

The emergence of forest herbs from dormancy is controlled largely by temperature. Most research on flowering phenology of spring herbs has shown that some sort of heat summation index usually is well correlated with blooming time (Jackson, 1966; Lindsey and Newman, 1956). In wide-ranging species there is evidence of phenological races that have different sums necessary for initiation of activity (Castelli, 1970). Less is known about the control of vegetative growth, but in *Trientalis bore-*

alis the emergence of shoots and production of tubers is controlled in part by temperature, with photoperiod also having an important influence on tuber formation (Anderson, 1970; Anderson and Loucks, 1973).

Herbs may be classified into phenological groups by the length and timing of their growth periods relative to the phenology of the canopy and stand microclimate (Mahall and Bormann, 1978; Sparling, 1964). These groups are:

1. Spring ephemerals ('light phase' species of Sparling and 'vernal photosynthetics' of Mahall and Bormann) complete all or most of their annual growth cycle in the brief period lasting a few weeks between snowmelt and closure of the tree canopy. Flowering may be at or after the period of vegetative growth.

2. Shade tolerants ('shade phase' species of Sparling and 'summer green' and 'late summer' species of Mahall and Bormann) start to grow at or before the time of canopy closure, and carry out most of their growth and photosynthesis under a closed canopy.

3. Evergreens (includes 'semievergreens' of Mahall and Bormann) retain all or most of their aboveground biomass for an entire year. They are not usually metabolically active at all seasons, having a season of dormancy during the winter months. Many evergreens retain a single cohort of leaves, usually produced in the spring, for a 12-month period. A few species have leaves living up to three years. Others have continuous leaf turnover during the growing season, with the leaves produced in late summer and fall surviving through the winter until growth resumes in the spring. Evergreens are not uniform in their flowering times although many species flower in the late spring or early summer.

Spring ephemerals are seasonal stress avoiders, while the latter two groups contain mostly stress tolerators. These three groups differ considerably in their physiology as will be described.

Spring ephemerals are active in periods of high light, water, and nutrient availability. Caldwell (1969) has pointed out the similarities in environment and growth response of ephemerals and alpine species. Both of these types must complete their annual growth within a short period of favorable conditions. Caldwell found that an alpine and a

Table 12.1 Photosynthetic characteristics of deciduous forest herbaceous species

Species	Compensation point*	Saturation point*	Maximum photosynthetic rate[†]	Leaf life span (week)	References
Spring ephemerals					
Erythronium americanum	1.0	25.0	20[‡]	11	T and P
	0.7	30.0	85.2	11	S
	1.0	~ 50.0	125	11	H
Allium tricoccum	0.5	20.0	82	10	S
	—	—	13.1[‡]	11	T and P
Dicentra canadensis	1.8	30.0	28.4	9	S
Claytonia caroliniana	1.8	15.0	26.5	11	S
Dentaria diphylla	0.4	3.8	36.6	10	S
Summer green					
Solidago flexicaulis	0.5	12.0	11[‡]	28	T and P
Trillium grandiflorum	0.5	14.0	10[‡]	22	T and P
	0.2	23.0	30.3	11	S
T. erectum	0.5	2.5	—	20	S
Podophyllum peltatum	0.5	2.5	27.1	18	S
	—	—	11.5[‡]	11	T and P
Parthenocissus quinquefolia	—	—	4.7[‡]	25	T and P
Caulophyllum thalictroides	1.0	30	49.2	20	S
Sanguinaria canadensis	0.2	5.0	70.	20	S
Dryopteris marginalis	0.2	3.0	19.6	30	S
Maianthemum canadense	0.1	2.5	31	22	S
	0.3	3.0	82.7	11	H
Polygonatum pubescens	0.1	2.5	12	~ 22	S
Hydrophyllum appendiculatum	1.0	21.0	5.7[‡]		M
Evergreens					
Mitchella repens	0.4	7.0	43	120	H
Dryopteris spinulosa	0.2	2.5	20.8	54	S
Pyrola elliptica	0.4	7.0	55.5	54	H
Hexastylis arifolia	0.2–1.5	24.0	11[‡]	50	G
Fragaria vesca	0.6–2.5	20–30	71(3.9)[‡]	14–20	C, J

*% of full sunlight
[†]$\mu mol\ CO_2\ g^{-1}\ s^{-1}$;   [‡]$\mu mol\ CO_2\ m^{-2}\ s^{-1}$

Notes

Compensation and saturation points are expressed as % full sunlight, because different authors have used different units. Full sunlight is considered to be $2000\ \mu E\ m^{-2}\ s^{-1}$ or $10\,000$ ftc. Maximum photosynthetic rates are maximum rates at light saturation and are expressed in units of $\mu mol\ CO_2\ m^{-2}\ s^{-1}$ for the data of Taylor and Pearcy and in $\mu mol\ CO_2\ g^{-1}\ s^{-1}$ for the data of Sparling and Hicks. These figures represent maxima over an entire leaf span and are typically found in fully expanded, though still young, leaves. The leaf life span is the period during which mature leaves are present. Data for leaf life spans are from Sparling (1964) for southern Ontario, Mahall and Bormann (1978) in New Hampshire, and unpublished observations of the authors in New York State. The last column indicates the source of the photosynthetic data: T and P = Taylor and Pearcy (1976), for upstate New York plants; S = Sparling (1967), for southern Ontario; H = unpublished data of D.J. Hicks for the Ithaca, New York area; M = Morgan (1971); G = Gonzales (1972). Only species for which more or less complete data were available are included.

deciduous forest species of *Erythronium* were similar in their response to temperature. Both could grow at low temperatures, but were also capable of facultative response to periods of warmer weather. The rapid flush of growth in ephemerals and alpine herbs is supported by rapid mobilization and translocation of stored materials (compare Kieckheffer, 1962 and Risser and Cottam, 1968 to Mooney and Billings, 1960). For short periods the growth rates can be very high. For example, *Erythronium americanum* in New Hampshire has an average rate of productivity of $0.67\,g\,m^{-2}\,d^{-1}$ for a 21-day period and *E.albidum* in Kentucky has a rate of $1.18\,m^{-2}\,d^{-1}$ over 36 days (calculated from Muller, 1978, 1979). The rapid growth of spring ephemerals when tree growth has hardly commenced reduces the loss of nutrients from the forest ecosystem in spring runoff (Muller, 1978).

Ephemerals are typically sun plants in their photosynthetic characteristics. They have relatively high light compensation and light saturation points and high absolute photosynthesis rates (Table 12.1). The high photosynthesis rates of ephemerals are due to higher RuBP-carboxylase concentrations and higher mesophyll and stomatal conductances (Taylor and Pearcy, 1976). These species also maintain high dark respiration rates $(0.6-0.9\,mol\,CO_2\,m^{-2}\,s^{-1})$, presumably related to the higher maintenance costs for larger amounts of mesophyll tissue and a larger pool of RuBP-carboxylase (Taylor and Pearcy, 1976). Spring ephemerals appear to lack the capacity to modulate photosynthetic and respiratory physiology in response to the decreased light levels that accompany canopy closure. The investment in leaf photosynthetic structure and light-absorbing pigments is not substantially altered with shading (Harvey, 1980). This lack of adaptability limits ephemerals from exploiting reduced light environments. Consequently, ephemerals revert to a dormant condition soon after the canopy species leaf out.

For woody tree and shrub species, the end of the dormant period in the spring is also controlled largely by temperature although precipitation has been implicated as well (Ahlgren, 1967). Tree species differ in their reaction to temperature so that bud break and development of forest canopy takes place over a period of a month or more. When dormancy is broken, stored carbohydrates are translocated to growing organs, and the tree experiences a period of negative carbon balance (McLaughlin and McConathy, 1979). In some species the carbon gain in the preceding year controls growth of the current flush (Kozlowski and Keller, 1966). This is also probably true for herbaceous species (Jurik, 1980).

The duration and timing of the growth flush is also correlated with the successional status and hydraulic architecture of the species. Early successional trees have a long period of shoot growth, often from frost to frost, and an indeterminate growth pattern. Such opportunistic growth allows maximum utilization of favorable weather conditions and high growth rates. These features allow successional species to efficiently use ephemeral resources. In contrast, late successional trees have a shoot growth period of only 4–5 weeks and are typically determinate. These two groups have contrasting patterns of energy allocation, with early successional trees allocating energy primarily to stems and late successional trees to leaves and roots. The conservative behavior of the late successional species is adaptive in the competitive, stressful conditions of the forest floor, but precludes response to favorable weather by additional shoot growth (Marks, 1975).

The hydraulics of water transport also affect growth phenology in the spring. Ring porous species must differentiate a new ring of conducting tissue before bud break and shoot growth can occur. Diffuse pore species have active xylem at all times (Kramer and Kozlowski, 1979; Zimmerman and Brown, 1971). As a result, ring porous species are later in starting their growth flush than are diffuse porous species (Federer, 1976).

12.4.3 THE CHANGING MICROCLIMATE

With the onset of growth in the spring, the climate within the forest increasingly comes under the control of the plants themselves. Many aspects of microclimate modification by the forest can be traced to interception of solar radiation by the tree canopy (Fig. 12.3). The degree of interception is

Figure 12.3 Microclimate sampling in the Northern hardwoods association (Hubbard Brook, NH). The tower is used to support meteorological instruments used in characterizing environmental changes through the forest canopy as well as providing access for monitoring photosynthesis and stomatal behavior. Principal species are beech (*Fagus grandiflora*) and sugar maple (*Acer saccharum*).

quite variable through the year (Fig. 12.4). During winter, when the trees are leafless, their branches and trunks can absorb 50–70% of incoming solar radiation (Federer, 1971; Hutchison and Matt, 1977; Vezina and Grandtner, 1965). In contrast, during the summer full leaf phase typically only 1–5% of incoming solar radiation penetrates the canopy to ground level. This amount is variable and depends on the density of tree and understory strata, with greater amounts of penetration in xeric and

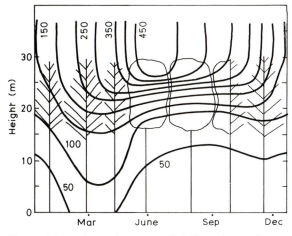

Figure 12.4 Annual course of daily solar radiation within a stand of tulip tree (*Liriodendron tulipifera*) in eastern Tennessee. The isopleths are in units of ly day^{-1} (from Hutchinson and Matt, 1977).

leaf and branch tissue between the ground and the sun varies. Under a patch of dense canopy or in the shade of a tree trunk, less than 0.5% of incoming radiation may reach the ground. At the other extreme, canopy gaps allow full or nearly full sunlight to penetrate to the surface at certain times of day (Minckler *et al.*, 1973). At a particular point on the forest floor light levels can vary by orders of magnitude as the sun briefly shines through a gap in the canopy (Fig. 12.5). The great spatial and temporal variation in light in the forest makes statistically adequate descriptions of the light regime quite difficult (Anderson, 1964; Reifsnyder *et al.*, 1971). Instantaneous point light measurements are especially subject to misinterpretation. Depending upon application, either daily integrated measurements or frequent instantaneous measurements at a large number of points in the forest provide the best characterizations of the light climate.

hydric forests with open canopies (Kittredge, 1948; Whittaker, 1966). The change from winter to summer conditions takes place over a period of a month or more. There is a similar period of light increase in the fall as leaves of different species lose chlorophyll and are shed.

Forest canopies are not homogeneous in density, so light levels on the forest floor are highly variable from point to point and time to time as amount of

Passage of sunlight through the fully leafed canopy also affect the spectral distribution of the radiation as specific wavelengths are differentially absorbed, reflected, or transmitted. Blue and red light are absorbed by the leaves of the canopy and are somewhat impoverished in the understory, while far red is greatly increased (Federer and Tanner, 1966; Vezina and Boulter, 1966). Alterations of the red/far red ratios have potential

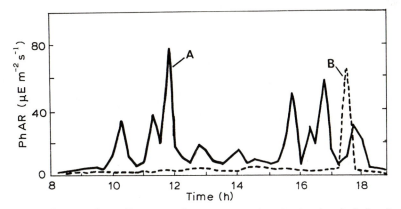

Figure 12.5 Variation in photosynthetically active radiation at two microsites in a hemlock–hardwood forest, Ithaca, NY. Site A is under hemlock (*Tsuga canadensis*) canopy but receives sidelight and direct beam light through adjacent deciduous trees with thinner crowns. Site B is directly under dense hemlock canopy and is also shaded by a patch of sugar maple (*Acer saccharum*) saplings. The day of measurement was sunny with midday maximum photosynthetically active radiation (PhAR) of 1800–2000 $\mu E\,m^{-2}\,s^{-1}$.

influence over phytochrome-mediated processes such as morphogenesis and seed germination (Smith, 1973).

Other aspects of microclimate are largely determined by the disposition of solar energy. The fully leafed forest can be divided into three microclimatic strata. The uppermost of these is the crown space where radiation exchange is most active. The greatest temperature extremes tend to occur in the crown space. Below this, in the trunk space, little radiation exchange occurs so that temperatures are nearly isothermal. The ground surface is a radiation exchange surface of secondary importance during the summer. Daily temperature changes are buffered by turbulent air exchange in the trunk space and storage of heat in the soil. With a sparse canopy or during the leafless period in the winter, the ground becomes an active radiation exchange surface, producing the greatest temperature extremes at this level.

The profiles of other microclimatic factors depend upon their source and interaction with the canopy. Windspeed declines from the bulk atmosphere at an approximately exponential rate with depth in the canopy (Fig. 12.6). In stands that are open at one or more edges, 'blowthrough' can occur with windspeeds high both above and below the canopy. Carbon dioxide and water vapor, in contrast to other microclimatic factors, have major sources and sinks in the soil and in the metabolizing leaves of the canopy. During the day when the canopy is fully leafed out, these elements typically decline from the ground surface up into the canopy, often with a secondary maximum of water vapor and minimum of CO_2 in the crown space. At night and in the winter both quantities decrease from the ground upward.

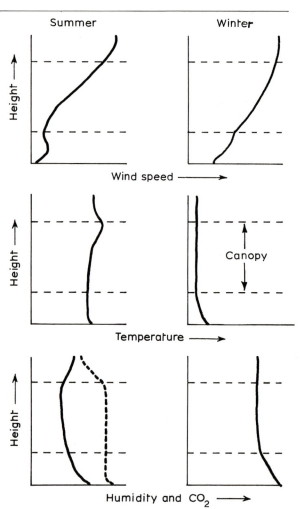

Figure 12.6 Idealized profiles of the microclimatic elements within a deciduous forest canopy. The contrast between conditions during the day in summer and winter is shown (sources: Christy, 1952; DeSelm, 1952; Garrett and Roberts, 1978; Geiger, 1965; Lee, 1978).

12.4.4 RESPONSE TO SHADING

Low light levels most restrict photosynthesis and growth rates of forest species. It is well known to foresters that tree species differ widely in their abilities to maintain populations of seedings in the shade of a closed, fully leafed canopy. Individuals of shade tolerance species, usually simply called 'tolerants' in the forestry literature, can persist as shade-suppressed saplings for long periods. Intolerants do not maintain sapling or seedling populations under closed canopies. Baker (1950) provides a widely accepted classification of shade tolerance in which many intermediate conditions are possible. Herb species also differ substantially in their ability to tolerate shade stress. In this section we discuss the mechanisms that allow such tolerance. These adaptations are expressed in three levels: within leaf (anatomical and physiological), within plant (canopy, architecture and energy allocation), and

Table 12.2 Photosynthetic properties of deciduous tree species

Type and species	Light saturated photosynthesis	Respiration	Light saturation level	Compensation point	Reference
Intolerant (early successional)					
Liriodendron tulipifera	6.7	0.25	25	—	W and D
	10.7	1.4	30	—	L
Quercus marilandica	9	—	85	—	Bo
Ulmus alata	10.9	1.1	53	4.0	Ba
Diospyros virginiana	9.5	1.1	53	5.5	Ba
Populus tremuloides		2.0	32	—	L
Sassafras albidum	6.2	0.7	30	~1	Bz
Populus deltoides	16.4	0.9	75	~1	Re
Midtolerant (mostly gap phase)					
Quercus rubra	5.0	—	40	—	Bo
Acer rubrum	3.8	0.4	22	—	W and D
Quercus rubra	4.5	0.5	14	—	L
Acer rubrum	2.8	1.1	8	—	L
Betula lutea	4	0.2	7	~0.5	Lg
Quercus alba	11.2	—	29	0.8	A
Tolerant (late successional)					
Cornus florida	2.6	0.2	25	—	W and D
Acer saccharum	4.4	0.3	30	—	L and K
Fagus grandifolia	4.4	0.9	17	~0.5	L

Notes

Metabolic rates are all expressed in units of μmol CO_2 m^{-2} s^{-1}. Saturation and compensation points are expressed as % full sun, *assuming full sun to be 2000* μE m^{-2} s^{-1}, or 10 000 ftc, or 50 000 lux. Where contrasting data were available, data for intolerants are for plants grown in high light (100% sun), those for tolerants and midtolerants for plants grown in low light (10–15% full sun). Rates at 25–30 °C. References: W and D, Wallace and Dunn (1980); L = Loach (1967); Bo = Bourdeau (1954); Ba = Bacone *et al.* (1976); Bz = Bazzaz *et al.* (1972); Re = Regehr *et al.* (1975); Lg = Logan (1970); A = Aubuchon *et al.* (1978); L and K = Logan and Krotkov (1968).

life cycle (seed and seedling ecology). Species growing exclusively in dense shade show the best development of the mechanisms of tolerance discussed below, but all forest species display these mechanisms to some extent.

Many adaptive properties within leaf level can be understood as aspects of photosynthetic differences of sun and shade leaves (Boardman, 1977). The number of characteristics which distinguish sun and shade leaves may be extensive and are summarized in Table 12.2. Leaves developing under low light intensities are usually thin with small mesophyll-cell volume per unit leaf area. Concentrations of the primary carboxylase enzyme, ribulose-1.5-diphosphate carboxylase oxygenase, are low and this

seems to be the primary cause of low rates of net photosynthesis indicated in Table 12.2 (Loach, 1967; Taylor and Pearcy, 1976). Because of the smaller pool of leaf protein, maintenance costs as expressed by the rate of dark respiration, are lower in tolerant species (Loach, 1967; Taylor and Pearcy, 1976). Low respiration rates in turn lead to reduced light compensation points which facilitates a positive carbon balance under dense shade. Chlorophyll concentration and photosynthetic unit size (chlorophyll/P700) may be higher in low-light grown leaves. In some species this simply represents a protection from photobleaching of light-harvesting pigments by high-light intensities. While light harvesting may be relatively more effective in shade leaves, intermediate electron transport pigments and Hill reaction activity are often reduced to match the reduced carboxylation capacities. Because of these complex cellular changes, a common measure of light-use efficiency, CO_2 fixed per quanta of light absorbed, may be higher, lower, or unchanged in high- versus low-light grown leaves (Wallace and Dunn, 1980). Stomatal behavior differs in such a way that tolerants can utilize the sun flecks that characterize the forest floor more efficiently than intolerant species. Stomates of tolerant species open in a few minutes when light levels increase, but intolerants need 30 min or more for maximum stomatal opening (Davies and Kozlowski, 1974; Woods and Turner, 1971; Pereira and Kozlowski, 1977). Stomatal behavior plus the rapid increase in photosynthesis under light fleck conditions (Gross and Chabot, 1979) may partly compensate for the lower biochemical efficiency of shade-tolerant species.

For most species growing in shade a substantial fraction of annual carbon gain is accumulated prior to canopy closure, at the end of the growing season, or during light flecks. Jurik (1980) presents data indicating that 45% of annual carbon supply in *Fragaria vesca* occurs when the tree canopy is leafless. Gross (1982) has calculated that when the canopy is in full leaf, about 50% of the daily carbon income for understory herbs occurs as a result of light flecks. Light gaps can have an important influence on plant density and reproductive output of herbs forest (Pitelka *et al.*, 1980). It is not surprising then to find that shade-adapted species retain an ability to exploit these high-light environments. In part, this is seen in photosynthesis rates which are not saturated at 'mean' light levels. Biochemical and photochemical capacity is usually in excess of that required for optimal photosynthetic performance at low light levels. Leaf phenology also plays an important role with some species, producing a succession of leaves with each set adapted to specific environmental conditions at the time of development. For example, *Fragaria vesca* maintains over-wintering leaves which initiate photosynthesis early in the spring. These leaves are replaced by new leaves produced in late March, more than 1–2 months prior to canopy leafout. When canopy

closure occurs, these new leaves are replaced by leaves better adapted to the shaded conditions which remain for the duration of the summer. These leaves senesce and are replaced by over-wintering leaves only after very severe frosts and usually well after the canopy trees have lost their leaves. Species classified as spring ephemerals or aestival species are phenological specialists with leaves adapted primarily to restricted portions of the season.

The canopy architecture of shade-tolerant species is organized to efficiently intercept low-light intensities with minimum shading within the canopy. Many of these species display leaves horizontally in a non-overlapping single layer. Intolerant species in the sun have a more nearly random distribution of leaves within the canopy. They have a branching pattern with many short shoots which leads to mutual shading by adjacent leaves. Also, intolerants have a multilayered canopy. In full sun, leaves are displayed at a considerable angle to the horizontal, minimizing absorption of energy and possible overheating (McMillen and McClennon, 1979; Whitney, 1976). Within plant level, intolerant species have a greater ability to modify leaf angle. McMillen and McClennon (1979) found that a group of intolerant species change their leaf angle by a factor of 3.5 between sun- and shade-produced leaves while tolerant species modified leaf angles by a factor of only 1.7. Wallace and Dunn (1980) also found that leaf angles changed in tree species that exploited gaps in the forest canopy. They attributed the ability to attain high growth rates in gaps partially to this modification.

Plants in the shade typically allocate more energy to producing a crop of thinner leaves, thus expanding the absolute photosynthetic surface area (Abrahamson and Gadgil, 1972; Loach, 1969). As light intensity increases, allocation shifts to roots and secondarily to stems (Loach, 1969). Such a shift in allocation pattern apparently allows better growth than competitors and more effective competition for light and water. The optimal allocation of resources is of obvious importance for forest species. It has only begun to be addressed formally with regard to reproduction and with respect to the utilization of fluctuating light environments. Optimal adaptation is implicit in the interpretation of physiological adaptations to various light intensities. However, there remains considerable room for explicit quantitative assessment of the optimality of various adaptations.

The above discussion, and most of the literature, is concerned with maintenance of positive carbon budgets in shade as a major aspect to shade tolerance. For example, the significance of high respiration rates of intolerant species in the shade has been emphasized by Grime (1965) and Loach (1967). Observations of carbon budgets and growth rates in shaded situations do not always confirm the primary role of carbon accumulation. Geis *et al.* (1971) made field measurements of carbon balance for several species in an old-growth forest in Illinois. Their measurements were integrated over entire days and compared carbon gain of species of different tolerance in a gap and under full canopy. In the gap, *Quercus macrocarpa*, an intolerant species, gained $63.6 \, \text{mmol CO}_2 \, \text{m}^{-2} \, \text{d}^{-1}$ and was superior to the semi-tolerant species, *Q. rubra* and *Q. alba* (38.6 and $31.8 \, \text{mmol CO}_2 \, \text{m}^{-2} \, \text{d}^{-1}$, respectively) and the tolerant *Acer saccharum* ($29.5 \, \text{mmol CO}_2 \, \text{m}^{-2} \, \text{d}^{-1}$). In the shade, the lowest rates of carbon gain were found in the semi-tolerant oaks (*Q. alba* 8.2, *Q. rubra* $6.8 \, \text{mmol CO}_2 \, \text{m}^{-2} \, \text{d}^{-1}$). However, *A. saccharum* and *Q. macrocarpa* did not differ statistically in their carbon-gaining capacity (14.8 and 11.6 mmol $\text{CO}_2 \, \text{m}^{-2} \, \text{d}^{-1}$, respectively). The experiments of Loach (1967) also contradict expectations based upon carbon accumulation. He grew species of contrasting tolerance under various levels of artifical shade. *Populus tremuloides*, the species considered most intolerant, indeed had the lowest growth rate in the shade, with negative rates in the deepest shade. However, *Liriodendron tulipifera*, also intolerant, had growth rates equal to or greater than those of tolerant species in the shade. The intolerants had the greatest rates of growth by far in full sun.

It appears that intolerants are competitively superior in gaps because of their higher rates of carbon gain and growth. On the other hand, the mechanisms that allow tolerants to be superior to intolerants in shade are not clear. As the exceptions in the above discussion show, no aspect can be singled out as the most important property allowing shade tolerance.

Further research is necessary in several areas. Long-term measurements of carbon budgets will be essential in identifying the significance of relative gain in shade tolerance. Short-term measurements during the summer do not provide an adequate basis for evaluating carbon budgets over an annual environmental cycle. Other aspects of the environment may also be important. The trenching experiments of the 1930s clearly showed the significance of competition between understory and canopy, but this has not been followed up with the methods of modern physiological ecology (Korstian and Coile, 1938). The role of fungal diseases has not been extensively investigated in the field but may be quite important (Grime, 1965; Vaartaja, 1962). The significance of herbivory has also been overlooked. Both of these factors may interact with the carbon balance since plants with inadequate carbon income may be particularly susceptible to disease and herbivory.

12.5 Other periodic stresses

In addition to seasonal changes in the light climate, moisture and temperature significantly affect the structure and physiology of the forest. Many of the responses of deciduous forest species to these changes are similar to those of plants in more extreme climates and more detailed descriptions can be found in chapters on desert and tundra communities. Information specifically important to understanding the deciduous forest will be covered here.

12.5.1 TEMPERATURE

Temperature climates are hardly 'temperate' in the sense meant by geographers (Bailey, 1964). Seasonal variation in several environmental factors, but especially temperature, is rather extreme. The highest and lowest mean monthly temperatures in a year differ by $17-33\,°C$ in the deciduous forest region. Winter invasion by arctic air masses and summer intrusions by tropical air masses push maximum and minimum temperatures to even more extreme values. Extreme temperatures limit physiological activity during the growing season. Sensitivity to low winter temperatures has been associated with the northernly distribution limits of a number of woody species (George *et al.*, 1974).

The most striking event of the year in a deciduous forest is the annual loss of leaves by the dominant trees. Within a period of a few weeks virtually all of the tree leaves lose their chlorophyll exposing accessory pigments to tint the leaves red, yellow, or orange. Stomates close with the loss of chlorophyll (Gee and Federer, 1972) and leaves are shed shortly after the change in color. Leaf abscission is a complex process controlled by many internal and environmental factors. Addicott and Lyon (1973) list over 20 environmental variables that influence the time and rate of leaf fall. The most important controls for deciduous forest species are the decreasing daylengths and temperatures of autumn.

Deciduous forest trees can be leafless for a majority of the year in the northern part of the region. Tree leaves are photosynthetically active for mean periods of 140 days in southern Ontario, about 175 days in Staten Island, NY and 200 days in the North Carolina Piedmont (Britton, 1978). The extent to which this leafless period of 150 to more than 225 days is adaptive is not well understood. The most generally accepted explanation, but with very little supporting data, is that metabolically expensive tissue is being shed to avoid a portion of the year when there are limited possibilities for positive photosynthetic gains and growth. This and other hypotheses are reviewed more fully in Chabot and Hicks (1982). Low temperatures reduce photosynthetic rates and make soil water unavailable, with a strong possibility of negative carbon and water budgets for the winter season. Seasonal loss of leaves restricts respiration and transpiration increasing the probability of survival and leaving the plant with the ability to take advantage of favorable conditions in the spring. Photosynthetic rates of deciduous species are usually higher than for evergreens and the growing season is long enough in the deciduous forest region that the costs of replacing leaves can be recovered and a carbon profit made as well.

That the vascular systems of deciduous and evergreen species frequently differ has not been overlooked. It is likely that the deciduous habit evolved as drought avoidance behavior. Freezing-induced cavitation is a likely possibility in large-vesseled broadleaved species. Separation of the xylem water column would disrupt water supply to the leaves and limit their usefulness during the winter. With a trachiid-based water conducting system, most gymnosperms are less vulnerable to cavitation.

Perennial tissues not shed in the fall must be able to withstand freezing temperatures which occur throughout the region. In northern regions of the deciduous forest, there is a high probability for freezing temperatures for more than half the year. Mechanisms of freezing tolerance remain an active area of investigation. Much attention is focused on changes in cell membrane properties. The proportion of unsaturated fatty acids in membrane lipids may have some role in determining how temperature alters membrane structure and, as a consequence, affects physiological processes dependent upon membrane-bound enzymes (Wilson and McMurdo, 1981). Most, if not all species of temperate deciduous forests are unlikely to be very sensitive to chilling injury (physiological damage at low, but non-freezing temperatures) such as has been studied extensively in crop plants. Indeed, some forest species, the spring ephemerals in particular, are capable of growth and physiological activity at temperatures below freezing (Muller, 1978). *Aplectrum hymenale* has its highest photosynthesis rates in winter-collected plants. Maximum photosynthesis rates drop substantially when plants are held at temperatures of 25 °C for 3 days.

Many plants avoid freezing until very cold temperatures are reached. The accumulation of sugars through starch – sugar conversion in the fall is a part of a cold acclimation process which produces a depression of the freezing point by 2–4 °C. Increase in osmotic concentration of the cytoplasm also occurs through tissue dehydration, a freezing avoidance behavior, particularly common in lichens. A further phenomen, which appears to be unique to woody tree and shrub species of the eastern deciduous forest is deep undercooling. With deep undercooling, tissue water remains unfrozen to temperatures of − 37° to − 40 °C. It is limited to isolated tissues, such as bud primordia, where ice nucleation in one group of cells will not propagate throughout the plant. Burke (1979) has hypothesized that deep undercooling is an intermediate freezing adaptation in species which had their primary evolution in warmer temperate regions.

Eventually tissues do freeze. Debate continues as to what causes the primary lesions. With deep undercooled and unacclimated tissue, it appears to be mechanical disruption due to ice crystal formation. Recently, Steponkus *et al.* (1981) has focused on the mechanical disruption due to cell shrinkage as water moves to ice crystals which develop initially in extracelluar spaces. During freeze-induced dehydration, cell membrane pieces may break off as the cells shrink and the membrane folds in on itself. With rehydration the volume reduction of the cell places increased mechanical stress on the membrane leading to cell lysis. Cold-hardening reduces sensitivity to this process. A major aspect of cold hardening is the quantitative augmentation of membrane lipids. Cold-induced increases in phospholipids have been observed in several woody and herbaceous species of the deciduous forest. The hypothesis is that in cold-hardened tissues the observed increase in the number of membrane fragments are available for substitution into the cell membrane as it contracts and expands.

Temperature extremes during the growing season seem to place little limitation on physiological activity. Jurik (1980) has estimated for *Fragaria virginiana* growing in both shaded and open situations that seasonal carbon income is reduced by 5–7% in response to temperatures above or below the optimum. Leaves within the canopy are substantially light limited and photosynthesis consequently is insensitive to temperature. Most forest species have rather broad thermal response curves under non-limiting light. They are able to maintain high photosynthesis rates under the normal daily and monthly temperature ranges of midsummer. Most species are capable of substantial physiological acclimation which shifts the overall thermal response of photosynthesis and respiration as seasonal temperatures change (Berry and Bjorkman, 1980; Chabot, 1978).

Temperatures do affect growth processes in pro-
nounced ways. This does not necessarily involve a linear increase throughout the normal temperature range. Seed germination, for example, frequently shows a temperature optimum between 20° and 30°C (Angevine and Chabot, 1979; Baskin and Baskin, 1971). Development of floral primordia and flowering in *Fragaria vesca* occurs at temperatures below 25°C while production of runners and a-sexual plantlets occurs at higher temperatures (Chabot, 1978). The requirement for 'stratification' at low temperatures in order to overcome strong innate dormancy is found in the seeds of a number of species of the deciduous forest (Baskin and Baskin, 1971; Ware and Quarterman, 1969).

12.5.2 MOISTURE

Deciduous forest communities that are not in mesic situations may be exposed to drought or flooding. Through the region there are pronounced gradients in both precipitation and evaporation. Annual precipitation ranges from more than 250 cm in the southern Appalachains to 60 cm at the western transition to grassland. Vapor-pressure deficits increase westward so that the most favorable precipitation/evaporation ratios are in the northeastern areas (Transeau, 1905). Early analyses by Transeau (1905) seemed to point to moisture as a major determinant of forest composition within the deciduous forest region. Mositure gradients have continued to receive emphasis in more recent vegetation studies (Bratton, 1976; Curtis, 1959; Goff and Cottam, 1967).

The majority of investigations examining water relations of plants focused on individual components of water transport, loss, and stress tolerance. The attempt was to discover limitations which explained growth or distribution. An example is the study of Wuencher and Kozlowski (1971) which examined five tree species occurring along a moisture gradient. They concluded that water-use efficiency, measured as a ratio of the H_2O/CO_2 diffusion resistances, was important to the habitat requirements of the species. The most efficient species occurred in the driest sites. Stomatal and mesophyll conductances varied complexly in response to light and temperature and individually could not be used to explain the ecological differences between species. Not examined was the role of plant water status in regulating stomatal conductance. Several other studies have found that more xeric species tend to keep their stomates open longer during a drought period (Bunce *et al.*, 1977; Hinckley *et al.*, 1979; Tobiessen and Kana, 1974; Zabadal, 1974). Such behavior would lead to decreased water-use efficiency and higher absolute rates of water loss under just those conditions where efficiency and water conservation might seem to be most important. However, when stomates close, carbon gain through photosynthesis ceases. In a competitive situation, maximizing carbon gain appears to be more important than minimizing water loss (Bunce *et al.*, 1977). Carbon gain maximization under water stress conditions requires a coordinated set of physiological traits. These include stomatal closure at more negative xylem water potentials, ability to limit nonstomatal (physiological) impairment to photosynthesis, ability to rapidly recover physiological capacity when water stress is relieved, and lower thresholds for lethal stress levels (Bunce, 1977; Bunce *et al.*, 1977; Hinckley *et al.*, 1979). Studies with several herbaceous species (Zabadal, 1974) showed that moist site species in comparison with dry site species tended to produce higher levels of abscissic acid and continue with closed stomates for longer periods following water stress events. Such behavior would substantially limit carbon gain in habitats where drought events are frequent.

Detailed studies on white oak (*Quercus alba*) have added to our understanding of characteristics in addition to those described previously which allow species to be successful in more xeric sites within the deciduous forest region. These characteristics include reduced dark respiration under stress conditions (which minimizes loss of carbon while photosynthesis is inhibited), an ability to acclimate photosynthesis to a wide range of leaf temperatures and continue photosynthesis under stress conditions, a deep and well developed root system, roots able to grow under reduced moisture conditions, and long duration of leaf area to maximize seasonal carbon gain (Dougherty and Hinckley, 1981. Dougherty *et al.*, 1979; Hinckley *et al.*, 1978). General growth patterns, especially increased car-

bon allocation to the root system, along with various physiological adjustments seem to play key roles in drought tolerance ability (McLaughlin and McConathy, 1979; McLaughlin *et al.*, 1980).

An increased emphasis on the integration of morphological and physiological traits relating to water use has appeared in several studies. The amount of leaf area which an individual can maintain relates to the total conducting volume in the stem, vascular architecture, and root system development and ultimately upon site water availability (Hinckley *et al.*, 1978). Reduction of water flow through vascular decreases can lead to a loss of leaf area (Talboys, 1968). The water requirement of broad leaves requires a vessel-based vascular system which, in turn, makes the plants more vulnerable to water stress and freezing-induced cavitation.

While drought affects many species and, as a consequence, has been well studied, there are many situations where excess moisture presents problems for plant growth. Such situations range from brief periods of flooding during the spring melt-off to swamps, bogs, lake and stream margins where standing water is more continuous. Flooding causes several deleterious changes in the soil environment including reduced oxygen and increased concentrations of reduced forms of elements such as iron, sulfur, or manganese. Most studies of plant adaptations to such conditions have occurred outside North America and are well reviewed by Armstrong (1975). A key element in avoiding these stresses involves provision for more effective oxygen transport to the root zone. This involves open lacunae and aerenchyma tissues which allows for the free diffusion of O_2 into the root system (Hook and Brown, 1973; Hook *et al.*, 1972). There may be an abundance of lenticels (small openings in the epidermis) or pneumataphores (air roots) which facilitate the entry of O_2. These increased O_2 levels permit aerobic metabolism in roots, but also provide for the oxidation and immobilization of toxic elements at the root surface.

When water levels rise and restrict O_2 entry, tolerance reactions occur in well adapted species. Anaerobic metabolism differs between flooding-tolerant and intolerant species. Normal anaerobic metabolism results in the accumulation of acetaldehyde and ethanol. Under prolonged flooding, these compounds may accumulate to high, toxic levels. The toxicity thresholds tend to be higher in flooding-tolerant species (Hook *et al.*, 1971). Blockage of the normal induction of alcohol dehydrogenase tends to limit ethanol accumulation in some species. Other flooding-tolerant species accumulate less toxic products such as malic, lactic, or shikimic acid.

12.6 Summary

In comparison with other major vegetation types, the deciduous forest occupies a climatic range where substantial periods of the year are favorable for high productivity. Nevertheless species of the forest have evolved to function under a combination of stresses which vary seasonally. Light intensities which are strongly limited by the developing tree canopy places a severe limitation on photosynthesis at a time of the year when moisture and temperature are most favorable. As a consequence, there is a significant vertical structure to the forest and a selection for species which specialize for optimal growth at well-defined times of the year. This is especially seen in the spring ephemerals which in many respects are similar to species of the tundra or to desert annuals in having to restrict growth to a narrow time period.

Adaptations which permit growth under low light intensities have been studied at the physiological level and are reasonably well understood relative to plant response to other environmental factors. Yet some challenges remain. One such challenge is the recent focus on how forest plants use fluctuating light and sunflecks. Another lies in understanding how whole plant maintenance costs can be reduced so as to maximize the effectiveness of photosynthetic income. Answers to such questions have broad application beyond the deciduous forest.

All species of the deciduous forest must accommodate to low temperatures both as a mechanism to extend the growing season and to survive below freezing temperatures of winter. There is substantial physiological acclimation on a seasonal basis in both evergreen and deciduous species. Specialized processes accompany the entry of dormancy. These changes are triggered by shortened photoperiod or by reduced temperatures.

We have come far in being able to detail the specific responses to isolated environmental factors. We have yet to fully appreciate how many traits and behaviors are integrated, how complex environments call for compromise in plant response, and how the cycles of plant activity are coordinated.

References

Abrahamson, W.G. and Gadgil, M. (1973) Growth form and reproductive effort in goldenrods (*Solidago*, Compositae). *American Midland Naturalist*, **107**, 657–61.

Addicott, F.T. and Lyons, J.L. (1973) Physiological ecology of abscission, in *Shedding of Plant Parts* (ed. T.T. Kozlowski), Academic Press, New York.

Ahlgren, C.E. (1967) Phenological observation of nineteen tree species in northeastern Minnesota. *Ecology*, **38**, 622–8.

Anderson, M.C. (1964) Light relations of terrestrial plant communities and their measurement. *Biological Review*, **39**, 425–86.

Anderson, R.C. (1970) The role of daylength and temperature in tuber formation and rhizome growth of *Trientalis borealis* Raf. *Botanical Gazette*, **131**, 122–8.

Anderson, R.C. and Loucks, O.L. (1973) Aspects of the biology of *Trientalis borealis*. *Ecology*, **54**, 798–808.

Angevine, M.W. and Chabot, B.F. (1979) Seed germination syndromes in higher plants, in *Topics in Plant Population Biology* (eds. O.T. Solbrig, S. Jain, G.B. Johnson and P.H. Raven), Columbia Univeristy Press, New York.

Armstrong, W. (1975) Waterlogged soils, in *Environment and Plant Ecology* (ed. J.R. Etherington), John Wiley, London.

Aubuchon, R.R., Thompson, D.R. and Hinckley, T.M. (1978) Environmental influences on photosynthesis within the crown of a white oak. *Oecologia*, **35**, 295–306.

Bacone, J., Bazzaz, F.A. and Boggess, W.R. (1976) Correlated photosynthetic responses and habitat factors of two successional tree species. *Oecologia*, **23**, 63–74.

Bailey, H.P. (1964) Toward a unified concept of the temperate climate. *Geographical Review*, **54**, 516–45.

Baker, F.S. (1950) A revised tolerance table. *Journal of Forestry*, **48**, 179–81.

Baskin, J.M. and Baskin, C.C. (1971) Germination ecology and adaptation to habitat in *Leavenworthia* spp. (Cruciferae). *American Midland Naturalist*, **85**, 22–35.

Bazzaz, F.A., Paape, V. and Boggess, W.R. (1972) Photosynthetic and respiratory rates of *Sassafras albidum*. *Forest Science*, **18**, 218–22.

Berry, J. and Bjorkman, O. (1980) Photosynthetic response and adaptation to temperature in higher plants. *Annual Review of Plant Physiology*, **31**, 491–543.

Boardman, N.K. (1977) Comparative photosynthesis of sun and shade plants. *Annual Review of Plant Physiology*, **28**, 355–77.

Bourdeau, P. (1954) Oak seedling ecology determining segregation of species in Piedmont oak forests. *Ecological Monographs*, **24**, 297–320.

Bratton, S.P. (1976) Resource division in an understory herb community. *American Naturalist*, **110**, 679–93.

Braun, E.L. (1950) *Deciduous Forests of Eastern North America*, Blakiston, Philadelphia.

Britton, N.L. (1978) When the leaves appear. *Bulletin of the Torrey Botanical Club*, **6**, 235–7.

Bunce, J.A. (1977) Nonstomatal inhibition of photosynthesis at low water potentials of intact leaves from a variety of habitats. *Plant Physiology*, **59**, 348–50.

Bunce, J.A., Miller, L.N. and Chabot, B.F. (1977). Competitive exploitation of soil reserves by five eastern North American tree species. *Botanical Gazette*, **138**, 168–73.

Cain, S.A. (1950) Life forms and phytoclimate. *Botanical Reviews*, **16**, 1–32.

Caldwell, M.L.H. (1969) *Erythronium*: comparative phenology of alpine and deciduous forest species in relation to environment. *American Midland Naturalist*, **82**, 543–69.

Castelli, M.R. (1970) Flowering behavior in a uniform garden of wide-ranging, spring-blooming woodland herbs of the eastern United States. *Castanea*, **35**, 260–77.

Chabot, B.F. (1978) Environmental influences on photosynthesis and growth in *Fragaria vesca*. *New Phytologist*, **80**, 87–98.

Chabot, B.F. and Hicks, D.J. (1982) The ecology of leaf life spans. *Annual Review of Ecology and Systematics*, **13**, 229–59.

Christy, H.R. (1952) Vertical temperature gradients in a beech forest in central Ohio. *Ohio Journal of Science*, **52**, 199–209.

Curtis, J.T. (1959) *The Vegetation of Wisconsin*, University of Wisconsin Press, Madison.

Daubenmire, R.F. (1978) *Plant Geography, With Special Reference to North America*, Academic Press, New York.

Davies, W.J. and Kozlowski, T.T. (1974) Stomatal responses of five woody angiosperms to light intensity and humidity. *Canadian Journal of Botany*, **52**, 1525–34.

Davis, M.B. (1969) Pleistocene biogeography of temperate deciduous forests *Geoscience and Man*, **13**, 13–26.

Delcourt, P.A. and Delcourt, H.R. (1979) Late Pleistocene and Holocene distributional history of the deciduous forest in the southeastern United States. *Veroffentilichungen des Geobotanischen Institutes der Eidgenossische Technische Hochschule, Stiftung Rubel*, **68**, 79–107.

De Selm, H.R. (1952) Carbon dioxide gradients in a beech forest in central Ohio. *Ohio Journal of Science*, **52**, 187–98.

Dougherty, P.M. and Hinckley, T.M. (1981) The influence of a severe drought on net photosynthesis in white oak (*Quercus alba* L.) *Canadian Journal of Botany*, **59**, 335–41.

Dougherty, P.M., Teskey, R.O., Phelps, J.E. and Hinckley, T.M. (1979) Net photosynthesis and early growth trends of a dominant white oak (*Quercus alba* L.). *Plant Physiology*, **64**, 930–5.

Federer, C.A. (1971) Solar radiation absorption by leafless deciduous forest. *Agricultural Meteorology*, **9**, 3–20.

Federer, C.A. (1976) Differing diffusive resistance and leaf development may cause differing transpiration among hardwoods in the spring. *Forest Science*, **22**, 359–64.

Federer, C.A. and Tanner, C.B. (1966) Special distribution of light in the forest. *Ecology*, **47**, 555–61.

Garrett, H.E., Cox, G.S. and Roberts, J.E. (1978) Spatial and temporal variations in carbon dioxide concentration in an oak-hickory forest ravine. *Forest Science*, **24**, 180–90.

Gee, G.W. and Federer, C.A. (1972) Stomatal resistance during senescence of hardwood leaves, *Water Resources Research*, **8**, 1456–60.

Geiger, R. (1964) *The Climate Near the Ground*, revised edition, Harvard University Press, Cambridge.

Geis, J.W., Tortorelli, R.L. and Boggess, W.R. (1971) Carbon dioxide assimilation of hardwood seedlings in relation to community dynamics in Illinois. I. Field measurements of photosynthesis and respiration. *Oecologia*, **7**, 276–89.

George, M.F., Burke, M.J., Pellett, H.M. and Johnson, A.G. (1974) Low temperature exotherms and woody plant distribution. *Horticultural Sciences*, **9**, 519–22.

Goff, F.G. and Cottam, G. (1967) Gradient analysis: the use of species and synthetic indices. *Ecology*, **48**, 793–806.

Gonzalez, V.C. (1972) The ecology of *Hexastylis arifolia*, an evergreen herb in the North Carolina deciduous forest.

PhD thesis, Duke University, Durham, North Carolina (*Dissertation Abstracts*, **33**, 5246–B).

Graham, A. (ed.) (1972) *Floristics of Asia and Eastern North America*, Elsevier, New York.

Grime, J.P. (1965) Shade tolerance in flowering plants. *Nature*, **208**, 161–163.

Gross, L.J. (1982) Photosynthetic dynamics in varying light environments: a model and its application to whole leaf carbon gain. *Ecology*, **63**, 84–93.

Gross, L.J. and Chabot, B.F. (1979) Time course of photosynthetic response to changes in incident light energy. *Plant Physiology*, **63**, 1033–1038.

Harvey, G.W. (1980) Seasonal alteration of photosynthetic unit size in three herb layer components of a deciduous forest community. *American Journal of Botany*, **67**, 293–299.

Hicks, D.J. (1980) Intrastand distribution patterns of southern Appalachian cove forest herbaceous species. *American Midland Naturalist*, **104**, 209–223.

Hinckley, T.M., Lassoie, J.P. and Running, S.W. (1978) Temporal and spatial variations in the water status of forest trees. *Forest Science Monograph 20*. American Society of Foresters.

Hinckley, T.M., Dougherty, P.M., Lassoie, J.P., Roberts, J.E. and Tesky, R.O. (1979) A severe drought: impact on tree growth phenology, net photosynthetic rate and water relations. *American Midland Naturalist*, **102**, 307–316.

Hook, D.D. and Brown, C.L. (1973) Root adaptations and relative flood tolerance of five hardwood species. *Forest Science*, **19**, 225–229.

Hook, D.D., Brown, C.L. and Kormanik, P.P. (1971) Inductive flood tolerance in swamp tupelo (*Nyssa sylvatica* var. *biflora* (Walt.) Sarg.). *Journal of Experimental Botany*, **22**, 78–79.

Hook, D.D., Brown, C.L. and Wetmore, R.H. (1972) Aeration in trees. *Botanical Gazette*, **113**, 443–454.

Hunt, C.B. (1974) *Natural Regions of the United States and Canada*. W.H. Freeman, San Francisco.

Hutchison, B.A. and Matt, D.R. (1977) The distribution of solar radiation within a deciduous forest. *Ecological Monographs*, **47**, 185–207.

Jackson, M.T. (1966) Effects of microclimate on spring phenology. *Ecology*, **47**, 407–415.

Jurik, T.W. (1980) Physiology, growth, and life-history characteristics of *Fragaria virginiana* Duchesne and *F. vesca* L. (Rosaceae). PhD Thesis, Cornell University, Ithaca, New York.

Keeley, J.E. (1979) Population differentiation along a flood frequency gradient: physiological adaptations to flooding in *Nyssa sylvatica*. *Ecological Monographs*, **49**,

89–108.

Kieckheffer, B.J. (1962) Correlation between phenology and caloric content in forest herbs. *Transactions of the Illinois Academy of Science*, **55**, 215–223.

Kittredge, J. (1948) *Forest Influences*, McGraw-Hill, New York.

Korstian, C.F. and Coile, T.S. (1938) Plant competition in forest stands. *Duke University Forestry Bulletin*, **3**.

Kozlowski, T.T. and Keller, T. (1966). Food relations of woody plants. *Botanical Reviews*, **32**, 293–382.

Kramer, P.J. and Kozlowski, T.T. (1979) *Physiology of Woody Plants*, Academic Press, New York.

Lee, R. (1978) *Forest Microclimatology*, Columbia University Press, New York.

Lindsey, A.A. and Newman, J.E. (1956) Use of official weather data in spring time–temperature analysis of an Indiana phenological record. *Ecology*, **37**, 812–23.

Loach, K. (1967) Shade tolerance in tree seedlings. I. Leaf photosynthesis and respiration in plants raised in artificial shade. *New Phytologist*, **66**, 607–21.

Loach, K. (1969) Shade tolerance in tree seedlings II, Growth analysis of plants raised under artificial shade. *New Phytologist*, **68**, 274–86.

Logan, K.T. (1970) Adaptations of the photosynthetic apparatus of sun- and shade-grown yellow birch (*Betula alleghaniensis* Britt.). *Canadian Journal of Botany*, **48**, 1681–8.

Logan, K.T. and Krotkov, G. (1968) Adaptations of the photosynthetic mechanism of sugar maple (*Acer saccharum*) seedlings grown in various light intensities. *Physiologia Plantarum*, **22**, 104–16.

Mahall, B.E. and Bormann, F.H. (1978) A quantitative description of the vegetative phenology of herbs in a northern hardwood forest. *Botanical Gazette*, **139**, 467–81.

Marks, P.L. (1975) On the relation between extension growth and successional status of deciduous trees of the northeastern United States. *Bulletin of the Torrey Botanical Club*, **102**, 172–7.

McLaughlin, S.B. and McConathy, R.K. (1979) Temporal and spatial patterns of carbon allocation in the canopy of white oak. *Canadian Journal of Botany*, **57**, 1407–13.

McLaughlin, S.B., McConathy, R.K., Barnes, R.L. and Edwards, N.T. (1980) Seasonal changes in energy allocation by white oak (*Quercus alba* L.). *Canadian Journal of Forest Research*, **10**, 379–88.

McMillen, G.C. and McClennon, J.H. (1979) Leaf angle: an adaptive feature of sun and shade leaves. *Botanical Gazette*, **140**, 437–42.

Minckler, L.S., Woerhide, J.D. and Schlesinger, R.C. (1973) *Light, soil, moisture and tree reproduction in hardwood forest openings*, US Department of Agriculture Forest Service Research Paper NC-89.

Mooney, H.A. (1972) Carbon balance of plants. *Annual Review of Ecology and Systematics*, **3**, 315–46.

Mooney, H.A. and Billings, W.D. (1960) The annual carbohydrate cycle of alpine plants as related to growth. *American Journal of Botany*, **47**, 594–8.

Morgan, M.D. (1971) Life history and energy relationships of *Hydrophyllum appendiculatum*. *Ecological Monograhs*, **41**, 329–49.

Muller, R.N. (1978) The phenology, growth, and ecosystem dynamics of *Erythronium americanum* in the northern hardwoods forest. *Ecological Monographs*, **48**, 1–20.

Muller, R.N. (1979) Biomass accumulation and reproduction in *Erythronium albidum*. *Bulletin of the Torrey Botanical Club*, **106**, 276–83.

Oosting, H.J. (1956) *The Study of Plant Communities*, W.H. Freeman, San Francisco.

Osmond, C.B., Bjorkman, O. and Anderson, D.J. (1980) *Physiological Processes in Plant Ecology*, Springer-Verlag, Berlin.

Pereira, J.S. and Kozlowski, T.T. (1977) Influence of light intensity, temperature and leaf area on stomatal aperture and water potential of woody plants. *Canadian Journal of Forest Research*, **7**, 145–53.

Pitelka, L.F., Stanton, D.S. and Peckenham, M.O. (1980) Effects of light and density on resource allocation in a forest herb, *Aster acuminatus*. *American Journal of Botany*, **67**, 942–8.

Regehr, D.L., Bazzaz, F.A. and Boggess, W.R. (1975) Photosynthetis, transpiration and leaf conductance of *Populus deltoides* in relation to flooding and drought. *Photosynthetica*, **9**, 52–61.

Reifsnyder, W.E., Furnival, C.M. and Horowitz, J.L. (1971) Spatial and temporal distribution of solar radiation beneath forest canopies. *Agricultural Meteorology*, **9**, 21–38.

Risser, P.G. and Cottam, G. (1968) Carbohydrate cycles in the bulbs of some spring ephemerals. *Bulletin of the Torrey Botanical Club*, **95**, 359–69.

Smith, H. (1973) Light quality and germination: ecological implications, in: *Seed Ecology* (ed. W. Heydecker), Pennsylvania State University Press, University Park.

Sparling, J.H. (1964) Ontario's woodland flora. *Ontario Naturalist*, **2**, 18–24.

Sparling, J.H. (1967) Assimilation rates of some woodland herbs in Ontario. *Botanical Gazette*, **128**, 160–8.

Steponkus, P.L., Wolfe, J. and Dowght, M.F. (1981) Stresses induced by contraction and expansion during a freeze-thaw cycle: a membrane perspective, in *Effects of Low Temperatures on Biological Membranes* (eds. G.J.

Morris and A. Clarke), Academic Press, London, pp. 307–22.

Stone, E.L., Jr. (1975) Windthrow influences on spatial heterogeneity in a forest soil. *Mitteilungen Eidgenossische Anstalt fur das Forstliche Versuchwesen*, **51**, 77–87.

Struik, G.J. and Curtis, J.T. (1962) Herb distribution in an *Acer saccharum* forest. *American Midland Naturalist*, **68**, 285–96.

Talboys, P.W. (1968) Water deficits and vascular disease, in *Water Deficits and Plant Growth, Vol. II, Plant Water Consumption and Use* (ed. T.T. Kozlowski), Academic Press, New York, pp. 255–311.

Taylor, R.J. and Pearcy, R.W. (1976) Seasonal patterns of CO_2 exchange characteristics of understory plants from a deciduous forest. *Canadian Journal of Botany*, **54**, 1094–103.

Thompson, J.N. (1980) Treefalls and colonization patterns of temperate forest herbs. *American Midland Naturalist*, **104**, 176–84.

Tobiessen, P. and Kana, T.M. (1974) Drought-stress avoidance in three pioneer tree species. *Ecology*, **55**, 667–70.

Transeau, E.N. (1905) Forest centres of eastern North America. *American Naturalist*, **39**, 875–89.

Vaartaja, O. (1962) The relationship of fungi to survival of shaded tree seedlings. *Ecology*, **43**, 547–9.

Vankat, J.L. (1979) *The Natural Vegetation of North America*, John Wiley, New York.

Vezina, P.E. and Boulter, D.W.K. (1966) The spectral composition of near ultraviolet and visible ratiation beneath forest canopies. *Canadian Journal of Botany*, **44**, 1267–83.

Vezina, P.E. and Grandtner, M.M. (1965) Phenological observations of spring geophytes in Quebec. *Ecology*, **46**, 869–72.

Wallace, L.L. and Dunn, E.L. (1980) Comparative photosynthesis of three gap-phase successional tree species. *Oecologia*, **45**, 331–40.

Walter, H. and Lieth, H. (1967) *Klimmadiagramm-Weltatlas*, VEB Gustav Fischer-Verlag, Jena.

Ware, S.A. and Quarterman, E. (1969) Seed germination in cedar glade *Talinum*. *Ecology*, **50**, 137–40.

Whitney, G.C. (1976) The bifurcation ratio as an indication of adaptive strategy in woody plant species. *Bulletin of the Torrey Botanical Club*, **103**, 67–72.

Whittaker, R.H. (1966) Forest dimensions and production in the Great Smoky Mountains. *Ecology*, **47**, 103–21.

Wilson, J.M. and McMurdo, A.C. (1981) Chilling injury in plants, in *Effects of Low Temperatures on Biological Membranes* (eds. G.J. Morris and A. Clarke), Academic Press, London, pp. 145–72.

Wolfe, J.N., Wareham, R.T. and Scofield, H.T. (1949)

Microclimates and macroclimate of Neotoma, a small valley in central Ohio, Ohio Biological Survey Bulletin 41.

Woods, D.B. and Turner, W.C. (1971) Stomatal response to changing light of four tree species of varying shade tolerance. *New Phytologist*, **70**, 77–84.

Wuencher, J.E. and Kozlowski, T.T. (1971) Relationship of gas-exchange resistance to tree seedling ecology. *Ecology*, **52**, 1016–23.

Zabadal, T.J. (1974) Ecophysiological aspects of drought resistance, PhD Thesis, Cornell University, Ithaca, New York.

Zimmerman, M.H. and Brown, C.L. (1971) *Tree Structure and Function*, Springer-Verlag, New York.

13

Tropical and subtropical forests

ROBERT W. PEARCY AND ROBERT H. ROBICHAUX

13.1 Introduction

A significant amount of information has accumulated over the last century on the ecology of tropical and subtropical forests (Golley, 1983a; Golley and Medina, 1975; Richards, 1952; Walter, 1971). However, as pointed out recently by Mooney *et al.* (1980), very little information is available on the physiological ecology of tropical and subtropical plants. Yet, knowledge of the physiological responses of these species is crucial for understanding the mechanisms by which the functional integrity of these forests is maintained. Obtaining such information has a particular urgency because of the rapid rate of conversion of tropical and subtropical forests for timber harvesting and agricultural purposes (Myers, 1983). This conversion may have significant global implications, since these forests may play a major role in the global cycling of carbon and water.

Tropical and subtropical forests also provide excellent opportunities for studying fundamental aspects of the physiological adaptation of higher plants to contrasting environments. This is particularly true for the forests of tropical and subtropical islands, whose endemic species frequently exhibit very distinctive patterns of evolutionary diversification.

Hence, in the present chapter, we briefly review the distributions of tropical and subtropical forests

and the limited information available on the physiological ecology of their component species. We then discuss the results of some of our recent studies with Hawaiian forest species, which illustrate the excellent opportunities afforded by working with species native to tropical and subtropical islands.

13.2 Distribution of tropical and subtropical forests

Tropical and subtropical forests have been classified in a variety of ways. In geographical terms, these forests occur between $23\frac{1}{2}°$N and $23\frac{1}{2}°$S latitude. According to the climatic classification of Koeppen, tropical and subtropical forests are found in regions delimited by the 18 °C and − 3 °C isotherms, respectively, for the coldest month of the year (Trewartha, 1954). Much narrower definitions have been provided by classification systems based on community structure (Richards, 1952; Walter, 1971; Webb, 1968; Wolfe, 1979). These systems have distinguished forest types on the basis of canopy structure, leaf size, physiographic position, and seasonality of leaf fall in relation to precipitation.

Lowland tropical rainforests usually occur within 10° to 15° of the equator, although they may extend to higher latitudes along eastern continental margins. These forests occur in areas where the mean annual temperature is high, the annual variation in temperature is small, and abundant rainfall is present throughout the year. They are characterized by their high species diversity and their structural complexity (Richards, 1952; Golley, 1983a). Canopy species have leaves that are typically moderate in size and evergreen. Deciduous species may be present, depending on the fertility of the soil. However, these species usually show no particular synchrony of leaf fall.

At higher elevations, montane tropical rainforests of simpler structure and lower stature are found (Grubb, 1977). These forests occupy sites that are cooler and frequently wetter than those of the lowland rainforests (Grubb and Whitmore, 1966). Indeed, most montane rainforests are characterized by a high incidence of low-level cloud cover, which promotes the development of an abundant epiphytic vascular flora (Grubb, 1977). Canopies are frequently two-layered, with the dominant tree species possessing small, evergreen leaves. While the evergreen leaves in both lowland and montane rainforests are usually coriaceous, there is a trend toward increasing leaf thickness with increasing elevation (Grubb, 1977).

The classification of forest communities at higher latitudes is not uniform. These communities are often classified as subtropical (Walter, 1971), although Wolfe (1979) refers to them as paratropical. The climate is usually more seasonal, particularly with respect to precipitation. Semi-evergreen and raingreen seasonal forests occur in areas with a pronounced dry season. These forests are typically two-layered, with deciduous species dominating the upper stratum and evergreen species dominating the lower stratum. Longer dry seasons result in forests composed almost entirely of deciduous species. While these dry forests are more common at higher latitudes, they also may be found at lower latitudes on the leeward sides of large mountain ranges.

Wet forests experiencing only limited seasonal variation in precipitation may occur at latitudes as high as 20° to 23°, particularly on subtropical islands such as the Caribbean and Hawaiian Islands. In these forests, rainfall from frontal storms is abundant in the winter, while convectional or orographic precipitation is significant during the summer. In many respects, such as in canopy structure, these wet subtropical forests are similar to the montane tropical rainforests of lower latitudes.

13.3 Physiological ecology of tropical and subtropical forest species

Mooney *et al.* (1980) have reviewed the literature available on the physiological ecology of tropical plants and have suggested areas in which future research is needed. In addition, data on the physiological characteristics of tropical trees relative to their successional status and regeneration potential have been summarized by Bazzaz and Pickett (1980). These excellent reviews have focused on lowland tropical rainforests and have not considered montane tropical rainforests or subtropical forests, where the information is even more limited. As a result, we have attempted to incorporate data on species from these latter habitats in the following brief review. For the sake of brevity, we have chosen to concentrate on studies of photosynthetic carbon assimilation and water relations. Information on patterns of nutrient acquisition and allocation may be found in the reviews of Grubb (1977) and Golley (1983b).

Much of the research in tropical and subtropical forests has focused on light as a limiting factor. The early investigations of the light environment did not specifically measure photosynthetically active radiation (Ashton, 1958; Evans, 1956; Evans *et al.*, 1960; Grubb and Whitmore, 1967; Whitmore, 1966). Nevertheless, they did provide considerable insight into the large spatial variability and important influences of cloudiness and fog on the light regime in the understory.

Measurements of photosynthetically active photon flux densities (PFD; 400–700 nm wavelengths) in a lowland tropical rainforest in Queensland, Australia, showed that diffuse light was generally less than 5–$10\,\mu\mathrm{mol\,m^{-2}\,s^{-1}}$, and that the average daily total for a 2 week period was only $0.21\,\mathrm{mol\,m^{-2}}$ in the understory as compared to $44\,\mathrm{mol\,m^{-2}}$ above the canopy (Björkman and Ludlow, 1972). Brief sunflecks contributed 62% of the total PFD in the understory on a clear day. Measurements in a Hawaiian forest understory gave higher daily totals (0.55–$1.38\,\mathrm{mol\,m^{-2}}$), of which sunflecks contributed about 40% of the total (Pearcy, 1983). Diffuse light is relatively uniform in the understory, whereas the direct beam component is highly variable (Reifsnyder *et al.*, 1971). This great variability is evident in the Hawaiian forest, where on relatively clear days, locations in the understory less than 3 m apart received sunflecks for as little as 7 min or for as much as 90 min per day.

Plants in these understory environments must be extremely frugal in terms of their carbon allocation, since most of the time the PFD barely exceeds the light compensation point for net photosynthesis. Mechanisms of adaptation to these extremely low light conditions have been elegantly demonstrated in studies of several herbaceous understory species in the lowland rainforest in Queensland (Björkman *et al.*, 1972; Boardman, 1977; Boardman *et al.*, 1972: Goodchild *et al.*, 1972). These plants appear to maximize their investment in the light harvesting membrane complex of the chloroplast while minimizing the concentrations of electron transport carriers and soluble proteins. High concentrations of the latter are required for high rates of carbon fixation, but are also costly in terms of respiratory energy for their synthesis and maintenance, particularly since they do not add to the net carbon gaining capacity of the leaf at extremely low PFD. As a result, these plants are characterized by very low light-saturated photosynthetic rates coupled with the capacity to maintain positive rates of carbon fixation at the very low PFD characteristic of their natural habitat.

Photosynthesis during sunflecks is also of great importance. Measurements by Björkman *et al.* (1972) showed that about 40% of the daily carbon gain of an *Alocasia macrorrhiza* leaf was due to several sunflecks that were incident on the leaf for only a few minutes total. Similar results were obtained by Pearcy and Calkin (1983) for tree seedlings in a Hawaiian forest (see Section 13.4). These results suggest an important role for sunflecks in the carbon balance of understory plants but at present little is understood about the mechanisms that determine the capacity of a leaf to utilize sunflecks for photosynthesis. There have been only a few studies published on the dynamics of the response of photosynthesis to changing light (e.g.

Gross and Chabot, 1979) and, to our knowledge, none with tropical forest species.

Photosynthetic measurements have been made for a few lowland rainforest tree species, but usually in much less detail and with less sophisticated techniques than in the studies of understory herbaceous species discussed above. Nonetheless, measured photosynthetic rates of tropical trees are quite similar to those of temperate trees. Relatively high rates ($15-20\ \mu\mathrm{mol\ m^{-2}\ s^{-1}}$) are found in pioneer species that colonize heavily disturbed sites. Some of these species persist as emergents in the mature forest and apparently continue to maintain high rates of CO_2 uptake (Kira, 1978). Other, more shade-tolerant species that develop under the canopy of the mature forest but eventually reach the emergent layer have lower photosynthetic capacities (Hozumi *et al.*, 1969). Understory trees and shrubs exhibit still lower photosynthetic capacities, as would be expected for plants growing exclusively under low light conditions (Boardman, 1977; Lugo, 1970).

Many, if not all lowland rainforests consist of a relatively high proportion of shade-intolerant species that depend on canopy gaps created by tree falls for successful regeneration (Hartshorn, 1980; Pires and Prance, 1977). Gap size is an important parameter; pioneer species require large gaps, while many other species require small gaps created by the fall of one or two trees for successful regeneration (Denslow, 1980). Surprisingly, species classified on the basis of their mode of regeneration as shade intolerant or moderately intolerant constitute a large proportion of the understory tree canopy at La Selva, Costa Rica (Hartshorn, 1980). These species may survive in a relatively suppressed state for many years, but reproduce sexually only after exposure to higher light levels following gap formation. Little is known about the physiological responses of these species to the higher light levels and higher temperatures characteristic of the gap environment. Clearly, studies of photosynthetic acclimation to changing light and temperature regimes would be very valuable in terms of understanding differences in the utilization of canopy gaps by different species.

Except for the data discussed in Section 13.4,

there are no reported measurements of photosynthesis in plants from montane tropical rainforests or wet subtropical forests. The presence of thick leaves and cuticles in these species, together with their large investment in leaf structural tissues such as the hypodermis (Grubb, 1977), suggest that their photosynthetic capacities are probably relatively low. However, the sclerophyllous nature of these leaves may promote greater longevity. This may be very important for nutrient conservation in these cool, wet environments where litter decay is slow (Grubb, 1977). According to Buckley *et al.* (1980), the sclerophyllous nature of these leaves does not result in any increase in their drought tolerance.

Very few photosynthetic measurements have been reported for dry tropical and subtropical forest species. Lugo *et al.* (1978) measured the diurnal course of photosynthesis in the deciduous species, *Exostema caribaeum*, in a dry forest in Puerto Rico. They found a six-fold increase in the daily leaf carbon gain of this species during the wet season relative to that during the dry season, suggesting a strong influence of drought stress on leaf carbon balance. Even less is known about the photosynthetic responses of evergreen species from dry forest habitats. However, the limited evidence suggests that they have lower photosynthetic rates than the deciduous species (Medina and Kinge, 1983). This difference is consistent with the results of studies on deciduous and evergreen species from dry habitats in temperate regions (Mooney and Dunn, 1970).

The water relations of tropical and subtropical forest species are poorly understood. There are very few reported measurements of water potentials or osmotic potentials for these species (Fetcher, 1979; Medina, 1983; Reich and Borchert, 1982; Robichaux, 1984, Robichaux and Pearcy, 1980a; Walter 1971). In addition, transpirational rates or stomatal conductances to water vapor have been measured in very few species (Fetcher, 1979; Robichaux and Pearcy, 1980a; Shreve, 1914; Weaver *et al.*, 1973; Whitehead *et al.*, 1981). Yet, water stress is of obvious significance in seasonally dry forests, and may even be of considerable significance in lowland and montane rainforests during brief dry periods. For example, Fetcher (1979) measured water potentials as low as -3.0 to -4.0 MPa at the end of a

brief dry period in a lowland tropical rainforest on Barro Colorado island, Panama. In dry tropical and subtropical forests, phenological events such as leaf development, leaf shedding, and flowering may be controlled primarily by water availability (Borchert, 1980; Medina, 1983; Opler *et al.*, 1976, 1980; Reich and Borchert, 1982). Reich and Borchert (1982) have shown that variation in the timing of seasonal development in *Tabebuia neochrysantha*, both within and between populations, is controlled almost exclusively by variation in tree water status.

In at least some lowland rainforest tree species, stomatal conductance to water vapor may be quite sensitive to the vapor-pressure deficit (VPD) between the leaf and the atmosphere (Fetcher, 1979; Whitehead *et al.*, 1981). During midday, relative humidities are often around 60% in the canopy and leaf temperatures are often 30–35 °C. As a result, the VPD may be as high as 2–4 kPa. Stomatal closure in response to these high values of VPD would strongly limit transpiration and would presumably limit photosynthesis during midday. In the understory, relative humidities are higher, midday leaf temperatures are lower (20–25 °C), and the VPD usually does not exceed 0.5 kPa. As a consequence, the degree of stomatal opening in understory species is probably influenced primarily by factors other than VPD.

13.4 Physiological ecology of Hawaiian forest species

The endemic flora of the Hawaiian Islands offers a series of unparalleled opportunities for studying the physiological adaptation of higher plants to contrasting environments. In part, this is a consequence of the tremendous diversity of habitats represented in the archipelago. On each of the five larger islands, for example, annual rainfall varies from less than 600 mm yr^{-1} to more than 8000 mm yr^{-1} (Blumenstock and Price, 1967). In addition, localized precipitation gradients can be extremely steep. Over 4 km near Mt Waialeale on Kauai, rainfall increases 1800 mm yr^{-1} km^{-1} (Mueller-Dombois *et al.*, 1981). Thus it is possible to find communities as different as arid scrub and wet subtropical forest in close proximity.

Figure 13.1 *Euphorbia forbesii* (with notebook on the trunk) as it appears in its native habitat of the tropical forest understory on the island of Oahu, Hawaii.

The Hawaiian Islands are also very isolated geographically. Indeed, geological evidence suggests that the entire archipelago, or at least its present major islands, has always been isolated by more than 3500 km of unbroken ocean from any other island group or continent (Carson and Kaneshiro, 1976). As a result, the modern flora has descended from a relatively small number of ancestral immigrants that crossed the ocean by long distance dispersal. It has been estimated that as few as 280 successful colonization events may have produced the entire Hawaiian flora (Fosberg, 1948). This represents approximately one colonization event per 20 000 years during the six million year history of the current major islands in the archipelago.

This combination of habitat diversity and geographical isolation has resulted in an unusual degree of evolutionary diversification in several major plant groups in Hawaii. In some groups, this diversification has occurred primarily at the infraspecific level. For example, *Metrosideros polymorpha* not only dominates the wet forests of Hawaii but also colonizes habitats as varied as new lava flows and bogs. This species also occurs over an elevational gradient from near sea level to approximately 2600 m in the subalpine zone (Mueller-Dombois *et al.*, 1981). In other groups, this diversification is also apparent at the specific and generic levels. As a result, very closely related species may often be found occupying markedly contrasting habitats. In particularly striking groups, such as in the genera *Euphorbia* and *Dubautia*, this diversification has produced species capable of occupying nearly every type of habitat in Hawaii. Indeed, these latter groups represent unusually well-developed examples of the phenomenon of insular adaptive radiation (Carlquist, 1974). As discussed below, these groups have provided us with excellent opportunities for studying plant performance in a variety of habitats, including mesic and wet subtropical forests.

13.4.1 PHOTOSYNTHETIC DIFFERENTIATION IN THE C_4 *EUPHORBIA* SPECIES

The fourteen endemic Hawaiian species of *Euphorbia* possess C_4 photosynthesis, yet range from arid coastal strand shrubs to rainforest trees (Pearcy and Troughton, 1975). Indeed, the tree species are the only known C_4 forest trees, and clearly occur in a most unusual habitat for plants possessing this pathway (Fig. 13.1). Fig. 13.2 illustrates the range of photosynthetic differentiation in these species. The photosynthetic light response curve of the dry scrub, drought-deciduous species, *E. celastroides*, is similar to that of many C_4 crop species and many other arid habitat C_4 shrubs (Pearcy *et al.*, 1982). In contrast, the photosynthetic light response curve of the mesic forest, evergreen tree species, *E. forbesii*, is similar to that of many C_3 trees. Comparison of the two response curves indicates that while *E. celastroides* has a significantly higher photosynthetic rate at high PFD, *E. forbesii* has a much lower light compensation point and a higher photosynthetic

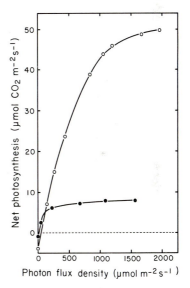

Figure 13.2 Response of net photosynthetic CO_2 uptake to incident photon flux density in *Euphorbia celastroides* (○) and *E. forbesii* (●). *Euphorbia celastroides* was grown in the open in a greenhouse (daily PFD = 15–35 mol m^{-2}), while *E. forbesii* was grown under 90% shade cloth (daily PFD = 0.7–1.8 mol m^{-2}). Mean daily maximum and minimum temperatures in the greenhouse were 24 and 17°C, respectively, for both species. Photosynthetic measurements were made at a leaf temperature of 30°C for *E. celastroides* and 22°C for *E. forbesii*. The ambient CO_2 pressure was 30–32 Pa and the vapor-pressure deficit was 0.4–0.7 kPa.

rate at the low PFD characteristic of its forest habitat. Although the photosynthetic light response curve of *E. forbesii* thus exhibits a striking convergence with that of C_3 tree species, this species still possesses a photosynthetic CO_2 response curve typical of C_4 plants, with a zero CO_2 compensation point and CO_2 saturation of photosynthesis at intercellular $p(CO_2)$ above 10–15 Pa (Fig. 13.3). The differences illustrated in Fig. 13.2 and 13.3 reflect both environmental and genetic effects, since each species was grown and measured under conditions similar to those in its respective native habitat. As will be discussed later, both of these effects are important in determining these differences.

We have also compared the photosynthetic responses of the C_4 *Euphorbia* species with those of the C_3 Hawaiian *Scaevola* species (Robichaux and Pearcy, 1984). Species in both of these genera grow in habitats ranging from dry scrub to wet forest. In the dry scrub habitat, where precipitation is $500 \, \mathrm{mm \, yr^{-1}}$, the small-leaved shrubs, *E. celastroides* and *S. gaudichaudii*, occur together. In the wet forest habitat, in contrast, where precipitation exceeds $5000 \, \mathrm{mm \, yr^{-1}}$, the large-leaved evergreen shrubs, *E. clusiaefolia* and *S. mollis*, grow together. In addition, the evergreen shrubs, *E. hillebrandii* and *S. gaudichaudiana*, occur in somewhat drier forests ($2500 \, \mathrm{mm \, yr^{-1}}$ precipitation), while the prostrate shrubs, *E. degeneri* and *S. coriacea*, grow in coastal strand sites ($500 \, \mathrm{mm \, yr^{-1}}$ precipitation).

As indicated in Fig. 13.4, the range of variation in photosynthetic rates among the C_4 *Euphorbia* species is significantly greater than among the C_3 *Scaevola* species. There is a three-fold range of variation in the photosynthetic rates of the *Euphorbia* species, with the mesic and wet forest species possessing much lower capacities for CO_2 fixation than the coastal strand and dry scrub species. In contrast, the range of variation among the *Scaevola* species is significantly less. This difference between the two genera is present regardless of whether photosynthesis is expressed per unit leaf area, per unit leaf dry weight, or per unit leaf nitrogen. As discussed by Robichaux and Pearcy (1984), this marked difference between the two genera is not related necessarily to constraints associated with the difference in their photosynthetic pathways. It may well be related instead to other constraints associated with the difference in their phylogenetic legacies.

The difference in the photosynthetic capacities of the wet forest species, *E. clusiaefolia* and *S. mollis*, is relatively small (Fig. 13.4). While this may represent evolutionary convergence, it is unclear whether selection in this habitat would have been for a lower photosynthetic capacity per se or for other characteristics, such as a greater leaf longevity. The latter characteristic, which has been shown to be inversely correlated with photosynthetic capacity (Chabot and Hicks, 1982), appears to be related to enhanced nutrient conservation in these extremely wet and cool habitats (Grubb, 1977).

Mesophyll conductances to CO_2 vary in parallel with photosynthetic rates expressed per unit leaf area in both the *Euphorbia* and *Scaevola* species (Fig. 13.4). However, because C_4 plants exhibit a greater intrinsic efficiency of CO_2 utilization at low intercellular $p(CO_2)$ than C_3 plants (Björkman, 1976), the C_4 *Euphorbia* species within each habitat has a significantly higher mesophyll conductance than the corresponding C_3 *Scaevola* species.

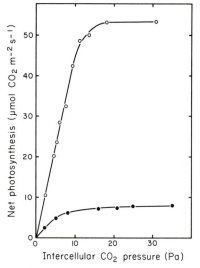

Figure 13.3 Response of net photosynthetic CO_2 uptake to intercellular CO_2 partial pressure in *Euphorbia celastroides* (○) and *E. forbesii* (●). Growth and measurement conditions were the same as in Fig. 13.2, except that CO_2 partial pressure was varied, while incident PFD was held constant at 1500–$1900 \, \mu\mathrm{mol \, m^{-2} \, s^{-1}}$.

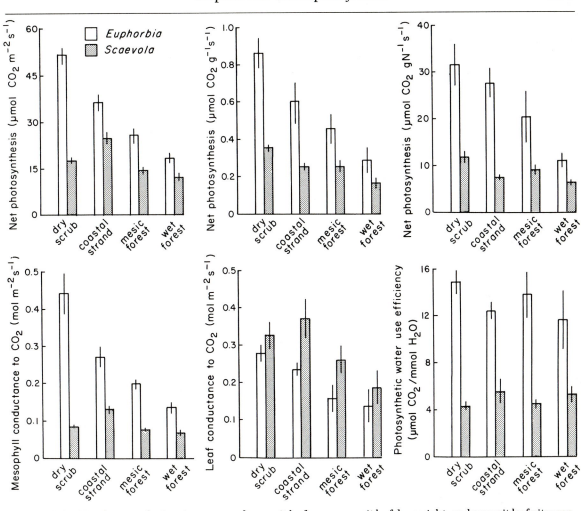

Figure 13.4 Net photosynthetic rates expressed per unit leaf area, per unit leaf dry weight, and per unit leaf nitrogen, mesophyll and leaf conductances to CO_2, and photosynthetic water-use efficiencies of C_3 *Scaevola* and C_4 *Euphorbia* species from dry scrub (*S. gaudichaudii* and *E. celastroides*), coastal strand (*S. coriacea* and *E. degeneri*), mesic forest (*S. gaudichaudiana* and *E. hillebrandii*), and wet forest habitats (*S. mollis* and *E. clusiaefolia*). Vertical lines indicate ± 1 s.d. for 3–9 replicates. The plants were grown under uniform environmental conditions in a greenhouse and measured at high incident PFD (1500–1800 μmol m^{-2} s^{-1}), optimal temperatures for net photosynthesis (27 °C for the *Scaevola* species and 30 °C for the *Euphorbia* species), normal atmospheric CO_2 partial pressures (30–32 Pa), and low water vapor-pressure deficits (0.3–0.9 kPa). All transpirational rates were calculated for a vapor-pressure deficit of 0.8 kPa (Robichaux and Pearcy, 1984).

Leaf conductances to CO_2 also vary in parallel with photosynthetic rates expressed per unit leaf area in both genera (Fig. 13.4). Indeed, this covariation in leaf conductance and photosynthesis is such that, within each genus, intercellular $p(CO_2)$ is relatively constant under typical ambient $p(CO_2)$ of 30–32 Pa. Intercellular $p(CO_2)$ is 12–15 Pa in the

Euphorbia species and 23–25 Pa in the *Scaevola* species. This covariation does not mean that the differences in photosynthetic capacity within each genus are determined by the differences in leaf conductance. Rather, the differences in photosynthetic capacity appear to be determined primarily by differences in the capacity of the mesophyll to fix

CO_2, with leaf conductance being regulated in such a manner as to maintain a relatively constant intercellular $p(CO_2)$ (Pearcy et al., 1982).

Viewed in this manner, leaf conductance appears to be a rather conservative characteristic. As a consequence, however, photosynthetic water-use efficiencies (or photosynthesis/transpiration ratios) are essentially constant within each genus. As indicated in Fig. 13.4, the C_4 *Euphorbia* species possess water-use efficiencies that are $2-3\frac{1}{2}$ times as high as those of the C_3 *Scaevola* species. This large difference is present in all of the habitats, despite the fact that annual precipitation along this habitat gradient varies by an order of magnitude. This difference is similar to that commonly measured in C_3 and C_4 plants (Robichaux and Pearcy, 1984). We have found an exception to this pattern in only one of the *Euphorbia* species, *E. remyi*, which occurs in a bog habitat (Pearcy et al., 1982). This latter species has a high leaf conductance relative to its photosynthetic capacity, a high intercellular $p(CO_2)$ under typical ambient conditions, and a low photosynthetic water-use efficiency in comparison with the other *Euphorbia* species.

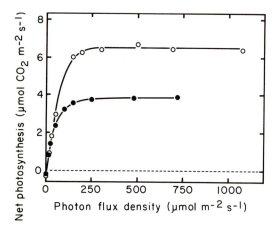

Figure 13.5 Response of net photosynthetic CO_2 uptake to incident photon flux density in *Euphorbia forbesii* ($\bigcirc$) and *Claoxylon sandwicense* ($\bullet$). The plants were growing in the shaded understory of a mesic evergreen forest on Oahu. Leaf temperatures were $23.0 \pm 0.5\,°C$, ambient CO_2 partial pressures were 32.0 ± 1.5. Pa, and vapor-pressure deficits were 0.6 ± 0.1 kPa. Light was provided by an artificial source (Pearcy and Calkin, 1983).

Because of its cool, shaded understory habitat and arborescent growth form, *E. forbesii* is certainly one of the most unusual C_4 plants yet described. Pearcy and Calkin (1983) conducted comparative field measurements of photosynthesis in this species and a co-occurring C_3 tree species, *Claoxylon sandwicense*, in the shaded understory of a mesic evergreen forest on Oahu. Fig. 13.5 shows the photosynthetic responses to PFD of *E. forbesii* and *C. sandwicense* measured *in situ* with a portable field gas exchange apparatus and an artificial light (Pearcy and Calkin, 1983). *Euphorbia forbesii* has a higher light-saturated photosynthetic rate than *C. sandwicense*, but the responses of the two species at low light are quite similar, with both species possessing very low light compensation points and low dark respiration rates. Moreover, quantum yields in these two species are equal at $22-23\,°C$, the prevailing temperature in this understory habitat (Robichaux and Pearcy, 1980b). Although the differences are very small, respiration rates and light compensation points are consistently higher in *E. forbesii*. This small difference is the only potentially disadvantageous photosynthetic characteristic possessed by *E. forbesii* relative to *C. sandwicense*. It is not known if this is somehow related to the differences in photosynthetic pathway or to other factors. Leaf conductances are lower in *E. forbesii* and, as a result, it has a higher water use efficiency than *C. sandwicense*, which is consistent with the difference in their photosynthetic pathways. It is difficult to imagine an advantage or disadvantage to this higher water use efficiency, however, since it is unlikely that water stress poses a significant limitation in this understory habitat (Robichaux and Pearcy, 1980a).

Measurements of photosynthetic gas exchange under natural light regimes illustrate the very dynamic nature of photosynthetic responses to sunflecks and the importance of sunflecks in daily leaf carbon gain (Fig. 13.6). The estimated percentage of the daily CO_2 uptake attributable to photosynthesis during sunflecks on these measurement days was about 40% and 60% for *E. forbesii* and *C. sandwicense*, respectively. The differences have more to do with the total duration of sunflecks received on the two days and at the two locations than with any species differences in capacity to respond to sun-

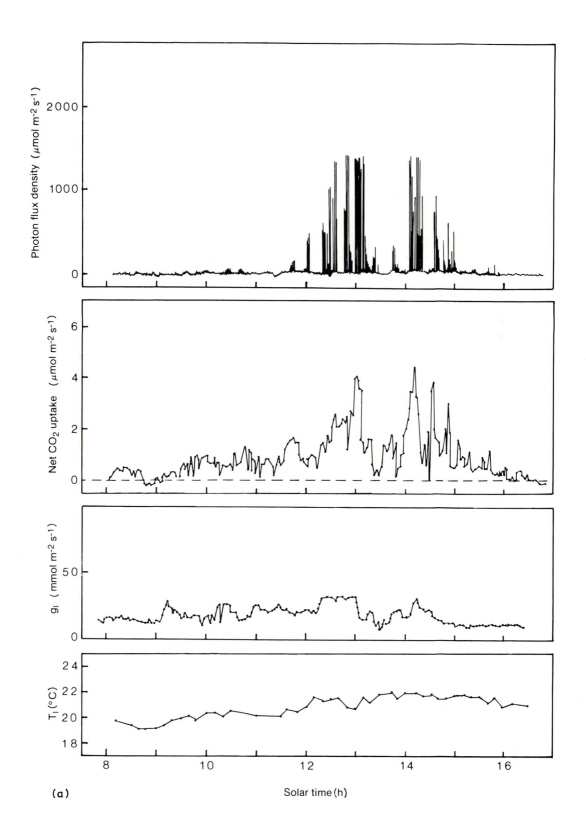

(a)

Solar time (h)

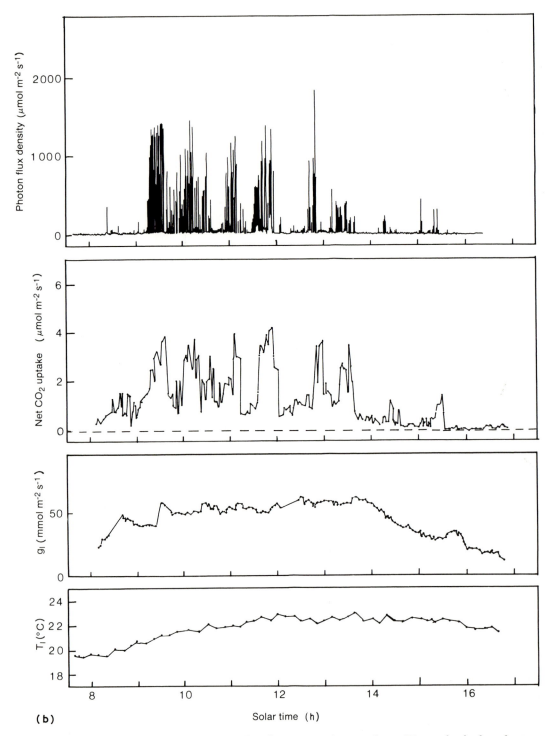

Figure 13.6 Diurnal courses of incident photon flux density, net photosynthetic CO_2 uptake, leaf conductance, and leaf temperature for *Euphorbia forbesii* on 16 July 1981 (a) and for *Claoxylon sandwicense* on 22 July 1981 (b) in Pahole Gulch, Oahu, Hawaii (Pearcy and Calkin, 1983).

flecks. The importance of sunflecks in leaf carbon gain in the understory also appears to carry over into growth rates. Pearcy (1983) found much higher relative growth rates for seedlings of both *E. forbesii* and *C sandwicense* at locations estimated on the basis of canopy photographs to receive more sunflecks than at locations estimated to receive fewer sunflecks.

The photosynthetic response to sunflecks is very fast. Indeed, the rapid changes in apparent photosynthetic rate shown in Fig. 13.6 are probably limited more by instrument lags than by the rate of the response of the leaf. Stomata appear to respond much more to light changes in *E. forbesii* than in *C. sandwicense*, but may limit photosynthetic response to sunflecks more in the former species, especially after periods when sunflecks are infrequent. Since sunflecks tend to be grouped, the limitation by stomata may only be apparent for the first few sunflecks in any series. The leaf conductances in *C. sandwicense* are high enough that intercellular $p(CO_2)$ never declines to levels that indicate a significant stomatal limitation of the photosynthetic response to sunflecks. In addition to the stomatal

responses, other factors such as differences in the ability to utilize short versus long sunflecks (cf. Gross and Chabot, 1979), which are technically difficult to resolve in field measurements such as these, may be of considerable importance. These all suggest a dependence of the response not only on the PFD of the sunfleck but also on the duration and timing relative to other sunflecks. Much more work will be required to resolve these very complex interactions.

Another important aspect of the physiological adaptation of forest trees is the response to changes in the light environment, such as those that occur when a tree grows from the understory into the canopy or when a gap forms. The higher PFD provides more energy for photosynthesis, but can also lead to photoinhibition of the photosynthetic apparatus, particularly if the PFD greatly exceeds that which can be utilized by the plant (Björkman, 1981). While there is little information on the responses of tropical and subtropical forest trees, it is likely that, as with temperate zone trees, most exhibit some photosynthetic light acclimation, whereby the rate of light-saturated photosynthesis increases when the plant is exposed to increased

Table 13.1 Photosynthetic characteristics and leaf chlorophyll contents of *Euphorbia forbesii* and *Claoxylon sandwicense* grown at low and high light intensities*

Characteristics	E. forbesii		C. sandwicense	
	Low light	High light	Low light	High light
Light-saturated photosynthetic rate (μmol CO_2 m^{-2} s^{-1})†	7.3 ± 1.2	12.5 ± 1.5	4.7 ± 0.4	5.9 ± 0.9
Dark respiration rate (μmol CO_2 m^{-2} s^{-1})	0.4 ± 0.1	0.9 ± 0.1	0.3 ± 0.1	0.7 ± 0.1
Light compensation point (μmol m^{-2} s^{-1})	5 ± 1	19 ± 1	4 ± 1	14 ± 3
Quantum yield (mol CO_2 mol photons^{-1})	0.061 ± 0.006	0.058 ± 0.005	0.052 ± 0.003	0.043 ± 0.008
Chlorophyll content (μg cm^{-2})	51 ± 6	63 ± 5	55 ± 7	35 ± 4

*The plants were grown in a greenhouse, either on an open bench (high light treatment) or under shade cloth transmitting approximately 10% of the incident light (low light treatment).
†All photosynthetic characteristics, except quantum yield, were measured at 22 °C. Quantum yield was measured at 30 °C. Values are means $\pm$ 1 s.d.

light during growth. Greenhouse experiments with *E. forbesii* and *C. sandwicense* illustrate the differences in response for these two species (Table 13.1). Both have the capacity to acclimate to the light environment, but the change in light-saturated photosynthesis is much greater in *E. forbesii*. Moreover, *C. sandwicense* appears to acclimate to the higher light and, at the same time, to be injured by it. The reduced chlorophyll contents and quantum yields in the high light grown plants are both symptoms of photoinhibition (Björkman, 1981). In contrast, there is no evidence for a significant photoinhibition in *E. forbesii*. As has been shown for herbaceous species (Björkman, 1981) the responses of these two trees are rather complex, and involve changes in photosynthetic capacity of the mesophyll, electron transport capacity, dark respiration rate, leaf conductance, leaf nitrogen content, chloroplast ultrastructure, and leaf anatomy.

The differences in the capacity for light acclimation appear to fit well with observed differences in the ecology of these two species. While both species are most abundant in the understory canopy layers, *E. forbesii* may be much more dependent on receiving high light for successful reproduction. The only individuals that we have observed reproducing successfully are small saplings in gaps or large individuals having at least one or two branches exposed to high light in the overstory canopy. This dependence on high light for reproduction has been described in several tropical tree species (Whitmore, 1978). In contrast, understory individuals of *C. sandwicense* commonly reproduce. *Claoxylon sandwicense* also occasionally occurs in the canopy, but it is not known if the photoinhibition observed in the greenhouse occurs under field conditions. However, it is fairly clear that while the differences in the behavior of these two species with respect to the light environment in the understory may seem rather small, there may be significant contrasts for large individuals that reach the overstory canopy.

While these comparisons involve a C_3 and a C_4 tree, which may accentuate some of the differences, particularly at high light, the contrasts in photosynthetic rates and behavior are not much different from the contrasts observed in comparisons of C_3 tree species from tropical or from temperate forests.

As further work proceeds on tropical and subtropical forest species, it is likely that a diversity of responses will be found that may prove useful for understanding differences in the ecology and life history of the component species of these forests.

13.4.2 VARIATION IN THE TISSUE WATER RELATIONS OF THE *DUBAUTIA* SPECIES

The twenty-one endemic Hawaiian species of *Dubautia* occupy habitats as varied as exposed lava, dry scrub, dry forest, mesic forest, wet forest, and bog (Carlquist, 1980; Carr and Kyhos, 1981). The range of moisture availability among these habitats is quite dramatic. For example, *D. linearis* occupies sites on the island of Hawaii that receive less than 400 mm rainfall yr^{-1}, while *D. waialealae* grows in an area on Kauai that receives more than 12 500 mm yr^{-1}. This latter area is one of the wettest terrestrial sites on the surface of the earth (Carlquist, 1980; Grosvenor, 1966).

A recent study by Robichaux (1984) has analyzed the extent and significance of variation in the tissue elastic and osmotic properties of two of these species and their hybrid recombinant. These two species grow sympatrically at approximately 2000 m elevation on the slopes of Mauna Loa, Hawaii (Carr and Kyhos, 1981). At this site of sympatry, a 1935 lava flow occurs over an older, prehistoric lava flow. The older substrate is exposed in several places at this site in the form of small 'islands' surrounded by the younger substrate. Both species of *Dubautia* are common at this locality. However, *D. scabra* is restricted exclusively to the younger substrate, while *D. ciliolata* is restricted to the 'islands' of older substrate.

The geographical distributions of these two species extend in different directions from this site of sympatry. The distribution of *D. scabra* extends into the montane rainforest belt at lower elevations, where annual rainfall is 4000–5000 mm yr^{-1}. In contrast, the distribution of *D. ciliolata* extends into the dry scrub zone at higher elevations, where annual rainfall, is 400–500 mm yr^{-1}. At the site of sympatry, annual rainfall is approximately 800–1000 mm yr^{-1}.

The diurnal water potentials experienced by these

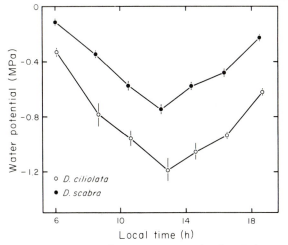

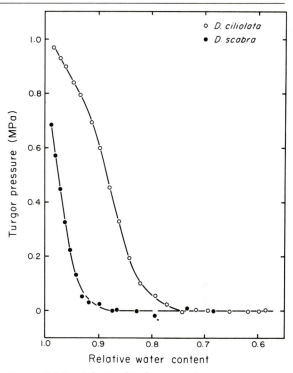

Figure 13.7 Diurnal water potentials for *Dubautia ciliolata* (○) and *D. scabra* (●) on 4 August 1982 near Puu Huluhulu, Saddle Road, Hawaii. Vertical bars indicate ± 1 s.d. for five replicates of *D. ciliolata* between 08.30 and 16.30 h and five replicates of *D. scabra* at 12.30 h. At other times, vertical bars indicate the range of variation for duplicates (Robichaux, 1984).

Figure 13.8 Relationship between tissue turgor pressure and tissue water content for *Dubautia ciliolata* (○) and *D. scabra* (●). See Robichaux (1984) for a detailed discussion of the methods by which these relationships were obtained.

two species at their site of sympatry differ markedly (Fig. 13.7). Water potentials in *D. ciliolata* are significantly lower than in *D. scabra* during the entire day. During the summer of 1982, the difference in the midday water potentials of these two species was typically 0.45–0.65 MPa. This difference is present in winter and summer, and may reach magnitudes of 0.8 MPa during particularly dry periods (Robichaux, 1984).

Significant differences are also present in the tissue elastic and osmotic properties of these two species. These differences are apparent in Fig. 13.8, where tissue turgor pressure is plotted as a function of tissue water content. The differences in the slopes of these two curves reflect differences in tissue elastic properties, while the differences in the *y*-intercepts reflect differences in tissue osmotic potentials at full turgor (Robichaux, 1984). *Dubautia ciliolata* exhibits both a lower osmotic potential at full turgor and a lower initial bulk elastic modulus than *D. scabra*. The effect of these differences on the relative abilities of the two species to maintain high and positive turgor pressures as tissue water contents decrease is quite striking. For example, at a relative water

content of 0.93, the difference in turgor pressure in these two species is 0.7 MPa. In addition, the relative water content at which turgor reaches zero is significantly lower in *D. ciliolata* than in *D. scabra*.

From the data in Fig. 13.8, it is possible to derive a relationship between tissue water potential and tissue turgor pressure for these two species (Robichaux, 1984). Knowing the respective diurnal patterns of water potential, one may then estimate with reasonable accuracy the diurnal patterns of turgor pressure experienced by *D. scabra* and *D. ciliolata* in the field. As indicated in Fig. 13.9, these two species appear to experience very similar values of tissue turgor pressure throughout the day. In addition, one may calculate diurnal values of turgor pressure for a hypothetical individual that exhibits the tissue elastic and osmotic properties of *D. scabra*, yet that experiences the diurnal water potentials of *D. ciliolata*. The possible functional significance of

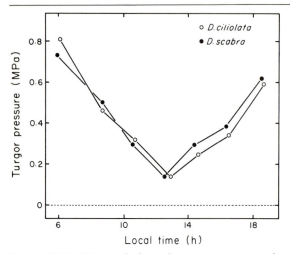

Figure 13.9 Estimated diurnal turgor pressures for *Dubautia ciliolata* (○) and *D. scabra* (●) on 4 August 1982. See text for details (Robichaux, 1984).

the modified tissue elastic and osmotic properties of *D. ciliolata* may then be estimated by comparing the actual diurnal turgor pressures for *D. ciliolata* to those of the hypothetical individual (Fig. 13.10). It is readily apparent from such a comparison that there is a significant potential benefit for *D. ciliolata* in terms of diurnal turgor maintenance. Not only are turgor pressures several tenths of an MPa higher

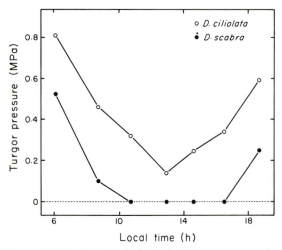

Figure 13.10 Estimated diurnal turgor pressures for *Dubautia ciliolata* (○) and a hypothetical individual of *D. scabra* (●) on 4 August 1982. See text for details (Robichaux, 1984).

in *D. ciliolata* than in the hypothetical individual throughout the day, but turgor pressures in the latter individual reach 0 MPa at 10.30 h in the morning and remain at 0 MPa until 16.30 h in the afternoon. This marked difference in diurnal turgor maintenance would presumably translate into a significant difference in growth, given that a number of important physiological processes, such as leaf expansion, depend intimately on turgor (Hsiao *et al.*, 1976).

We are currently examining the extent of variation in these properties in a large number of the other *Dubautia* species. This should provide us with an indication of the degree to which these properties vary in closely related species growing in a wide variety of habitats.

13.5 Summary

It is clear that we have a great deal to learn about the physiological responses of tropical and subtropical plants. Indeed, as emphasized in the excellent review by Mooney *et al.* (1980), our knowledge of the physiological ecology of tropical plants is so limited that virtually any research effort will result in valuable new information. These authors discuss a number of areas, in particular, in which research is urgently needed. These include: mechanisms of carbon, water, and nutrient acquisition and allocation, mechanisms of predator protection, capacities of different growth forms for regeneration in response to different levels of disturbance, and relationships between resource levels and seasonality of flowering, leaf ageing, and dormancy (Mooney *et al.*, 1980). The urgency of such research stems from the need to develop sound management policies with regard to the exploitation of tropical and subtropical forests. Unfortunately, these forests are currently being converted at an alarming rate despite the very limited amount of quantitative information for predicting the long-term impact of different kinds of disturbance (Myers, 1983).

The results of our recent studies with Hawaiian forest species also suggest that research on tropical and subtropical plants may provide valuable information on physiological mechanisms of adaptation to contrasting environments in higher plants. In this regard, we emphasize the importance of studying very closely related species when the goal is a fundamental one of understanding mechanisms of adaptation. Only by comparing very closely related species can one minimize the potential influence of phylogenetic constraint and maximize the possibility of eventually understanding the genetic basis of those adaptive mechanisms.

Acknowledgement

This research was supported by NSF grants DEB 7921826 to RWP and DEB 82-06411 to RHR.

References

Ashton, P.G. (1958) Light intensity measurements in rainforest near Santarem, Brazil. *Journal of Ecology*, **46**, 65–70.

Bazzaz, F.A. and Pickett, S.T.A. (1980) The physiological ecology of tropical succession: a comparative review. *Annual Review of Ecology and Systematics*, **11**, 287–310.

Björkman, O. (1976) Adaptive and genetic aspects of C_4 photosynthesis, in *CO_2 Metabolism and Plant Productivity* (eds R.H. Burris and C.C. Black). University Park Press, Baltimore.

Björkman, O. (1981) Responses to different quantum flux densities, in *Encyclopedia of Plant Physiology, New Series*, Vol 12A, *Physiological Plant Ecology I* (eds O.L. Lange, P.S. Nobel, C.B. Osmond and H. Ziegler) Springer-Verlag, Berlin.

Björkman, O. and Ludlow, M.M. (1972) Characterization of the light climate on the floor of a Queensland rainforest. *Carnegie Institution of Washington Yearbook*, **71**, 85–94.

Björkman, O., Ludlow, M.M. and Morrow, P.A. (1972) Photosynthetic performance of two rainforest species in their native habitat and analysis of their gas exchange. *Carnegie Institution of Washington Yearbook*, **71**, 94–107.

Blumenstock, D.I. and Price, S. (1967) *Climates of the States, Hawaii*, US. Dept. of Commerce, 27 pp.

Boardman, N.K. (1977) Comparative photosynthesis of sun and shade plants. *Annual Review of Plant Physiology*, **28**, 355–57.

Boardman, N.K., Anderson, J.M., Thorne, S.W. and Björkman, O. (1972) Photochemical reactions of the chloroplasts and components of the electron transport chain in two rainforest species. *Carnegie Institution of Washington Yearbook*, **71**, 107–14.

Borchert, R. (1980) Phenology and ecophysiology of a tropical tree, *Erythrina peoppigiana* O.F. Cook, *Ecology*, **61**, 1065–71.

Buckley, R.C., Corlett, R.T. and Grubb, P.J. (1980) Are the xeromorphic trees of tropical upper montane rainforests drought resistant? *Biotropica*, **12**, 124–36.

Carlquist, S. (1974) *Island Biology*, Columbia University Press, New York.

Carlquist, S. (1980) *Hawaii, A Natural History*, Pacific Tropical Botanical Garden, Lawai, Hawaii.

Carr, G.D. and Kyhos, D.W. (1981) Adaptive radiation in the Hawaiian silversword alliance (Compositae: Madiinae). I. Cytogenetics of spontaneous hybrids. *Evolution*, **35**, 543–56.

Carson, H.L., and Kaneshiro, K.Y. (1976) *Drosophila* of Hawaii: systematics and ecological genetics. *Annual Review of Ecology and Systematics*, **7**, 311–45.

Chabot, B.F. and Hicks, D.F. (1982) The ecology of leaf life spans. *Annual Review of Ecology and Systematics*, **13**, 229–59.

Denslow, J.S. (1980) Gap partitioning among tropical rainforest trees, in *Tropical Succession* (ed. J.C. Ewel), Supplement to *Biotropica*, **12**, 47–80.

Evans, G.C. (1956) An area survey method of investigating the distribution of light intensity in woodlands, with particular reference to sunflecks. *Journal of Ecology*, **44**, 391–428.

Evans, G.C., Whitmore, T.C. and Wong, T.K. (1960) The distribution of light reaching the ground vegetation in a tropical rainforest. *Journal of Ecology*, **48**, 193–204.

Fetcher, N. (1979) Water relations of five tropical tree species on Barro-Colorado Island, Panama. *Oecologia*, **40**, 229–33.

Fosberg, F.R. (1948) Derivation of the flora of the Hawaiian Islands, in *Insects of Hawaii* (ed. E.C. Zimmerman), University of Hawaii Press, Honolulu.

Golley, F.B. (ed.) (1983a) *Tropical Rain Forest Ecosystems: Structure and Function*, Elsevier, Amsterdam.

Golley, F.B. (1983b) Nutrient cycling and nutrient conservation, in *Tropical Rain Forest Ecosystems: Structure and Function* (ed. F.B. Golley), Elsevier, Amsterdam.

Golley, F.B. and Medina, E. (ed.) (1975) *Tropical Ecological Systems: Trends in Terrestrial and Aquatic Research*, Ecological Studies 11, Springer-Verlag, Berlin.

Goodchild, D.J., Björkman, O. and Pyliotis, N.A. (1972) Chloroplast ultrastructure, leaf anatomy, and content of chlorophyll and soluble protein in rainforest species. *Carnegie Institution of Washington Yearbook*, **71**, 102–7.

Gross, L.J. and Chabot, B.F. (1979) Time course of photosynthetic response to changes in incident light energy. *Plant Physiology*, **63**, 1033–8.

Grosvenor, M.B. (1966) *National Geographic Atlas of the World*, National Geographic Society, Washington, D.C.

Grubb, P.J. (1977) Control of forest growth on wet tropical mountains: with special reference to mineral nutrition. *Annual Review of Ecology and Systematics*, **8**, 83–107.

Grubb, P.J. and Whitmore, T.C. (1966) A comparison of montane and lowland rainforest in Ecuador. II. The climate and its effects on distribution and physiognomy of the forests. *Journal of Ecology*, **54**, 303–33.

Grubb, P.J. and Whitmore, T.C. (1967) A comparison of montane and lowland rainforest in Ecuador. III. The light reaching the ground vegetation. *Journal of Ecology*, **55**, 33–57.

Hartshorn, G.S. (1980) Neotropical forest dynamics, in *Tropical Succession* (ed. J.C. Ewel), supplement to *Biotropica*, **12**, 23–30.

Hozumi, K., Yoda, K. and Kira, T. (1969) Production ecology of tropical rainforests in southwestern Cambodia. II. Photosynthetic production in an evergreen seasonal forest. *Nature and Life in Southeast Asia, Tokyo*, **6**, 57–81.

Hsiao, T.C., Acevedo, E., Fereres, E. and Henderson, D.W. (1976) Water stress, growth, and osmotic adjustment. *Philosophical Transactions Royal Society London B*, **273**, 479–500.

Kira, T. (1978) Community architecture and organic matter dynamics in tropical lowland rainforest of Southeast Asia with special reference to Pasoh Forest, West Malaysia, in *Tropical Trees as Living Systems* (ed. P.B. Tomlinson and M.H. Zimmermann), Cambridge University Press, Cambridge.

Lugo, A. (1970) Photosynthetic studies on four species of rainforest seedlings, in *A Tropical Rainforest* (ed. H.T. Odum), US Atomic Energy Commission, Washington.

Lugo, A., Gonzalez-Libby, J.A. and Dugger, K. (1978) Structure, productivity, and transpiration of a subtropical dry forest in Puerto Rico. *Biotropica*, **10**, 278–91.

Medina, E. (1983) Adaptations of tropical trees to moisture stress, in *Tropical Rain Forest Ecosystems: Structure and Function* (ed. F.B. Golley), Elsevier, Amsterdam.

Medina, E. and Kinge, H. (1983) Tropical forests and tropical woodlands, in *Encyclopedia of Plant Physiology, New Series*, Vol. 12D, *Physiological Plant Ecology IV* (eds O.L. Lange, P.S. Nobel, C.B. Osmond, and H. Ziegler), Springer-Verlag, Berlin.

Mooney, H.A. and Dunn, E.L. (1970) Photosynthetic systems of mediterranean climate shrubs and trees of California and Chile. *American Naturalist*, **194**, 447–53.

Mooney, H.A., Björkman, O., Hall, A.E., Medina, E. and Tomlinson, P.B. (1980) The study of the physiological ecology of tropical plants – current status and needs. *Bioscience*, **30**, 22–6.

Mueller-Dombois, D., Bridges, K.W. and Carson, H.L. (1981) *Island Ecosystems: Biological Organization in Selected Hawaiian Communities*, Hutchinson Ross, Stroudsburg.

Myers, N. (1983) Conversion rates in tropical moist forest, in *Tropical Rain Forest Ecosystems: Structure and Function* (ed. F.B. Golley), Elsevier, Amsterdam.

Opler, P.A., Frankie, G.W. and Baker, H.G. (1976) Rainfall as a factor in the release, timing, and synchronization of anthesis by tropical trees and shrubs. *Journal of Biogeography*, **3**, 321–36.

Opler, P.A., Frankie, G.W. and Baker, H.G. (1980) Comparative phenological studies of treelet and shrub species of tropical wet and dry forests in the lowlands of Costa Rica. *Journal of Ecology*, **68**, 167–88.

Pearcy, R.W. (1983) The light environment and growth of C_3 and C_4 tree species in the understory of a Hawaiian forest. *Oecologia*, **58**, 19–25.

Pearcy, R.W. and Calkin, H.C. (1983) Carbon dioxide exchange of C_3 and C_4 tree species in the understory of a Hawaiian forest. *Oecologia*, **58**, 26–32.

Pearcy, R.W. and Troughton, J. (1975) C_4 photosynthesis in tree form *Euphorbia* species from Hawaiian rainforest sites. *Plant Physiology*, **55**, 1054–6.

Pearcy, R.W., Osteryoung, K. and Randall, D. (1982) Carbon dioxide exchange characteristics of C_4 Hawaiian *Euphorbia* species native to diverse habitats. *Oecologia*, **55**, 333–41.

Pires, J.M. and Prance, G.T. (1977) The Amazon forest: A natural heritage to be preserved, in *Extinction is Forever: The Status of Threatened and Endangered Plants of the Americas* (eds G.T. Prance and T.S. Elias), New York Botanical Garden, New York.

Reich, P.B. and Borchert, R. (1982) Phenology and ecophysiology of the tropical tree, *Tabebuia neochrysantha* (Bignoniaceae). *Ecology*, **63**, 294–9.

Reifsnyder, W.E., Furnival, G.M. and Horowitz, J.L. (1971) Spatial and temporal distribution of solar radiation beneath forest canopies. *Agricultural Meteorology*, **9**, 21–37.

Richards, P.W. (1952) *The Tropical Rainforest*, Cambridge University Press, Cambridge.

Robichaux, R.H. (1984) Variation in the tissue water relations of two sympatric Hawaiian *Dubautia* species and their natural hybrid. *Oecologia*, in press.

Robichaux, R.H. and Pearcy, R.W. (1980a) Environmental characteristics, field water relations and photosynthetic responses of C_4 Hawaiian *Euphorbia* species from contrasting habitats. *Oecologia*, **47**, 99–105.

Robichaux, R.H. and Pearcy, R.W. (1980b) Photosynthetic responses of C_3 and C_4 species from cool shaded habitats in Hawaii. *Oecologia*, **47**, 106–9.

Robichaux, R.H. and Pearcy, R.W. (1984) Evolution of C_3 and C_4 plants along an environmental moisture gradient: patterns of photosynthetic differentiation in Hawaiian *Scaevola* and *Euphorbia* species. *American Journal of Botany*, **71**, 121–9.

Shreve, F. (1914) *A Montane Rain-forest: A Contribution to the Physiological Plant Geography of Jamaica*, Carnegie Institution of Washington, Washington, D.C.

Trewartha, G. (1954) *An Introduction to Climate*, McGraw-Hill, New York.

Walter, H. (1971) *Ecology of Tropical and Subtropical Vegetation*, Oliver and Boyd, Edinburgh.

Weaver, P.L., Byer, M.D. and Bruck, D.L. (1973) Transpiration rates in the Luquillo Mountains of Puerto Rico. *Biotropica*, **5**(2), 123–33.

Webb, L.J. (1968) Environmental relationships of the structural types of Australian rainforest vegetation. *Ecology*, **49**, 296–311.

Whitehead, D., Okali, D.U.U. and Fasechun, F.E. (1981) Stomatal response to environmental variables in two tropical forest species during the dry season in Nigeria. *Journal Applied Ecology*, **18**(2), 571–87.

Whitmore, T.C. (1966) A study of light conditions in forests in Ecuador with some suggestions for further studies in tropical forests, in *Light as an Ecological Factor*, (eds R. Bainbridge, G.C. Evans and O. Rackham), Blackwell, Oxford.

Whitmore, T.C. (1978) Gaps in the forest canopy, in *Tropical Trees as Living Systems* (eds P.B. Tomlinson and M.H. Zimmermann) Cambridge University Press, Cambridge.

Wolfe, J.A. (1979) *Temperature parameters of humid to mesic forests of eastern Asia and relation to forests of other regions of the northern hemisphere and Australasia*, US Geological Survey Professional Paper 1106, 37 pp.

14

Marine beach and dune plant communities

MICHAEL G. BARBOUR, THEODORE M. DE JONG
AND BRUCE M. PAVLIK

14.1 Introduction

The biology of beach and dune plants along North American shores has received scientific attention for nearly 100 years. Cowles founded his dynamic view of plant communities on examples of succession in coastal dunes (albeit Great Lakes dunes; Cowles, 1891). Within 20 years, Harshberger (1900, 1902, 1903, 1908, 1914) had contributed many field observations from Atlantic and Gulf coasts and Cooper (1936) had begun a long-term study of Pacific coast strand plant distribution.

Early ecologists may have been attracted by the dynamism of the strand's ecosystem, the simplicity of the flora, the presumed (though poorly measured) dramatic gradients of salt and wind, the easy access to study sites, and the aesthetic quality of the setting. The latter two attributes have also attracted many tourists and home owners, leading to a reduction in the ecological quality of many beach fronts along all coasts (US Army Corps of Engineers, 1973).

Although the span of ecological interest has been long, its thrust has been diffuse and scattered. Thus, a single research theme has seldom been applied throughout a region and a given area has seldom been examined from many perspectives. The result is a collection of excellent but isolated studies, such as those on mineral cycling by van der Valk (1974b), morphological adaptations to a few stresses (Purer, 1936b), measurement of and response to salt spray (Boyce, 1954; Oosting, 1945), zonation of species (Sauer, 1967), germination (Seneca, 1969), water relations (Martin and Clements, 1939), and revegetation (Dolan *et al.*, 1973; McLaughlin and Brown, 1942).

Especially rare are ecophysiological studies, which use modern approaches to gas exchange and water balance, and demographic studies. Around the turn of the century, world-scale works by Schimper (1903) and Warming (1909) triggered a strong interest in the physiological adaptations of plants to their habitats. However, techniques for making physiological and micro-environmental measurements were limited, so strand research at first relied primarily on deductions from anatomical traits (e.g. Coker, 1905; Harshberger, 1908, 1909; Kearney 1900, 1904; Purer, 1936b; Snow, 1902).

One major theme of strand research has been a documentation of species zonation back from tide line and an increasingly refined correlation of that zonation with gradients in the environment. Oosting and Billings (1942), for example described gradients of soil moisture, soil temperature, soil salinity, soil pH, air temperature, relative humidity, evaporation, and salt spray on North Carolina outer banks, and concluded that salt spray was the only factor which could account for observed patterns of species zonation. Other investigators have emphasized the importance of sand movement (Martin, 1959; van der Valk, 1974a). But many other factors have been relatively overlooked. Thus, even though early investigators described anatomical and morphological features which could be interpreted as adaptations to a xeric habitat with high light intensity, high surface temperature, low soil nitrogen, high soil chlorinity, and high winds with desiccating and abrasive consequences, we still today have few data concerning physiological responses to these factors. Sauer (1976) has summarized the situation well: 'Past workers have suffered from a persistent delusion that broad generalization could be drawn from measuring casually chosen parameters in arbitrarily delimited regions. Coastal vegetation research has barely begun.' In short, the North American strand literature does not lend itself very well to the kind of synthetic review which could take us to a new level of understanding, especially if it is to be a review with a physiological ecology bias. Nevertheless, we do here attempt such a review. First, we deal with demonstrated or presumed plant adaptations to such environmental factors as light, temperature, water, salt spray, sand movement, soil salinity, and nutrient availability, then we analyse floristic lists in order to develop life-form, leaf-form, and metabolic (C_3, C_4, CAM) spectra as a base from which to extrapolate further autecological conclusions.

We have limited our review to plants which characterize beaches and unstabilized dunes which face open, maritime coasts. We excluded shingle beaches, coastal bluffs, perched dunes of recent sand atop consolidated substrate, stabilized (grey) dunes with a high per cent of plant cover, fresh water strands, and protected marine strands facing bays, sounds, harbors, and other quiet bodies of water not exposed to the open ocean. The locations of some major areas which have been studied are shown in Fig. 14.1, with some descriptive information in Table 14.1.

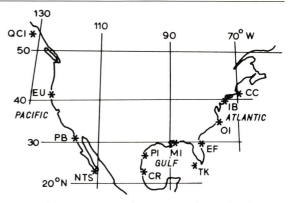

Figure 14.1   Location of 12 regions chosen for climatic and floristic summaries. Climatic data appear in Table 14.1; growth-form and leaf-form spectra appear in Table 14.6.

Beach is that strip of sandy substrate which extends from mean tide line to the top of the foredune (littoral dune) or, in the absence of a foredune, to the furthest inland reach of storm waves. The beach is characterized by a maritime climate, high exposure to salt spray and sand blast, and a shifting, sandy substrate with low water-holding capacity and low organic matter content. Open *dunes* extend from the lee side of the foredune to inland or dune climax vegetation on stabilized substrate. 'White' or 'yellow' dunes have little accumulation of organic matter and relatively open cover, in contrast to 'grey' dunes. The foredune which separates beach from dune is created by the sand-stilling abilities of herbaceous perennials in temperate and arid zones of the North American coast, and by woody species in tropical zones. When the dominants are temperate zone rhizomatous grasses, the foredune tends to be a continuous ridge several metres high (Fig. 14.2); otherwise, the foredunes may be broken up into separate hillocks. *Strand* will be a general term, meaning beach and/or dune.

The vegetated portion of the beach is considerably narrower than the beach itself, and the leading edge of vegetation is subject to seasonal disturbance such that its position moves toward shore in times of calm weather when beaches prograde, and back toward the foredune in times of stormy weather when beaches retrograde. In exceptional times, the fore-

Table 14.1 Selected environmental data for the 12 regions shown in Fig. 14.1. R = annual solar radiation (kcal cm^{-2} yr^{-1}); T = mean annual temperature; A = range of temperature, warmest – coldest month. From many sources.

Region (code)	Latitudinal and Longitudinal limits or center	R	Temperatures (°C)		Precipitation	Winds	Waves and Tides
			T	A			
Queen Charlotte Islands (QCI)	53° 15'N 131° 49'W	< 100	7.9	12.6	125 cm, incl. some snow; peak in fall – winter	Mainly SE; 92 km h^{-1}Dec. average, 34 km h^{-1}July average; gusts to 192 km h^{-1}	High energy waves, gale-force winds; 4–6 m tidal range
Eureka (EU)	40° 45'N 124° 10'W	100–150	11.2	5.7	102 cm, peak Nov. – Mar., summer fog	BF 4 peak in July, BF 8 peak in Jan.	High energy, but gale-force winds rare; 2–4 m tidal range
Punta Banda (PB)	31° 53'N 116° 10'W	150–200	17.0	7.0	20–28 cm, peak Nov. – Mar., summer fog	Mainly NW, BF 4$^+$ not common, but tropical cyclones do occur	Mostly low energy waves; 2–4 m tidal range
North Todos Santos (NTS)	23° 25'N 110° 15'W	150–200	22.0	9.0	17–26 cm, peak July – Oct.	Mainly NW, BF 4$^+$ not common, but tropical cyclones do occur	Mostly low energy waves; 2–4 m tidal range
Mexican Coast, mainly Cabo Rojo (CR)	21–26° N 97–98° W	150–200	23.0	12.0	80–130 cm, increasing south; May – Oct. peak	39 km h^{-1}ann. average; nortes with gale force winds may occur 12 + times a year	Low energy; diurnal range 0–3 m
Padre Island (PI)	26° 15'N 97° 10'W	150–200	22.0	15.0	61 cm; April – Oct. peak	Mainly SE, generally mild	Low energy; diurnal range < 1 m
Mississippi Islands (MI) (Ship, Cat, Horn I.)	30° N 89° W	150–200	20.0	15.0	147 cm; April – Oct. peak	Mainly SE, generally mind	Low energy; diurnal range < 1 m
Tortugas Keys (TK)	24° 30'N 81–82° W	150–200	25.0	3.0	99 cm; April – Oct. peak	Mainly E, ann. average 15 km h^{-1}; mild	Low energy; mixed semi-diurnal; < 1 m range
Eastern Florida (EF), Ponte Vedra Beach and Jupiter I.	27° 02'– 30° 13'N 80° 05'– 81° 05'W–	150–200	21.2	12.9	133 cm; summer peak	Still within range of tropical storms, but generally mild	Low energy; 5–10% of waves < 2.6 m; 2–4 m spring tide range

298

Table 14.1 *Contd.*

Region (code)	Latitudinal and Longitudinal limits or center	R	Temperatures (°C)		Precipitation	Winds	Waves and Tides
			T	A			
Ocracoke Island (OI)	35°N 76°10'W	150–200	16.8	17.5	138 cm; modest summer peak	Moderate to strong winds; 15–30% Jan. winds BF 4+	Moderate energy, 10–15% of waves > 2.6 m; 2–4 m spring tide range
Island Beach State Park (IB)	39°55'N 74°05'W	100–150	12.2	22.4	109 cm; evenly distributed	Generally mild; 0–15% BF 4+ winds all year	Moderate energy, 10–15% of waves > 2.6 m; 2–4 m spring tide range
Cape Cod (CC)	41°55'N 70°15'W	100–150	9.9	22.8	110 cm; evenly distributed	Generally mild; 0–15% BF 4+ winds all year	High energy storm waves; 10–15% of waves > 2.6 m; 4–6 m spring range

dune itself may be eroded, and all beach vegetation is lost.

14.2 Light and temperature

14.2.1 ENVIRONMENTAL PATTERNS

As one moves south along the coastlines of North America between 60° and 20°N latitude, there is an increase in annual solar radiation, annual net radiation, mean annual temperature, and length of the growing season (Table 14.1). Frequency of frost decreases along the same transect, although freezing temperatures may extend much further south on the east coast (see Davies, 1977). Gulf Coast stations exhibit little differentiation in regards to climate and are typically mild temperate or subtropical. Sites along the west coast tend to be warmer than comparable latitudes on the east coast, with desert-like extremes in southern California and Mexico.

Microenvironmental data are scant and scattered in the literature. Data in Fig. 14.3 illustrate light and temperature patterns within stands of the grasses *Elymus mollis* and *Ammophila arenaria* at Point Reyes, California. On days with clear skies and light breezes, air and sand surface temperatures are quite warm. The grass canopies decrease turbulent exchange between the sand and the air (see also Fig. 14.4), and attenuate incident radiation. However, significant differences in canopy structure of typical *Elymus* and *Ammophila* stands result in very different microclimates. *Ammophila* allocates dry matter to the production of a dense, taller, more vertically-oriented sward of leaves (Pavlik, 1983b). Such an architecture permits a higher leaf area index and a more efficient utilization of direct and diffuse radiation (de Wit, 1965; Monsi *et al.*, 1973). On overcast and foggy days, however, the extinction of direct beam radiation is greater and only the uppermost leaf surfaces are significantly illuminated. Planophilic, low canopies near or within *Ammophila* stands would be at a competitive disadvantage with regard to light (Pavlik, 1982). Additional data on the microclimate of coastal dune grasses at Cape Hatteras, North Carolina, have been presented by van der Valk (1977a), and for several beach plants near *Veracruz*, Mexico by Moreno-Casasola (1982).

Leaf temperatures of several species at Bodega

Figure 14.2 Continuous dune line developing under a cover of rhizomatous grasses. Bodega Bay, California

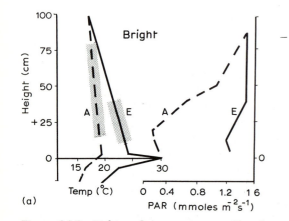

Figure 14.3 Light and temperature profiles through canopies of *Ammophila arenaria* (A, dashed lines) and *Elymus mollis* (E, solid lines) on (a) bright and (b) cloudy days, at Kehoe Beach, Point Reyes National Seashore, California. Measurements for a bright day were taken at 3 p.m., 27 April 1978; those for a cloudy day were taken at 3:30 p.m., 9 May 1978. Redrawn from Pavlik (1982). Shaded region shows maximum canopy development.

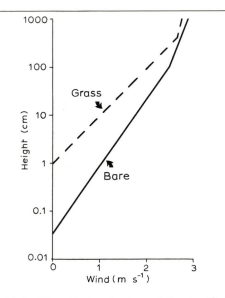

Figure 14.4 The effect of *Ammophila breviligulata* canopy on wind profiles over a Great Lakes sand dune. Canopy height was about 1 m, but the wind profile was affected for several meters above that. Redrawn from Olson (1958b) and Goldsmith (1978).

Bay, a cool, wet, northern site, were generally 1–2 °C higher than air temperatures 10 cm above the sand surface (De Jong and Barbour, 1979). The same taxa at a warmer and drier southern locality showed no consistent trend with respect to air temperatures, although leaves of *Atriplex leucophylla* were usually 1 °C cooler than ambient. The succulent, entire leaves of *Abronia maritima* were warmer than the pubescent, white-scurfy, or dissected leaves of *Ambrosia chamissonis*, *Atriplex leucophylla*, and *Cakile maritima* (De Jong, 1979), but it is surprising that leaf characteristics which promote shortwave reflectance and turbulent heat transfer are not more common in beach and dune floras from southern latitudes (see Table 14.6).

14.2.2 GERMINATION

The responses of imbibed seeds to light can be very different among strand plants. For example, light enhances germination of *Physalis viscosa*, *Erigeron canadensis*, *Iva imbricata*, and *Solidago sempervirens* and can substitute for a cold pretreatment or allow better germination under otherwise unfavorable temperature regimes (van der Valk, 1974a). Most species are light-neutral and do not exhibit either enhancement or inhibition (Kumler, 1969; van der Valk, 1974a). However, studies on *Cakile maritima* (Barbour, 1970a) and *Abronia maritima* (Johnson, 1978) showed that these species have light-inhibited germination mechanisms.

Seeds of *Iva imbricata*, *Croton punctatus*, *Uniola paniculata*, and *Ammophila breviligulata* require stratification at 4 °C for maximum germination (Seneca, 1969; van der Valk, 1974a; Wagner, 1964). Seedlings of these plants appear in the late winter and spring, whereas species lacking a stratification requirement (e.g. *Physalis viscosa*, *Erigeron canadensis*) may germinate in the fall and early winter. However, populations of *Uniola* from Florida have little or no cold treatment requirement when compared to populations in Virginia and North Carolina (Seneca, 1972). A similar cline was found for *Ammophila breviligulata* between Lake Michigan and North Carolina, populations from the latter not requiring stratification (Seneca and Cooper, 1971).

Laboratory germination studies of strand plants show that optimal constant temperatures (in the dark) tend to lie between 20 and 30 °C. Thermoperiods of alternating temperatures are required by some beach plants (e.g. *Iva ibricata*, *Uniola paniculata*) for any germination. Even if not strictly required, alternating temperatures will enhance germination over constant temperatures in *Ammophila breviligulata*, *Solidago sempervirens*, *Abronia maritima*, and *Calystegia soldanella* (Johnson, 1973; Kumler, 1969; Seneca and Cooper, 1971; van der Valk, 1974a). Fluctuating temperature regimes enhance germination regardless of the absolute temperature values (within 15–35 °C), the range of the alternation, or the relative time spent at either high or low temperature. This may be an artifact of the small number of time–temperature permutations investigated.

It is difficult to assign ecological significance to such a wide array of responses. On the one hand, light inhibition may ensure proximity of the seed to more constant levels of moisture or prevent germination of imbibed seeds exposed by sand movement (Koller, 1956; Mayer and Poljakoff-Mayber, 1975).

Table 14.2 Anatomical, morphological and physiological traits of selected west coast strand plants. Data from Purer (1936b, 1942), and zonation indices calculated from Barbour *et al.* (1976).

Species	Thickness (mm) Leaf and trichomes	Cuticle (ad/ab)	Trichomes (% leaf)	Stomatal Distrib. and index (adax/adax mm^{-2})	Architecture of mesophyll	Psn mode	Root system	Zonation index[‡]
Atriplex leucophylla	0.75	0.0025/ 0.0016	66	100/110*	Kranz-type, compact	C_4	Tap[‡]	0.39
Cakile maritima	3.0	0.0025	0	64/96	Undifferentiated, with air spaces	C_3	Tap	0.58
Ambrosia chamissonis	0.90	0.005/ 0.003	33	153/150[†]	Isolateral, multi-seriate palisade, compact	C_3	Tap	0.60
Abronia maritima	3.4	0.005/ 0.005/	17	21/20*	Isolateral, uniseriate palisade, compact	C_3	Tap	0.66
Ammophila arenaria	0.56	0.008/ 0.002	12 (abaxial)	0/12[†]	Undifferentiated, compact	C_3	Fibrous	0.68
Calystegia soldanella	0.50	0.004/ 0.003	0	155/150*	Isolateral multi-seriate palisade, compact	C_3	Fibrous	0.71
Camissonia cheiranthifolia	0.64	0.003^{+}/ 0.003	44	170–250/ 285–354	Isolateral, multi-seriate palisade, with air spaces	C_3	Tap	0.72
Mesembryanthemum chilense	10.10	0.004/ 0.004	0	90/86[†]	Isolateral, multi-seriate, compact	C_3	Fibrous	0.76
Abronia umbellata	1.10	0.003	12	50/60*	Isolateral, multi-seriate palisade, compact	C_3	Tap	Dune
Lupinus chamissonis	1.10	0.003/ 0.002	68	150/150	Isolateral, multi-seriate palisade	C_3	Tap	Dune
Happlopappus ericoides	0.56	0.005	0	60	Isolateral, uni – or multiseriate palisade, compact	C_3	Tap	Dune
Eriogonum parvifolium	0.65	0.002/ 0.001	54 (abaxial)	0/282*	Isolateral, multi-seriate palisade, compact	C_3	Tap	Dune
Rhus integrifolia	0.75	0.007/ 0.004	0	0/255	Dorsiventral to iso-lateral, multiseriate palisade, compact	C_3	Tap	Dune

*Slightly sunken.   [†]Well-sunken stomata.   [‡]Where 0 = leading edge and dune is > 1.0.
[‡]Diffuse according to De Jong (1979).

Burial by only 2–3 mm of coarse and can attenuate light to 1% of surface intensity (Wooley and Stoller, 1978) and may remove the inhibition. Alternatively, light enhancement may prevent copious germination at sand depths too great to allow seedling emergence. Ruderal species commonly exhibit light-induced germination (Sauer and Struik, 1964; Wesson and Wareing, 1969). More importantly, it is usually difficult to correlate optimal germination in the laboratory with soil temperatures and seedling emergence observed in the field because relevant data on the micro-environment around the seed usually are not gathered.

14.2.3 CARBON METABOLISM

The leaves of beach and dune plants seem anatomically adapted to high light intensities (Table 14.2, and see also Harshberger, 1908, 1909; Kearney, 1901). Like typical sun leaves, they are thick, exhibit a well-developed cuticle, can have high stomatal indices, and possess an isolateral, multiseriate, and compact palisade mesophyll (unless C_4 or grass-like). Leaf surfaces may be quite pubescent, glabrous, or glaucous, and are usually amphistomatous, with the stomata slightly sunken or in furrows. There is no apparent correlation of the leaf anatomical traits shown in Table 14.2 with type of root system or zonation along the strand.

Beach plants are fully capable of utilizing high light intensities for carbon fixation. When grown in the laboratory, rates of CO_2 uptake of Californian C_3 grasses and a C_4 dicot ranged between 25 and $40 \, \mu mol \, m^{-2} \, s^{-1}$ at photon fluxes above $1.6 \, mmol \, m^{-2} \, s^{-1}$ (De Jong 1978a; Pavlik, 1983a). The C_3 species approached light saturation near $1.0–1.2 \, mmol \, m^{-2} \, s^{-1}$ whereas C_4 *Atriplex leucophylla* did not saturate until $2.0 \, mmol \, m^{-2} \, s^{-1}$ (essentially full sun). Studies using $^{14}CO_2$ have shown that field plants exhibit similar light responses although the absolute rates of photosynthesis are much lower (Pavlik, 1982). De Jong and Barbour (1979) could find no evidence of ecotypic differentiation of the light responses of *Atriplex* from northern and southern localities which differed in total annual solar radiation (Table 14.1).

In the laboratory, De Jong (1978a) found that *Atriplex* had a photosynthetic temperature optimum 6–8 °C above those of *Cakile maritima* (22 °C), *Ambrosia chamissonis* (20 °C), and *Abronia maritima* (20 °C), all C_3 species. Leaf conductances of *Atriplex* were much lower and unaffected by 15–40 °C temperatures whereas the other species closed their stomata as temperatures increased. Even though its photosynthetic optimum was usually above naturally occurring leaf temperatures in the stand habitat, De Jong demonstrated that *Atriplex* maintained a higher water use efficiency at the temperature optima of C_3 taxa.

Dubois (1977) could find no significant differences in the photosynthetic temperature optima of C_4 and C_3 strand plants in the field. Over a single summer season in Georgia, the C_3 species *Heterotheca subaxillaris, Croton punctatus*, and *Iva imbricata* exhibited broad temperature responses and moderate rates of CO_2 fixation ($13–19 \, \mu mol \, CO_2 \, m^{-2} \, s^{-1}$) even at 40 °C. The C_4 species *Atriplex arenaria, Uniola paniculata*, and *Panicum amarum* had slightly higher temperature optima and fixation rates ($13–19 \, \mu mol \, CO_2 \, m^{-2} \, s^{-1}$), but the differences were not statistically significant. Differences in water and nitrogen use efficiency between C_4 and C_3 taxa, however, were dramatically accentuated at higher temperatures (see Section 14.6).

There are very few temperature acclimation data for plants of the strand. Populations of *Atriplex leucophylla* from northern and southern sites differ in their ability to acclimate (De Jong and Barbour, 1979). A coastal (although not strand) race of *Atriplex lentiformis* lacked the physiological flexibility of a desert race, perhaps reflecting the modestly fluctuating maritime climate (Pearcy, 1976). However, recent work on *Heliotropium curassavicum* from coastal sites has shown that some maritime plants have maintained a significant capacity for temperature acclimation (Mooney, 1980).

The relationship between latitude, temperatures during the growing season, and certain aspects of carbon catabolism have been investigated in *Lathyrus japonicus* of the east coast strand. Simon (1979) found that plants from warmer southern sites possessed a more thermostable form of NAD malate dehydrogenase, with higher activation energies and higher levels of activity. These correlations suggest a modification of the kinetic properties of this enzyme in response to gradients in the growing season temperature. He also found that some acclimation ability was associated with all populations of *Lathyrus* in regard to activation energy. Along the same latitudinal gradient, Lechowicz *et al.* (1980) demonstrated genetically based increases in total dark respiration with latitude, including both maintenance and growth components. They reasoned that increases in maintenance respiration reflected additional energy expended for nutrient uptake at lower temperatures, and that increased growth respiration rates (hence growth rates) would be adaptive in areas with a short growing season.

14.2.4 LATITUDINAL GRADATION OF C_3 AND C_4 BIOMASS

Using identical methods Barbour and Robichaux (1976) and Hall and Eleuterius (unpublished) developed regressions of plant biomass on absolute cover for West and Gulf Coast strand plants. With their equations we made estimates of total plant biomass and the percentage contributed by C_4 taxa using three sites along a latitudinal gradient. Cover data for each site were the combination of three transect surveys found in the literature. Although the number of sites is small, Table 14.3 suggests that total plant biomass of the strand is most closely related to precipitation and its pattern (winter peak in California, summer peak in Texas). The contribution of C_4 plants, however, increases abruptly towards the south where annual solar radiation input and overall temperatures are higher (Table 14.1). This pattern emulates the data in our floristic analysis (Table 14.6) and supports the conclusions of Teeri and Stowe (1976) and Doliner and Jolliffe (1979). It seems somewhat anomalous, though, that C_4 taxa should extend so far north along the east coast where temperatures are cooler and frost more common than in the west. The differences between the two coasts may reflect their floristic histories, the winter-dormant nature of many eastern C_4 plants,

and the strikingly different patterns of oceanic currents which modify adjacent onshore environments.

14.3 Water relations

14.3.1 ENVIRONMENTAL PATTERNS

Many early botanists were impressed by the 'xerophytic' nature of strand vegetation (Coker, 1905; Harshberger, 1900, 1902, 1903; Kearney, 1900, 1901; Snow, 1902). There appear to be three sources of non-saline water for strand vegetation: atmospheric precipitation (including fog drip), a water table within the root zone of tap-rooted species, and 'internal dew' formation. The latter two will be treated in the next section.

Precipitation falling directly on the beach and dune has an important role in leaching salts from the soil, beyond its role as major water supply for seedling establishment and survival of shallow-rooted species. However, soil moisture storage is rapidly exhausted during periods of drought because of the low water-holding capacity of coarse-textured sands (Oosting, 1954; Salisbury, 1952). Patterns of precipitation, then, may be as or more important than total annual precipitation.

Atlantic coast sites receive 109–133 cm yr^{-1} (Table 14.1). All four experience significant precipitation every month, but the southern two receive

Table 14.3 Estimates of total plant biomass and the contribution of C_4 taxa along a latitudinal gradient. Sites are from transect data of Barbour *et al.* (1976) and Judd *et al.* (1977)

Site	Mean annual Temp. (°C)	Mean annual Ppt (cm)	Dominants	Total biomass (g m^{-2}) Range	Total biomass (g m^{-2}) Average	C_4 biomass (%) Range	C_4 biomass (%) Average
California, 40° N	12	80	*Elymus mollis* (C_3) *Ammophila arenaria* (C_3) *Ambrosia chamissonis* (C_3)	30–211	121	0–0	0
California, 33° N	16	30	*Ambrosia chamissonis* (C_3) *Abronia maritima* (C_3) *Camissonia cheiranthifolia* (C_3)	20–174	98	0–6	3
Texas, 26° N	22	61	*Uniola paniculata* (C_4) *Panicum amarulum* (C_4) *Ipomoea stolonifera* (C_3)	103–211	149	51–83	70

more in summer than winter. Rains tend to be gentle and prolonged; winds are moderate throughout the year. Gulf sites continue the trend of summer peak in precipitation, but yearly totals are variable: the Tortugas Keys, the Mississippi Islands, and the Mexican coast near Cabo Rojo receive 99–147 cm (that is, similar to the Atlantic sites), but Padre Island is much drier, with only 61 cm. Fall hurricanes with heavy, intense rain and gale-force winds strike these sites at infrequent intervals; otherwise, winds are calm. Pacific coast sites are the most variable, ranging from an evenly distributed 125 cm at the Queen Charlotte Islands, to 102 cm at Eureka with a short summer drought moderated by extensive summer fog, to less than 25 cm at the southern two sites with pronounced (4–9 mon) droughts. Winds are high at the Queen Charlotte Islands and moderate elsewhere, but tropical cyclones with intense rain and high winds can occur at the two southern sites.

14.3.2 PLANT WATER POTENTIALS

Few strand species have had their internal water status measured. At a semi-arid southern California site (39 cm ppt yr^{-1}), the mean dawn plant water potential showed minor fluctuation throughout the year, from -0.75 to -1.25 MPa for *Abronia maritima* and from -0.3 to -0.7 MPa for *Ambrosia chamissonis* (De Jong, 1979). Mean mid-day plant water potential was rarely more negative than -1.8 MPa for either. Both are tap-rooted species and obtain moisture from a fresh water table (700 ppm salt) about 3 m below the surface.

In contrast, mean dawn water potential of *Atriplex leucophylla*, located at the same beach but in the more saline swash zone, and of the shallow-rooted annual *Cakile maritima*, ranged from -0.3 to -1.8 MPa seasonally and fluctuated by as much as 2.0 MPa during a given day. These plants did not tap the water table. At a wetter northern site (82 cm ppt yr^{-1}), mean dawn water potentials of the same four species were dampened, never being more negative than -1.1 MPa, but daily fluctuations in water potential of *Atriplex* were still large, often as great as 1.5 MPa.

On the Georgia coast mean dawn potentials

ranged between -0.4 and -0.87 MPa for several species (Dubois, 1977). Mean afternoon values for the two C$_3$ herbaceous species (*Heterotheca subaxillaris* and *Croton punctuatus*) were -1.2 and -1.0 MPa, respectively, and -1.8 MPa for the C$_3$ shrub *Iva imbricata*. Three C$_4$ plants, *Atriplex arenaria*, *Uniola paniculata*, and *Spartina patens* all had mean afternoon xylem pressure potentials between -2.0 and -2.2 MPa. Au (1969) reported water potentials for North Carolina herbaceous beach and dune species that ranged between -0.5 and -1.5 MPa. In that study the C$_4$ species also tended to have lower mean water potentials but the differences were not significant.

Another source of moisture available to shallow-rooted species other than direct precipitation may be 'internal dew'. This is the condensation of water vapor into liquid on the surface of sand grains as a result of diurnal temperature fluctuations. The zone of condensation corresponds with the rooting depth of shallow-rooted plants. Measurements of increases in soil moisture during periods of extended drought have been made in several strands (De Jong, 1979; Olsson-Seffer, 1909; Salisbury, 1952). The water vapor in the soil atmosphere may come from diffusion from the air above the soil or from deeper layers in the sand.

De Jong (1979) found at least three strategies of water use in the species he examined: tap-rooted species which utilize the water table; shallow-rooted perennials with a diffuse root system, such as *Atriplex leucophylla*, which utilize precipitation and internal dew; and annuals such as *Cakile maritima* which complete their life cycle before the onset of drought. In addition, *A. leucophylla* has the C$_4$ mode of photosynthesis (De Jong, 1978a), with a substantially higher water use efficiency (26.2 mg CO$_2$ gH$_2$O^{-1}) than the other tap-rooted or annual C$_3$ species (17.–18.7).

If we can extrapolate this pattern elsewhere, then we should expect shallow, fibrous-rooted taxa to predominate in regions of high precipitation or where conditions for internal dew formation are favorable (e.g. a humid coast with high insolation and clear skies at night). Tap-rooted species may be expected to predominate in drier areas underlain with fresh water tables. Indeed, the strand of the north Pacific coast and the Atlantic coast are dominated by fibrous-rooted grasses (Barbour *et al.*, 1976; Martin, 1959; Oosting and Billings, 1942; Stalter, 1974), whereas tap-rooted species dominate the dry strands of southern and Baja California (Barbour *et al.*, 1976; Johnson, 1977). The semi-arid Padre Island strand, however, exhibits shared dominance by both categories of plants and does not seem to fit the pttern (see Table 14.6).

We would also expect shallow, fibrous-rooted taxa to exhibit more pronounced xeromorphies than tap-rooted taxa because the amount of water available is more limited. To some degree, this is borne out in the literature as we now interpret it (e.g. papers by Harshberger, 1908, 1909; Kearney, 1901). Additionally, as in a Georgia strand (Dubois, 1977), many fibrous-rooted species are C_4 grasses with water use efficiencies (14.9–36.0 mg CO_2 g H_2O^{-1} at 22–28 °C, 7.0–15.5 at 32–38 °C) much higher than associated C_3 tap-rooted species (6.6–8.6 over both temperature regimes).

14.4 Salt spray

14.4.1 NATURAL SPRAY INTENSITIES

Gradients of salt spray on the beach and from beach to dune have been measured by several investigators and invoked as one of the principal controlling factors in the distribution of North American coastal plants (Barbour and De Jong, 1977; Boyce, 1954; Martin, 1959; Oosting and Billings, 1942; van der Valk, 1974a; Wells, 1939). Boyce (1951a, 1954) demonstrated that the droplets of salt spray had a mixed salt content no different from that of sea water, and that in fact they were ejected by breaking bubbles of sea water in the zone of heavy surf. By the time they reach the beach, brought in by on-shore wind, the spray droplets are generally 25–75 μm diameter.

The intensity of salt spray within the vegetated portion of the beach is correlated with wind speed, distance back from the tide line, height above the ground, and site microtopography (Barbour, 1978). At about 15 cm above the ground – just at or above average canopy height – salt spray deposits are 1–200 mg salt dm^{-2} vertical surface d^{-1}. Leaves which project higher than this are subjected to higher spray intensities; leaves which are horizontal may receive an order of magnitude less spray than this (Barbour, 1978); leaves which are broad may receive less salt per unit area than linear leaves (Woodcock, 1953). Within the vegetated portion of the beach, salt spray intensity at any time may vary 2 to 20-fold (Barbour, 1978), and dune locations 600 m inland may receive only 1–5% of the salt spray load of the beach (Barbour *et al.*, 1973; Martin, 1959).

14.4.2 HYPERTROPHY AND SPRAY TOLERANCE

A typical response of spray-tolerant plants to salt spray is leaf hypertrophy (enlargement of cells), resulting in a doubling or tripling of leaf thickness (Boyce, 1951b; Martin and Clements, 1939). The leaf cells which enlarge are either leucoparenchyma or chlorenchyma, depending on the species, and the ion most responsible for the enlargement is chloride.

Generally, in dicots, the degree of salt tolerance seems positively correlated with the ability to undergo hypertrophy and to become succulent. None of the dune grasses, however, exhibit hypertrophy. In the presence of high levels of salt spray, epidermal lesions and tip burns appear. A thick cuticle may be an adaptation which minimizes such injuries. Succulent leaves contain 4–6 times the chloride concentration of protected, non-succulent leaves of the same species. Most of the chloride enters through lesions in the leaves, rather than through the root system. Chloride may be translocated within a leaf to the tip and the margin, but very little is translocated out of the leaf. Intercellular space inside succulent leaves is reduced, compared to the non-succulent state. There are no data in the literature regarding the effects of salt spray-induced hypertrophy on photosynthetic gas exchange. Leaf modifications which may reduce the load of salt spray reaching the epidermis include: broad shape, horizontal (rather than vertical) orientation, position close to the ground surface, and dense pubescence.

In general, tolerance to salt spray does correlate with species zonation relative to tide line: those most tolerant are characteristically closest to shore, and those less or least tolerant are furthest from shore. In the two most extensive studies, however, important anomalies were reported. In Island Beach State Park, New Jersey, Martin (1959) pointed out that some back-dune species exhibited very high salt spray resistance, and some foredune species (e.g. *Hudsonia tomentosa*) exhibited very low resistance. Along the California coast, Barbour and De Jong (1977) found similar exceptions: *Ambrosia chamissonis*, for example, exhibited nearly the same level of spray intolerance as the control glycophyte *Phaseolus vulgaris*. In both studies, low tolerance was coincident with low or prostrate morphology, pubescent leaves, or occurrence in favorable microsites.

14.5 Soil salinity

14.5.1 STRAND SOIL SALINITY

Schimper (1903) and Warming (1909) considered soil salinity to be an important feature in the strand, while Kearney (1904) questioned whether the salt content was sufficient to label the resident plants halophytes. After taking soil samples on both Atlantic and Pacific coasts, Kearney concluded that even at Long Beach, CA, where the greatest concentrations of salt were measured, salt content did not '...exceed the maximum occurring in ordinary cultivated soils'. He suggested that halophytic features described by other investigators were probably xerophytic adaptations instead. Oosting and Billings (1942), Boyce (1954), and Barbour *et al.* (1973) reported a maximum strand topsoil salinity of 0.1% and were unable to show much correlation between changes in soil salinity and species distribution as one moves inland through the strand. Barbour and De Jong (1977) did show that beach salinity in the root zone (10–30 cm depth) at the leading edge of vegetation could reach 0.3% following storm surges, but this is still a relatively non-haline environment (oligohaline to mesohaline; see Waisel, 1972). Even if one were to express the above concentrations in terms of salt dissolved in soil water at field capacity (instead of expressed as per gram of soil), and one assumed that 10% by weight of the strand was occupied by water at that point, salinity would be typically less than 1% except briefly following storm surges. Further, some temperate zone foredune species, such as *Ammophila breviligulata*, are found in nature growing on both maritime and fresh water shores with equal vigor. In his 1970c review, Barbour concluded that '...strand species appear to be intolerant halophytes and cannot be described as either facultative or obligate halophytes'.

Tropical and arid strands, however, may be more saline. Davis (1942) reported 2–3% salt in Florida strands when calculated in relation to the actual amount of water present in the sand. Sea-water around the Florida Keys where he was working was about 3.5% salt. Similarly, Johnson (1977) reported soil water salinities (corrected to a water-holding capacity of 5%) ranging from 0.35 to 4.55% with a mean of about 2.1% for the beach and dunes along arid Baja California. De Jong (1979) reported seasonal soil osmotic potentials based on soil water and salt contents for two locations on the California coast. At both sites mean calculated soil osmotic potentials ranged from approximately -0.4 to -2.0 Mpa at a depth of 30 cm and -0.2 to -0.7 MPa at a depth of 100 cm.

14.5.2 PLANT RESPONSE

Knowledge about the salt tolerance of beach and dune species is based on relatively few experiments. Oosting and Billings (1942) transplanted *Uniola paniculata*, *Andropogon littoralis*, and *Spartina patens* from the dunes of North Carolina to large pots and watered them daily with full strength sea-water for 3 wk. At the end of the treatment *Uniola* plants 'looked unhealthy and pale' three of four *Andropogon* plants were dead, and *Spartina* was unaffected compared to controls watered with fresh water. Wagner (1964) reported that watering *Uniola paniculata* three times a week for 6 weeks with sea-water resulted in the death of many plants.

Barbour and De Jong (1977) attempted to simulate tidal inundation by flooding trays of eleven California beach taxa with sea-water and subsequently leaching them with fresh water, over a period of 3 weeks. All beach taxa tested were substantially more tolerant to this treatment than was the control glycophyte (*Phaseolus vulgaris*). With a few exceptions this study also showed that the species which were the most tolerant are those that have distributions nearest to the shoreline. Barbour (1970a) showed that root growth of *Cakile maritima* was unaffected by NaCl concentrations up to 28% sea-water but declined at higher concentrations.

De Jong (1978) measured photosynthetic gas exchange and growth responses of two California beach taxa *Abronia maritima* and *Atriplex leucophylla*. Growth and water-use efficiency of *Abronia maritima*, a succulent C_3 species, was actually stimulated by 17% sea-water solution; however, photosynthesis, leaf conductance to water vapor, and mesophyll conductance were all negatively affected at the same sea-water concentrations. Higher sea-water concentrations caused declines in all responses measured. Relative growth rate of the C_4 species *Atriplex leucophylla* was unaffected by up to 35% sea-water concentration. Mesophyll conductances were stimulated by 17% sea-water but leaf photosynthesis declined with all salinity treatments because of decreases in leaf conductance. Water-use efficiency of *A. leucophylla* increased with increasing salinity up to 70% sea-water.

Other evidence which suggests that some strand plants may be adapted to soil salinity is that several species occur in salt marshes as well: *Iva imbricata* (Colosi and McCormick, 1978), *Spartina patens* (Silander, 1979) and *Atriplex leucophylla* (De Jong and Barbour, 1979), among others. Osmotic water potentials of some dune species may be appreciable, between −1.0 and −1.5 MPa (Martin and Clements, 1939).

Germination studies with beach and dune species indicate that of those tested the germination of most species is relatively sensitive to salinity. Germination of *Cakile maritima* is inhibited by NaCl concentration above 0.1% (Barbour, 1970a). Maximum tolerance limits for germination of *Ammophila breviligulata* and *Uniola paniculata* to NaCl are between 1.0 and 1.5% (Seneca, 1969). The upper limit for germination of *Panicum amarulum* is between 1.5 and 2.0% and for *Spartina patens* is as high as 4%. Seneca concluded that the germination inhibition is primarily due to an osmotic effect and not to chloride toxicity.

In general, studies involving the effects of salinity on germination and growth of beach and dune species indicate that they are neither as tolerant to salinity as some typical salt marsh halophytes, nor as sensitive as some glycophytes.

14.6 Soil nutrients

14.6.1 INPUTS

Studies of nutrient cycling conducted on the east coast (van der Valk, 1974a,b; Art *et al.*, 1974) and west coast (Clayton, 1972; Holton, 1980) have shown that:

1. meteorologic inputs greatly exceed inputs from substrate weathering;

2. aerosols in spray and fog comprise the largest meteorologic input, and concentrations of Na, Mg, Ca, K, N, and P reflect their relative abundances in sea-water;

3. rapid leaching through a coarse substrate lacking clay and organic matter prevents high soil salinity and nutrient status;

4. annual ion input slightly exceeds the output through leaching, such that:

5. storage in plant biomass is small and the proportions of ions in plant tissues reflect selective uptake and/or exclusion.

Regarding input from aerosols, it has been shown that the morphology, topographic position, and exposure of shoots greatly influence the interception of nutrients in beach and dune habitats (Barbour, 1978; Boyce, 1954; van der Valk, 1974b; Woodcock, 1953). In the perennial dune grasses *Uniola paniculata* and *Ammophila breviligulata*, however, van der Valk (1977b) concluded that nutrients contained in spray were indirectly available through interception and stem flow and not via foliar absorption. In addition to aerosols, the onshore movement of nutrients may be affected by bulk precipitation, sea-water inundation, detritus deposition, and sand movement (Azevedo and Morgan, 1974; Barbour and De Jong, 1977; Holton, 1980; Olson 1958a,b; Ranwell, 1972; Wagner, 1964; Wilson, 1959).

All of the known macro- and micronutrients regarded as essential for plant growth are present in sea-water (Epstein, 1972; Weast, 1977). Many are at such high concentrations (i.e. Na, Cl, Mg, Ca) that plant requirements for these ions are readily met or detrimentally exceeded by meteorologic inputs. Others are in very low concentrations (i.e. N, P, K) compared to levels required for optimum plant growth and may impose major restrictions on coastal plant productivity (Morris *et al.*, 1974; Pavlik, 1982, 1983a; Woodhouse and Hanes, 1966). Supplementing beach stands with fertilizer results in significantly greater biomass production (Augustine *et al.*, 1964; Brown and Hafenrichter, 1948; Willis, 1965). Excesses of Na and Cl have been already discussed so we will now look at the relation of N–P–K deficits to the physiological ecology of beach and dune plants.

14.6.2 NITROGEN

The concentration of organic and inorganic nitrogen in sea-water ranges between 0.036 and 1.000 ppm, the dominant fraction being $N-NO_3$ (Wagner, 1964; Weast, 1977). Concentrations of nitrogen in bulk precipitation and fog vary between 1 and 7 ppm along the northern California coast, mostly in the form of $N-NH_4$ (Azevedo and Morgan, 1974; Holton, 1980).

At Point Reyes, California, Holton found that inorganic nitrogen pools in beach and dune sand could range between 0.5 and 2.0 ppm, equally divided into $N-NO_3$ and $N-NH_4$ fractions. The distribution of soil organic nitrogen was shown to be highest near the tidemark (0.01%) and within the stabilized dunes (0.04–0.2%), as compared to more exposed areas between tidemark and foredune (0.003–0.006%). Wave-deposited detritis and litter fall from dune species (several of which are nodulated with nitrogen-fixing bacteria) were thought to be responsible for the pattern. Despite the low meteorologic input and small soil reserve, tissue nitrogen contents of the dominant taxa (1.4–2.8%) were at levels adequate for plant growth (1.5% according to Epstein, 1972). A similar though wider range of 1.0–3.0% occurred in seven beach and dune species along the Georgia coast (Dubois, 1977).

Sclerophylly (hence pienomorphy, Loveless, 1962) is rare among North American strand plants (see Table 14.6). What adaptations, then, do these plants possess which enable them to maintain a favorable nitrogen economy, given the seemingly low pool of available nitrogen?

Possibly, episodic deposits of detritus represent an important additional source of nitrogen to the strand (Ranwell, 1972). It is known that strand plants can utilize the decay products of algae and respond positively to them. For example, Holton (1980) provided greenhouse-grown plants of *Cakile maritima* and *Atriplex leucophylla* with either full nutrient solution, nutrient solution lacking nitrogen, distilled water, or distilled water plus 6 g brown algae *Macrocystis pyrifera* and *Egregia mentiesii*. After 3 mon, the shoot dry weight of seaweed-innoculated plants was 70–74% of those given full nutrients as compared to the 9–39% achieved by plants under the other regimes. Tissue nitrogen of the algae was between 1 and 2% dry weight and it was readily released during decomposition.

In the field, detritus is deposited above the tidemark and would be available primarily to plants at the leading edge of vegetation, such as *Cakile* and *Atriplex* along the California coast. It has been shown that higher nitrate levels in tidemark (beach) sands induce nitrate reductase activity in *Cakile maritima*, and partially compensate for the inhibition of the enzyme by salinity (Garcia Novo, 1976; Holton, 1980). Garcia Novo (1976) also noted that the less salt-tolerant dune grass *Elymus arenarius* did not exhibit significant nitrate reductase induction. The correlation of *Cakile* individuals with enriched nitrogen microsites was suggested by Johnson and York (1915) and Veldkamp (1971) and measured by Barbour (1972, who held that the differences were ecologically insignificant, however). The C_4 *Atriplex leucophylla* maintains higher rates of CO_2 fixation at lower leaf conductances caused by salinity (De Jong, 1978b) and perhaps utilizes detrital nitrogen from wave deposition. De Jong found that the more salt-tolerant plants in his study, *A. leucophylla*, *C. maritima*, and *Abronia maritima*, had higher leaf protein contents but lower rates of carbon fixation than the salt-sensitive *Ambrosia chamissonis*. This may indicate that the additional protein (nitrogen) is used in developing salt tolerance rather than photosynthetic capacity (cf. Pavlik, 1983a; Stewart *et al.*, 1979; Storey *et al.*, 1977).

Nitrogen fixation by rhizosphere bacteria ('associative symbiosis') has been demonstrated for marine angiosperms (Patriquin and Knowles, 1972), some tropical and temperate dune grasses (Ahmad and Neckelmann, 1978; Day *et al.*, 1975), and mangroves in Florida (Zuberer and Silver, 1975). However, the importance of rhizosphere nitrogen fixation to the nutrition of most North American beach plants remains in question. Using acetylene reduction, Holton (1980) could not detect fixation in sand and root samples from *Ammophila arenaria* and *Elymus mollis* stands at Point Reyes, California. Nor could he culture nitrogen-fixing bacteria in the lab from dune sand inoculum. This conflicts with European studies of *Ammophila* (Abdel Wahab, 1975; Abdel Wahab and Wareing, 1980; Hassouna and Wareing, 1964), which showed low but measureable rhizosphere fixation rates of $2 \, g \, g^{-1}$ root d^{-1}, and growth was stimulated when sterile root media were innoculated with *Azotobacter*. High nitrogenase activity has been associated with C_4 grasses (Ahmad and Neckelmann, 1978; Day *et al.*, 1975), so rhizosheath fixation may be more important on low-latitude beaches where such grasses predominate (see Table 14.6).

Nodulated (symbiotic) fixation has not been widely investigated for North American strand plants. Legumes in beach and dune floras are rare along the east coast (two species), increasing along the western (nine species) and Gulf (22 species) margins. Holton (1980) demonstrated significant rates of acetylene reduction by nodules of *Lupinus chamissonis* and *L. arboreus* ($0.05-0.60 \, \mu m \, C_2H_4 \, g^{-1}$ fresh weight h^{-1}) collected from stabilized dune sites at Point Reyes. Nodules of the strand legume *Lathyrus littoralis* exhibited the highest nitrogenase activities from October to December ($0.25-2.83 \, \mu m \, C_2H_4 \, g^{-1}$ fresh weight h^{-1}) with much lower activities in late summer. Holton and Johnson (1979) discussed the synecological importance of these nodulated legumes to dune scrub communities. Non-legumes, such as *Myrica pennsylvanica* of east coast dunes, also have nodules with significant acetylene reduction rates ($0.07-5.03 \, \mu m \, C_2H_4 \, g^{-1}$ fresh weight $^{-1}$, to which a portion of their success on duen sites

has been attribued (Morris *et al.*, 1974).

Many plants, regardless of natural habitat, respond to nitrogen fertilization with increased dry matter production. It should not be surprising that plants of beaches and dunes would do the same; thus, one can infer little about their nutritional ecology from augmentation studies. Comparative studies of physiological response to nitrogen deficiency by strand and non-strand taxa might be more productive. Holton (1980) performed such stress experiments on *Cakile maritima* and Pavlik (1983a,b,c) similarly treated *Elymus mollis* and *Ammophila arenaria*, all from California beaches. They showed that low-nitrogen growth regimes decreased: total leaf area, total plant dry weight, leaf area ratio, relative growth rate, dark respiration rate, and tissue nitrogen content. The ratio of root: shoot dry weight increased significantly with decreasing available soil nitrogen.

There were, however, some differences in response to low nitrogen between the annual *Cakile* and the two perennial grasses. All three species exhibited lower mesophyll conductances when nitrogen-stressed, but *Cakile* maintained its net photosynthetic rate by increasing leaf conductance (Table 14.4). Although this served to increase the gaseous flux of CO_2 into the leaf, it also resulted in a large increase in transpiration and a considerable decrease in an already low water-use efficiency (a reduction of 54%). This induction of higher leaf conductances and increased water expenditure by nitrogen deficiency may be the cause of *Cakile* senescence in the field. When supplied with adequate moisture *Cakile* is not an obligate annual but will continue to flower and set seed (see Barbour, 1970b). In contrast, the perennial grasses responded to low nitrogen supply by lowering leaf conductances and maintaining a high water use efficiency. Presumably, rapid depletion of water in the root zone would more severely affect the establishment, survival, and reproductive output of a perennial psammophyte than a rapidly growing and precocious annual. Other, non-psammophytic annuals may not respond like *Cakile*; for example, Medina (1970) found that the C_3 plant *Atriplex patula* ssp. *hastata* has gas exchange responses to low nitrogen that parallel those of *Ammophila* and *Elymus*.

Table 14.4 Gas exchange characteristics of leaves of greenhouse grown *Cakile maritima, Ammophila arenaria*, and *Elymus mollis* given high (+) or low (−) supplies of nitrogen. Statistical comparison of treatment means is valid only within each species. Values within a row followed by the same letter are not significantly different at $p < 0.05$ (ANOVA). Data from Holton (1980) and Pavlik (1983a)

	Cakile[*]		*Ammophila*[†]		*Elymus*[†]	
	+	−	+	−	+	−
Net photosynthesis (μmol m^{-2} s^{-1})	23.3[a]	20.5[a]	31.2	19.4	26.7	14.5
Mesophyll conductance (mm s^{-1})	3.4	2.0	5.1	3.1	3.2	2.1
Leaf conductance to H_2O (mm s^{-1})	9.0	13.6	8.1	5.5	8.1	4.1
Transpiration (mg m^{-2} s^{-1})	93.0	155.0	59.0	39.9	59.1	30.5
Water use efficiency (g CO_2 mg^{-1} H_2O)	11.0	5.8	23.1[b]	24.3[b]	18.5[c]	24.6[c]

[*]Leaf temperature $= 22°$ C, quantum flux $= 1.2$ mmol m^{-2} s^{-1}.
[†]Leaf temperature $= 25°$ C, quantum flux $= 1.7$ mmol m^{-3} s^{-1}, 1.0 KPa vapor-pressure deficit.

Differences in photosynthetic nitrogen use efficiency (Brown, 1978) have been demonstrated between species that are restricted to beach and dune habitats. Pavlik (1983a,b,c) showed that under laboratory and field conditions the introduced *Ammophila arenaria* fixed more carbon per unit of blade nitrogen than the native *Elymus mollis*. *Ammophila's* superiority was correlated with the production of more live blade area and vegetative buds, slower rates of blade senescence, and reduced dry matter and nitrogen allocations to roots regardless of nitrogen supply; its superiority may explain why it has aggressively supplanted *Elymus* along portions of the Pacific Coast (Barbour *et al.*, 1976; Slobochikoff and Doyen, 1977). Comparative studies have also shown that C_4 grasses as a whole are able to fix more carbon per unit of assimilated nitrogen than C_3 grasses (Bolton and Brown, 1980). In Georgia, the C_4 dune grasses *Uniola paniculata*, *Spartina patens*, and *Panicum amarum* had significantly higher nitrogen use efficiencies than three C_3 dicots (especially at leaf temperatures of 32–38 °C) and a slightly higher efficiency than the C_4 dicot *Atriplex arenaria* (Dubois, 1977). De Jong (1978a), however, provided some evidence to suggest that the salt-sensitive C_3 strand dicot *Ambrosia chamissonis* had a higher efficiency than the salt-tolerant C_4 dicot *Atriplex leucophylla*. We need further studies, comparing plants with similar leaf morphotypes over a range of temperatures, before reaching major conclusions on the ecological significance of nitrogen use efficiency to plants from beach and dune habitats. Whether or not such taxa are more productive per unit of assimilated nitrogen than those from other, more fertile, habitats has not been addressed.

14.6.3 POTASSIUM AND PHOSPHORUS

At Cape Hatteras National Seashore, North Carolina, van der Valk (1974b) concluded that potassium '...could limit the growth of the vegetation if adequate levels of all other requirements for growth were present', because meteorologic inputs and soil storage of potassium were the smallest of the four cations investigated (Na, Mg, Ca, and K), and yet potassium had the highest concentration in the vegetation. A similar relationship was also found by Art *et al.* (1974) on Fire island, New York, by Clayton (1972) along the central California coast, by Moreno-Casasola *et al.* (1982) near Veracruz, Mexico, and by Poggie (1963) at calcareous sites near Tampico, Mexico. In Britain, however, potassium deficiencies in calcareous dunes are not regarded as being as critical as other nutrient limitations (Etherington, 1967; Willis, 1965; Willis and Yemm, 1963).

Exchangeable potassium in beach sands has a measured range of 13–40 ppm (Moreno-Casasola, 1982; Poggie, 1963; Wagner, 1964). Epstein (1969, 1972) noted that plants require potassium in amounts similar to that of nitrogen (about 1% dry weight), yet we know of no studies on the potassium nutrition of beach and dune plants conducted on this continent. This is particularly surprising in the light of what is known about the mechanisms of potassium uptake in the presence of high sodium concentrations (cf. Rains and Epstein, 1967). However, Eshel *et al.* (1974) demonstrated that when grown under conditions of high sodium, *Cakile maritima* utilized sodium ions instead of the usual potassium ions in effecting stomatal opening (Table 14.5).

Even less is known about phosphorus in beach and dune communities. Not only is the element dilute in sea water, but as phosphate it would be readily leached. Its availability would be further decreased in calcareous sands of high pH such as occur at the tip of Baja, California, and Florida, and along portions of the Gulf coast of Mexico. Wagner (1964) measured a range of 73–133 ppm in North

Carolina; Johnson (1963) assayed beach and dune sands near Arcata, California for phosphoric anhydride (P_2O_5) and detected 0.3–7.8 ppm at 30 cm depth.

14.7 Sand movement

14.7.1 RATES OF SAND ACCUMULATION

The rate at which sand is blown onto the strand is determined by such obvious factors as sand sediment load on to shore from off-shore currents, wind speed, grain size, and relative air humidity (Bagnold, 1941; Davies, 1977). The threshold velocity needed to move average dune sand of mixed diameter is 4 m s^{-1} (9 mph), or 3 on the Beaufort Scale (8–12 mph, 'gentle breeze'). Our 12 coastal sites (Table 14.1) exhibit three basic wind regimes: strong, frequency of Beaufort Force 4^+ in January and July 15–30%; moderate, frequency 5–15%; calm 0–5% (Davies, 1977, in part). Sites QCI and OI have strong regimes, EU, PB, NTS, CC, IB, and EF moderate, and the Gulf sites TK, MI, PI and CR are calm (see Fig. 14.1 for abbreviations).

Other factors determine the rate of settling out and sand retention, once the sand reaches the strand: primarily these factors are topographic roughness and plant cover (Fig. 14.3, see also Au, 1969). Any object extending above the ground increases the surface roughness and can cause the wind velocity to fall below that necessary to move sand, causing sand to pile up around the object. As the object is buried its roughness diminishes until wind velocity again attains a point where it can move sand past the object; however, if the object is a growing plant which can continue to project above the piling sand, a dune will be built.

The morphologies of different beach plants apparently can induce different rates of sand accumulation, even on the same beach during the same time period. At Point Reyes National Seashore, California, for example, Barbour *et al.* (1974–77) erected permanent transects through nearby foredunes dominated by *Ammophila arenaria*, *Elymus mollis*, and *Abronia umbellata*. Over the course of 19 months, 0.9 m accumulated within the *Ammophila*, 0.7 m accumulated within the *Abronia*, and 0.5 m ac-

Table 14.5 An X-ray microanalysis of the guard cells of *Cakile maritima* grown under high and low salt regimes. (Data of Eshel *et al.*, 1974).

| Growth medium | Element | X-ray yields (average net cpm $\pm$ 1 S.D.) | |
		Open stomates	Closed stomates
Hoagland's solution	Na	14 ± 12	16 ± 9
	K	$1115 \pm 277^*$	$534 \pm 195^*$
Hoagland's + 100 mm NaCl	Na	688 ± 465	371 ± 283
	K	25 ± 19	46 ± 72

*·Significance at the 0.01 level between open and closed, stomates.

cumulated within the *Elymus*. In the same time period, unvegetated portions of beach typically lost 0.5–1.0 m sand. At Padre Island National Seashore, Texas, *Panicum amarum* was able to accumulate 0.8–1.2 m sand yr^{-1}, and the authors (Dahl *et al.*, 1975) concluded, 'The main limitation of dune growth was not the trapping capacity of the plants, but the amount of blowing sand available'. At Island Beach State Park, New Jersey, Martin (1959) measured maximum rates of accumulation around foredune stands of *Ammophila breviligulata* of 34–53 cm yr^{-1}. *A. breviligulata* plantings along lake Michigan trapped an average 30 cm sand yr^{-1} (Olson 1958a, b). These rates compare very favorably with those provided by inanimate objects such as sand fences, which on the Outer Banks of North Carolina trap 60 cm sand yr^{-1} or less (Savage and Woodhouse, 1969; Woodhouse *et al.*, 1968).

14.7.2 PLANT RESPONSE

Apparently, beach and dune species not only accumulate different amounts of sand, but their ability to grow rapidly enough to keep ahead of piling sand differs. Most of the evidence for this correlation comes from England. *Ammophila arenaria* can tolerate yearly depositions of 60–90 cm, *Elymus arenarius* can tolerate about 30 cm, and *Agropyron junceum* about 20 cm (Ranwell, 1972). Burial stress experiments have rarely been done for North American species. Johnson (1978) subjected greenhouse plants of the Pacific coast species *Abronia maritima* to sand burial for 3–4 mon and found an ecotypic response. Collections from windier sites with finer sand (presumably, then, subject to greater burial stress) grew 18 cm mon^{-1}, while those from Gulf sites with considerably lower burial stress grew 5 cm mon^{-1}. *Sesuvium portulacastrum*, collected from one windy site grew 22 cm mon^{-1}. De Jong and Barbour (1979) described morphological ecotypes of *Atriplex leucophylla* along the Pacific coast which may have evolved in response to differential rates of sand movement.

It is not clear how many beach species merely tolerate burial in contrast to others which require burial for maximum growth and completion of the life cycle. Three grasses are known to decline in vigor and to reproduce vegetatively only when sand deposition declines: *Ammophila arenaria*, *A. breviligulata*, and *Uniola paniculata*. Several theories proposed to explain how sand burial might interact with metabolism have been summarized by Marshall (1965) and Eldred and Maun (1982). They involve competition, plant age, soil pH, allelopathy, soil microflora, soil aeration, the rate of root growth, and nutrient inputs. Senescent stands can be rejuvenated by nitrogen fertilizer alone (Wagner, 1964) or by burial with sterile sand (Hope-Simpson and Jeffries, 1966), so it is unlikely that a single factor is responsible. Competition with other species is probably not a major factor (Eldred and Maun, 1981). Possibly, physical pressure could exert an effect. In some simple field experiments, Purer (1936a) showed that *Calystegia soldanella* rhizomes appeared to 'seek' a 10 cm depth, independent of light level. As sand level increased around the above-ground portions, the rhizomes ascended to the usual 10 cm level. Deflation had the opposite effect; complete natural deflation exposing the rhizome tip caused the tip to arch back into the sand. Do we have a 'new' tropism here, which could be called barostropism (Gr. *baros*, weight, pressure + Gr. *tropism*, a turning)? Or could CO_2 concentration be factor?

Sand accumulation not only affects growth and a balance between sexual and vegetative reproduction; it affects seedling establishment. Beach and foredune species appear to have larger seeds than dune species, implying an ability to successfully send up a seedling from a deeply buried seed. Johnson (1978) compared seed weights of 11 common California beach herbs with that of 39 backdune herbs, and reported that foredune seeds were 73% heavier, on the average. Van der Valk (1974a) found a significant, curvilinear relationship between seed weight and maximum depth of burial for six species of the North Carolina strand; to his data we added *Abronia maritima* and *Cakile maritima* from the Pacific coast to yield the linear relationship shown in Fig. 14.5. Platt (unpublished) has shown that the different seed weights of *Ipomoea stolonifera* and *I. pes-caprae* along the Gulf coast correlate with sand deposition rate: rapidly prograding beaches tend to be invaded by *I. pes-caprae*, which has a heavier seed and can germinate at greater depth.

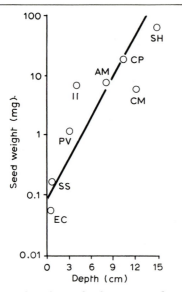

Figure 14.5 The relationship between seed weight and maximum depth of burial for successful seedling establishment, for eight forbs: SH = *Strophostyles helvola*, CP = *Croton punctatus*, AM = *Abronia maritima*, II = *Iva imbricata*, CM = *Cakile maritima*, PV = *Physalis viscosa*, SS = *Solidago sempervirens*, EC = *Erigeron canadensis*. *Abronia* data from Johnson (1978), *Cakile* data from Barbour (1970a, 1972), the rest from van der Valk (1974a).

14.8 Deductions from floristic analyses

Regional summaries of various floristic spectra are shown in Table 14.6. Latitudinal gradients are apparent in a few categories. For example, as one moves south along the Pacific coast, nanochamaephytes, chamaephytes, shrubs, armed taxa, prostration, broad leaves, succulence, and C$_4$ metabolism all increase, while geophytes and dissected-linear leaves decline. As one moves south on the east coast, only the following patterns appear: chamaephytes, shrubs, and the C$_4$ mode increase, while annuals and dissected-linear leaves decline. Thus only four trends repeat themselves on the two coasts: increasing frequency of chamaephytes, shrubs, and C$_4$ plants, and decreasing frequency of linear-dissected leaves with lower latitudes.

In comparison to Raunkiaer's world norm, these beaches as a whole are overrepresented with annuals, geophytes, and hemicryptophytes, and are under-represented with phanerophytes. Leaf morphology is typically broad or grass-like. On average,

species with linear, dissected, or reduced leaves account for less than 22% of the flora. Leaf texture is quite variable in the Mexican region, but usually mesic; that is, 42% of the species have entire, non-pubescent, non-sclerophyllous, and broad leaves. Sclerophyllous or leathery-leaved species account for less the 16% of the flora, on average, indicating that such leaves are not a typical evolutionary solution to the beach environment.

Additional patterns emerge when the sites are grouped climatically rather than regionally. For example, five sites can be grouped into a temperate/wet climate (QCI, EU, CC, IB, OI), four others into a warm/wet climate (MI, EF, CR, TK), and three others into a warm/dry climate (PI, NTS, PB). Moving along this complex gradient from cool/wet to warm/dry, nanochamaephytes, chamaephytes, shrubs, reduced/absent-leaved taxa, and C$_4$ taxa increase, while annuals, geophytes, vines, linear-dissected-leaved taxa, and those with sclerophyllous or leathery leaves decline.

We can also arrange the sites into exposure categories based on wind regimes: strong (QCL, OI), moderate (EU, PB, NTS, CC, IB, EF), and calm (TK, MI, PI, CR). As exposure decreases, chamaephytes, trees, and C$_4$ taxa increase while geophytes, sclerophyllous or leathery-leaved taxa, and those with pubescent leaves decline.

From all of the above aspects we may conclude that the array of growth forms, leaf traits, and metabolic pathways are a result of many factors, and that the importance of various factors differs from coast to coast. The prostrate growth habit does not seem to be related to exposure, while the number of taller forms is inversely related to exposure. Leaf morphologies do not seem to be related to heat load. For example, pubescence is not related to heat stress gradients and dissected- or linear-leaved taxa decline as warmth/aridity climbs. Most beach taxa have broad, unspecialized, mesic leaves. Some of them may be oriented vertically, which may reduce salt spray deposition or heat load, but in other ways they are not specialized. Succulence is common but not dominant (floristically); sclerophylly may be related to exposure, but it is not a common strategy on any beach.

When beach vegetation is sampled with replicate

Table 14.6 Life-form, leaf-form and metabolic spectra of selected strand floras.[1]

Trait	QCI	EU	PB	NTS	CR	PI	MI	IK	CC	LI	OI	EF
						Location						
Flora:												
No. Species	30	22	17	13	49	86	52	63	18	16	26	23
No. Families	14	10	10	10	25	25	20	26	11	10	14	16
% Legumes (L)	13	4	6	8	8	12	4	11	5	13	0	0
% Composites (C)	16	45	18	15	12	13	14	13	22	19	15	9
% Grasses (G)	20	9	0	15	12	24	30	27	11	19	27	17
L + C + G	49	58	24	38	32	49	48	51	28	50	42	26
Life Forms (%)												
Annuals	20	27	24	0	11	29	26	18	44	44	31	17
Geophytes	43	32	0	8	11	22	20	13	22	44	35	9
Hemicryptophytes	23	27	12	31	16	26	12	14	33	12	12	17
Nanochamaephytes	13	9	24	31	9	14	14	9	0	6	8	22
Chamaephytes	0	4	12	15	9	5	9	10	0	0	4	17
Shrubs	0	0	29	15	25	2	10	20	6	6	15	26
Trees	0	0	0	0	15	0	7	7	0	0	0	4
Vines	7	0	0	0	4	2	2	9	0	6	8	0
Armed	0	0	12	15	4	3	3	4	6	6	27	9
Prostrate	36	32	84	70	18	20	21	19	44	56	31	35
Leaf Morphology (%):												
Broad, entire	57	64	71	77	76	55	45	60	50	50	46	52
Linear or dissected	20	27	18	8	4	10	18	7	39	31	19	26
Grass-like	23	9	0	15	16	30	36	32	11	19	27	22
Reduced or absent	0	0	12	0	4	5	1	1	0	0	8	0
Leaf Texture (%):												
Fleshy or succulent	37	32	59	46	21	13	23	23	50	44	19	56
Mesic	43	59	29	38	27	52	58	51	39	37	46	22
Sclerophyllous or leathery	20	9	0	15	23	5	10	9	11	19	35	22
Pubescent	47	73	65	54	23	21	17	20	44	50	54	30
Physiology (%):												
C_4	0	0	24	39	33	54	63	70	17	25	35	30
C_3	100	100	71	61	67	45	36	29	83	75	61	66
CAM	0	0	6	0	0	1	1	1	0	0	4	4

[1] Major references: Bowman (1918), Calder and Taylor (1968), Davis (1943), Gillespie (1976), Kurz (1942), Lonard *et al.* (1978), Macdonald and Barbour (1974), Martin (1959), Miller (1975), Moreno-Casasola (1982), Orme (1973), Parker (1974), Penfound and O'Neill (1934), Pessin and Burleigh (1941), Stalter (1974), Fry *et al.* (1978).

transects which run from the leading edge of vegetation to the foredune top (eg. Barbour *et al.*, 1976) it is usually apparent that the pattern of species presence and cover is zoned into bands parallel to the shore. In order to generate correlations between species zonation and environmental factors or biological traits, we used such transect data to calculate a 'zonation index' for each species. The zonation index is the average relative position on the beach for the average unit biomass of a given species. The index scale ranges from 0 (the leading edge of vegetation) to 1 (top of the foredune). We were able to find several studies on each coast with published data on plant distribution detailed enough for us to calculate zonation indices. The complete list of species, their zonation indices, and our method of calculating the indices are available on request.

Table 14.7 Zonation indices for plant categories on four Gulf sites: Indices range from 0 (the leading edge of vegetation) to 1 (top of foredune)

Category	Index	Category	Index
Annuals	0.72	Succulent	0.65
Geophytes	0.73	Pubescent	0.82
Hemicryptophytes	0.78	Linear or dissected	0.83
Nanochamaephytes	0.69	Sclerophyllous or leathery	0.92
Chamaephytes	0.70	Grass-like	0.90
Shrubs	0.91	Mesic	0.72
Trees	1.00	Reduced or absent	0.99
Vines	0.79	C_4	0.82*
Armed	0.94	C_3	0.87[†]
Prostrate	0.82		

*Based on 28 taxa.
[†]Based on 17 taxa.

Few of the life form, leaf form, or metabolic categories showed any separation in terms of beach zonation at any site or group of sites. Table 14.7, for example, summarizes average zonation indices for the Gulf coast sites. None of the differences was statistically significant at the 95% level of confidence, but a few patterns stand out. Annuals, geophytes, and vines with mesic unspecialized leaves did tend to have lower index numbers; shrubs, trees, and armed species, with sclerophyllous, leathery, or reduced leaves tended to have higher index numbers. Phanerophytes, especially if sclerophyllous and/or armed, had the highest (most inland) indices (0.91–1.00) and succulent nanochamaephytes (often prostrate) had the lowest indices (0.65–0.69). Succulence, pubescence, linear/dissected leaves, and a prostrate habit did not accumulate in any zone. There was no zonation difference between C_3 and C_4 taxa. Another comparison between zonation index and leaf anatomy was shown in Table 14.2 for a southern California beach. This table has already been discussed from the standpoint of adaptation of strand taxa to high light intensity; an additional point here is that there were no patterns of anatomical change which correlated with zonation index. The only C_4 plant, however, did have the lowest zonation index. Clearly, microenvironmental gradients within a given beach or group of related beaches are not sufficient to strongly affect the plant and leaf traits we examined.

14.9 Summary

Although historically and regionally extensive, the literature on strand plant autecology is too diffuse to permit a significant synthetic review at this time. Furthermore, certain modern approaches to gas exchange, water balance, and plant demography have rarely been applied to strand taxa. The review we have provided highlights major patterns when possible and suggests directions of further research in a few cases.

The strand experiences high light intensities, and summer temperatures near the sand surface can be warm to hot. Generally, leaf temperatures are within a few degrees Celsius of ambient. Germination, growth, and photosynthesis have broad temperature optima, generally between 12 and 36 °C. Although leaf anatomy of beach plants typically shows certain heliophytic traits as dense mesophyll, multiple palisade parenchyma layers, and isolateral symmetry, other xeromorphies such as sclerophylly, small or dissected leaves, and dense pubescence are not common. Where measured, photosynthetic temperature optima of C_4 taxa are not consistently greater than those of C_3 taxa. Most germination experiments are so artificial and unrelated to the actual microenvironment in nature that it is difficult to extrapolate from them. However, it appears that few strand species are light inhibited, thus seed burial is not essential for germination. Some taxa require stratification (e.g.

Uniola paniculata) but only in the temperate part of their range. Many species respond to a thermoperiod for germination.

Tap-rooted herbaceous perennials may reach a water table beneath the strand, and then stem pressure potential may be seldom more negative than $-1.8\,\text{MPa}$ at midday. Fibrous or shallow-rooted annuals and perennials experience greater water stress but may be able to take advantage of 'internal dew' – water vapor which condenses in the upper soil horizons. We should expect the latter plants to exhibit more pronounced xeromorphies and improved water use effciencies; they may additionally possess the C_4 mode of photosynthesis more frequently than tap-rooted species.

Salt spray load 15 cm above the beach ranges from 1 to 200 mg dm^{-2} vertical surface d^{-1}. Broad leaves receive less spray than linear ones, horizontal leaves receive an order of magnitude less spray than vertical leaves, and in general the lower the canopy the lower the spray load. Hypertrophy is a typical response of salt-tolerant forbs to chloride ion accumulations. We can hypothesize that gas exchange is reduced in hypertrophic leaves because of anatomical changes, but measurements have not been made. Salt spray tolerance does appear to be correlated with species zonation relative to distance back from shoreline, but some taxa (e.g. *Ambrosia chamissonis*) have surprisingly low salt spray tolerances.

Temperate strand soil salinity, even when expressed on a basis of soil solution at field capacity, is typically less than 1%, except following storm surges. Tropical strands may exhibit 2–3% salt on the same basis. Most beach taxa appear to be more tolerant of constant or acute osmotic shock than glycophytes, in terms of germination and growth. Some beach taxa also occur in salt marshes. Photosynthesis typically declines with increasing soil salinity for both C_3 and C_4 taxa.

Meteorologic inputs of N, P, and K come mainly from salt spray and fog droplets, and their concentrations reflect their relative abundances in seawater. Their levels in strand soils are low compared to levels required for optimum growth of inland mesophytes, yet tissue content of N (1–3%) is as high as that of some crop plants. Utilization of detritus decay products and symbioses with N-fixing micro-organisms may be additional sources of nitrogen. Some soil nitrogen, such as that close to the tidemark, may only be available to the most salt-tolerant strand plants. Nitrogen-stressed plants show decreased net photosynthesis, dark respiration, mesophyll conductance, relative growth rate, and altered patterns of resource allocation. The gas exchange responses of annual and perennial psammophytes to low nitrogen may be quite different; the former maintain photosynthetic rates by increasing leaf conductance while the latter reduce leaf conductance and photosynthesis but maintain a high water use efficiency. Variations in photosynthetic nitrogen use efficiency occur between littoral species, but additional studies that compare similar leaf morphotypes over a range of temperature need to be done. Little is known about potassium nutrition of strand plants, but in *Cakile maritima* sodium has been reported to replace potassium in effecting guard cell activity.

Sand accumulation is very important to the germination, growth, vigor, and even presence/absence of some strand species. The mechanism by which depth of sand burial or rate of sand accumulation affects metabolism is unresolved. Some taxa (e.g. rhizomes of *Calystegia soldanella*) exhibit movement relative to burial which we term 'barostropism'. Others (*Abronia maritima, Atriplex leucophylla*) may have evolved ecotypes in response to different regimes of sand movement.

In comparison to Raunkiaer's world norm, beach floras are over-represented with annuals, geophytes, and hemicryptophytes, and under-represented with phanerophytes. Leaves are typically broad or grass-like and mesic; sclerophylly, reduced leaves, and liner-dissected leaves are not common features. In general, as one moves from cold/wet to warm/dry beaches, chamaephytes, shrubs, and C_4 species increase, while annuals, geophytes, and taxa with linear-dissected leaves decrease. Exposed (windy) beaches have higher proportions of geophytes and taxa with sclerophyllous or pubescent leaves than calm beaches, but fewer chamaephytes, trees, and C_4 taxa. Such leaf traits as pubescence and dissection or linearity are not related to heat or aridity gradients, nor is prostration related to any environmental gradient that we examined. It appears further that, on a given beach, microenvironmental gradients back from tideline do not play a major role in sorting out growth forms, leaf forms, or metabolic traits.

References

Abdel Wahab, A.M. (1975) Nitrogen fixation by *Bacillus* strains isolated from the rhizosphere of *Ammophila arenaria*. *Plant and Soil*, **42**, 703–8.

Abdel Wahab, A.M. and Wareing, P.F. (1980) Nitrogenase activity associated with the rhizosphere of *Ammophila arenaria* L. and effect of inoculation of seedlings with *Azotobacter*. *New Phytologist*, **84**, 711–21.

Ahmad, M.H. and Neckelmann, J. (1978) N$_2$-fixation by roots and rhizosphere of sand dune plants. *Zeitschrift Fuer Pflanzenernahrung, und Bodenkunde*, **141**, 171–21.

Art, H.W., Bormann, F.H., Voigt, G.K. and Woodwell, G.M. (1974) Barrier Island forest ecosystem: role of meteorologic inputs. *Science*, **184**, 60–2.

Au, S.-F. (1969) Vegetation and ecological processes on Shackleford Bank, North Carolina, PhD dissertation, Duke University, Durham, North Carolina.

Augustine, M.T., Thornton, R.B., Sanborn, J.M. and Leiser, A.T. (1964) Response of American Beachgrass to fertilizer. *Journal of Soil and Water Conservation*, **19**, 112–15.

Azevedo, J. and Morgan, D.L. (1974) Fog precipitation in coastal California forests. *Ecology*, **55**, 1135–41.

Bagnold, R.A. (1941) *The Physics of Blown Sand and Desert Dunes*, Methuen, London.

Barbour, M.G. (1970a) Germination and early growth of the strand plant *Cakile maritima*. *Bulletin of the Torrey Botanical Club*, **97**, 13–22.

Barbour, M.G. (1970b) Seedling ecology of *Cakile maritima* along the California coast. *Bulletin of the Torrey Botanical Club*, **97**, 280–9.

Barbour, M.G. (1970c) Is any angiosperm an obligate halophyte? *American Midland Naturalist*, **84**, 105–20.

Barbour, M.G. (1972) Seedling establishment of *Cakile maritima* at Bodega Head, California. *Bulletin of the Torrey Botanical Club*, **99**, 11–16.

Barbour, M.G. (1978) Salt spray as a microenvironmental factor in the distribution of beach plants at Point Reyes, California. *Oecologia*, **32**, 213–24.

Barbour, M.G. and DeJong, T.M. (1977) Response of west coast beach taxa to salt spray, seawater inundation, and soil salinity. *Bulletin of the Torrey Botanical Club*, **104**, 29–34.

Barbour, M.G. and Johnson, A.F. (1979) Beach and dune, in *Terrestrial vegetation of California*, (eds. M.G. Barbour and J. Major), Wiley–Interscience, New York.

Barbour, M.G. and Robichaux, R.H. (1976) Beach phytomass along the California coast. *Bulletin of the Torrey Botanical Club*, **103**, 16–20.

Barbour, M.G., Craig, R.B., Drysdale, F.R. and Ghiselin, M.T. (1973) *Coastal Ecology: Bodega Head*, University of California Press, Berkeley, CA.

Barbour, M.G., DeJong, T.M. and Johnson, A.F. (1976) Synecology of beach vegetation long the Pacific Coast of the United States of America: a first approximation. *Journal of Biogeography*, **3**, 55–69.

Barbour, M.G., Johnson, A.F. and Holton, B. (1974–1977) Management of beach and dune vegetation, project R/CZ-22, annual reports, *University of California Sea Grant College Program, Publications* **47**, 41–2; **57**, 27–9; **61**, 41–3.

Bolton, J.K. and Brown, R.H. (1980) Photosynthesis of grass species differing in carbon dioxide fixation pathways. V. Response of *Panicum maximum*, *Panicum milioides*, and tall fescue (*Festuca arundinacea*) to nitrogen nutrition. *Plant Physiology*, **66**, 97–100.

Bowman, H.H.M. (1918) *Botanical Ecology of the Dry Tortugas*, Carnegie Institution of Washington, Publication No. 252.

Boyce, S.G. (1951a) Source of atmospheric salts. *Science*, **113**, 620–1.

Boyce, S.G. (1951b) Salt hypertrophy in succulent dune plants. *Science*, **114**, 544–5.

Boyce, S.G. (1954) The salt spray community. *Ecological Monographs*, **24**, 29–67.

Brown, R.H. (1978) A difference in nitrogen use efficiency in C$_3$ and C$_4$ plants and its implications in adaptation and evolution. *Crop Science*, **18**, 93–8.

Brown, R.L. and Hafenrichter, A.L. (1948) Factors influencing the production and use of beachgrass and dunegrass clones for erosion control: III. Influence of kinds and amounts of fertilizer on production. *Agronomy Journal*, **40**, 677–84.

Clayton, J.L. (1972) Salt spray and mineral cycling in two California coastal ecosystems. *Ecology*, **53**, 74–81.

Calder, J.A. and Taylor, R.L. (1968) *Flora of the Queen Charlotte Islands, Part I* Research Branch, Canada Department of Agriculture, Monograph No. 4, Ottawa.

Coker, W.C. (1905) Observations on the flora of the Isle of Palms, Charleston, S.C. *Torreya*, **5**, 135–45.

Colosi, J.C. and McCormick, J.F. (1978) Population structure of *Iva imbricata* in five coastal habitats. *Bulletin of the Torrey Botanical Club*, **105**, 175–86.

Cooper, W.S. (1936) The strand and dune flora of the Pacific coast of North America, in *Essays in Geobotany* (ed. T.H. Goodspeed), University of California Press, Berkeley.

Cowles, H.C. (1891) The ecological relations of the vegetation on the sand dunes of Lake Michigan. *Botanical Gazette*, **27**, 95–117.

Dahl, B.E., Fall, B.A., Lohse, A. and Appan, S.G. (1975) *Construction and stabilization of coastal foredunes, with vegetation: Padre Island, Texas*, United States Army Crops of Engineers, Coastal Engineering Research Centre, Fort Belvoir, Virginia.

Davies, J.L. (1977) *Geographical Variation in Coastal Development*, Longman, New York.

Davis, J.H., Jr (1942) The ecology of the vegetation and topography of the Sand Keys of Florida. *Papers from the Tortugas Laboratory*, **33**, 113–95.

Davis, J.H., Jr (1943) The natural features of southern Florida. *Florida Department of Conservation. Geological Survey, Geological Bulletin*, **25**, 1–311.

Day, J.M., Neves, M.C.P. and Dobereiner, J. (1975) Nitrogenase activity on the roots of tropical forage grasses. *Soil Biology and Biochemistry*, **7**, 107–12.

De Jong, T.M. (1978a) Comparative gas exchange of four California beach taxa. *Oecologia*, **34**, 343–51.

De Jong, T.M. (1978b) Comparative gas exchange and growth responses of C_3 and C_4 beach species grown at different salinities. *Oecologia*, **36**, 59–68.

De Jong, T.M. (1979) Water and salinity relations of Californian beach species. *Journal of Ecology*, **67**, 647–63.

De Jong, T.M. and Barbour, M.G. (1979) Contributions to the biology of *Atriplex leucophylla*, a C_4 Californian beach plant. *Bulletin of the Torrey Botanical Club*, **106**, 9–19.

De Wit, C.T. (1965) *Photosynthesis of leaf canopies*, Agricultural Research Report 663, Institute for Biological and Chemical Research on Field Crops and Herbage, Wageningen, pp. 1–57.

Dolan, R., Godfrey, P.J. and Odum, W.E. (1973) Man's impact on the Barrier Islands of North Carolina. *American Scientist*, **61**, 152–162.

Doliner, L.H., Jolliffe, P.A. (1979). Ecological evidence concerning the adaptive significance of the C_4 dicarboxylic acid pathway of photosynthesis. *Oecologia* (Berlin), **38**, 23–34.

Dubois, S.J. (1977) Comparative Ecophysiology of C_3 and C_4 Sand Dune Plant Species of the Georgia Coast, MS thesis, University of Georgia, Athens, GA.

Eldred, R.A. and Maun, M.A. (1982) A multivariate approach to the problem of decline in vigor of *Ammophila. Canadian Journal of Botany*, **60**, 137–80.

Epstein, E. (1969) Mineral metabolism of halophytes, in *Ecological aspects of the mineral nutrition of plants* (ed. I.H. Rorison), Blackwell, London.

Epstein, E. (1972) *Mineral nutrition of plants: principles and perspectives*, Wiley, New York.

Eshel, A., Waisel, Y. and Ramati, A. (1974) The role of sodium in stomatal movements of a halophyte: a study by X-ray microanalysis, in *Plant analysis and fertilizer problems, 7th International Colloquium*, Hanover, Germany.

Etherington, J.R. (1967) Studies of nutrient cycling and productivity in oligotrophic ecosystems. I Soil potassium and wind-blown seaspray in a South Wales dune grassland. *Journal of Ecology*, **55**, 743–52.

Fry, B., Jeng, W., Scanlan, R.S., Parker, P.L. and Baccus, J. (1978) ^{13}C food web analysis of a Texas sand dune community. *Geochimica et Cosmochimica Acta*, **42**, 1299–302.

Galtsoff, P.S. (ed.) (1954) *Gulf of Mexico: its origin, waters, and marine life*, Fishery Bulletin 89 and Fishery Bulletin of the US Fish and Wildlife Service, Vol. 55.

Garcia Novo, F. (1976) Ecophysiological aspects of the distribution of *Elymus arenarius* and *Cakile maritima* on the dunes of Tents-Muir point (Scotland). *Oecologia Plantarium*, **11**, 13–24.

Gillespie, T.S. (1976) The flowering plants of Mustang Island, Texas – an annotated checklist. *Texas Journal of Science*, **27**, 131–48.

Goldsmith, V. (1978) Coastal dunes, in *Coastal sedimentary environments* (ed. R.A. Davis, Jr), Springer-Verlag, New York.

Harshberger, J.W. (1900) An ecological study of the New Jersey strand flora. *Academy of Natural Sciences of Philadelphia, Proceedings*, **1900**, 623–71.

Harshberger, J.W. (1902) Additional observations on the strand flora of New Jersey. *Academy of Natural Sciences of Philadelphia, Proceedings*, **1902**, 642–69.

Harshberger, J.W. (1903) Notes on the strand flora of Great Inagua, Haiti, and Jamaica. *Torreya*, **3**, 67–70.

Harshberger, J.W. (1908) The comparative leaf structure of the sand dune plants of Bermuda. *Proceedings of the American Philosophical Society*, **47**, 97–110.

Harshberger, J.W. (1909) Comparative leaf structure of the strand plants of New Jersey. *American Philosophical Society, Proceedings*, **48**, 72–89.

Harshberger, J.W. (1914) *The vegetation of south Florida,*

Transactions of the Wagner Free Institute of Science, Philadelphia.

Hassouna, M.G. and Wareing, P.F. (1964) Possible role of rhizosphere bacteria in the nitrogen nutrition of *Ammophila arenaria*. *Nature*, **202**, 467– 9.

Holton, B. (1980) Some aspects of the nitrogen cycle in a northern California coastal dune-beach ecosystem, with emphasis on *Cakile maritima*, PhD dissertation, University of California, Davis, CA.

Holton, B. and Johnson, A.F. (1979) Dune scrub communities and their correlation with environmental factors at Point Reyes National Seashore, California. *Journal of Biogeography*, **6**, 317–28.

Hope-Simpson, J.F. and Jeffries, R.L. (1966) Observations relating to vigour and debility in marram grass (*Ammophila arenaria* (L.) Link). *Journal of Ecology*, **54**, 271–4.

Johnson, A.F. (1973) A survey of the strand and dune vegetation along the Pacific coast of Baja California, Mexico, MS thesis, University of California, Davis, CA.

Johnson, A.F. (1977) A survey of the strand and dune vegetation along the Pacific and southern gulf coasts of Baja California, Mexico. *Journal of Biogeography*, **7**, 83–99.

Johnson, A.F. (1978) Some aspects of the autecology of *Abronia maritima* Nutt. ex Wats, PhD dissertation, University of California, Davis, CA.

Johnson, A.F. (1982) Dune vegetation along the eastern shore of the Gulf of California. *Journal of Biogeography*, **9**, 317–30.

Johnson, D.S. and York, H.H. (1915) *The relation of plants to tidelevels*, Carnegie Institution of Washington, Publication 206.

Johnson, J.W. (1963) Ecological study of dune flora, Humboldt Bay, MS thesis, Humboldt State University, Arcata.

Judd, F.W., Lonard, R.I. and Sides, S.L. (1977) The vegetation of South Padre Island, Texas in relation to topography. *The Southwestern Naturalist*, **22**, 31–48.

Kearney, T.H. (1900) The plant covering of Ocracoke Island; a study in the ecology of the North Carolina strand vegetation. *United States National Herbarium, Contributions*, **5**, 261–319.

Kearney, T.H. (1901) Report on a botanical survey of the Dismal Swamp Region. *United States National Herbarium, Contributions*, **5**, 321–585.

Kearney, T.H. (1904) Are plants of sea beaches and dunes true halophytes? *Botanical Gazette*, **37**, 424–36.

Koller, D. (1956) Germination regulating mechanisms in some desert seeds. III. *Calligonum comosum* L'Her. *Ecology*, **37**, 430–3.

Kumler, M.L. (1969) Plants succession on the sand dunes of the Oregon coast. *Ecology*, **50**, 695–704.

Kurz, H. (1942) Florida dunes and scrub, vegetation and ecology. *Florida Department of Conservation, Geological Bulletin*, **23**, 1–154.

Lechowicz, M.J., Hellens, L.E., and Simon, J. -P. (1980) Latitudinal trends in the responses of growth respiration and maintenance respiration to temperature in the beach pea, *Lathyrus japonicus*. *Canadian Journal of Botany*, **58**, 1521–4.

Lonard, R.I., Judd, F.W. and Sides, S.L. (1978) Annotated check-list of the flowering plants of South Padre Island, Texas. *The Southwestern Naturalist*, **23**, 497–510.

Loveless, A.R. (1962) Further evidence to support a nutritional interpretation of sclerophylly. *Annals of Botany*, **26**, 551–61.

Macdonald, K.B. and Barbour, M.G. (1974) Beach and salt marsh vegetation along the Pacific coast, in *Ecology of halophytes*, (eds. R.J. Reimold and W.H. Queen), Academic Press, New York.

Marshall, J.K. (1965) *Corynephorus canescens* (L.) P. Beum. as a model for the *Ammophila* problem. *Journal of Ecology*, **53**, 447–63.

Martin, W.E. (1959) The vegetation of Island Beach State Park. *Ecological Monographs*, **21**, 1–46.

Martin, E.V. and Clements, F.E. (1939) *Adaptation and origin in the plant world. I. Factors and functions in coastal dunes*, Carnegie Institution of Washington, Publication 521.

Mayer, A.M. and Poljakoff-Mayber, A. (1975) *The germination of seeds*, Pergamon Press, McMillan company, New York.

McLaughlin, W.T. and Brown, R.L. (1942) *Controlling coastal sand dunes in the Pacific Northwest*, USDA Circular No. 660, USGPO, Washington, D.C.

Medina, E. (1970) Effect of nitrogen supply and light intensity during growth on the photosynthetic capacity and carboxydismutase activity of leaves of *Atriplex patula* ssp. *hastata*. *Carnegie Institution Yearbook*, **70**, 551–9.

Miller, G.R. (1975) Vegetation dynamics of Ship Island, Mississippi, PhD dissertation, University of Georgia, Athens, GA.

Monsi, M., Uchihima, Z. and Oikawa, T. (1973) Structure of foliage canopies and photosynthesis. *Annual Review of Ecology and Systematics*, **1**, 301–27.

Mooney, H.A. (1980) Photosynthetic plasticity of populations of *Heliotropium curassavicum* L. originating from differing thermal regimes. *Oecologia*, **45**, 372–6.

Moreno-Casasola, P. (1982) Ecologia de la vegetacion de

dunas costeras: factores fisicos. *Biotica*, **7**, 577–602.

Moreno-Casasola, P. *et al.* (1982) Ecologia de la vegetation de dunas costeras: estructura y composicion en el morro de la mancha, Ver. I. *Biotica*, **7**, 491–526.

Morris, M., Eveleigh, D.E., Riggs, S.C. and Tiffney, W.N. (1974) Nitrogen fixation in the bayberry (*Myrica pennsylvanica*) and its role in coastal succession. *American Journal of Botany*, **61**, 867–70.

Olson, J.S. (1958a) Rates of succession and soil changes on southern Lake Michigan sand dunes. *Botanical Gazette*, **119**, 125–70.

Olson, J.S. (1958b) Lake Michigan dune development. *Journal of Geology*, **56**, 254–63, 345–51, 413–83.

Olsson-Seffer, P. (1909) Hydrodynamic factors influencing plant life on sandy seashores. *New Phytologist*, **8**, 39–49.

Oosting, H.J. (1945) Tolerance to salt spray of plants of coastal dunes. *Ecology*, **26**, 85–9.

Oosting, H.J. (1954) Ecological processes of vegetation of the maritime strand in the southeastern United States. *Botanical Review*, **20**, 226–62.

Oosting, H.J. and Billings, W.D. (1942) Factors affecting vegetational zonation on coastal dunes. *Ecology*, **23**, 131–42.

Orme, A.R. (1973) *Coastal dune systems of Northwest Baja California, Mexico*. Office of Naval Research, Technical Report 0–73–1.

Parker, J. (1974) Coastal dune systems between Mad River and Little River, Humboldt County, California, MA thesis, Humboldt State University, Arcata, CA.

Patriquin, D. and Knowles, R. (1972) Nitrogen fixation in the rhizosphere of marine angiosperms. *Marine Biology*, **16**, 49–58.

Pavlik, B.M. (1982) Nutrient and productivity relations of the beach grasses, *Ammophila arenaria* and *Elymus mollis* at Point Reyes, California, PhD thesis, University of California, Davis, CA.

Pavlik, B.M. (1983a) Nutrient and productivity relations of the dune grasses *Ammophila arenaria* and *Elymus mollis*. I. Blade photosynthesis and nitrogen use efficiency in the laboratory and field. *Oecologia (Berlin)*, **57**, 227–32.

Pavlik, B.M. (1983b) Nutrient and productivity relations of the dune grasses *Ammophila arenaria* and *Elymus mollis*. II. Growth and patterns of dry matter and nitrogen allocation as influenced by nitrogen supply. *Oecologia (Berlin)*, **56**, **57**, 238–38.

Pavlik, B.M. (1983c) Nutrient and productivity relations of the dune grasses *Ammophila arenaria* and *Elymus mollis*. III. Spatial aspects of clonal expansion with reference to rhizome growth and the dispersal of buds. *Bulletin of the Torrey Botanical Club*, **110**, 271–79.

Pearcy, R.W. (1976) Temperature responses of growth and photosynthetic CO_2 exchange rates in coastal and desert races of *Atriplex lentiformis*. *Oecologia (Berlin)*, **26**, 245–55.

Penfound, W.T. and O'Neill, M.E. (1934) The vegetation of Cat Island, Mississippi. *Ecology*, **15**, 1–16.

Pessin, L.J. and Burleigh, T.D. (1941). Notes on the forest biology of Horn Island, Mississippi. *Ecology*, **22**, 70–81.

Poggie, (1963) *Coastal pioneer plants and habitat in the Tampico region, Mexico*, Louisiana State University Studies, Coastal Studies Series, No. 6.

Purer, E.A. (1936a) Growth behavior in *Convolvulus soldanella* L. *Ecology*, **17**, 541–50.

Purer, E.A. (1936b) Studies of certain coastal and sand dune plants of southern California. *Ecological Monographs*, **6**, 1–87.

Purer, E.A. (1942) Anatomy and ecology of *Ammophila arenaria* Link. *Madrono*, **6**, 167–71.

Rains, D.W. and Epstein, E. (1967) Preferential absorption of potassium by leaf tissue of the mangrove, *Avicenna marina*: an aspect of halophytic competence in coping with salt. *Australian Journal of Biological Science*, **20**, 847–57.

Ranwell, D.S. (1972) *Ecology of salt marshes and sand dunes*, Chapman and Hall, London, UK.

Raunkiaer, C. (1934) *The life forms of plants and statistical plant geography*, Clarendon Press, Oxford, UK.

Salisbury, E.J. (1952) *Downs and dunes, their plant life and environment*, Bell, London, UK.

Sauer, J. (1967) *Geographic reconnaissance of seashore vegetation along the Mexican Gulf coast*, Louisiana State University Press, Coastal Studies Series, No. 21, Baton Rouge.

Sauer, J.D. (1976) Problems and prospects of vegetational research in coastal environments. *Geoscience and Man* **14**, (June), 1–16.

Sauer, J. and Struik, G. (1974) A possible ecological relation between soil disturbance, light flash and seed germination. *Ecology*, **46**, 884–886.

Savage, R.P. and Woodhouse, W.W. Jr (1969). Creation and stabilization of coastal barrier dunes, in *Proceedings 11th Conference on Coastal Engineering, London, UK Sept. 1968.* American Society of civil Engineers, United Engineering Centre, New York.

Schimper, A.F.W. (1903) *Plant geography upon a physiological basis*, Clarendon Press, Oxford, UK.

Seneca, E.D. (1969) Germination response to temperature

and salinity of four dune grasses from the Outer Banks of North Carolina. *Ecology*, **50**, 44–53.

Seneca, E.D. (1972) Germination and seedling response of Atlantic and Gulf coasts populations of *Uniola paniculata*. *American Journal of Botany*, **59**, 290–6.

Seneca, E.D. and Cooper, A.W. (1971) Germination and seedling response to temperature, daylength, and salinity by *Ammophila breviligulata* from Michigan and North Carolina. *Botanical Gazette*, **132**, 203–15.

Silander, J.A. (1979) Microevolution and clone structure in *Spartina patens*. *Science*, **203**, 658–60.

Simon, J. -P. (1979) Adaptation and acclimation of higher plants at the enzyme level: Latitudinal variations of thermal properties of NAD malate dehydrogenase in *Lathyrus japonicus* Willd. (Leguminosae). *Oecologia (Berlin)*, **39**, 273–87.

Slobodchikoff, C.N. and Doyen, J.T. (1977) Effects of *Ammophila arenaria* on sand dune arthropod communities. *Ecology*, **58**, 1171–5.

Snow, L.M. (1902) Some notes on the ecology of the Delaware coast. *Botanical Gazette*, **34**, 284–306.

Stalter, R. (1974) Vegetation in coastal dunes of South Carolina. *Castanea*, **39**, 95–103.

Stewart, G.R., Larher, F., Ahmad, I. and Lee, J.A. (1979) Nitrogen metabolism and salt-tolerance in higher plant halophytes, in *Ecological processes in coastal environments* (eds R.L. Jeffries and A.J. Davy), Blackwell, London, UK.

Storey, R., Ahmad, N. and Wyn Jones, R.G. (1977) Taxonomic and ecological aspects of the distribution of glycinebetaine and related compounds in plants. *Oecologia (Berlin)*, **27**, 319–32.

Teeri, J.A. and Stowe, L.G. (1976) Climatic patterns and the distribution of C_4 grasses in North America. *Oecologia (Berlin)*, **23**, 1–12.

US Army Corps of Engineers (1973) *National shoreline study*, Vols, 1–5 USGPO, Washington, D.C.

van der Valk, A.G. (1947a) Environmental factors controlling the distribution of forbs on foredunes in Cape Hatteras National Seashore. *Canadian Journal of Botany*, **52**, 1057–73.

van der Valk, A.G. (1974)a Environmental factors con-foredune plant communities in Cape Hatteras National Seashore. *Ecology*, **55**, 1349–58.

van der Valk, A.G. (1977a) The macroclimate and microclimate of coastal foredune grasslands in Cape Hatteras National Seashore. *International Journal of Biometeorology*, **21**, 227–37.

van der Valk, A.G. (1977b) The role of leaves in the uptake of nutrients by *Uniola paniculata* and *Ammophila breviligulata*. *Chesapeake Science*, **18**, 77–9.

Veldkamp, J.F. (1971) Enige opmerkingen over de aanpassing van zeeraket (*Cakile maritima* Scop.) aan het strand. *Gorteria*, **5**, 227–31. (English summary.)

Wagner, R.H. (1964) The Ecology of *Uniola paniculata* in the dune strand habitat of North Carolina. *Ecological Monograph*, **34**, 79–96.

Waisel, Y. (1972) *Biology of halophytes*, Academic Press, New York.

Warming, J.E.B. (1909) *Oecology of plants*, Oxford University Press, London, UK.

Weast, R.C. (ed.) (1977) *Handbook of chemistry and physics*, 57th edn, Chemical Rubber Company Press, Chicago.

Wells, B.W. (1939) A new forest climax: the salt spray climax of Smith Island, North Carolina. *Bulletin of the Torrey Botanical Club*, **66**, 629–34.

Wesson, G. and Wareing, P.F. (1969) The induction of light sensitivity in weed seeds by burial. *Journal of Experimental Botany*, **20**, 414–25.

Willis, A.J. (1965) The influence of mineral nutrients on the growth of *Ammophila arenaria*. *Journal of Ecology*, **53**, 735–45.

Willis, A.J. and Yemm, E.W. (1961) Braunton Burrows: mineral nutrient status of the dune soils. *Journal of Ecology*, **49**, 377–90.

Wilson, A.T. (1959) Surface of the ocean as a source of airborne nitrogenous material and other plant nutrients. *Nature*, **184**, 99–101.

Woodcock, A.H. (1953) Salt nuclei in marine air as a function of altitude and wind force. *Journal of Meteorology*, **10**, 362–71.

Woodhouse, W.W. and Hanes, R.E. (1966) *Dune stabilization with vegetation on the Outer Banks of North Carolina*, Soils Infertility Series 8, North Carolina State University, Raleigh, N.C.

Woodhouse, W.W., Seneca, E.D., Jr and Cooper, A.W. (1968) Use of sea oats for dune stabilization in the southeast. *Shore and Beach*, **36(2)**, 15–21.

Wooley, J.T. and Stoller, E.W. (1978) Light penetration and light-induced seed germination in soil. *Plant Physiology*, **61**, 597–600.

Zuberer, D.A. and Silver, W.S. (1975) Mangrove associated nitrogen fixation, in *Proceedings of the First International Symposium on Biology and Management of Mangroves*, US Government Printing Office, Washington.

15

*Coastal marshes**

B.L. HAINES AND E.L. DUNN

15.1 Introduction

For North American coastal marshes we review habitat characteristics, geographic distribution, microtopographic distribution, the environmental stresses of flooding, salinity, low redox potential, and finally limitations of nitrogen and phosphorus. Physiological and morphological responses and adaptations to these stresses are described. Coastal marsh ecosystem properties of primary production and nutrient cycling as affected by plant ecophysiology are explored. The implications of plant ecophysiology for the management of oil spills, heavy metals, biocides, municipal wastes, and dredging are treated. Finally, we offer a list of general conclusions, a conceptual model of processes controlling production in a southeastern *Spartina alterniflora* dominated salt marsh, and a list of future research needs especially regarding belowground processes.

15.1.1 HABITAT CHARACTERISTICS

Plants in salt marshes are exposed to unique combinations of environmental variables. The mixing of fresh and sea-waters creates temporal and spatial salinity gradients. The rise and fall of tides produce anaerobic soils. Changes in tidal inundation time with elevation above mean sea-level and evaporation from the marsh surface result in salt accumulation and soil water potentials more negative than sea water. In northern marshes

winter ice may shear off plant stems. Rivers bring new sediments to the marsh while tidal movements and ocean currents constantly redistribute the sediments in which the plants are rooted. Toxic wastes including heavy metals, herbicides, and pesticides are carried by rivers through marshes where they might accumulate. Petroleum exploration and recovery, the loading and unloading of oil tankers, tanker wrecks, and refineries may occur in or near marshes and release petroleum into them. Dredging to maintain navigable waterways, and marsh filling for housing, industry, and diking for agriculture, mariculture, or for salt works can change the aereal extent of the habitat.

15.1.2 GEOGRAPHIC DISTRIBUTION

Marsh development depends on the interaction of coastal physiography, tidal amplitude, tidal energy, sediment availability and dispersal ability, salt-tolerance and flood-tolerance of plants. Coastal salt marshes are associated with off-shore barrier islands, with estuaries or with shelter provided by spits (Chapman, 1974). Salt marshes occur (Fig. 15.1) at many points from Barrow Alaska to the tip of Baja California on the Pacific Coast (MacDonald and Barbour, 1974). Salt marshes occur almost continuously along the Gulf and Atlantic coasts except in parts of Florida where they are replaced by mangroves (Cooper, 1974). Salt marshes also occur along the Arctic Ocean.

* Contribution No. 452 from the University of Georgia Marine Institute, Sapelo Island Ga.

Figure 15.1 Approximate distribution of North American coastal salt marshes (adapted from Chapman, 1977; Kurz and Wagner, 1957; MacDonald and Barbour, 1974; Miller and Egler, 1950).

The most extensive coastal marshes occur in the southeastern United States especially in Georgia and South Carolina where a unique combination of high tidal amplitude, low tidal energy, and a coastline protected behind barrier islands has fostered their development. Along the Gulf Coast with a lower tidal amplitude, similar vegetation patterns and extensive marsh development occur over a smaller elevational gradient. Pacific coastal marshes are less extensive because of a steeper shoreline, higher sea energy, fewer protected areas, and heavy impact by humans. Descriptions and classifications of North American salt marshes are given by Chapman (1977) and by Frey and Basan (1978).

Atlantic and Gulf coastal marshes are floristically

rich with 347 vascular plant species in 177 genera in 75 families (Duncan, 1974). Pacific coast marshes are less rich with 78 vascular plant species between Barrow, Alaska, and Cabo San Lucas, Baja California (MacDonald and Barbour, 1974). Only 28 vascular plant species are reported from a Hudson Bay salt marsh (Kershaw, 1976). Atlantic and Gulf Coast marshes are dominated by few species, mostly *Spartina alterniflora* Loisel., *Spartina patens* (Ait.) Muhl., *Salicornia* sp., *Iva* sp., and *Juncus* sp. Pacific coast salt marshes of California are dominated by *Spartina foliosa* Trin., *Salicornia virginica* L., and *Distichlis spicata* (L.) Greene (MacDonald and Barbour, 1974). *Puccinellia phryganodes* (Trin.) Scrib. and Merr. and *Carex subspathacea* Wormsk dominate in the Hudson Bay (Kershaw, 1976). Most studies of physiological ecology have focused on one or more of these dominant species.

15.1.3 LOCAL ZONATION – MICROTOPOGRAPHIC PATTERNS

North American salt marshes have long captured the interest of ecologists because of the distinct zonation of species in relation to presumed environmental gradients and the high plant productivity. For examples of these species zonation patterns see MacDonald and Barbour (1974) for the Pacific Coast, Kurz and Wagner (1957) and Eleuterius and Eleuterius (1979) for the Gulf Cost, Kurz and Wagner (1957), Adams (1963), and Niering and Warren (1980) for the Atlantic Coast, Smith *et al.* (1980) for Nova Scotia; and Ringius (1980) and Kershaw (1976) for the Hudson Bay region. Species zonation for the same marsh in relation to selected environmental parameters is shown in Fig. 15.2. For individual species, gradients in height also occur. For example in Georgia, *Borrichia frutescens* (L.) D.C. and *Batis maritima* L. plants may change 40 cm in height over a distance of 1 m or less. *Spartina alterniflora* grows tallest (> 2 m) on tidal creek banks and gradually decreases in stature with distance and elevation toward the land to a short form < 0.3 m. Since height and standing crop are positively related in *Spartina alterniflora* (Broome *et al.*, 1975a; Nixon and Oviatt, 1973; Williams and

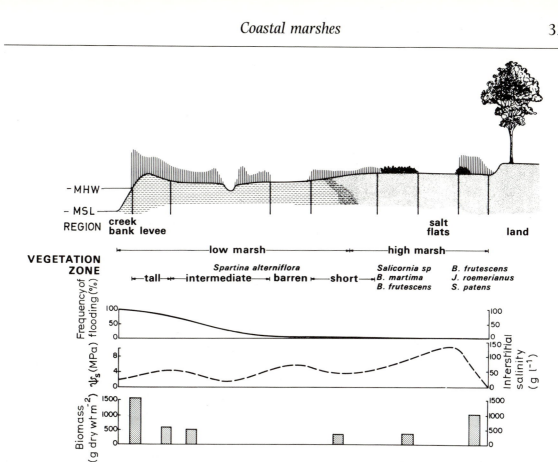

Figure 15.2 Schematic representation of typical zonation and standing biomass (g dry wt m⁻²) of plants in a Georgia salt marsh in relation to tidal inundation (MHW, mean high water, MSL, mean sea level), water potential Ψ_s ($-$MPa), and interstitial salinity (gl⁻¹). (Adapted from Antlfinger and Dunn, 1979; Letzsch and Frey, 1980; Nestler, 1977; Pomeroy and Wiegert, 1981; Reimold *et al.*, 1973.)

Murdoch, 1969), the widespread existence of height gradients in *Spartina alterniflora* suggests the general occurrence of productivity gradients in coastal marsh (see Turner, 1976). With the highly predictable tidal rhythm imposing environmental gradients and the strong gradients of size forms within species and zonation among species, coastal salt marshes provide opportunities to study the physiological and genetic basis of plant adaptation.

15.2 Environmental stress

The general patterns of environmental stresses that change along gradients in salt marshes are related to the dominant influence of tidal flooding by estuarine water and include the frequency and duration of flooding, salinity changes, and effects of flooding on soil chemistry.

15.2.1 FLOODING

Frequency and duration of flooding depend on the elevation at which a plant is growing and the tidal amplitude in a specific location (see Fig. 15.2). The range of flooding conditions to which vascular plants have adapted vary from twice daily flooding with up to 17h partial daily submergence for *Spartina alterniflora* at its lower limit (Johnson and York, 1915) to flooding only by spring tides each month followed by evaporation and salt concentration or precipitation and land runoff and salt dilution (Gardner, 1973; Kurz and Wagner, 1957). Some areas at the extreme upper edge of the marsh are only flooded by the highest annual tides or by unpredictable storm-related events.

15.2.2 SALINITY

Associated with the pattern of tidal flooding are great differences in soil and interstitial water salinities. There are great differences among marsh systems in the salinity gradients that develop. In the regularly flooded salt marshes along the Atlantic Coast of the Southeastern US, strong interstitial salinity gradients develop (Fig. 15.2), with values ranging from estuarine concentrations of $20-25 \, gl^{-1}$ along the creekbanks to $35-40 \, gl^{-1}$ in the upper regions of the high marsh where the short height form of *Spartina alterniflora* occurs (Antlfinger and Dunn, 1979; Gardner, 1973; Nestler, 1977). In the high marsh areas flooded only by spring tides, salt flats or salt pans develop, with interstitial salinities exceeding $100 \, gl^{-1}$ (Antlfinger and Dunn, 1979; Kurz and Wagner, 1957). In some New England and Gulf Coast marshes, however, species distribution patterns and gradations in the height of *Spartina alterniflora* occur without gradients in salinity (Howes *et al.*, 1981; Mendelssohn *et al.*, 1981; Niering and Warren, 1980).

15.2.3 SEDIMENT CHEMISTRY

Probably the most important effects of flooding on vascular halophytes are in the soil chemical environment. Several sediment factors are associated with flooding, including oxygen availability, redox potential (Eh), and solubility of various elements both beneficial and toxic (Armstrong, 1975; Delaune *et al.*, 1976; Patrick and DeLaune, 1977; Ponnamperuma, 1972). Redox potentials vary from > $+400 \, mV$ in the upper few centimetres of frequently flooded and drained areas to < $-200 \, mV$ at depths of 30–50 cm and in areas with little or no water movement (Howes *et al.*, 1981; Teal and Kanwisher, 1961). Related to redox potential as a measure of the degree of anaerobiosis, the basic sequence of chemical changes that occur are free oxygen disappearance from $+600-500 \, mV$,

nitrate disappearance over the same range, manganese reduction and increased solubility from $+600-400 \, mV$, iron reduction and increased solubility from $+350-180 \, mV$, and sulfate reduction from 0 to $-190 \, mV$ (Patrick and DeLaune, 1977; Phung and Fiskell, 1972). There are also interactions among the different ionic forms since these processes overlap along a redox potential gradient. With sulfate reduction and the release of free sulfide, insoluble precipitates of FeS form, thus lowering the concentration of soluble iron and preventing high concentrations of free sulfide (King *et al.*, 1982; Ponnamperuma, 1972). The formation of FeS also releases phosphate from insoluble ferric phosphate (Gooch, 1968).

The dominant form of inorganic nitrogen is NH_4 under typical redox conditions in the marsh. Higher NH_4 concentrations are often found in the high marsh, short height zones of *Spartina alterniflora* than in the low marsh, tall height zones (Chalmers *et al.*, 1976; Mendelssohn, 1979). Nitrate may occur where there are significant inputs from land runoff or groundwater seepage (Valiela and Teal, 1979). Small concentrations of NO_3 and NO_2 are also found in marsh sediments at redox potentials below which these ions should occur, indicating microzones of oxidation by O_2 diffusing out of roots and rhizomes (Chalmers *et al.*, 1976; Howes *et al.*, 1981).

15.3 Plant responses and adaptations

15.3.1 INTRA- AND INTERSPECIFIC CHANGE IN PLANTS ALONG ENVIRONMENTAL GRADIENTS

Gradients in plant size and in plant species composition have been studied in many North American marshes. In Georgia (Fig. 15.2) *Spartina alterniflora* plants growing on tidal creek banks are taller, greener and have wider leaves than plants growing at higher elevations in the marsh. In the intermediate marsh *Salicornia* spp. also show gradients in plant size. In the higher marsh where *Borrichia frutescens* and *Batis maritima* border on the barren salt flats or pans, the plants are stunted to a few cm in height at the edge of the salt pan but increase in stature to 50 cm taller inside the stand. These patterns of plant change along with knowledge of gradients in tidal inundation, substrate salinity, and substrate redox potential have resulted in studies to determine cause and effect relationships. Three questions have been raised. Within species, do size differences reflect ecotypic differentiation or phenotypic plasticity? Are there phenological ecotypes? Do species segregate across the marsh because of difference in tolerance to some combination of environmental variables?

(a) *Ecotypes versus ecophenes*

Following a transplant study, Stalter and Batson (1969) suggested that *Spartina alterniflora* might occur in two varieties, a tall form and a short form with the short form being more salt and flood tolerant than the tall form. A series of genetic and physiological studies were conducted by others to explore this ecotype hypothesis.

Seedling growth was found to be independent of height form of parent plants, thus Mooring *et al.* (1971) concluded that height forms are best described as ecophenes. The phenotypic plasticity of *Spartina alterniflora* in response to artificial gradients in NaCl and in NH_4 in solution culture (Haines and Dunn, 1976) supports the ecophene interpretation. The most definitive study is that of Shea *et al.* (1975) in which tall and short forms were investigated for enzyme polymorphism in the laboratory and were reciprocally transplanted in the field. Gel electrophoretic banding patterns of total soluble protein, peroxidase, and the dehydrogenases of malic, lactic and isocitric acid showed some seasonal changes in pattern but were the same for both tall and short forms of *Spartina alterniflora*. After three growing seasons, reciprocally transplanted *Spartina alterniflora* were '... virtually indistinguishable from the surrounding, undisturbed populations'. In another study of protein polymorphism, Valiela *et al.* (1978) compared the electrophoretic bands of α, β esterase, leucine aminopeptidase, alkaline phosphatase, acid phosphatase, malate dehydrogenase, and tetrazolium oxidase in tall and short plants of *Spartina alterniflora* but found no differences. The ecotype–ecophene controversy is reviewed by Anderson and Treshow (1980). Most of the evidence supports the ecophene rather than the ecotype interpretation of the tall and short forms of *Spartina alterniflora*.

Spartina patens has a broad ecological amplitude, occurring along the Atlantic coast from 50°N in Canada to 16°N lat. in Central America (Silander and Antonovics, 1979). South of about 25°N lat. it is found on beaches and low coastal dunes while north of about 43° lat. it is restricted to salt marshes. At the center of its geographic distribution in North Carolina it is found in marsh, swale and dune habitats. Silander (1979) used isozyme analysis to investigate the genetic basis of this ecological amplitude. Plants were collected along a transect from marsh, through swale to dune and grown in a greenhouse for 24 mon before electrophoretic separations. Of the 346 plants surviving transplantation to the greenhouse he identified 101 genotypes with only three common between dune and swale populations, only two between dune and marsh and only six between swale and marsh. Dune populations were comprised of a small number of common genotypes and had high allocation to sexual reproduction. Swale and marsh populations were composed of large numbers of rare genotypes with a smaller allocation to sexual reproduction (Silander, 1979). In further experimental studies dune plants were found to have higher salt tolerance and drought stress tolerance than swale or marsh populations but were less responsive to nutrient supplementation. Further, in both common garden and in field studies, flowering times were found to differ among the populations thus restricting gene flow (Silander and Antonovics, 1979). Thus *Spartina patens*, in contrast to *Spartina alterniflora*, clearly exhibited ecotypic differentiation in response to local environmental gradients.

Microscale genetic differentiation in *Borrichia frutescens* (Compositae) with respect to salinity occurs in a Georgia salt marsh (Antlfinger, 1979). These high marsh salt flats or pans (Fig. 15.2) are dotted with mixed species patches of one to many metres in diameter comprised of *Borrichia frutescens, Salicornia virginica* L., *Salicornica bigelovii* Torr., *Distichlis spicata* and *Juncus roemerianus* Scheele. Plants at the centers of the patches are taller than those at the periphery. The salinity can vary from $15 \, gl^{-1}$ at the center of the path to $45 \, gl^{-1}$ at the edge of the patch. Individuals of *Borrichia frutescens* at the edges of patches are $\frac{1}{4}$ as tall as those in the center, have smaller leaves, more negative xylem water potentials (-3.8 compared to -2.7 MPa) and more negative leaf solute potentials (-3.3 compared to -3.0 MPa). Differences in xylem water potential and leaf solute potential were shown to be hereditable (Antlfinger, 1979).

In a comparative study of morphology, height, color, flowering time and length of growing period of 12 *Spartina alterniflora* populations from Atlantic and Gulf coasts, Seneca (1974a) concluded that intraspecific variations result from some combination of genetic differences in relation to north–south latitudinal gradients in thermoperiods and photoperiods and from differences in preconditioning of seeds.

The effect of seed source of *Spartina alterniflora* upon flowering phenology was also studied by Somers and Grant (1981) who harvested seeds from nine marshes from Maine to Virginia and grew them in a Delaware marsh. Flowering occurred progressively later from northern to southern populations. The flowering times of Maine and Virginia populations did not overlap, thus they could not cross pollinate. They found Delaware populations to be mostly outcrossing. These studies of seed germination and of flowering time also indicate ecotypic variation within *Spartina alterniflora* in response to latitudinal gradients in photoperiod. Related studies of genetic structure in salt marsh macrophyte populations in Britain are reviewed by Gray *et al.* (1979).

(b) Ecotones

A complete understanding of environment and plant attributes causing species zonation within a given marsh has yet to be achieved. The characteristic patterns and species distributions in relation to environmental factors in the coastal salt marshes have been described in many previous studies and reviewed by Ranwell (1972), Chapman (1974), Cooper (1974), and Niering and Warren (1980). For example, species distribution patterns have been reported in New England salt marshes (Niering and Warren, 1980), and in southeastern salt marshes in

North Carolina (Adams, 1963; Reed, 1947), South Carolina (Kurz and Wagner, 1957; Stalter, 1973), Georgia (Gallagher and Reimold, 1974), Florida (Jackson, 1952; Kurz and Wagner, 1957), Louisiana (Penfound and Hathaway, 1938) and Mississippi (Grabriel and de la Cruz, 1974).

In the Eastern intertidal marshes, the major environmental factors suggested to be determining these distribution patterns were elevation or inundation frequency (Chapman, 1974; Hinde, 1954; Jackson, 1952; Kurz and Wagner, 1957), salinity (Adams, 1963; Kurz and Wagner, 1957; Miller and Egler, 1950; Reed, 1947), nutrients (Adams, 1963; Pigott, 1969), soil drainage and aeration (Howes *et al.*, 1981; King *et al.*, 1982; Mendelssohn *et al.*, 1981; Purer, 1942; Reed, 1947).

In a Western intertidal marsh, Mahall and Park (1976) investigating plant standing stock across an ecotone between *Spartina foliosa* and *Salicornia virginica*, concluded that the physical environment and physiological properties of the species, rather than interspecific competition, were determinants of plant performance. The higher salinity in the higher elevations of the marsh decreased the water use efficiency of *Spartina foliosa* but increased the water use efficiency of *Salicornia virginica*. Thus, higher salinity excluded *Spartina foliosa* from the upper elevations in the marsh. Rates of diffusion of oxygen into the soil did not vary along transects through the ecotone. Flooding experiments showed that tidal immersion had little effect on *Spartina foliosa*, but could decrease the seaward advance of *Salicornia virginica*.

15.3.2 PHYSIOLOGICAL AND MORPHOLOGICAL RESPONSES AND ADAPTATIONS

(a) Gas exchange responses

Early estimates of primary production in the C_4 grass *Spartina alterniflora* in a Georgia marsh indicated high rates of gross photosynthesis and high rates of respiration equalling 77% of gross photosynthesis (Teal, 1962). Interest in the high productivity of salt marshes generated by Teal's work and the discovery of C_4 photosynthesis led to several comparative studies of photosynthesis and respiration of C_4 and

Table 15.1 Summary of typical growing season gas exchange rates of salt marsh species measured in the field under natural environmental conditions compared with herbaceous land plants

Species	Location	Net photosynthesis — Maximum rates per leaf area ($\mu mol\, m^{-2}s^{-1}$)	Net photosynthesis — Daily rates per land area ($g\, C\, m^{-2}\, d^{-1}$)	Night time respiration* — Daily photosynthesis	Transpiration ($\mu mol\, H_2O\, m^{-2}\, s^{-1}$)	Reference
Salt marsh plants						
S. alterniflora	NY		1.4–3.2	0.51		Houghton and Woodwell (1980)
S. alterniflora tall	NC		4.4	0.17–0.61(0.43)		Blum et al. (1978)
S. alterniflora short	NC		2.5	0.01–1.00(0.32)		Blum et al. (1978)
S. alterniflora tall	GA	8–17	21–44	0.07–0.13(0.11)	1900–2600	Giurgevich and Dunn (1979, 1982)
S. alterniflora short	GA	4–11	4–8	0.08–0.56(0.26)	1200–1700	Giurgevich and Dunn (1979, 1982)
S. alterniflora	LA	14–18	3.5–3.6	0.24–0.36		Gosselink et al. (1977)
S. patents/Distichlis spicata	MD	4.6	7.2	0.3		Drake and Reed (1981)
J. roemerianus	NC		1.70	0.32–0.61(0.48)		Blum et al. (1978)
J. roemerianus	GA	9–15	9–12	0.07–0.23(0.14)	2500–4300	Giurgevich and Dunn (1982, 1978)
Land plants – phanaerogams, herbaceous						
mixed agriculture		4.9–5.4				Houghton and Woodwell (1980)
C_4		19–50				Larcher (1980)
C_3 crop plants		13–50				
C_3 heliophytes		13–25				

*Night time respiration as a fraction of daily photosynthesis, range (mean).

C_3 marsh species in relation to environmental variables. Gas exchange measurements for salt marsh halophytes have ranged from whole ecosystem responses using aerodynamic approaches to individual plant responses using small cuvettes. Results from both field and laboratory measurements of gas exchange responses of salt marsh halophytes have shown rates of photosynthesis, dark respiration and transpiration similar to species from many other environments (Table 15.1).

Measurements of photosynthesis and respiration at the community and ecosystem levels have been made in North American salt marshes in an effort to detemine rates, efficiencies and environmental factors affecting net ecosystem productivity (see (b) below). Houghton and Woodwell (1980) used the aerodynamic flux technique to measure CO_2 exchange rates over Flax Pond, a 57 ha salt marsh on Long Island, NY dominated by *S. alterniflora* (Table 15.1). Over the summer growing season, daytime uptake was $618\, g\, C\, m^{-2}$ and nighttime loss $342\, g\, C\, m^{-2}$ or 55% of daytime fixation. On an annual basis, uptake was $814\, g\, C\, m^{-2}$, loss was $512\, g\, C\, m^{-2}$ (63%) resulting in a net fixation of $302\, g\, C\, m^{-2}\, yr^{-1}$. Even though photosynthesis in *Spartina alterniflora* dominated carbon balance most of the year, photosynthesis by intertidal algae maintained net CO_2 uptake by the marsh during the winter and early spring (Brinkhuis et al., 1976; Houghton and Woodwell, 1980).

Drake and Read (1981) used $1\, m^2$ chambers to enclose two brackish marsh communities in Maryland, one composed of the C_4 species, *Spartina patens* and *Distichlis spicata*, and the other dominated by a C_3 species, *Scirpus olneyi* Gray, but including the two C_4 species. They followed diurnal CO_2 exchange rates for 125 days over a period of six years (Table 15.1). Over the growing season, CO_2 assimilation was $744\, g\, C\, m^{-2}$ and respiratory loss by the community, $432\, g\, C\, m^{-2}$ or 740 g dry weight. The efficiency of community assimilation was 1.9% of photosynthetically active radiation (PAR) or 0.9% of total solar radiation and that of net ecosystem production was 0.8% of PAR or 0.4% of incident solar radiation. Seasonal changes in canopy structure from an erect – to horizontal – leaved display was partially responsible for a decline in photosynthetic efficiency in the fall (Turitzin and Drake, 1981).

Blum et al. (1978) in North Carolina, and Gosselink et al. (1977) in Louisiana, used smaller chambers covering individual plants as well as an area of marsh surface. CO_2 exchange rates included both plant and sediment responses and efforts were made to separate the two components (Table 15.1). Total ecosystem respiratory losses including belowground biomass and soil components ranged from 65 to 84% of photosynthesis and averaged 70% for the entire marsh. Maximum efficiencies of photosynthesis per unit of solar radiation were 0.7% and 1.1% for *Juncus roemerianus* and *Spartina alterniflora* (Blum *et al.*, 1978).

Giurgevich and Dunn (1982) measured both CO_2 and water-vapor exchange from leaves or parts of shoots enclosed in cuvettes without interference from sediment surfaces (Table 15.1). The tall form of *Spartina alterniflora* had higher rates of photosynthesis, higher water use efficiencies, higher leaf area indices, and lower respiration losses, resulting in much higher values of net primary production. The aereal extent of the tall form of *Spartina alterniflora* in Georgia marshes is only about 10% of that of the short form, with the physiological characteristics of the short form of *Spartina alterniflora* determining the production of the marsh. The response pattern of *Juncus roemerianus* differed from *Spartina alterniflora*. A large reasonably constant leaf area index combined with moderate rates of photosynthesis and respiration in *Juncus roemerianus* resulted in net primary productivity rates between those of the tall and short height forms of *Spartina alterniflora*. The transpiration rate of *Juncus roemerianus* was higher than *Spartina alterniflora* resulting in the lower values of water use efficiency of photosynthesis. Since water loss from salt marsh plants growing in sediments with high interstitial salinities requires some mechanism to cope with salt in, or excluded from, the transpirational stream, the differences in water use efficiencies of photosynthesis of *Spartina alterniflora* and *Juncus roemerianus* reflect one of the major adaptive advantages of C_4 photosynthesis in the salt marsh (Giurgevich and Dunn, 1982).

In order to explain these patterns of net primary production, several field and laboratory studies of gas exchange responses of salt marsh species have attempted to evaluate the effects of the major environmental factors in controlling rates of photosynthesis and transpiration.

The *in situ* temperature and light dependence of photosynthesis of *Juncus roemerianus* and *Spartina alterniflora* were compared in a Georgia salt marsh (Giurgevich and Dunn, 1978, 1979, 1981). *Juncus roemerianus* had higher photosynthetic rates in early

spring and fall than in the summer with a temperature optimum of photosynthesis at moderate temperatures (20 to 25 °C). Leaf conductances to CO_2 and H_2O vapor were higher at cooler temperatures, resulting in a higher water use efficiency of photosynthesis below 25 °C, and a sharp decline in water use efficiency at increasing temperature. The light response of photosynthesis in *Juncus roemerianus* showed light saturation only at very high photosynthetically active radiation (PAR). In contrast, the photosynthetic responses in *Spartina alterniflora* showed greater differences between the tall and the short height forms within the same species than between the C_3 and C_4 species (Giurgevich and Dunn, 1978, 1979). The temperature optima in both height forms of *Spartina alterniflora* were between 30 and 35 °C, but plants of *Spartina alterniflora* maintained net positive photosynthesis all year long, even during the cool winter. Photosynthesis in tall forms of *Spartina alterniflora* did not show light saturation even at full sunlight while plants of the short height form showed light saturation during most seasons. Leaf conductances to CO_2 and water vapor also decreased with increasing temperature. Resulting values of water use efficiency were much higher in the tall form of *Spartina alterniflora*, even at high temperatures, than in either short *Spartina alterniflora* or in *Juncus roemerianus*. The water use efficiency in short *Spartina alterniflora* was similar to that in *Juncus roemerianus*.

The results of an analysis of stomatal versus internal metabolic resistance to CO_2 uptake showed that metabolic limitations to photosynthesis were more important than stomatal limitations in *Juncus roemerianus* and in the short form of *Spartina alterniflora* (Giurgevich and Dunn, 1978, 1979). Photosynthesis in the tall form of *Spartina alterniflora* was capable of inherently higher rates and was more responsive to changes in stomatal conductance than in the short form. A similar analysis showed that photosynthesis in *Spartina cynosuroides*, a C_4 species from a lower salinity marsh habitat, was also controlled more by stomatal activity than by internal (metabolic) resistances to CO_2 uptake (Giurgevich and Dunn, 1981).

Field measurements of CO_2 exchange in *Spartina alterniflora* and *Juncus roemerianus* in North Carolina

(Blum *et al.*, 1978) and in *Spartina alterniflora* in Louisiana (Gosselink *et al.*, 1977) showed qualitatively similar responses of photosynthesis to light and temperature (Table 15.1), but precise quantitative comparisons are difficult due to differences in chamber design and experimental tenchniques. Similar responses were observed in *Spartina townsendii* H. and J. Groves (*sensu lato*) even in the cool temperate marshes of England (Long and Incoll, 1979).

Laboratory measurements of photosynthesis of several marsh species have also attempted to partition the effects of various environmental factors. The effect of salinity and growth illumination on photosynthesis in *Spartina alterniflora* was evaluated by Longstreth and Strain (1977) in phytotron grown individuals. Plants grown at low illumination showed decreased stomatal conductance and net photosynthesis with increasing salinity. Plants grown at high illumination showed little difference in photosynthesis and stomatal conductance with respect to increasing salinity. Significant increases in specific leaf weight were observed in response to increasing salinity, which may explain some of the differences in photosynthetic rates observed in this study.

An extensive series of gas exchange measurements on salt-free, laboratory-grown individuals of *Spartina townsendii* showed typical C_4 photosynthetic responses to light intensity and CO_2 concentration, but lower than average temperature optima and less detrimental effects of low temperature on photosynthesis in this cool temperate species (Dunn *et al.*, 1981; Long and Incoll, 1979; Long and Woolhouse 1978a,b). Internal resistance to CO_2 uptake was low and comparable to stomatal resistance in these salt free plants. In *Spartina anglica* C.E. Hubbard, net photosynthesis was reduced greatly at high light and high temperature when plants were grown in salt-free compared to estuarine salinities (Mallot *et al.*, 1975).

There is much less information on the field photosynthetic responses of other less dominant salt marsh species. DeJong *et al.* (1982) compared the photosynthetic responses of two C_4 species, *Spartina patens* and *Distichlis spicata*, with those of *Scirpus olneyi*, a C_3 species, from the brakish marshes of the Chesapeake Bay. Under field conditions, all three

species had more similar rates of photosynthesis and light saturation values than were observed in greenhouse-grown individuals. The C_4 species did have higher water use efficiencies of photosynthesis and higher internal compared to stomatal conductances than the C_3 species.

Antlfinger and Dunn (1979) measured the gas exchange responses of three salt marsh succulents, *Borrichia frutescens*, *Batis maritima*, and *Salicornia virginica*, under natural field conditions in a Georgia marsh. The succulents were photosynthetically active all year long and exhibited typical C_3 photosynthetic responses with no evidence of CAM metabolism. The water use efficiency values in these plants were lower than in *Juncus roemerianus* and *Spartina alterniflora*, even though the succulent species occupy a higher salinity habitat in the marsh. The CO_2 compensation point of photosynthesis in these succulents was exceptionally high and may partially explain the low water use efficiency values.

The short-term effects of increasing salinity on photosynthesis in greenhouse-grown individuals from marshes in southern California showed little or no reduction in photosynthesis in *Limonium californicum* (Boiss.) Heller when flooded with sea-water (Woodell and Mooney, 1970) or in *Batis maritima* and *Salicornia europea* L. subjected to Ψ_s of -2.6 MPa (Kuramoto and Brest, 1979). Photosynthesis was reduced in *Spartina foliosa* at Ψ_s of -2.6 MPa and in *Distichlis spicata*, severe reduction in net photosynthesis occurred at osmotic potentials below -1.6 MPa (Kuramoto and Brest, 1979).

Early work on the effects of salinity on glycophytes showed an increased respiration or 'salt respiration' as salt uptake or salinity increased (reviewed by Rains, 1972). Subsequent work on whole plant responses in salt marsh halophytes has not shown exceptionally high or variable dark respiration rates in any one species in either field measurements under natural conditions (Antlfinger and Dunn, 1979; Giurgevich and Dunn, 1978, 1979, 1982) or in the laboratory under manipulated salinity conditions (Kuramoto and Brest, 1979). However, when comparing the respiration rates of species along a natural gradient of salinity and other environmental factors in a Georgia marsh, dark respiration rates were higher and were a greater fraction of daily photosynthetic rates in the species at the higher end of the salinity gradient (Pomeroy *et al.*, 1981). Estimates of photorespiration or respiratory release in the light in the same species along this gradient showed dramatic differences. The highest estimates of photorespiration occurred in the succulents inhabiting the areas of highest salinity (Pomeroy *et al.*, 1981).

In summary, photosynthesis in *Spartina alterniflora* is reduced by cooler temperatures of winter and spring, even though this species extends to > 45° N latitude (Hatcher and Mann, 1975). Under some conditions of flooding, nutrient supply, and sediment characteristics, *Spartina alterniflora* is capable of utilizing full sunlight intensities and photosynthesis is limited by solar radiation and canopy characteristics. In the high marsh, photosynthesis in the short height form of *Spartina alterniflora* is limited by some combination of flooding and sediment characteristics. In contrast photosynthesis in *Juncus roemerianus* is reduced by high summer temperatures and excessive water loss rate. This species is also able to utilize full sunlight.

(b) *Responses to flooding*

Aerenchyma or air conducting passages found in some salt marsh plants are adaptations to flooded environments. In *Spartina alterniflora*, lower stems are aerenchymatous and the rhizome is a hollow cylinder while plants of *Spartina patens* are less aerenchymatous (Anderson, 1974). Aerenchyma is reported for *Limonium* spp. and for *Juncus roemerianus* but not for *Distichlis spicata, Salicornia virginica* or *Aster tenuifolius* L. in a North Carolina salt marsh (Anderson, 1974). Aerenchyma functions as a pathway for diffusion of gases between stems and roots (Gleason and Dunn, 1982; Howes *et al.*, 1981; Teal and Kanwisher, 1961, 1966).

The plant-microbe-soil interaction is the most important and least understood part of the marsh system. Salt marsh microbial ecology was reviewed by Christian *et al.* (1981) and by Wiebe *et al.* (1981). In general soils of the tidal creek banks supporting tall *Spartina alterniflora* are thought to be least anaerobic, soils in the high marsh supporting

short chlorotic *Spartina alterniflora* are thought to be more anaerobic, and soils of dieback areas where *Spartina alterniflora* degenerates and disappears are thought to be most anaerobic. The relationships between plant vigor, drainage, aeration, redox potential (Eh), H_2S and nutrients have been investigated in a number of field and greenhouse studies. Height of *Spartina alterniflora* in the field was positively related to soil drainage where drainage was defined as distance from marsh surface to water table at low tide (Howes *et al.*, 1981; Mendelssohn and Seneca, 1980). Profiles of sediment Eh showed higher values in soils occupied by tall *Spartina alterniflora* than soils occupied by the short form (Howes *et al.*, 1981; Mendelssohn and Seneca, 1980). In greenhouse studies where soils were either flooded twice daily or continuously flooded, height, aerial biomass, and root biomass of *Spartina alterniflora* increased with increased aeration and the resulting increased Eh and decreased sulfide concentration (Linthurst, 1979).

Soil drainage had a positive effect on the growth of *Spartina alterniflora* in field studies but a negative effect in greenhouse studies. In the field studies when drainage was experimentally impeded beneath tall, medium and short *Spartina alterniflora*, growth of tall and medium forms of *Spartina alterniflora* was significantly decreased relative to unimpeded controls (Mendelssohn and Seneca, 1980). Impeded drainage was shown to decrease the Eh. In greenhouse experiments in which tall, medium and short *Spartina alterniflora* were exposed to twice daily tidal inundation of soil, but were completely drained, half drained or not drained, total aboveground biomass from tall and medium height plant cores were significantly greater in undrained soil treatments than for medium drained treatments which were in turn greater than the completely drained treatments (Mendelssohn and Seneca, 1980). The mechanisms underlying these seemingly contradictory results are unknown and need further investigation.

While aeration promotes growth of *Spartina alterniflora* in the greenhouse, oxygen diffusion through the roots to sediments promotes growth in the field. In a field study of sediment profiles (Howes *et al.*, 1981), Eh values increased from plant-free sites, to short *Spartina alterniflora* sites, to tall *Spartina alterniflora* sites. Eh values were highest in the rooted zone. Fertilization increased plant size and sediment Eh. The authors propose a positive feedback model in which increased nutrient availability increases nutrient uptake thereby increasing plant growth which in turn increases O_2 diffusion through the plants to the sediments. Oxidation in the rhizosphere increases nutrient availability and further increases nutrient uptake by the plants. Tall and short *Spartina alterniflora* had about the same stem cross-sectional area per unit marsh area, thus about the same rates of O_2 diffusion into the sediments via plant stems and roots was expected. Because tall *Spartina alterniflora* oxidized the sediments more than did the short form, Howes *et al.* (1981) propose an additional sediment oxidizing mechanism, namely the production of H_2O_2 and other oxidizing compounds by root respiration.

Anaerobic respiration is another adaptation to flooding. Diffusion of oxygen from stems to roots is insufficient to maintain an aerobic rhizosphere in some of the marsh. On a transect from tall creek-bank through 'intermediate' to 'die-back' short *Spartina alterniflora*, Mendelssohn *et al.* (1981) evaluated aboveground standing crop, stem density, plant height, soil Eh, root alcohol dehydrogenase activity (ADH) and concentrations of ethanol, malate, ATP and total adenine nucleotide in roots. Eh and aboveground plant parameters decreased from creek bank to 'die-back' areas. At the creek bank, relatively low malate, relatively low ADH and high ATP suggested that root respiration was mostly aerobic. Accumulated malate and decreased ATP in the intermediate plants suggested anaerobic metabolism. Further inland toward the 'die-back' region, ADH, ATP, total adenine and the adenylate energy charge ratio [(ATP + 0.5ADP)/(ATP + ADP + AMP)] increased in roots indicating a different anaerobic metabolism where ethanol was the main end product. In the die back zone, activities and concentrations of materials in the roots all declined. Here, the accumulation of phytotoxins, low energy status of the roots, and a potential carbon deficit evidently led to decline in growth and to die back.

The aerenchyma tissue through which O_2 diffuses to the roots also serves as a pathway for the internal transport of respiratory CO_2 by roots and refixation by leaves (Gleason and Dunn, 1982) as well as release of dimethylsulfide to the atmosphere (Carlson, 1980).

(c) *Responses to salinity*

The salinities of the inundating tidal water and of the interstitial soil water create problems of water stress and ionic balance for normal plants (glycophytes) as well as for halophytes. Halophytes, plants which grow and reproduce in saline environments (Jefferies, 1981) cope with salinity by some combination of quantitative or qualitative salt exclusion at the roots, salt excretion from leaves, and osmoregulation with compartmentalization of salts within cells.

Quantitative exclusion of salt was inferred from field measurements of water vapor loss, CO_2 uptake, and interstitial water salinity in three marsh succulents, *Salicornia virginica*, *Borrichia frutesoens* and *Batis maritima*. Daily rates of carbon fixation, water loss, potential salt accumulation, and salt to carbon ratios showed that a significant quantity of salt in interstitial water must be excluded from the transpiration stream as water passes through the plants (Antlfinger, 1976; Antlfinger and Dunn, 1983). Quantitative exclusion was also inferred from the sewage sludge application study of Chalmers (1979). Application of sludge to the short *Spartina* marsh stimulated aboveground biomass and increased transpiration thereby increasing interstitial salinity compared to adjacent controls. Similar increases in salinity resulting from the growth of *Spartina alterniflora* are reported by Smart and Barko (1978, 1980).

Qualitative or selective exclusion of Na in favor of K was demonstrated for the mangrove *Avicennia marina* (Forsk.) Vierh. by Rains and Epstein (1967) and for *Spartina alterniflora* by Smart and Barko (1980). More recently Yeo (1981) found that Na ions were transported out of vacuoles of *Suaeda maritima* (L.) Dumort, about four times more rapidly than were K ions, indicating selective transport.

Salt excreting glands occur in the leaves of *Spartina alterniflora*, *Spartina patens*, *Limonium* sp. and in *Distichlis spicata* (Anderson, 1974). Their function was inferred from anatomical studies and from the appearance of moisture droplets and salt crystals on leaves of plants grown in a relatively dry greenhouse atmosphere (Anderson, 1974). In the same study salt glands were absent in *Salicornia virginica*, *Juncus roemerianus* and in *Aster tenuifolius*. Salt glands also occur in some genera of the graminae which are not halophytes (Liphschitz and Waisel, 1974). The inducibility of salt excretion by *Limonium* was reviewed by Lüttge (1975). Supplying Cl^- to plants growing in Cl^--free nutrient solution stimulated salt excretion in a rate versus time relation which was sigmoidal. Decreased salt in the medium decreased the rate of salt pumping. Rates of salt excretion by *Spartina alterniflora* leaves are reported by McGovern *et al.* (1979) to be 1.5×10^{-4} g salt secreted g^{-1} dry wt h^{-1}. They also found excreted salts to be depleted in Ca, Mg and SO_4 relative to sea-water. What salt excretion means to plant energetics is not clear. Studies of salt inputs and outputs coupled with measures of respiration over a range of salinities are needed.

Osmoregulation is an important process because water uptake requires that the plant leaf water potential be more negative than water potentials of the rooted substrate. The solute potential of sea-water at a salinity of $35 \, gl^{-1}$ at $20°C$ is about $-2.5 \, MPa$ (Cox, 1965). In a Georgia salt marsh the interstitial water potentials ranged from about $-1.4 \, MPa$ in the tidal creek to -9.0 in the salt flats (Fig. 15.2). In this marsh, xylem pressure potentials ranged from -1.8 to $-4.3 \, MPa$ and leaf solute potentials ranged from -1.5 to -3.6 in *Spartina alterniflora*, *Juncus roemerianus*, *Batis maritima*, *Salicornia virginica* and *Borrichia frutescens* with the species on the higher end of the salinity gradient showing more negative water potentials (Antlfinger and Dunn, 1983; Giurgevich and Dunn, 1979). Leaf water potentials ranged from -2.9 to $-4.2 \, MPa$ and leaf solute potentials ranged from -4.2 to $-5.2 \, MPa$ in *Salicornia europea*, *Batis maritima*, *Spartina foliosa* and *Distichlis spicata* in a southern California marsh (Kuramoto and Brest, 1979). In a San Francisco, California, marsh, interstitial water solute potentials ranged from -0.4 to $-3.4 \, MPa$ while mid-day xylem pressure potentials of *Spartina foliosa*, *Scirpus robustus* Pursh. and *Salicornia virginica* ranged from -2.3 to -4.6 (Ustin *et al.*,

1982). In a Chesapeake tidal marsh, soil solute potentials ranged from about -0.1 to $-0.8 \, MPa$ while plant pressure potentials for five vascular plant species ranged from -0.5 to $-1.5 \, MPa$ (De Jong and Drake, 1981).

Osmoregulation is accomplished by accumulation of inorganic and organic solutes in cells. Halophytes, especially in the Chenopodiaceae, have higher inorganic ion levels than non-halophytes growing in comparable conditions. In a Georgia salt marsh, the salt concentrations in *Borrichia frutescens*, *Salicornia virginica* and *Batis maritima* were 0.27, 0.37, and $0.48 \, g \, g^{-1}$ dry wt, respectively (Antlfinger and Dunn, 1979). *Salicornia europea* accumulated Na up to $0.15 \, g \, g^{-1}$ dry wt in a Moncton, New Brunswick, Canada salt marsh (Poulin *et al.*, 1978). *Suaeda maritima* accumulated $0.45 \, g$ salts g^{-1} dry wt (Yeo and Flowers, 1980).

Salt accumulation is accomplished at an unknown energetic cost, but halophytic plant cells expend more energy than glycophytic plant cells (Dainty, 1979). Enzymes of halophytes are not particularly salt tolerant (Flowers *et al.*, 1977). Apparently, salt toxicity is avoided by compartmentalization of inorganic ions in the vacuole, which accounts for 90% of the cell volume, and of organic osmotica to the same solute potential in the smaller volume of cytoplasm (Epstein, 1980). Compartmentation of K and Na between cytoplasm and vacuoles in *Suaeda maritima* was demonstrated by Yeo (1981) using radioactive tracers. Compartmentation of Na, K, Cl in cytoplasm, cell walls, vacuoles and chloroplasts of leaf mesophyll cells in *Suaeda maritima* was also demonstrated by use of X-ray dispersive microanalysis with a scanning electron microscope (Harvey *et al.*, 1981).

Organic solutes including organic acids, carbohydrates, and nitrogen compounds accumulate in halophytes in response to water stress (Jefferies *et al.*, 1979). Proline accumulation was examined in 8 Atlantic salt marsh species in growth chamber and field studies (Cavalieri & Huang 1979). Three general patterns appeared:

1. *Limonium carolinianum* and *Juncus roemerianus* started accumulating proline when grown at 0.25 M NaCl;

2. the threshold salinity for proline accumulation

of the C_4 grasses *Spartina alterniflora, Spartina patens* and *Distichlis spicata* was around 0.5 M NaCl;

3. the succulent plants *Salicornia bigelovii, Salicornia virginica* and *Borrichia frutescens* did not accumulate proline until the NaCl was 0.7 M.

When interstitial water salinities and tissue proline concentrations were determined for these species in the field, similar results were obtained. When plants were grown in solutions containing either NaCl or polyethylene glycol 4000 (PEG) at Ψ_s of -2.0 MPa, patterns of proline accumulation were qualitatively similar. Quantitatively, for all species except *Distichlis spicata* and *Salicornia bigelovii*, proline production was greater in the NaCl than in the PEG treatment. In *Distichlis spicata* and in *Salicornia bigelovii* the greatest proline accumulation occurred in the PEG treatment. Proline production is controlled both by an osmotic effect and a specific salt effect and the response varies with the species. The authors conclude that proline accumulation is important to *Limonium carolinianum* and *Juncus roemerianus*, sometimes important to the C_4 grasses when the salinity temporarily increases, and is of little importance to the succulent plants. In a study of tall and short *Spartina alterniflora*, Cavalieri and Huang (1981) found that NH_4NO_3 fertilization increased growth, proline and glycinebetaine concentrations, and decreased water potentials of the short but not the tall plants. They suggest that because short *Spartina alterniflora* may be N limited, it has a decreased capacity to allocate N to osmoregulation and other N-requiring metabolism with a resulting decreased growth.

Comparing the production of organic osmotica by the halophytes *Plantago maritima* L., *Triglochin maritima* L., *Limonium vulgare* P. Mill. and *Halimione portulacoides* (L.) Aell grown in sea-water and in PEG 6000 at comparable solute potentials, Jefferies *et al.* (1979) found that sorbitol, proline, reducing sugars, quarternary ammonium compounds and alpha amino nitrogen compounds accumulated in tissues depending on the species. They found that the accumulations of osmotica were very similar in the sea-water and in the PEG treatments and ascribed accumulation of organics to osmotic effects and not to salinity *per se*.

Decreased growth of halophytes at high salinities has been reported by many investigators (e.g.

Haines and Dunn, 1976; Nestler, 1977; Parrondo, *et al.*, 1978; Yeo and Flowers, 1980). Mechanisms involved in this growth retardation are still not clear but could involve the costs of osmotic regulation.

(d) *Response to nutrient limitation*

Gradients in height of *Spartine alterniflora* have been attributed to salinity (Mooring *et al.*, 1971), iron limitation, (Adams, 1963; Nixon and Oviatt, 1973) and to nitrogen limitation (Broome *et al.*, 1975b; Gallagher, 1975; Mendelssohn, 1979; Smart and Barko, 1980; Sullivan and Daiber, 1974). The salinity–Fe–nitrogen limitation hypotheses were subjected to a factorial solution culture experiment (Haines and Dunn, 1976). Increasing NH_4 promoted growth except at the highest salinity. Shoot: root ratios indicated decreased carbon allocation to root production with increases in both NH_4 and NaCl. Fe did not affect dry weight or height.

The possibility of phosphorus limitation was tested by measuring growth of short *Spartina alterniflora* in plots supplied with Ca $(H_2PO_4)_2$ compared to plots receiving NH_4NO_3, or to control plots. While N addition promoted growth, phosphorus addition did not (Sullivan and Daiber, 1974).

In a study of nitrogen uptake kinetics by *Spartina alterniflora* grown in half-strength sea-water enriched with K, P, and chelated Fe, Morris (1980) found Michaelis–Menten half saturation constants in mg l^{-1} for NH_4–N of 0.057 ± 0.016 and for NO_3–N of 0.124 ± 0.034. Because NH_4–N in interstitial water beneath short *Spartine alterniflora* was about 3.0 mgNl^{-1} but the plants showed growth responses to NH_4 fertilizer, Morris (1980) suggest that some combination of anaerobiosis, salinity, and H_2S could increase the apparent K_m for NH_4 uptake. Further, soil resistance to NH_4 diffusion might decrease NH_4 concentration at the root surface well below the 3.0 mg Nl^{-1} in the interstitial water.

15.4 Ecosystem properties affected by plant ecophysiology

15.4.1 PRIMARY PRODUCTIVITY

Coastal salt marshes are among the most productive of natural ecosystems (Odum, 1959). While gradients in production occur across marshes, for reasons already discussed, there are also latitudinal gradients in marsh productivity. Production in fresh water marshes and in Gulf and in Atlantic coastal marshes has been reviewed by several authors (Keefe, 1972; Reimold, 1977; Turner, 1976). Turner (1976) found a south to north decline in production parallelling a gradient of decreased solar energy input at a 0.2–0.35% net conversion efficiency. Production in Pacific coast marshes has been less studied. Eilers (1979) reported higher production at West Coast sites than at corresponding latitudes on the east coast and suggested that greater production on the Northwest Coast may result from the absence of large summer–winter temperature changes, little snow cover, and rarity of ice rafting, all of which become more pronounced with increasing latitude on the eastern seaboard. Furthermore, on the west coast rainfall decreases from north to south such that in southern Oregon and California drought could result in increased rooting zone salinity and decreased production.

On the Atlantic seaboard, production has also been positively related to tidal amplitude over a mean tide range of 0.7–2.5 m (Steever *et al.*, 1976). This supports the 'tidal energy subsidy' hypothesis of Odum (1974), Odum and Fanning (1973) but does not explain the mechanism of subsidy.

15.4.2 RELATION OF GAS EXCHANGE MEASUREMENTS TO PRIMARY PRODUCTIVITY

While at least 31 studies of aboveground production by *Spartina alterniflora* using harvest methods are reported (Turner, 1976) and several have appeared since, very few gas exchange studies have been coupled with modelling efforts to provide mechanistic explanations for production (Blum *et al.*, 1978; Cammen *et al.*, 1982; Giurgevich and Dunn, 1982). Morris (1982) modelled dry matter growth for *Spartina alterniflora* and was able to analyse the relative effects of climatic variables such as PAR, temperatures, and lengths of growing season on net primary production in the salt marsh. Morris *et al.* (unpublished) converted dry matter growth rates to CO_2 exchange rates and used this model to predict maximum and probable rates of belowground production.

15.4.3 NUTRIENT CYCLING

Nutrient cycling in the ecosystem is affected by the physiology of the plants. In coastal salt marshes the interactions of microbial processes, soil drainage, and tidal flux may be quantitatively more important in determining nutrient flux than are plant processes.

(a) Nitrogen fixation

Nitrogen fixing activity (acetylene reducing activity, ARA) has been demonstrated in excised roots of angiosperms in a coastal salt marsh in Nova Scotia (Patriquin and Keddy, 1978). Among the 33 species investigated, 13 without root nodules had rates in excess of $100 \, nmol \, C_2H_4 \, g^{-1} \, h^{-1}$. There was an inverse relation between nitrogen fixing activity and the concentration of NH_4 in interstitial water. A nitrogen fixing *Campylobacter* sp. has been isolated from surface sterilized roots of *Spartina alterniflora* (McClung and Patriquin, 1980). It metabolized organic and amino acids but not carbohydrates. It might be exploiting root exudates from *Spartina alterniflora*. In another study where ARA was localized both inside and outside of roots of *Spartina alterniflora*, Boyle and Patriquin (1980) suggested the existence of a primitive symbiosis.

(b) Denitrification

Potential denitrification (N_2O gas production) was greater in sediments supporting *Spartina alterniflora* than sediments without plants. Oxidized microzones around *Spartina alterniflora* roots are postulated to allow nitrifying bacteria to produce NO_3 and NO_2. The NO_3 and NO_2 would then diffuse into adjacent more anaerobic zones where they would serve as terminal electron acceptors for the denitrifying bacteria (Sherr and Payne, 1979).

(c) Nitrogen uptake

Nitrogen uptake by plants has been quantified as the product of annual net primary productivity and tissue nitrogen concentration (Chalmers, 1979; Haines, 1979a; Patrick and DeLaune, 1977; Valiela and Teal, 1979). Nitrogen uptake by plants coincides with the decreased export of N from the marsh (Valiela and Teal, 1979). For nitrogen uptake, the kinetic properties of the Michaelis – Menten half saturation constants for both NH_4 and NO_3 were much too low compared to levels of these ions found in the field (Morris, 1980), suggesting that edaphic factors such as anaerobism, metabolic phytotoxins such as H_2S or competitive inhibition of NH_4-uptake by other ions may account for the apparent nitrogen limitation in the short *Spartina alterniflora*.

(d) *Phosphorus*

Spartina alterniflora translocates phosphorus from the marsh sediments to its leaves. (Patrick and DeLaune, 1976). Release from the leaves during tidal inundation provides an important source of phosphorus for coastal waters. Seasonal rates of phosphorus pumping follow seasonal patterns of *S. alterniflora* production (Reimold, 1972).

(e) *Sulfur*

Aerenchyma tissue in *S. alterniflora* is a pathway for the diffusion of atmospheric oxygen to the root surface. The uptake of the reduced sulfur species, H_2S, by roots of *S. alterniflora* (Carlson and Forrest, 1982) and measurement of emission of the other reduced sulfur gases, CS_2, COS, and dimethyl sulfide, over *S. alterniflora* (Aneja *et al.*, 1981) suggest that aerenchyma in this plant is a pathway for sulfur movement from anoxic sediments into the atmosphere. The possible reduction of SO_4^{-2} by marsh plants themselves needs investigation since *Glycine max* L. Merr. (Winner *et al.*, 1981) and *Picea abies* L. Karsten (Spaleny, 1977) have been shown to reduce SO_4^{-2} to H_2S, and marine algae have recently been shown to reduce SO_4^{-2} to dimethyl sulfide (Barnard *et al.*, 1982). The oxidation of these sulfur gases in the atmosphere and subsequent wash out by rain could contribute to rainfall acidity (Galloway *et al.*, 1982; Haines, 1983).

(f) *Detrital carbon and energy flow*

Spartina alterniflora, with a high annual productivity and C_4 photosynthesis with resulting ^{13}C: ^{12}C ratios of -12.3 to -13.7 compared to C_3 plants with ratios of -23 to -26, and having a high annual productivity, produces stable isotopically labelled detritus in large quantities which has been successfully used to trace carbon and energy flow through marsh and estuarine food webs (Haines, 1977; Haines and Montague, 1979). Recent results of such studies have stimulated a reevaluation of the contribution of coastal marshes to the carbon budgets of estuaries and nearshore waters (Haines 1979a, b).

15.5 Management implications

15.5.1 PETROLEUM SPILLS

Oil may enter salt marshes from exploratory and production wells so common on the Gulf Coast, from oil refineries at sea ports, and from oil tanker breakups. General effects of oils are known from numerous studies in which oils were applied to plants as herbicides or insecticides and from a few studies of oil spills (Baker, 1970). Oils vary in toxicity according to composition. The higher the concentrations of low boiling point compounds the more toxic the oil to plants. Oils may penetrate plants, travel through intercellular spaces and damage cell membranes. Oils decrease transpiration and photosynthesis and may either increase or decrease respiration (Baker, 1970).

Spartina alterniflora production was unaffected by Louisiana crude oil largely because a low tidal amplitude of 0.3 m did not distribute the oil over the leaves (DeLaune *et al.*, 1979). In a Massachusetts salt marsh after a spill of No. 2 fuel oil, plant growth was inversely related to oil concentration in sediment (Burns and Teal, 1979). *Spartina alterniflora* was killed at concentrations of 1 to 2 mg oil g^{-1} sediment. In the Chesapeake Bay, *Spartina alterniflora* receiving the most No. 2 fuel oil became chlorotic and died, while plants receiving sub-lethal amounts exhibited delayed spring development, increased stem density and reduced stem weight (g cm^{-2}) in comparison to *Spartina alterniflora* in a control marsh (Hershner and Lake, 1980). At some locations where *Spartina alterniflora* was killed, sediments were subsequently eroded away to an elevation below which *Spartina alterniflora* can grow; thus oil spills may have long-term implications for marsh productivity beyond simple toxicity to plants.

15.5.2 HEAVY METALS

The role of *Spartina alterniflora* in the flow of lead, cadmium and copper through the salt marsh was investigated by Dunstan *et al.* (1975). They determined concentrations of the metals in *Spartina alterniflora* and in underlying sediments derived from six southeastern rivers. Mean concentrations were usually lower in the plants than in sediment. The potential toxicity of these metals was investigated by growing *Spartina alterniflora* seedlings in nutrient solution amended to contain 100 ppm of each of the metals. Copper killed all the plants. Lead caused a 50% mortality and those surviving attained a height of about half that of controls, but these stunted plants contained less lead than field-grown plants. Cadmium had no effect on growth and concentrations in tissue were as high as 150 times the amounts found in field grown plants. The annual uptake by *Spartina alterniflora* of cadmium and copper from nine southeastern estuaries was estimated to be less than 4% of the metal input from rivers. The authors conclude that increased levels of cadmium and copper in the river would have little effect on *Spartina alterniflora*.

15.5.3 BIOCIDES

Biocides may be subjected to 'biological magnification' or show a systematic increase in concentration with increasing trophic level (Woodwell *et al.*, 1967). Biocides manufactured and used on land may be carried to the salt marsh where they can accumulate.

The herbicide atrazine used for selective weed control in sugar cane, corn, and sorghum, is an inhibitor of the Hill reaction in photosynthesis and possibly of protein synthesis (McEnerney and Davis, 1979). About 1.2% of atrazine applied to a corn field in Maryland (Wu *et al.*, 1977) and 2.4% of atrazine applied to soils in Pennsylvania (Hall *et al.*, 1972) were lost in runoff. For the Maryland site the weekly volume weighted concentration at the stream gages ranged from 0 to 0.05 ppm. This is half the atrazine

concentration found to decrease dry weight accumulation in roots and in shoots of *Spartina alterniflora* (Pillai *et al.*, 1977). Runoff stream water would be diluted by estuarine water before reaching the salt marsh; thus the effect of atrazine in the marsh is not clear.

Toxaphene, an insecticide used against a variety of cotton plant pests is manufactured adjacent to a Brunswick, GA, salt marsh. Toxaphene concentrations in the top 10 cm of sediments near the waste water outfall from the manufacturing plant ranged from 35 to 1860 ppm (Durant and Reimold, 1972). Where concentrations were 32 ppm in sediment, toxaphene concentrations in roots, rhizomes, leaves, and seed heads of *Spartina alterniflora* were 1.9, 1.2, 36 and 5 ppm, respectively (Reimold and Durant, 1974). When toxaphene was injected at a depth of 1 cm into soil producing a concentration of 3 ppm, *Spartina alterniflora* translocated the material both upward and downward in the plants and concentrated it in tissues as much as 300 times background (Gallagher *et al.*, 1979). While *Spartina alterniflora* clearly accumulates toxaphene, the effects on the plant are unknown.

In a Long Island, New York, salt marsh dominated by *Spartina patens*, Woodwell *et al.* (1967) showed biological magnification of DDT residues from 5×10^{-5} ppm in water to 0.33 and 2.80 ppm in shoots and roots of *Spartina patens* to 75 ppm in a ringbilled gull. Organic detritus in sediments in that marsh had DDT residue concentrations ranging from 3 to 50 times greater than concentrations of living shoots of *Spartina patens*, which were the ultimate sources of the detritus (Odum *et al.*, 1969). Whether DDT had accumulated in the tissue before or after it became detritus was not determined.

15.5.4 MUNICIPAL WASTES

The possible renovation of municipal wastes such as sewage sludge was investigated in salt marshes of New England (Valiela *et al.*, 1975) and of Georgia (Chalmers, 1979; Haines, 1979a). In both studies application of sewage sludge increased plant production by overcoming nitrogen deficiency. Negative effects of heavy metals and chlorinated hydrocarbons in sludge were found for the New

England marsh. Increased interstitial water salinity with increased plant production was thought to be the result of increased transpiration by larger *Spartina alterniflora* on sludge amended plots compared to plants outside the plots (Chalmers, 1979). Because increased salinity overcomes the growth promoting effects of high NH_4 (Haines and Dunn, 1976), the water balance of the high marsh sets limits to its capacity to trap and retain added nitrogen in the biomass. If large scale renovation of waste water or sludge by the salt marsh were contemplated, the saturation kinetics of the marsh would need to be determined. Because most of the plant biomass becomes detritus, the effect of increased detritus production, followed by increased detrital decomposition, upon the biological oxygen demand of the estuarine water and associated fauna would also have to be evaluated.

15.5.5 DREDGING

Navigable waterways such as the intracoastal waterway along the Gulf and Atlantic coasts pass through salt marshes. Dredge spoil from maintenance of these waterways buries existing marshes and sometimes provides substrate for development of new marsh. The need to stabilize dredge spoil to minimize its return to the ship channel has provided support for studies of seed production and seed germination in *Spartina alterniflora* as well as studies of edaphic factors controlling establishment, survival, and production of this dominant species (Broome *et al.*, 1975a, b; Seneca, 1974b; Smart and Barko, 1978, 1980). The depth of spoil placed on the marsh at Barnegat Bay, New Jersey, was found to influence rates of colonization by *Spartina*. Addition of less than 5 cm spoil permitted return to 60–90% cover in the first year. Addition of more than 5 cm spoil allowed only about 5% cover in the first year and resulted in changes in species composition during later years (Burger and Shisler, 1983).

15.6 Conclusions and future research needs

Salt marshes occur on all coasts of North America. Subject to periodic inundation by salt water, marsh plants are exposed to combinations of reducing and salinity conditions not found in other habitats. The result is a distinct zonation of production rates and plant height within species and a segregation of species into zones across the marsh. These zonations provide a unique opportunity to study plant performance along environmental gradients.

Coastal salt marsh plants cope with salinity by some combination of qualitative or quantitative salt exclusion at the root, osmoregulation with compartmentalization of salts within cells, and salt excretion from leaves.

Most of the ecophysiological investigations of coastal salt marsh plants have been performed with few taxa, notably *Spartina, Salicornia, Juncus, Borrichia*, or *Batis* along the east or California coasts. In these marshes little work has been done with the minor species and very few process studies have been conducted with plants from the northwestern and arctic coasts.

Ecotypic differentiation in *Spartina patens* and *Borrichia frutescens* has been demonstrated within individual marshes. Height forms in *Spartina alterniflora* reflect phenotypic plasticity rather than ecotypic differentiation within marshes, but clearly ecotypic differentiation occurs along latitudinal gradients.

We have summarized the plant–soil–microbial processes and interactions which contribute to the variations in height and productivity of *Spartina alterniflora*, the most studied of North American coastal halophytes, in a conceptual model (Fig. 15.3). Many of the processes have been reviewed and documented above, while others are less well documented or are still hypothetical. We believe that similar processes to some extent control horizontal patternings and plant performance in other species in other coastal marshes. Tidal flushing appears to be the major driving force.

The high amplitude of tidal flushing at low elevations in the marsh, the habitat of tall *Spartina alterniflora* (see Fig. 15.2), brings in new nutrients such as NH_4 and Fe, replaces partly anoxic interstitial water with new water thereby displacing toxic wastes such as sulfide, possibly metabolites and accumulated salts. Salinity is relatively stable and low, resulting in less inhibition of nutrient uptake by plant roots thus less NH_4 accumulates in

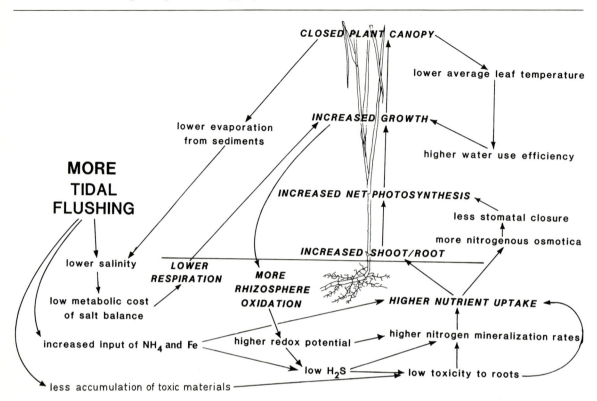

Figure 15.3 Conceptual summary of plant–soil–microbial processes and interactions which result in tall and short forms of *Spartina alterniflora* in Southeastern marshes. Processes indicated in capitals are well established. Other processes are less well documented or hypothetical.

interstitial water. Lower salinity also decreases the metabolic cost of maintaining salt balance, resulting in lower root respiration and lower whole plant respiration. Lower respiration favors a higher net photosynthesis. The higher redox potentials result in less sulfide production and less sulfide toxicity to plant roots. Higher rates of N mineralization make more N available for uptake by plants and allow production of nitrogen containing osmotica important in maintaining water balance. High rates of nutrient uptake by roots promote higher shoot:root ratios and greater overall plant growth. Increased shoot growth causes more diffusion of oxygen down through stems to the roots and into the soil, promoting higher redox potentials. Increased shoot growth leads to canopy closure, and a higher leaf area index. A high leaf area index rather than inherently high CO_2 fixation rates per leaf area is responsible for high production rates in the tall

Spartina alterniflora. The high leaf area index leads to shelf shading, decreased mean leaf temperature, increased water use efficiency of photosynthesis, decreased evapotranspiration per m² of marsh surface and finally to light limitation of photosynthesis.

The low amplitude of tidal flushing at the higher elevations in the marsh where the short *Spartina alterniflora* occurs (see Fig. 15.2) results in decreased inputs of new nutrients and toxic accumulations of materials such as sulfides and salts. Lower redox potentials favor high sulfide and toxicity to roots. Lower redox potentials and higher salinity result in decreased rates of decomposition, notably N mineralization. Higher salinities result in decreased uptake of mineralized N by plant roots and possibly of other elements such that NH_4 accumulates to higher concentrations here than in the region of the tall form. Decreased rates of nutrient uptake lead to decreased shoot:root ratios and to decreased overall growth. Decreased plant growth decreases oxygen diffusion into the sediment, promoting lower redox potentials. Decreased plant growth results in a more open canopy, higher average leaf temperature, decreased water use efficiency of photosynthesis, higher direct evaporation of water from sediment surface; all leading to higher interstitial salinity.

At the land–sea interface, marsh–estuarine systems receive wastes such as heavy metals and biocides from the land and petroleum products lost from production and transportation facilities in coastal areas. Live marsh plants appear to have low rates of pesticide accumulation while plant detritus has a higher affinity for both pesticides and heavy metals. Petroleum has little affect on marsh macrophytes when it coats the marsh sediments but kills macrophytes when it coats their leaves.

Many of the root–soil–microbial interaction processes are as yet poorly understood. Processes needing further investigation include:

1. mechanisms by which tidal flushing affects root–soil–microbial interactions. Paradoxically, while tidal flusing in the field promotes the production of tall *Spartina alterniflora*, simulated tidal flushing in the laboratory produced shorter *Spartina alterniflora* than is obtained from stagnant or unflushed controls;

2. the direct and interactive effects of H_2S, salinity, and anaerobism on the processes of root growth, root respiration, absorption of nutrients by roots, and upon the mineralization of nutrients out of plant remains.

References

Adams, D.A. (1963) Factors influencing vascular plant zonation in North Carolina salt marshes. *Ecology*, **44**, 445–56.

Anderson, C.E. (1974) A review of structure in several North Carolina salt marsh plants, *Ecology of Halophytes* (eds R.J. Reimold and W.H. Queen), Academic Press, New York, pp. 307–44.

Anderson, C.M. and Treshow, M. (1980) A review of environmental and genetic factors that affect height in *Spartina alterniflora* Loisel. (salt marsh cord grass). *Estuaries*, **3**, 168–76.

Aneja, V.P., Overton, J.H. and Aneja, A.P. (1981) Emission survey of biogenic sulfur flux from terrestrial surfaces. *Journal of the Air Pollution Control Association*, **31**, 256–8.

Antlfinger, A.E. (1976) Seasonal photosynthetic and water-use patterns in three salt marsh plants, MS thesis, University of Georgia, Athens, GA.

Antlfinger, A.E. (1979) The genetic basis of physiological and morphological variation in natural and experimental populations of *Borrichia frutescens* in relation to salinity, Doctoral dissertation, University of Georgia, Athens, 111 pp.

Antlfinger, A.E. and Dunn, E.L. (1979) Seasonal patterns of CO_2 and water vapor exchange of three salt marsh succulents. *Oecologia*, **43**, 249–60.

Antlfinger, A.E. and Dunn, E.L. (1983) Water use and salt balance in three salt marsh succulents, *American Journal of Botany* **70**, 561–7.

Armstrong, W. (1975) Waterlogged soils, *Environment and Plant Ecology* (ed. J.R. Etherington), J. Wiley and Sons, New York, pp. 181–218.

Baker, J.M. (1970) The effects of oils on plants. *Environmental Pollution*, **1**, 27–44.

Barnard, W.R., Andreae, M.O., Watkins, W.E., Bingemer, H. and Georgii, H.W. (1982) The flux of dimethylsufide from the oceans to the atmosphere. *Journal of Geophysical Research*, **87**, (C11) 8787–93.

Blum, U., Seneca, E.D. and Stroud, L.M. (1978) Photosynthesis and respiration of *Spartina* and *Juncus* salt-marshes in North Carolina: some models. *Estuaries*, **1**, 228–38.

Boyle, C.D. and Patriquin, D.G. (1980) Endorhizal and exorhizal acetylene-reducing activity in a grass (*Spartina alterniflora* Loisel.)–diazotroph association. *Plant Physiology*, **66**, 276–80.

Brinkhuis, B.H., Tempel, N.R. and Jones, R.F. (1976) Photosynthesis and respiration of exposed salt-marsh fucoids. *Marine Biology*, **34**, 349–59.

Broome, S.W., Woodhouse, W.W., Jr and Seneca, E.D. (1975a) The relationship of mineral nutrients to growth of *Spartina alterniflora* in North Carolina. I. Nutrient status of plants and soils in natural stands. *Soil Science Society of America Proceedings*, **39**, 295–301.

Broome, S.W., Woodhouse, W.W. Jr and Seneca, E.D. (1975b) The relationship of mineral nutrients of growth of *Spartina alterniflora* in North Carolina. II. The effects of N,P, and Fe fertilizers. *Soil Science Society of America Proceedings*, **39**, 301–7.

Burger, J. and Shisler, J. (1983) Succession and productivity on perturbed and natural *Spartina* salt-marsh areas in New Jersey. *Estuaries.* **6**, 50–6.

Burns, K.A. and Teal, J.M. (1979) The West Falmouth oil spill: hydrocarbons in the salt marsh ecosystem. *Estuarine Coastal Marine Science*, **8**, 349–60.

Cammen, L.M., Blum, U., Seneca, E.D. and Stroud, L.M. (1982) Energy flow in a North Carolina salt marsh: A. synthesis of experimental and published data. *Association of Southeastern Biologists Bulletin*, **29**, 111–34.

Carlson, P.R., Jr (1980) Oxygen diffusion from the roots of *Spartina alterniflora* and the role of *Spartina* in the sulfur cycle of salt marsh sediments, Doctoral dissertation, University of North Carolina, Chapel Hill, N.C.

Carlson, P.R., Jr. and Forrest, J. (1982) Uptake of dissolved sulfide by *Spartina alterniflora*: evidence from natural sulfur isotope abundance ratios. *Science*, **216**, 633–5.

Cavalieri, A.J. and Huang, A.H.C. (1979) Evaluation of proline accumulation in the adaptation of diverse species of marsh halophytes to the saline environment. *American Journal of Botany*, **66**, 307–12.

Cavalieri, A.J. and Huang, A.H.C. (1981) Accumulation of proline and glycine betaine in *Spartina alterniflora* in response to sodium chloride and nitrogen in the marsh. *Oecologia*, **49**, 224–8.

Chalmers, A.G. (1979) The effects of fertilization on nitrogen distribution in a *Spartina alterniflora* salt marsh. *Estuarine Coastal Marine Science*, **8**, 327–38.

Chalmers, A.G., Haines, E.B. and Sherr, B.F. (1976) Capacity of a *Spartina* salt marsh to assimilate nitrogen from secondarily treated sewage. Technical Completion Report USDI/OWRT no. A-057-GA, University of Georgia Marine Institute, Sapelo Island, GA, and Environmental Resources Center, Georgia, Institute of Technology, Atlanta, GA.

Chapman, V.J. (1974) *Salt Marshes and Salt Deserts of the World*, 2nd edn., Verlag von J. Cramer, Bremeshaven.

Chapman, V.J. (1977) *Wet Coastal Ecosystems*, Elsevier Scientific, New York. pp. 1–29.

Christian, R.R., Hanson, R.B., Hall, J.R. and Wiebe, W.J. (1981) Aerobic microbes and meiofauna, in *The Ecology of a Salt Marsh* (eds L.R. Pomeroy and R.G. Wiegert), Springer-Verlag, New York, pp. 113–35.

Cooper, A.W. (1974) Salt Marshes, (eds H.T. Odum, B.J. Copeland and E.A. McMahon), *Coastal Ecological Systems of the United States*, The Conservation Foundation, Washington, D.C., pp. 567–611.

Cox, R.A. (1965) The physical properties of sea water, *Chemical Oceanography* (eds J.P. Riley and G. Skirrow), Academic Press, New York, pp. 73–121.

Dainty, J. (1979) The ionic and water relations of plants which adjust to a fluctuating saline environment, in *Ecological Processes in Coastal Environments*, (eds R.L. Jefferies and A.J. Davy), Blackwell Scientific Publications, Oxford, UK, pp. 201–9.

De Jong, T.M. and Drake, B.G. (1981) Seasonal patterns of plant and soil water potential on an irregularly flooded tidal marsh. *Aquatic Botany*, **11**, 1–9.

De Jong, T.M., Drake, B.G. and Pearcy, R.W. (1982) Gas exchange responses of Chesapeake Bay tidal marsh species under field and laboratory conditions. *Oecologia*, **52**, 5–11.

DeLaune, R.D., Patrick, W.H., Jr and Brannon, J.M. (1976) *Nutrient transformations in Louisiana salt marsh soils*, Center for Wetland Resources, Louisiana State University, Baton Rouge, LA, 70803.

DeLaune, R.D., Patrick, W.H., Jr and Buresh, R.J. (1979) Effect of crude oil on a Louisiana *Spartina alterniflora* salt marsh. *Environmental Pollution*, **20**, 21–32.

Drake, B.G. and Read, M. (1981) Carbon dioxide assimilation, photosynthetic efficiency, and respiration of a Chesapeake bay salt marsh. *Journal of Ecology*, **69**, 405–23.

Duncan, W.H. (1974) Vascular halophytes of the Atlantic and Gulf coasts of North America north of Mexico, in *Ecology of Halophytes* (eds R.J. Reinold and W.H. Queen), Academic Press, New York, pp. 23–50.

Dunn, R., Long, S.P. and Thomas, S.M. (1981) The effect of temperature on the growth and photosynthesis of the temperate C_4 grass *Spartina townsendii*, in *Plants and their atmospheric environment* (eds J. Grace. E.D. Ford., and P.G. Jarvis), Blackwell Scientific, Oxford, UK, pp. 303–11.

Dunstan, W.M., Windom, H.L. and McIntire, G.L. (1975) The role of *Spartina alterniflora* in the flow of lead, cadmium, and copper through the salt marsh ecosystem, in *Mineral cycling in Southeastern ecosystems* (eds F.G. Howell, J.B. Gentry and M.H. Smith), US Energy Research and Development Administration CONF-740713, pp. 250–6.

Durant, C.J. and Reimold, R.J. (1972) Effects of estuarine dredging of toxaphene contaminated sediments in Terry Creek, Brunswick, GA, 1971. *Pesticides Monitoring Journal*, **6**, 94–6.

Eilers, H.P. (1979) Production ecology in an Oregon USA coastal salt marsh. *Estuarine Coastal Marine Science*, **8**, 399–410.

Eleuterius, L.N. and Eleuterius, C.K. (1979) Tide levels and salt marsh zonation. *Bulletin of Marine Science*, **29**, 394–400.

Epstein, E. (1980) Responses of plants to saline environments, in *Genetic engineering of osmoregulation.*

Symposium on genetic engineering of osmoregulation: impact on plant productivity for food, chemicals, and energy, (eds D.W. Rains, R.C. Valentine and A. Hollaender) Plenum Press, New York, pp. 7–12.

Flowers, T.J., Troke, P.F. and Yeo, A.R. (1977) The mechanism of salt tolerance in halophytes. *Annual Review of Plant Physiology,* **28,** 89–121.

Frey, R.W. and Basan, P.B. (1978) Coastal salt marshes, in *Coastal sedimentary environments* (ed. R.A. Davis, Jr), Springer-Verlag, New York, pp. 101–69.

Gabriel, B.C. and Cruz, de la A.A. (1974) Species composition, standing stock, and net primary production of a salt marsh community in Mississippi. *Chesapeake Science,* **15,** 72–7.

Gallagher, J.L. (1975) Effect of an ammonium nitrate pulse on the growth and elemental composition of natural stands of *Spartina alterniflora* and *Juncus roemerianus. American Journal of Botany,* **62,** 644–8.

Gallagher, J.L. and Reimold, R.J. (1974) Tidal marsh plant distribution and productivity patterns from the sea to fresh water – a challenge in resolution and discrimination, in *IV Biennial Workshop on color aerial photography in the plant sciences and related fields,* American Society of Photogrammetry, Washington, DC, pp. 165–83.

Gallagher, J.L., Robinson, S.E., Pfeiffer, W.J. and Seliskar, D.M. (1979) Distribution and movement of toxaphene in anaerobic saline marsh soils. *Hydrobiologia,* **63,** 3–10.

Galloway, J.N., Likens, G.E., Keene, W.C. and Miller, J.M. (1982) The composition of precipitation in remote areas of the world. *Journal of Geophysical Research,* **87,** (11), 8771–86.

Gardner, L.R. (1973) The effect of hydrologic factors on the pore water chemistry of intertidal marsh sediments. *Southeastern Geology,* **15,** 17–28.

Giurgevich, J.R. and Dunn, E.L. (1978) Seasonal patterns of CO_2 and water vapor exchange of *Juncus roemerianus* Scheele in a Georgia salt marsh. *American Journal of Botany,* **65,** 502–10.

Giurgevich, J.R. and Dunn, E.L. (1979) Seasonal patterns of CO_2 and water vapor exchange of the tall and short forms of *Spartina alterniflora* Loisel. in a Georgia salt marsh. *Oecologia,* **43,** 139–56.

Giurgevich, J.R. and Dunn, E.L. (1981) A comparative analysis of the CO_2 and water vapor responses of two *Spartina* species from Georgia coastal marshes. *Estuarine, Coastal and Shelf Science,* **12,** 561–8.

Giurgevich, J.R. and Dunn, E.L. (1982) Seasonal patterns of daily net photosynthesis, transpiration and net primary productivity of *Juncus roemerianus* and *Spartina alterniflora* in a Georgia salt marsh. *Oecologia,* **52,**

404–10.

Gleason, M.L. and Dunn, E.L. (1982) Effects of hypoxia on root and shoot respiration of *Spartina alterniflora, Esturina Comparisons* (ed. V. Kennedy), Academic Press, New York, pp. 243–53.

Gooch, E.L. (1968) Hydrogen sulfide production of its effect on inorganic phosphate release from the sediments of the Canary Creek Marsh, MA thesis, University of Delaware, Newark.

Gosselink, J.G. Hopkinson, C.S., Jr. and Parrondo, R.T. (1977) *Common marsh plant species of the Gulf Coast Area Vol II: Growth Dynamics,* US Army Corps of Engineers, US Waterways Experiment Station, Vicksburg, MS Tech Report D77-44, Vol. 2.

Gray, A.J., Parsell, R.J. and Scott, R. (1979) The genetic structure of plant populations in relation to the development of salt marshes, in *Ecological Processes in Coastal Environments* (eds R.L. Jefferies and A.J. Davy), London, UK, pp. 43–64.

Haines, B.L. (1983) Forest ecosystem SO_4–S input–output discrepancies and acid rain: are they related? *Oikos,* (**41**), 139–43.

Haines, B.L. and Dunn, E.L. (1976) Growth and resource allocation responses of *Spartina alterniflora* Loisel. to three levels of NH_4–N, Fe and NaCl in solution culture. *Botanical Gazette,* **137,** 224–30.

Haines, E.B. (1977) The origins of detritus in Georgia salt marsh estuaries. *Oikos,* **29,** 254–60.

Haines, E.B. (1979a) Growth dynamics of cord grass, *Spartina alterniflora* Loisel., on control and sewage sludge fertilized plots in a Georgia salt marsh. *Estuaries,* **2,** 50–3.

Haines, E.B. (1979b) Interactions between Georgia salt marshes and coastal waters: a changing paradigm, in *Ecological Processes in Coastal and Marine Systems* (ed. R.J. Livingston), Plenum Press, New York, pp. 35–46.

Haines, E.B. and Montague, C.L. (1979) Food sources of estuarine invertebrates analyzed using $^{13}C/^{12}C$ ratios. *Ecology,* **60,** 48–56.

Hall, J.K., Pawlus, M. and Higgins, E.R. (1972) Losses of atrazine in run off water and soil sediment. *Journal of Environmental Quality,* **1,** 172–6.

Harvey, D.M.R., Hall, J.L., Flowers, T.J. and Kent, B. (1981) Quantitative ion localization within *Suaeda maritima* leaf mesophyll cells. *Planta,* **151,** 555–60.

Hatcher, B.G. and Mann, K.H. (1975) Above-ground production of marsh cordgrass (*Spartina alterniflora*) near the northern end of its range. *Journal of the Fisheries Research Board of Canada,* **32,** 83–7.

Hershner, C. and Lake, J. (1980) Effects of chronic oil pollution on a salt marsh grass community. *Marine*

Biology, **56**, 163–73.

Hinde, H.P. (1954) The vertical distribution of salt marsh phanerograms in relation to tide levels. *Ecological Monographs*, **24**, 207–27.

Houghton, R.A. and Woodwell, G.M. (1980) The flax pond ecosystem study: Exchanges in carbon dioxide between a salt marsh and the atmosphere. *Ecology*, **61**, 1434–45.

Hopkinson, C.S., Jr. and Schubauer, J.P. (1984) Static and dynamic aspects of nitrogen cycling in the salt marsh graminoid *Spartina alterniflora* Loisel. *Ecology*, **65**, 961–9.

Howes, B.L., Howarth, R.W., Teal, J.M. and Valiela, I. (1981) Oxidation reduction potentials in a salt marsh: Spatial patterns and interactions with primary production. *Limnology and Oceanography*, **26**, 350–60.

Jackson, C.R. (1952) Some topographic and edaphic factors affecting distribution in a tidal marsh. *Quarterly Journal of the Florida Academy of Sciences*, **15**, 137–46.

Jefferies, R.L. (1981) Osmotic adjustment and the response of halophytic plants to salinity. *Bioscience*, **31**, 42–6.

Jefferies, R.L., Rudmik, T. and Dillon, E.M. (1979) Responses of halophytes to high salinities and low water potentials. *Plant Physiology*, **64**, 989–94.

Johnson, D.S. and York, H.H. (1915) *The relation of plants to tide levels*, Carnegie Institution of Washington Publications, 206.

Keefe, C.W. (1972) Marsh production: a summary of the literature. *Contributions in Marine Science (Texas)*, **16**, 163–81.

Kershaw, K.A. (1976) The vegetational zonation of the East Pen Island salt marshes, Hudson Bay. *Canadian Journal of Botany*, **54**, 5–13.

King, G.M., Klug, M.J., Wiegert, R.G. and Chalmers, A.G. (1982) The relationship of soil water movement, sulfide concentration, and *Spartina alterniflora* production in a Georgia salt marsh. *Science*, **218**, 61–3.

Kuramoto, R.T. and Brest, D.E. (1979) Physiological response to salinity by four salt marsh plants. *Botanical Gazette*, **140**, 295–8.

Kurz, H. and Wagner, K. (1957) *Tidal marshes of the Gulf and Atlantic coasts of northern Florida and Charleston, South Carolina*, Florida State University Studies No. 24, pp. 1–168. The Florida State University, Tallahassee.

Larcher, W. (1980) *Physiological Plant Ecology*, 2nd edn, Springer-Verlag, Berlin, pp. 94–5.

Letzsch, W.S. and Frey, R.W. (1980) Deposition and erosion in a holocene salt marsh, Sapelo Island, Georgia. *Journal of Sedimentary Petrology*, **50**, 529–42.

Linthurst, R.A. (1979) The effect of aeration on the growth of *Spartina alterniflora* Loisel. *American Journal of Botany*, **66**, 685–91.

Liphschitz, N. and Waisel, Y. (1974) Existence of salt glands in various genera of the gramineae. *New Phytologist*, **73**, 507–13.

Long, S.P. and Incoll, L.D. (1979) The prediction and measurement of photosynthetic rate of *Spartina townsendii (sensu lato)* in the field. *Journal of Applied Ecology*, **16**, 879–91.

Long, S.P. and Woolhouse, H.W. (1978a) The responses of net photosynthesis to vapor pressure deficit and CO_2 concentration in *Spartina townsendii (sensu lato)*, a C_4 species from a cool temperate climate. *Journal of Experimental Botany*, **29**, 567–77.

Long, S.P. and Woolhouse, H.W. (1978b) The response of net photosynthesis to light and temperature in *Spartina townsendii (sensu lato)*, a C_4 species from a cool temperate climate. *Journal of Experimental Botany*, **29**, 803–14.

Longstreth, D.J. and Strain, B.R. (1977) Effect of salinity and illumination on photosynthesis and water balance of *Spartina alterniflora* Loisel. *Oecologia*, **31**, 191–9.

Lüttge, U. (1975) Salt glands, in *Ion transport in plant cells and tissues* (eds D.A. Baker and J.L. Hall), North-Holland, New York, pp. 336–76.

MacDonald, K.B. and Barbour, M.G. (1974) Beach and salt marsh vegetations of the North American Pacific Coast, in *Ecology of Halophytes* (eds R.J. Reimold and W.H. Queen), Academic Press, New York, pp. 175–233.

Mahall, B.E. and Park, R.B. (1976) The ecotone between *Spartina foliosa* Trin. and *Salicornia virginica* L. in salt marshes of Northern San Francisco Bay I. Biomass and production; II Soil water and salinity: III Soil aeration and tidal immersion. *Journal of Ecology*, **64**, 421–33, 793–809, 811–19.

Mallott, P.G., Davy, A.J., Jefferies, R.L. and Hutton, M.J. (1975) Carbon dioxide exchange in leaves of *Spartina anglica* Hubbard. *Oecologia*, **20**, 351–8.

McClung, C.R. and Patriquin, D.G. (1980) Isolation of a nitrogen fixing *Campylobacter*-sp from the roots of *Spartina alterniflora*. *Canadian Journal of Microbiology*, **26**, 881–6.

McEnerney, J.T. and Davis, D.E. (1979) Metabolic fate of atrazine in the *Spartina alterniflora* detritus *Uca pugnax* food chain. *Journal of Environmental Quality*, **8**, 335–8.

McGovern, T.A., Laber, L.J. and Gram, B.C. (1979) Characteristics of the salt secreted by *Spartina alterniflora* and their relation to estuarine production. *Estuarine and Coastal Marine Science*, **9**, 351–6.

Mendelssohn, I.A. (1979) Nitrogen metabolism in the

height forms of *Spartina alterniflora* in North Carolina. *Ecology*, **60**, 574–84.

Mendelssohn, I.A. and Seneca, E.D. (1980) The influence of soil drainage on the growth of salt marsh cordgrass *Spartina alterniflora* in North Carolina. *Estuarine and Coastal Marine Science*, **11**, 27–40.

Mendelssohn, I.A., McKee, K.L. and Patrick, W.H., Jr (1981) Oxygen deficiency in *Spartina alterniflora* roots: metabolic adaptation to anoxia. *Science*, **214**, 439–41.

Miller, W.R. and Egler, F.E. (1950) Vegetation of the Wequetequock-Pawcatuck tidal-marshes, Connecticut. *Ecological Monographs*, **20**, 143–72.

Mooring, M.T., Cooper, A.W. and Seneca, E.D. (1971) Seed germination responses and evidence for height ecophenes in *Spartina alterniflora* from North Carolina. *American Journal of Botany*, **58**, 48–55.

Morris, J.T. (1980) The nitrogen uptake kinetics of *Spartina alterniflora* in culture. *Ecology*, **61**, 1114–21.

Morris, J.T. (1982) A model of growth responses by *Spartina alterniflora* to nitrogen limitation. *Journal of Ecology*, **70**, 25–42.

Nestler, J. (1977) Interstitial salinity as a cause of ecophenic variation in *Spartina alterniflora*. *Estuarine and Coastal Marine Science*, **5**, 707–14.

Niering, W.A. and Warren, R.S. (1980) Vegetation patterns and processes in New England salt marshes. *Bioscience*, **30**, 301–7.

Nixon, S.W. and Oviatt, C.A. (1973) Analysis of local variation in the standing crop of *Spartina alterniflora*. *Botanica Marina*, **16**, 103–9.

Odum, E.P. (1959) *Fundamentals of Ecology*, 2nd edn, Saunders, Philadelphia.

Odum, E.P. (1974) Halophytes, energetics, and ecosystems, in *Ecology of Halophytes* (eds R.J. Reimold and W.H. Queen), Academic Press, New York, pp. 599–602.

Odum, E.P. and Fanning, M.E. (1973) Comparison of the productivity of *Spartina alterniflora* and *Spartina cynosuroides* in Georgia coastal marshes. *Bulletin of the Georgia Academy of Sciences*, **31**, 1–12.

Odum, W.E., Woodwell, G.M. and Wurster, C.F. (1969) DDT residues absorbed from organic detritus by fiddler crabs. *Science*, **164**, 576–77.

Parrondo, R.T., Gosselink, J.G. and Hopkinson, C.S., Jr. (1978) Effects of salinity and drainage on the growth of three salt marsh grasses. *Botanical Gazette*, **139**, 102–7.

Patrick, W.H. Jr and DeLaune, R.D. (1976) Nitrogen and phosphorus utilization by *Spartina alterniflora* in a salt marsh in Barataria Bay, Louisiana. *Estuarine and Coastal Marine Science*, **4**, 59–64.

Patrick, W.H., Jr and DeLaune, R.D. (1977) Chemical and

biological redox systems affecting nutrient availability in the coastal wetlands. *Geoscience and Man*, **18**, 131–7.

Patriquin, D.G. and Keddy, C. (1978) Nitrogenase activity (acetylene reduction) in a Nova Scotian salt marsh: its association with angiosperms and the influence of some edaphic factors. *Aquatic Botany*, **4**, 227–44.

Penfound, W.T. and E.S. Hathaway. (1938) Plant communities in the marshlands of southeastern Louisiana. *Ecological Monographs*, **8**, 1–56.

Phung, H.T. and Fiskell, J.G.A. (1972) A review of redox reactions in soils. *Soil and Crop Science Society of Florida*, **32**, 141–5.

Pigott, C.D. (1969) Influence of mineral nutrition on the zonation of flowering plants in coastal salt marshes, in *Ecological Aspects of Mineral Nutrition in Plants* (ed. I.I. Rorison), 9th Symposium of the British Ecological Society, Blackwell Scientific Publications, Oxford, pp. 25–35.

Pillai, C.G.P., Weete, J.D. and Davis, D.E. (1977) Metabolism of atrazine by *Spartina alterniflora*. 1. Chloroform-soluble metabolites. *Journal of Agricultural Food and Chemistry*, **25**, 852–5.

Pomeroy, L.R. and Wiegert, R.G. (eds) (1981) *The Ecology of a Salt Marsh*, Springer-Verlag, New York.

Pomeroy, L.R., Darley, W.M., Dunn, E.L., Gallagher, J.L., Haines, E.B. and Whitney, D.M. (1981) Primary production, in *The Ecology of a Salt Marsh* (eds L.R. Pomeroy and R.G. Wiegert), Springer-Verlag, New York, pp. 39–67.

Ponnamperuma, F.N. (1972) The chemistry of submerged soils. *Advances in Agronomy*, **24**, 29–96.

Poulin, G., Bourque, D., Eid, S. and Jankowski, K. (1978) Composition chimique de *Salicornia europaea* L. *Naturaliste Canadien*, **105**, 473–8.

Purer, E.A. (1942) Plant ecology of the coastal salt marshlands of San Diego County, California. *Ecological Monographs*, **12**, 81–111.

Rains, D.W. (1972) Salt transport by plants in relation to salinity. *Annual Review Plant Physiology*, **23**, 367–88.

Rains, D.W. and Epstein, E. (1967) Preferential absorption of potassium by leaf tissue of the mangrove, *Avicennia marina*: an aspect of halophytic competence in coping with salt. *Australian Journal of Biological Sciences*, **20**, 847–57.

Ranwell, D.S. (1972) *Ecology of Salt Marshes and Sand Dunes*, Chapman and Hall, London, UK, 258 pp.

Reed, J.F. (1947) The relation of the *Spartinetum glabrae* near Beaufort, North Carolina to certain edaphic factors. *American Midland Naturalist*, **38**, 605–14.

Reimold, R.J. (1972) The movement of phosphorous

through the salt marsh cord grass, *Spartina alterniflora* Loisel. *Limnology and Oceanography*, **17**, 606–11.

Reimold, R.J. (1977) Mangals and salt marshes of Eastern United States, in *Ecosytems of the World 1. Wet Coastal Ecosystems* (ed. V.J. Chapman), Elsevier Scientific, New York, pp. 157–66.

Reimold, R.J. and Durant, C.J. (1974) Toxaphene content of estuarine fauna and flora before, during, and after dredging toxaphene contaminated sediments. *Pesticide Monitoring Journal*, **8**, 44–9.

Reimold, R.J., Gallagher, J.L., and Thompson, D.E. (1973) Remote sensing of tidal marsh. *Photogrammetric Engineering* **39**, 477–88.

Ringius, G.S. (1980) Vegetation survey of a James Bay coastal marsh. *Canadian Field-Naturalist*, **94**, 110–20.

Seneca, E.D. (1974a) Germination and seedling response of Atlantic and Gulf coasts populations of *Spartina alterniflora*. *American Journal of Botany*, **61**, 947–56.

Seneca, E.D. (1974b) Stabilization of coastal dredge spoil with *Spartina alterniflora*, in *Ecology of Halophytes* (eds. R.J. Reimold and W.H. Queen), Academic Press, New York, pp. 525–9.

Shea, M.L., Warren, R.S. and Niering, W.A. (1975) Biochemical and transplantation studies of the growth form of *Spartina alterniflora* on Connecticut salt marshes. *Ecology*, **56**, 461–6.

Sherr, B.F. and Payne, W.J. (1979) Role of the salt marsh grass *Spartina alterniflora* in the response of soil denitrifying bacteria to glucose enrichment. *Applied Environmental Microbiology*, **38**, 747–8.

Silander, J.A. (1979) Microevolution and clone structure in *Spartina patens*. *Science*, **203**, 658–60.

Silander, J.A. and Antonovics, J. (1979) The genetic basis of the ecological amplitude of *Spartina patens*. 1. Morphometric and physiological traits. *Evolution*, **33**, 1114–27.

Smart, R.M. and Barko, J.W. (1978) Influence of sediment salinity and nutrients on the physiological ecology of selected salt marsh plants. *Estuarine and Coastal Marine Science*, **7**, 487–95.

Smart, R.M. and Barko, J.W. (1980) Nitrogen nutrition and salinity tolerance of *Distichlis spicata* and *Spartina alterniflora*. *Ecology*, **61**, 630–8.

Smith, D.L., Bird, C.J., Lynch, K.D. and McLachlan, J. (1980) Angiosperm productivity in two salt marshes of Minas basin. *Proceedings of the Nova Scotian Institute Science*, **30**, 109–18.

Somers, G.F. and Grant, D. (1981) Influence of seed source on phenology of flowering of *Spartina alterniflora* Loisel. and the likelihood of cross pollination. *American Journal of Botany*, **68**, 6–9.

Spaleny, J. (1977) Sulphate transformation to hydrogen sulphide in spruce seedlings. *Plant and Soil*, **48**, 557–63.

Stalter, R. (1973) Factors influencing the distribution of vegetation of the Cooper River estuary. *Castanea*, **38**, 18–24.

Stalter, R. and Batson, W.T. (1969) Transplantation of salt marsh vegetation, Georgetown, South Carolina. *Ecology*, **50**, 1087–9.

Steever, E.Z., Warren, R.S. and Niering, W.A. (1976) Tidal energy subsidy and standing crop production of *Spartina alterniflora*. *Estuarine and Coastal Marine Science*, **4**, 473–78.

Sullivan, M.J. and Daiber, F.C. (1974) Response in production of cord grass, *Spartina alterniflora*, to inorganic nitrogen and phosphorus fertilizer. *Chesapeake Science*, **15**, 121–3.

Teal, J.M. (1962) Energy flow in the salt marsh ecosystem of Georgia. *Ecology*, **43**, 614–24.

Teal, J.M. and Kanwisher, J.W. (1961) Gas exchange in a Georgia salt marsh. *Limnology and Oceanography*, **6**, 388–99.

Teal, J.M. and Kanwisher, J.W. (1966) Gas transport in the marsh grass, *Spartina alterniflora*. *Journal of Experimental Botany*, **17**, 355–61.

Turitzin, S.N. and Drake, B.G. (1981) The effect of seasonal change in canopy structure on the photosynthetic efficiency of a salt marsh. *Oecologia*, **48**, 79–84.

Turner, R.E. (1976) Geographic variations in salt marsh macrophyte production: a review. *Contributions in Marine Science (Texas)* **20**, 47–68.

Ustin, S.L., Pearcy, R.W. and Bayer, D.E. (1982) Plant water relations in a San Francisco bay salt marsh. *Botanical Gazette* **143**, 368–73.

Valiela, I. and Teal, J.M. (1974) Nutrient limitation in salt marsh vegetation, in *Ecology of Halophytes* (eds R.J. Reimold and W.H. Queen), Academic Press, New York, pp. 547–63.

Valiela, I. and Teal, J.M. (1979) The nitrogen budget of a salt marsh ecosystem. *Nature*, **280** (5724), 652–6.

Valiela, I., Teal, J.M. and Sass, W.J. (1975) Production and dynamics of salt marsh vegetation and the effects of experimental treatment with sewage sludge. *Journal of Applied Ecology*, **12**, 973–81.

Valiela, I., Teal, J.M. and Deuser, W.G. (1978) The nature of growth forms in the salt marsh grass *Spartina alterniflora*. *American Naturalist*, **112**, 461–70.

Wiebe, W.J., Christian, R.R., Hansen, J.A., King, G., Sherr, B. and Skyring, G. (1981) Anaerobic respiration and fermentation, in *The Ecology of a Salt Marsh* (eds. L.R. Pomeroy and R.G. Wiegert), Springer-Verlag, New York, pp. 138–59.

Williams, R.B. and Murdoch, M.B. (1969) The potential importance of *Spartina alterniflora* in conveying zinc, manganese and iron into estuarine food chains, in *Symposium on Radioecology* (eds D.J. Nelson and F.C. Evans), US Atomic Energy Commission CONF 670503 TID 4500, pp. 431–9.

Winner, W.E., Smith, C.L., Koch, G.W., Mooney, H.A. Bewley, J.D. and Krouse, H.R. (1981) Rates of emission of H_2S from plants and patterns of stable sulphur isotope fractionation. *Nature*, **289**, 672–3.

Woodell, S.R.J. and Mooney, H.A. (1970) The effect of sea water on carbon dioxide exchange by the halophyte *Limonium californicum* (Bioss.). Heller. *Annuals of Botany*, **34**, 117–21.

Woodwell, G.M., Wurster, C.F., Jr. and Isaacson, P.A. (1967) DDT residues in an east coast estuary: a case of biological concentration of a persistent insecticide. *Science*, **156**, 821–24.

Wu, T.L., Mick, N.J. and Fox, B.M. (1977) Runoff studies of the agricultural herbicides alachlor and atrazine from the Rhode river watershed during the 1976 growing season, in *Watershed Research in Eastern North America* (ed. D.L. Correll), Chesapeake Bay Center for Environmental Studies, Smithsonian Institution, Edgewater, Maryland, pp. 707–26.

Yeo, A.R. (1981) Salt tolerance in the halophyte *Suaeda maritima* L. Dum: Intracellular compartmentation of ions. *Journal of Experimental Botany*, **32**, 487–97.

Yeo, A.R. and Flowers, T.J. (1980) Salt tolerance in the halophyte *Suaeda maritima* L. Dum.: Evaluation of the effect of salinity upon growth. *Journal of Experimental Botany*, **31**, 1171–83.

Index